NUCLEAR SPECTROSCOPY

Part B

This is Volume 9 in
PURE AND APPLIED PHYSICS
A Series of Monographs and Textbooks
Consulting Editors: H. S. W. Massey and Keith A. Brueckner
A complete list of titles in this series appears at the end of this volume

NUCLEAR SPECTROSCOPY

Edited by

FAY AJZENBERG-SELOVE

Department of Physics, Haverford College
Haverford, Pennsylvania

Part B

1960

ACADEMIC PRESS   New York and London

COPYRIGHT © 1960, BY ACADEMIC PRESS, INC.
ALL RIGHTS RESERVED
NO PART OF THIS BOOK MAY BE REPRODUCED IN ANY FORM,
BY PHOTOSTAT, MICROFILM, RETRIEVAL SYSTEM, OR ANY
OTHER MEANS, WITHOUT WRITTEN PERMISSION FROM
THE PUBLISHERS.

ACADEMIC PRESS, INC.
111 Fifth Avenue, New York, New York 10003

United Kingdom Edition published by
ACADEMIC PRESS, INC. (LONDON) LTD.
Berkeley Square House, London W1X6BA

LIBRARY OF CONGRESS CATALOG CARD NUMBER: 59-7675

Second Printing, 1969

PRINTED IN THE UNITED STATES OF AMERICA

Contributors to Part B

M. K. BANERJEE, *Saha Institute of Nuclear Physics, Calcutta, India*

L. C. BIEDENHARN, *The Rice Institute, Houston, Texas*

A. BOHR, *Nordisk Institut for Teoretisk Atomfysik, Copenhagen, Denmark*

HERMAN FESHBACH, *Massachusetts Institute of Technology, Cambridge, Massachusetts*

J. B. FRENCH, *University of Rochester, Rochester, New York*

D. KURATH, *Argonne National Laboratory, Lemont, Illinois*

R. D. LAWSON, *University of Chicago, Chicago, Illinois*

C. A. LEVINSON, *Princeton University, Princeton, New Jersey*

WILLIAM M. MACDONALD, *University of Maryland, College Park, Maryland*

B. R. MOTTELSON, *Nordisk Institut for Teoretisk Atomfysik, Copenhagen, Denmark*

M. E. ROSE, *Oak Ridge National Laboratory, Oak Ridge, Tennessee*

D. STROMINGER, *University of California, Berkeley, California*

D. H. WILKINSON, *Oxford University, Oxford, England*

Preface

The rapid development of the field of nuclear spectroscopy makes desirable a survey which will provide an up-to-date account of present knowledge. This is all the more important for those entering the field who are without extensive contact with the flood of literature which has appeared in recent years.

This book has been prepared both for the use of graduate students preparing for experimental research in nuclear spectroscopy and for specialists in one area of the field who wish to acquire a broader understanding of the entire field. In general, the treatment is at a level which should be accessible to a student who has completed graduate-level courses in quantum mechanics and nuclear physics.

The contributions to this volume are concerned with the ways in which experimental data may be analyzed to furnish information about nuclear parameters, and with the nuclear models in terms of which the data are interpreted. Some of the techniques discussed here have not been previously presented in detail in the printed literature.

Because of the complexity of the field of nuclear spectroscopy, there are many contributors to this book. This, of course, creates problems in level and depth of coverage, in duplication of material, and in the use of notation, but it insures that the author of a given section has expert knowledge of the material he discusses. The notation in a section is that most commonly used in the relevant literature.

The editor wishes to express her deep indebtedness to T. Lauritsen, R. F. Christy, W. Selove, B. T. Feld, M. E. Rose, and J. W. M. DuMond and to many others of her colleagues for their criticisms, suggestions, and advice.

This book is dedicated by the editor to two theorists who have supplied both illumination and inspiration to the practitioners of nuclear spectroscopy:

        Professors E. P. Wigner and V. F. Weisskopf.

                                FAY AJZENBERG-SELOVE

Haverford, Pennsylvania
March, 1960

ERRATA

"Nuclear Spectroscopy," Part B

Page 658, line 2 below Eq. (115). *For* "to measure Q_s and R_s (see Fig. 8a for Q_s and Fig. 8b for R_s)" *read* "to measure a linear combination of Q_s and R_s (see Fig. 8 for examples)."

Page 705, Eq. (11).
The K_F in the denominator should not be boldface.

Page 708, Eq. (24).
Insert T^2 in the right-hand side of the equation

$$\left[\text{i.e., } \mu_i\mu_f T^2 \left(\frac{1}{2\pi}\right)^2\right].$$

Page 710, Eq. (35a).
For: $j^{l_{AB}}(q_A r)$
read: $j_{l_{AB}}(q_A R)$.

Page 722, Eq. (49).
The right-hand side of the equation should contain the factor μ_f [i.e., $R_f(q_c)\mu_f$].

Page 723, line 2 of Eq. (50).
The denominator should contain μ_{BC}^2 instead of μ_{BC}.

Contents of Part B

CONTRIBUTORS TO PART B	v
PREFACE	vii
CONTRIBUTORS TO PART A	xiii
CONTENTS OF PART A	xv

V. Theoretical Analysis of the Data

V. A. The Compound Nucleus	625
By HERMAN FESHBACH	
1. Elastic and Inelastic Potential Scattering Amplitudes	629
2. Resonance Amplitudes for Isolated Resonances	631
3. Properties of the Widths	631
4. Total Elastic Cross Section	634
5. Reaction Cross Sections	639
6. Total Cross Section	645
7. Angular Distribution for Reactions	646
8. Angular Distribution for Elastic Scattering of Neutral Particles	646
9. Angular Distribution for the Elastic Scattering of Charged Particles	649
10. Gamma Rays in Nuclear Reactions	652
11. Polarization	654
12. Overlapping Levels	660
13. Many Overlapping Levels. Statistical Theory	661
14. Density of Levels	668
V. B. Direct Interactions	670
1. Direct Interactions in Inelastic Scattering	670
By C. A. LEVINSON	
2. The Theory of Stripping and Pickup Reactions	695
By M. K. BANERJEE	
V. C. Angular Correlations in Nuclear Spectroscopy	732
By L. C. BIEDENHARN	
1. The Angular Correlation Process	736
2. Application of Angular Correlations to Specific Cases	769
3. Appendix: Summary of Formulas and Notations	800

CONTENTS OF PART B

V. D. Analysis of Beta Decay Data 811
By M. E. ROSE

    1. The Role of Beta Decay in Nuclear Physics 813
    2. Outline of the Theory 815
    3. The Energy Distribution and Angular Correlation: Allowed Transitions 819
    4. Analysis of the Shapes of Allowed Spectra 820
    5. Forbidden Spectra 821
    6. Corrections to the Spectral Shapes 824
    7. ft Values 825
    8. Failure of Parity Conservation 829
    9. Orbital Capture 830

V. E. Analysis of Internal Conversion Data 834
By M. E. ROSE

    1. Definition of the Process 834
    2. The Selection Rules. Mixtures of Multipoles 836
    3. The Calculation of Internal Conversion Coefficients 838
    4. Numerical Values of Internal Conversion Coefficients 850

V. F. Analysis of Gamma Decay Data 852
By D. H. WILKINSON

    1. Modes of De-excitation 852
    2. General Selection Rules for Electromagnetic Transitions 853
    3. Internal Conversion 857
    4. Nomenclature 857
    5. Isotopic Spin Selection Rules 858
    6. Units of Transition Strength 858
    7. The Independent-Particle Model 863
    8. Methods of Observation 865
    9. Empirical Data: the Light Elements 867
    10. Empirical Data: the Heavy Elements 872
    11. Special Selection Rules 879
    12. Branching Ratios 882
    13. Sum Rules 882

V. G. The Analysis of Reduced Widths 890
By J. B. FRENCH

    1. Single-Particle Widths 891
    2. Selection Rules 894
    3. Quantitative Analysis via the Shell Model 901
    4. Miscellaneous Considerations 915

V. H. Isotopic Spin Selection Rules 932
By WILLIAM M. MacDONALD
1. Isotopic Spin Vector Operator 933
2. Isotopic Spin of Nuclear States 934
3. Coulomb Perturbation of Isotopic Spin States 940
4. Nuclear Reactions 946
5. Validity of the Isotopic Spin Quantum Number 957

VI. Nuclear Models

VI. A. The Nuclear Shell Model 963
By R. D. LAWSON
1. Evidence for Magic Numbers 964
2. The Shell Model 968

VI. B. Nuclear Coupling Schemes 983
By D. KURATH
1. Fundamental Coupling Procedure 984
2. Many-Particle Spectra 997
3. Intermediate Coupling in Light Nuclei 1002

VI. C. Collective Motion and Nuclear Spectra 1009
By A. BOHR and B. R. MOTTELSON
1. Vibrational Spectra 1012
2. Deformed Nuclei 1014
3. A Survey of the Low-Energy Nuclear Spectra 1023

VI. D. The Complex Potential Model 1033
By HERMAN FESHBACH
1. The Average Cross Sections and Average Wave Amplitudes 1034
2. Empirical Determination of the Complex Potential 1045

Appendixes

Appendix I. Physical Constants 1065

Appendix II. Table of Isotopes 1066
By D. STROMINGER

AUTHOR INDEX (PART B) 1105

SUBJECT INDEX (PARTS A AND B) 1114

Contributors to Part A

D. E. ALBURGER, *Brookhaven National Laboratory, Upton, New York*

H. H. BARSCHALL, *University of Wisconsin, Madison, Wisconsin*

G. A. BARTHOLOMEW, *Atomic Energy of Canada Ltd., Chalk River, Ontario, Canada*

L. M. BOLLINGER, *Argonne National Laboratory, Lemont, Illinois*

T. W. BONNER, *The Rice Institute, Houston, Texas*

W. W. BUECHNER, *Massachusetts Institute of Technology, Cambridge, Massachusetts*

L. CRANBERG, *Los Alamos Scientific Laboratory, Los Alamos, New Mexico*

S. DEVONS, *The University, Manchester, England*

H. FRAUENFELDER, *University of Illinois, Urbana, Illinois*

C. GEOFFRION, *Laval University, Quebec, Canada*

H. E. GOVE, *Atomic Energy of Canada, Ltd., Chalk River, Ontario, Canada*

WILLIAM F. HORNYAK, *University of Maryland, College Park, Maryland*

A. E. LITHERLAND, *Atomic Energy of Canada Ltd., Chalk River, Ontario, Canada*

H. T. RICHARDS, *University of Wisconsin, Madison, Wisconsin*

L. ROSEN, *Los Alamos Scientific Laboratory, Los Alamos, New Mexico*

W. SELOVE, *University of Pennsylvania, Philadelphia, Pennsylvania*

R. M. STEFFEN, *Purdue University, West Lafayette, Indiana*

F. S. STEPHENS, *University of California, Berkeley, California*

WILLIAM E. STEPHENS, *University of Pennsylvania, Philadelphia, Pennsylvania*

N. S. WALL, *Massachusetts Institute of Technology, Cambridge, Massachusetts*

WARD WHALING, *California Institute of Technology, Pasadena, California*

C. S. WU, *Columbia University, New York, New York*

Contents of Part A

INTRODUCTION
　BY FAY AJZENBERG-SELOVE

I. The Spectroscopy of Charged Particles

I. A. The Interactions of Charged Particles with Matter
　　1. The Interaction of Nuclear Particles with Matter
　　　BY WARD WHALING
　　2. The Interaction of Beta Particles with Matter
　　　BY C. S. WU

I. B. Charged-Particle Detectors
　　BY N. S. WALL

I. C. Measurement of Spectra
　　1. The Measurement of the Spectra of Charged Nuclear Particles
　　　BY W. W. BUECHNER
　　2. The Measurement of Beta-Ray Spectra
　　　BY C. S. WU AND C. GEOFFRION

I. D. Charged Particle Reactions
　　BY H. T. RICHARDS

I. E. Radioactive Decay Schemes
　　1. The Study of Nuclear States Observed in Beta Decay
　　　BY C. S. WU
　　2. The Study of Nuclear States Observed in Alpha Decay
　　　BY F. S. STEPHENS

II. Gamma Ray Spectroscopy

II. A. The Interaction of Gamma Rays with Matter
　　BY WILLIAM F. HORNYAK

II. B. The Detection of Gamma Rays and the Measurement of Gamma-Ray Spectra
　　BY D. E. ALBURGER

II. C. The Study of Nuclear States
　　1. The Gamma Decay of Bound Nuclear States
　　　BY D. E. ALBURGER

2. Gamma Rays from Unbound Nuclear States Formed by
   Charged-Particle Capture
   BY H. E. GOVE AND A. E. LITHERLAND
3. Neutron-Capture Gamma Rays
   BY G. A. BARTHOLOMEW

III. Neutron Spectroscopy

III. A. The Interactions of Neutrons with Matter
   BY W. SELOVE

III. B. Techniques of Slow Neutron Spectroscopy
   BY L. M. BOLLINGER

III. C. Measurement of Fast Neutron Spectra
   BY L. CRANBERG AND L. ROSEN

III. D. The Study of Bound Nuclear States
   BY L. CRANBERG AND L. ROSEN

III. E. The Study of Unbound Nuclear States
   1. Slow Neutron Resonances
      BY L. M. BOLLINGER
   2. Fast Neutron Resonances
      BY H. H. BARSCHALL

III. F. The Neutron Threshold Method
   BY T. W. BONNER

IV. Other Topics

IV. A. Photonuclear Reactions
   BY WILLIAM E. STEPHENS

IV. B. The Measurement of Very Short Lifetimes
   BY S. DEVONS

IV. C. The Measurement of Electromagnetic Moments of Nuclear States
   BY H. FRAUENFELDER AND R. M. STEFFEN

AUTHOR INDEX (PART A)

SUBJECT INDEX (PART A)

V.

Theoretical Analysis of the Data

V. A. The Compound Nucleus

by HERMAN FESHBACH

1. Elastic and Inelastic Potential Scattering Amplitudes.................... 629
2. Resonance Amplitudes for Isolated Resonances.......................... 631
3. Properties of the Widths... 631
4. Total Elastic Cross Section.. 634
5. Reaction Cross Sections... 639
6. Total Cross Section... 645
7. Angular Distribution for Reactions..................................... 646
8. Angular Distribution for Elastic Scattering of Neutral Particles........... 646
9. Angular Distribution for the Elastic Scattering of Charged Particles....... 649
10. Gamma Rays in Nuclear Reactions..................................... 652
11. Polarization.. 654
12. Overlapping Levels.. 660
13. Many Overlapping Levels. Statistical Theory........................... 661
14. Density of Levels... 668
    References.. 668

Resonance reactions and scattering occur at those energies for which the compound system, incident particle plus target nucleus, form a nearly stable state; that is a state of a relatively long lifetime. These resonance energies correspond to the energy levels of a physical entity, the compound nucleus, with the consequence that the collision of the incident bombarding particle, a, and the target nucleus, X, may be correctly described as resulting in the formation of a compound nucleus, C^*,

$$a + X \to C^*. \tag{1}$$

As indicated by the star on C the compound nuclear states are highly excited states of the compound system. However, by virtue of their long lifetime, these states have precise properties such as energy, total angular momentum J, parity Π, magnetic moment, and so on. The decay of C^* can occur in a variety of ways as set forth below:

$$\begin{aligned} a + X \to C^* &\to a + X & \text{elastic scattering} \\ &\to a + X^* & \text{inelastic scattering} \\ &\to \gamma + C & \text{radiative capture} \\ &\to b + Y & \text{transmutation} \end{aligned} \tag{2}$$

In this chapter we shall be concerned with the properties of these various reactions in the energy range in which resonances occur. The total cross sections for each of the processes above, as well as the angular distribution and polarization of the emergent particles a, b, and γ are sensitive to the quantum numbers of the levels of the compound nucleus C^*, and of the possible residual nuclei X, X^*, C, and Y. It is, of course, just this sensitivity which makes the investigation of resonance reactions an important tool for nuclear spectroscopy.

We start by quoting the general expression for the cross section describing a process leading from a state of the colliding system specified by α to a final state specified by α'. The index α denotes all the numbers specifying the initial state: such as E, the center of mass energy; i, the spin of the incident particle; I, the angular momentum of the target nucleus; Π, the parity. The same quantities when referred to the final state have prime superscripts. Since total angular momentum is preserved in any process it is useful to decompose the cross section according to possible values of the total angular momentum J. J is composed of the sum of the intrinsic spin of the target and projectile together with the orbital angular momentum l of the system:

$$\mathbf{J} = \mathbf{l} + \mathbf{I} + \mathbf{i}. \tag{3}$$

It is convenient to introduce an intermediate quantity, the channel spin $\mathbf{s}$ defined by the following equation:

$$\begin{aligned}\mathbf{s} &= \mathbf{I} + \mathbf{i} \\ \mathbf{J} &= \mathbf{l} + \mathbf{s}.\end{aligned} \tag{4}$$

For unpolarized beams and (or) unpolarized targets, $\mathbf{s}$ is random so that evaluation of the cross section involves averaging over all possible directions of $\mathbf{s}$. Wigner and Eisenbud (8),* Blatt and Biedenharn (9) corrected by Huby (10) have derived the result we need in terms of these quantities:

$$\frac{d\sigma(\alpha'|\alpha)}{d\Omega} = \frac{\lambda^2}{4} \sum \frac{(-)^{s'-s}}{(2I+1)(2i+1)} \bar{Z}(l_1 J_1 l_2 J_2; sL) \bar{Z}(l_1' J_1 l_2' J_2; s'L) \cdot \\ \operatorname{Re} \mathfrak{J}(\alpha' l_1' s' | \alpha l_1 s | J_1 \Pi_1) \mathfrak{J}^*(\alpha' l_2' s' | \alpha l_2 s | J_2 \Pi_2) P_L(\cos \vartheta). \tag{5}$$

In this formula ϑ is the angle between the emergent particle and the direction of the incident beam. $\mathfrak{J}(\alpha' l' s' | \alpha l s | J\Pi)$ is the transition matrix describing the transition from state α channel spin s to state α' channel spin s', the compound system having a total angular momentum J and parity Π, the incident particle an orbital angular momentum l, the emergent particle an orbital angular momentum l'. The $\bar{Z}$ factor is related to Blatt-Biedenharn's Z by

* The reference list for Section V.A. begins on page 668.

$$Z(l_1J_1l_2J_2;sL) = i^{L-l_1+l_2}\bar{Z}(l_1J_1l_2J_2;sL).$$

In terms of the Racah W coefficients

$$\bar{Z}(l_1J_1l_2J_2;sL) = [(2l_1+1)(2J_1+1)(2l_2+1)(2J_2+1)]^{1/2}$$
$$\cdot (l_1l_200|L0)W(l_1J_1l_2J_2;sL)$$

where $(l_1l_200|L0)$ is the Clebsch-Gordan coefficient for the coupling of two states l_1 ($m_1 = 0$) with l_2 ($m_2 = 0$) to obtain a state of angular momentum L ($M = 0$). The sum in Eq. (5) is taken over all angular momentum and parity quantum numbers but not over α or α' or I and i.

The total cross section for the process α into α' is

$$\sigma(\alpha'|\alpha) = \pi\lambda^2 \sum \frac{(2J+1)}{(2I+1)(2i+1)} |\mathfrak{z}(\alpha'l's'|\alpha ls|J\Pi)|^2, \tag{6}$$

while the total cross section for the target in state α is

$$\sigma_T(\alpha) = \sum_{\alpha'} \sigma(\alpha'|\alpha). \tag{7}$$

The $\bar{Z}$ factors are kinematical in the sense that they do not depend upon the nature of nuclear interactions but only upon the details of the decomposition of the initial plane wave into states of a given J, Π, l and s and upon the decay of this system into various possible final values of l' and s'. $\bar{Z}(l_1J_1l_2J_2;sL)$ describes the probability of forming a state of total angular momentum J_1 from l_1 and s or J_2 from l_2 and s. J_1 and J_2 are two possible J-values and l_1 and l_2 are two possible values for the orbital angular momentum of the incident beam which combine with s to form J_1 and J_2, respectively. Thus the triads (J_1,l_1,s) and (J_2,l_2,s) must form the sides of a triangle. If they do not, the factor $\bar{Z}(l_1J_1l_2J_2;sL)$ vanishes. Of course, a similar result is valid for the triads (J_1,l_1',s') and (J_2,l_2',s') describing the decomposition of the intermediate compound state J into the possible final values l_1', s', or l_2', s'. Note that $(s'-s)$ is an integer. Some further triangular triads are contained in the Z factors which come not so much from the conservation of angular momentum but simply from the fact that Eq. (5) gives the square of a matrix element which has been decomposed into the Legendre polynomials P_L. For example, the amplitude for the compound system to decay with the emitted particle having an orbital angular momentum l_1' will interfere with the amplitude for final angular momentum l_2'. In order for the interference of these two amplitudes to result in an angular distribution which contains a P_L component it is necessary that

$$l_1' + l_2' = L, \tag{8}$$

or (l_1',l_2',L) must form the sides of a possible triangle. The same condition holds for the triads (l_1,l_2,L) and (J_1,J_2,L). The following important result is a consequence:

$$L_{\max} \leq 2l_{\max}, \qquad L_{\max} \leq 2l'_{\max}. \tag{9}$$

In words, the maximum order of the Legendre polynomial which appears in the description of the angular distribution of the final state is twice (this factor appears because we are dealing with the cross section not the amplitude) the maximum value of the initial orbital angular momentum (or final orbital angular momentum whichever happens to be smaller) that enters appreciably into the reaction. Another property of the $\bar{Z}$ factors is again expected from the indicated origin of Eq. (8). The parity of the state L should be determined by the joint parities of l_1' and l_2'. Hence

$$l_1' + l_2' + L \quad \text{and} \quad l_1 + l_2 + L \text{ are even.} \tag{10}$$

So much for the Z factors.

The dynamics of the problem are contained in the $\mathfrak{J}$ factors. We note here some general properties of the $\mathfrak{J}$'s and the cross sections and shall then proceed to particular cases. The $\mathfrak{J}$'s are symmetric:

$$\mathfrak{J}(\alpha'l's'|\alpha ls|J\Pi) = \mathfrak{J}(\alpha ls|\alpha'l's'|J\Pi). \tag{11}$$

From Eq. (6) it immediately follows that

$$\sigma(\alpha'|\alpha) = \frac{\lambda^2}{\lambda'^2} \frac{(2I'+1)(2i'+1)}{(2I+1)(2i+1)} \sigma(\alpha|\alpha'). \tag{12}$$

The identical results hold for the angular distribution [Eq. (5)]; that is, Eq. (12) is a correct relation between $d\sigma(\alpha'|\alpha)/d\Omega$ and $d\sigma(\alpha|\alpha')/d\Omega$ when ϑ is suitably defined for each situation. Reciprocity relation (12) has on occasion been employed to measure a spin when three of the four spins I, i, I', i' are known (for example, spin of the π meson from the reaction $p + p \rightleftarrows \pi^+ + D$).

The $\mathfrak{J}$'s can be broken up into two terms one of which varies smoothly with energy, the other fluctuating rapidly, as is necessary for the description of resonances, namely:

$$\mathfrak{J}(\alpha'l's'|\alpha ls|J\Pi) = \mathfrak{J}_p(\alpha'l's'|\alpha ls|J\Pi) + \mathfrak{J}_R(\alpha'l's'|\alpha ls|J\Pi). \tag{13}$$

When $\alpha = \alpha'$, $\mathfrak{J}_p$ is the potential scattering term. When $\alpha \neq \alpha'$; that is, for reactions, $\mathfrak{J}_p$ is just the nonresonant smoothly varying part of the reaction amplitude. It represents the effects of all levels other than those contained in $\mathfrak{J}_R$. Upon inserting Eq. (13) into Eqs. (5) and (6), we can divide the cross section into three parts, the potential (which we use here in a generalized sense so as to include the effects of the nonresonant interactions) subscript, p, the resonant part subscript R, the interference between the two, subscript I:

$$\frac{d\sigma(\alpha'|\alpha)}{d\Omega} = \frac{d\sigma_p(\alpha'|\alpha)}{d\Omega} + \frac{d\sigma_R(\alpha'|\alpha)}{d\Omega} + \frac{d\sigma_I(\alpha'|\alpha)}{d\Omega}, \qquad (14)$$

where

$$\frac{d\sigma_R(\alpha'|\alpha)}{d\Omega} = \frac{\lambda^2}{4} \sum \frac{(-)^{s'-s}}{(2I+1)(2i+1)} \bar{Z}\bar{Z}' \operatorname{Re} \mathfrak{J}_R^{(1)}(\alpha'|\alpha)\mathfrak{J}_R^{(2)*}(\alpha'|\alpha), \qquad (15)$$

where a number of abbreviations which should be obvious have been adopted. The equation for $d\sigma_p(\alpha'|\alpha)/d\Omega$ is obtained from Eq. (15) simply by replacing the subscript R by p. The interference term is

$$\frac{d\sigma_I(\alpha'|\alpha)}{d\Omega} = \frac{\lambda^2}{4} \sum \frac{(-)^{s'-s}}{(2I+1)(2i+1)} \bar{Z}\bar{Z}' \cdot$$
$$\operatorname{Re} [\mathfrak{J}_R^{(1)}(\alpha'|\alpha)\mathfrak{J}_p^{(2)*}(\alpha'|\alpha) + \mathfrak{J}_p^{(1)}(\alpha'|\alpha)\mathfrak{J}_R^{(2)*}(\alpha'|\alpha)]. \qquad (16)$$

Decomposition (14) is pertinent when resonances are narrow and widely spaced. In the region between the resonances the slowly varying nonresonant cross section $d\sigma_p/d\Omega$ is the major term of the three while right at the resonance it forms a background upon which the structure of the resonance is superposed. The latter can be seriously affected by the interference term. These interference phenomena can occur in both elastic and inelastic scattering. Of course the magnitude of the effect will be large only when the magnitudes of the resonant and nonresonant amplitudes are comparable.

1. Elastic and Inelastic Potential Scattering Amplitudes

Consider elastic scattering first. The simplest situation occurs when the spin of the target nucleus is zero. Then the channel spin s is $1/2$ and because of the conditions imposed by parity and angular momentum conservation, l, the orbital angular momentum is a constant of the motion. Therefore,

$$\mathfrak{J}_p(\alpha l's'|\alpha ls|J\Pi) = 2i\delta_{ss'}\delta_{ll'}e^{i\delta(\alpha ls|J\Pi)} \sin \delta(\alpha ls|J\Pi), \qquad (17)$$

where $\delta(\alpha ls|J\Pi)$ is a phase shift. This quantity is obtained from the solution of the two-body Schroedinger equation for the target nucleus plus incident particle in which the interaction energy is given by a real potential. When the resonances are narrow and well separated in energy this potential is, to a good approximation, given by the real part of the optical potential (or more precisely by the shell model potential suitably extrapolated to the energy range of interest) for which empirical evaluations are available (see Section VI.D). Note that this definition of δ differs from the repulsive sphere phase shift assumed by Blatt and Biedenharn.

When the target nucleus spin differs from zero the description of elastic scattering becomes considerably more involved because l and s are no longer constants of the motion. It is then possible to "flip the channel spin." We illustrate with a simple case where $I = 1/2$ and $i = 1/2$. Then the channel spin can be either 0 or 1. The possible l', s', and J are shown in Table I for $l = 0$ and $l = 1$. For the $l = 0$, $s = 1$ case

TABLE I

l	s	J	l'	s'
0	0	0	0	0
	1	1	0, 2	1
1	0, 1	1	1	0, 1
	1	0	1	1
	1	2	1, 3	1

two values $l' = 0$ and $l' = 2$ are possible. This is, of course, just the familiar mixing of $^3S_1$ and $^3D_1$ states. When $l = 1$, l' can be either 1 or 3 corresponding to the mixing of the $^3P_2$ and $^3F_2$ states. In addition, when $l = 1$, $J = 1$ it is possible for the channel spin to change and we then have a mixing of the $^3P_1$ and $^1P_1$ states. When the target nucleus has a spin greater than $(1/2)$, more than two states can be coupled.[a]

Optical potentials which can lead to these complications are described in Section VI.D. However, in view of the fact that no evidence has been produced for their existence we shall assume that for potential elastic scattering changes in l and s do not occur with any appreciable probability. As a consequence Eq. (17) holds generally. We shall *not* make this assumption for the resonant scattering and this is our principal reason for discussing the possibility of channel spin and orbital angular momentum scattering.

[a] The channel spin representation we employ here suffers from the fact that the channel spin has no direct physical significance. A representation which is more directly connected with the experimental situation has been devised by Wick and Jacob (*11*). In it the channel spin quantum number is replaced by the chirality.

Potential reaction scattering (nonresonant reactions) is described elsewhere in this volume in the chapter on Direct Interactions. It is found that this process can be described in terms of a set of coupled Schroedinger equations the number corresponding to the various possible open channels. Since in this energy range the nonresonant reaction terms seem to be small it is useful to use the phase shift notation. Thus in the presence of nonresonant reactions

$$\mathfrak{J}_p(\alpha ls|\alpha ls|J\Pi) = 2ie^{i\delta(\alpha ls|J\Pi)} \sin \delta(\alpha ls|J\Pi) \qquad (18)$$
$$\mathfrak{J}_p(\alpha' l's'|\alpha ls|J\Pi) = 2ie^{i[\delta(\alpha ls|J\Pi)+\delta(\alpha' l's'|J\Pi)]} A_p(\alpha' l's'|\alpha ls|J\Pi). \qquad (19)$$

Note that in these equations the δ's are complex. A_p is symmetric.[b]

2. Resonance Amplitudes for Isolated Resonances

This discussion will be limited to the case of isolated nonoverlapping resonances. Overlapping resonances will be considered in a separate subsection. The position of the resonance is denoted by $E_{J\Pi}$ the total width of the resonance by $\Gamma_{J\Pi}$. Then we find (8)

$$\mathfrak{J}_R(\alpha' l's'|\alpha ls|J\Pi) = -ie^{i[\delta(\alpha ls|J\Pi)+\delta(\alpha' l's'|J\Pi)+\phi]} \frac{g(\alpha' l's'|J\Pi)g(\alpha ls|J\Pi)}{E - E_{J\Pi} + i\Gamma_{J\Pi}/2}. \qquad (20)$$

Here ϕ is a phase which depends on all the quantum numbers of the problem. It vanishes for diagonal elements; that is, when $\alpha = \alpha'$, $l = l'$, $s = s'$ and for all elements in the absence of nonresonant reaction processes. The factors g are real. They are related to the partial width $\Gamma(\alpha ls|J\Pi)$ as follows:

$$\Gamma(\alpha ls|J\Pi) = g^2(\alpha ls|J\Pi) \qquad (21)$$

while

$$\Gamma_{J\Pi} = \Sigma\Gamma(\alpha ls|J\Pi) + \Gamma(\gamma|J\Pi). \qquad (22)$$

The sum is taken over all possible values of α, l, s which can lead to the decay of the compound nuclear state $J\Pi$. $\Gamma(\gamma|J\Pi)$ is the radiation width.

3. Properties of the Widths

Obviously some qualitative understanding of the properties of the widths Γ and the amplitude factors g would be very helpful. Their energy

[b] The complex phase shifts may be avoided by employing the method of eigenphases. The latter has the advantage that the relations between the phase shifts which result from current conservation, for example, unitarity of the matrix are automatically contained. Blatt and Biedenharn (9) have discussed the method of eigenphases.

dependence for low energy neutrons is well known (*12*):

$$\Gamma(\alpha ls|J\Pi) \sim k^{2l+1}, \qquad k \to 0 \qquad (23)$$

while

$$g(\alpha ls|J\Pi) \sim k^{l+\frac{1}{2}}, \qquad k \to 0 \qquad (24)$$

Further information is contained in the equation relating the average width and the imaginary part of the phase shift, η, obtained from the complex potential model [Eq. (37'), Ch. VI.D]

$$\left\langle \frac{\Gamma(\alpha ls|J\Pi)}{D_{J\Pi}} \right\rangle = \frac{1}{2\pi} T_{ls}^{(J)} \simeq \frac{1}{2\pi} (1 - e^{-4\eta_{ls}^{(J)}}) \qquad (25)$$

where $D_{J\Pi}$ is the level spacing. This equation is valid when

$$\pi \left\langle \frac{\Gamma(\alpha ls|J\Pi)}{D_{J\Pi}} \right\rangle \ll 1$$

and when there are no nonresonant processes of importance. The transmission factors $T_{ls}^{(J)}$ are obtainable from the complex potential model calculations so that if some knowledge of $D_{J\Pi}$ is available an estimate of Γ can be made. We shall give an estimate of D later on in this chapter. We also observe that Eq. (23) is consistent with Eq. (25). In fact for incident particles of charge z and velocity v incident on a target of charge Z, Eq. (25) implies

$$\Gamma(\alpha ls|J\Pi) \sim k^{2l+1} C_l(\mu), \qquad (26)$$

where

$$C_l(\mu) = \frac{2\pi\mu}{e^{2\pi\mu} - 1} \frac{2^{(2l)}[(l^2 + \mu^2)((l-1)^2 + \mu^2) \cdots (1 + \mu^2)]}{(2l!)^2} \qquad (27)$$

and

$$\mu = (zZe^2)/\hbar v.$$

To continue further we require some of the properties of the transmission factor. This quantity should be proportional to the penetrability and should exhibit a single-particle resonance structure. In fact for the square well complex potential (*13*), it follows that

$$T_{ls}^{(J)} \simeq 4 s_l(kr) \frac{\hbar^2}{mR^2} \frac{\Delta}{\Delta^2 + (E - E_{sp})^2} \qquad (28)$$

E_{sp} is the single-particle resonance energy; the width of the resonance, Δ, is proportional to the imaginary part W of the optical potential; R is the radius of the square well. This equation is valid as long as $s_l \ll W/(\hbar^2/mR^2)$. The function s_l is related to the regular solution $F_l(kR)$ and the irregular solution $G_l(kR)$ of the equation satisfied by $r\psi$

where ψ is the wave function for the incident particle in the region outside of the square well. For neutrons

$$F_l(kr) \to \sin\left(kr - \frac{l\pi}{2}\right)$$
$$G_l(kr) \to \cos\left(kr - \frac{l\pi}{2}\right) \qquad kr \gg l \qquad (29)$$

while for protons:

$$F_l(kr) \to \sin\left(kr - \frac{l\pi}{2} - \mu \ln(2kr) + \sigma_l\right)$$
$$G_l(kr) \to \cos\left(kr - \frac{l\pi}{2} - \mu \ln(2kr) + \sigma_l\right) \qquad e^{2i\sigma_l} = \frac{\Gamma(l+1+i\mu)}{\Gamma(l+1-i\mu)}. \qquad (30)$$

The function s_l is then given by

$$s_l(kR) = \frac{kR}{F_l^2(kR) + G_l^2(kR)}. \qquad (31)$$

The function $s_l/kR \equiv v_l$ is tabulated by Blatt and Weisskopf (1,14). Equation (28) will be approximately valid for potentials which are more realistic than the square well since the transmission factors for these also show penetration effects as well as single-particle resonances. The best value of R is not easily determined but it presumably is of the order of the nuclear radius.

From Eq. (28) we immediately see that $(<\Gamma/D>)$ will be larger in the neighborhood of the single-particle resonance. Moreover with the aid of Eq. (28) we may now derive a sum rule. It is necessary to recall the meaning of the average in Eq. (25):

$$\left\langle \frac{\Gamma(\alpha ls|J\Pi)}{D_{J\Pi}} \right\rangle = \frac{1}{\Delta E} \sum \Gamma(\alpha ls|J\Pi), \qquad (32)$$

where the sum is taken over all the compound nuclear resonances of the type described by the argument of Γ contained in a small energy interval ΔE. The quantity ΔE satisfies the inequality:

$$\Delta \gg \Delta E \gg \Gamma. \qquad (33)$$

Hence from Eq. (25) the following may be obtained:

$$\sum \Gamma(\alpha ls|J\Pi) = \frac{1}{2\pi} \int T_{ls}^{(J)} dE, \qquad (34)$$

where now the sum is extended to include all the resonances within the single-particle resonance of Eq. (28). Inserting this last equation into

Eq. (34) we obtain

$$\sum \Gamma(\alpha l s | J\Pi) \simeq \frac{2\hbar^2}{mR^2} \overline{s_l(kR)}, \qquad (35)$$

where the bar indicates that the appropriate energy average over the resonance has been taken. This is the sum rule of Wigner and Teichmann (15) and obviously sets an upper limit to Γ:

$$\Gamma(\alpha l s | J\Pi) < \frac{2\hbar^2}{mR^2} \overline{s_l(kR)}. \qquad (36)$$

Inequality (36) is confirmed by experimental data (15,16) while Eq. (25) (see Chapter VI.D) provides a good fit if the appropriate complex potential is chosen. The empirical values of Γ and D are discussed elsewhere in this volume. For our present purposes we need only the order of magnitude. For neutrons of 1 ev energy (and therefore orbital angular momentum zero) $\Gamma(\alpha l s | J\Pi)/D$ is of the order of 10^{-4}. (See Fig. 1 in Section VI.D.) Order of magnitude values for other energies and orbital angular momenta can be obtained from this figure by taking the appropriate rates of penetration factors s_l. We ignore the resonance structure of (28) in making this estimate. This concludes this subsection on the properties of widths. We shall return to this subject when we discuss the statistical aspects of nuclear reactions in which the frequency distribution of widths obviously plays an important role.

4. Total Elastic Cross Section

It is now possible to discuss various aspects of the general formulae (5)–(7) since we can now substitute the parametric representation of the S matrix as given in Eqs. (18)–(20) and since we have a rough description of the properties of these parameters. Of course, in practice these parametric forms are employed to determine the widths empirically. Calculations of the latter for a specific case directly from a fundamental theory of nuclear structure are not yet feasible. In the next few sections we shall discuss the energy dependence of the total elastic scattering, of the total reaction cross section and of their sum, the total cross section.

Consider first the total elastic cross section $\sigma(\alpha|\alpha)$ given by Eq. (6). When (1) the target nucleus has a zero spin or (2) the energy is so low that only one orbital angular momentum (usually $l = 0$) can enter significantly into the formation or decay of a compound nucleus state with given J and Π, that is, l and consequently s are good quantum numbers, the nondiagonal J's in Eq. (6) can be either omitted or neglected and the cross section $\sigma(\alpha|\alpha)$ can be broken up into partial cross sections specified by J and l; namely,

$$\sigma(\alpha|\alpha) = \Sigma \sigma_{Jls}(\alpha|\alpha)$$

$$\sigma_{Jls} \simeq \pi \lambda^2 \frac{2J+1}{(2I+1)(2i+1)} |\mathfrak{Z}(\alpha ls|\alpha ls|J\Pi)|^2. \tag{37}$$

Inserting Eq. (18) and Eq. (20) we obtain

$$\sigma_{Jls} = 4\pi \lambda^2 \frac{2J+1}{(2I+1)(2i+1)} |e^{2i\delta(\alpha ls|J\Pi)}[\sin \delta(\alpha ls|J\Pi)e^{-i\delta(\alpha ls|J\Pi)}$$
$$- \Lambda \sin \gamma_{J\Pi} e^{-i\gamma_{J\Pi}}]|^2 \tag{38}$$

where

$$\Lambda(\alpha ls|J\Pi) = \Gamma(\alpha ls|J\Pi)/\Gamma_{J\Pi} \tag{39}$$

and

$$E - E_{J\Pi} + \tfrac{1}{2}i\Gamma_{J\Pi} = [(E - E_{J\Pi})^2 + (\tfrac{1}{2}\Gamma_{J\Pi})^2]^{\frac{1}{2}} e^{i\gamma_{J\Pi}}. \tag{40}$$

The simplest situation, δ real and $\Lambda = 1$, which can occur for light nuclei and low energies is instructive. Then

$$\sigma_{Jls} = 4\pi \lambda^2 \frac{2J+1}{(2I+1)(2i+1)} |\sin \delta e^{-i\delta} - \sin \gamma e^{-i\gamma}|^2 \quad \delta \text{ real}, \Lambda = 1 \tag{41}$$

where we have dropped the identifying notation for δ and γ. We see immediately that σ_{Jls} will have a minimum with the value of zero when $\delta = \gamma$:

$$(\sigma_{Jls})_{\min} = 0 \quad \text{when } \delta(\alpha ls|J\Pi) = \gamma_{J\Pi}. \tag{42}$$

The maximum of σ_{Jls} occurs when

$$\gamma_{J\Pi} = \delta(\alpha ls|J\Pi) + \frac{\pi}{2} \tag{43}$$

with

$$(\sigma_{Jls})_{\max} = 4\pi \lambda^2 \frac{2J+1}{(2I+1)(2i+1)}. \tag{44}$$

In terms of $(E - E_{J\Pi})$ and $\Gamma_{J\Pi}$ we have

$$\frac{2(E - E_{J\Pi})}{\Gamma_{J\Pi}} = \cot \delta(\alpha ls|J\Pi), \quad \text{when } \sigma_{Jls} \text{ has a minimum;} \tag{45}$$

$$= -\tan \delta(\alpha ls|J\Pi), \text{ when } \sigma_{Jls} \text{ has a maximum.} \tag{46}$$

For small energies, δ is negative so that the minimum occurs at an energy E below the resonance energy while the maximum will occur at an energy greater than the resonance energy. Since δ is of the order of $k^{(2l+1)}$ as k goes to zero we see that the maximum will generally be much closer to $E_{J\Pi}$ than the minimum. Since the minimum is a consequence of the interference between potential and resonance scattering it will be unobservable if the potential scattering is very small since at and near the minimum the resonance scattering amplitude will be correspondingly small. (See Fig. 1 and Fig. 2.)

More generally the maximum and minimum occur on opposite sides of $E_{J\Pi}$. The minimum occurs below $E_{J\Pi}$ if the potential phase shift is negative and above if it is positive. If $\delta = (n + \tfrac{1}{2})\pi$, n an integer, there is no maximum only a minimum at $E_{J\Pi}$, while if $\delta = n\pi$ there is a maximum at $E_{J\Pi}$ but no minimum.

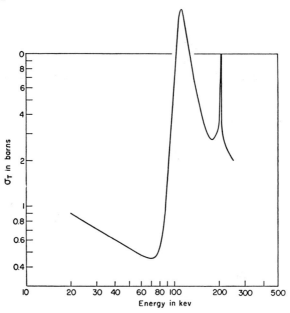

Fig. 1. Neutron resonances in sulfur. The resonance at 111 ± 2 kev shows an interference minimum at about 70 kev. [Curve taken from D. J. Hughes and R. B. Schwartz, *Neutron Cross-Sections*, 2nd ed. (Brookhaven National Lab., 1958), p. 129.]

These results are modified when other processes, such as radiative capture, as well as elastic scattering are possible. Then Λ is not unity. Again other modifications result if direct nonresonant reactions as well as resonant reactions can occur, since then δ is not real:

$$\delta(\alpha ls|J\Pi) = \xi(\alpha ls|J\Pi) + i\eta(\alpha ls|J\Pi). \tag{47}$$

Qualitatively however the results are the same, that is there is an interference minimum and an interference maximum. A significant difference is that the minimum cross section is not zero and the maximum is less than the maximum of Eq. (44). The minimum and maximum occur at energies obtained from the solution of the following equation:

$$\cot 2\gamma = \cot 2\xi - e^{-2\eta}(1 - \Lambda) \cos 2\xi. \tag{48}$$

There are two solutions for 2γ differing by π, one corresponding to the minimum of $\sigma_{J\text{II}}$, the other to the maximum. The values of the cross sections are

$$[\sigma_{J\text{II}}(\alpha|\alpha)]_{\substack{\max\\\min}} = \frac{\pi\lambda^2(2J+1)}{(2I+1)(2i+1)}$$
$$\{[1 + (1-\Lambda)^2 e^{-4\eta} - 2(1-\Lambda)e^{-2\eta}\cos 2\xi]^{\frac{1}{2}} \pm \Lambda e^{-2\eta}\}^2. \tag{49}$$

It is important to note from these formulae that when Λ is much smaller

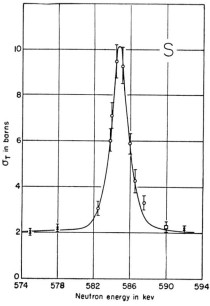

FIG. 2. Neutron resonance in sulfur. The resonance at 585 ± 3 kev shows no interference minimum. This together with the height at maximum and the order of magnitude of the width indicate that the compound nucleus has spin (3/2) and is formed by p wave neutrons. [Curve taken from R. E. Peterson, H. H. Barschall and C. K. Bockelman, Phys. Rev. **79**, 593 (1950).]

than 1 there is a negligible difference between the maximum and minimum cross section. This is no surprise as can be seen by examining $\mathfrak{I}_R(\alpha ls|\alpha ls|J\text{II})$, Eq. (20), which, for Λ very small is correspondingly small, even at its maximum. Thus for $\Lambda \ll 1$ the interference minimum and for that matter the interference maximum will not be easily observable in the elastic scattering. The competition provided by reactions of various kinds is much too strong so that the resonance will make its appearance only in the reaction and total cross sections. We shall discuss the parameter Λ in more detail later. For the present it is sufficient to state that the interference of potential and resonance scattering will be important when the resonance energy is so high that the width $\Gamma(\alpha ls|J\Pi)$ for the incident particle is comparable with the width of the possible emergent particles. This will generally be the case for light nuclei although some other possible examples can occur among the heavier target nuclei near the magic numbers.

We have so far included in our discussion only the partial cross section $\sigma_{Jls}(\alpha|\alpha)$. It is of course possible that many combinations of l and s can contribute to a given resonance. In that event the total contribution for a given J is

$$\sigma_J(\alpha|\alpha) = 4\pi\lambda^2 \frac{2J+1}{(2I+1)(2i+1)} \sum_{ls} |\sin \delta(\alpha ls|J\Pi)e^{-i\delta(\alpha ls|J\Pi)} - \Lambda(\alpha ls|J\Pi) \sin \gamma_{J\Pi}e^{i\gamma_{J\Pi}}|^2, \quad (50)$$

where we have taken δ to be real. Another form is obtained by squaring and substituting the values of Λ and γ:

$$\sigma_J(\alpha|\alpha) = 4\pi\lambda^2 \frac{2J+1}{(2I+1)(2i+1)} \sum_{ls} \Big\{ \sin^2 \delta(\alpha ls|J\Pi)$$
$$+ \frac{\Gamma(\alpha ls|J\Pi)}{4[(E-E_{J\Pi})^2 + \tfrac{1}{4}\Gamma_{J\Pi}^2]} [(\Gamma(\alpha ls|J\Pi) - \Gamma_{J\Pi}) + \Gamma_{J\Pi} \cos 2\delta(\alpha ls|J\Pi)$$
$$- 2(E - E_{J\Pi}) \sin 2\delta(\alpha ls|J\Pi)] \Big\}. \quad (51)$$

Each of the individual terms in the sum will have a minimum, but since these do not occur at identical values of E the minimum of $\sigma_J(\alpha|\alpha)$ will be neither as narrow nor as deep as it is in the special cases discussed above when only one term of the sum is significant.

We conclude with the equation for the total elastic cross section including all values of J:

$$\sigma(\alpha|\alpha) = \sigma_p + \frac{\pi\lambda^2}{(E-E_{J\Pi})^2 + (\tfrac{1}{2}\Gamma_{J\Pi})^2} \frac{2J+1}{(2I+1)(2i+1)}$$
$$\sum_{ls} \Gamma(\alpha ls|J\Pi)[(\Gamma(\alpha ls|J\Pi) - \Gamma_{J\Pi}) + \Gamma_{J\Pi} \cos 2\delta(\alpha ls|J\Pi)$$
$$- 2(E - E_{J\Pi}) \sin 2\delta(\alpha ls|J\Pi)], \quad (52)$$

where σ_p is the potential scattering cross section.

5. Reaction Cross Sections

We shall omit the effects of nonresonant reaction processes in the discussion to follow. Note, however, that if nonresonant terms are present, the reaction cross sections would exhibit interference maxima and minima in much the same fashion as the elastic scattering cross section.

In the absence of nonresonance terms the cross section of interest $\sigma(\alpha'|\alpha)$ for an isolated resonance when l and s are good quantum numbers is

$$\sigma(\alpha'l's'|\alpha ls|J\Pi) = \pi\lambda^2 \frac{2J+1}{(2I+1)(2i+1)} \frac{\Gamma(\alpha'l's'|J\Pi)\Gamma(\alpha ls|J\Pi)}{(E-E_{J\Pi})^2 + (\tfrac{1}{2}\Gamma_{J\Pi})^2}. \quad (53)$$

Radiative capture reactions follow a similar law

$$\sigma(\gamma|\alpha ls|J\Pi) = \pi\lambda^2 \frac{2J+1}{(2I+1)(2i+1)} \frac{\Gamma(\gamma|J\Pi)\Gamma(\alpha ls|J\Pi)}{(E-E_{J\Pi})^2 + (\tfrac{1}{2}\Gamma_{J\Pi})^2}. \quad (54)$$

If we sum over α' to obtain the total reaction cross section, σ_r, we obtain

$$\sigma_r(\alpha) = \sum_{\alpha' \neq \alpha} \sigma(\alpha'l's'|\alpha ls|J\Pi) = \pi\lambda^2 \frac{2J+1}{(2I+1)(2i+1)}$$
$$\cdot \frac{[\Gamma_{J\Pi} - \Gamma(\alpha ls|J\Pi)]\Gamma(\alpha ls|J\Pi)}{(E-E_{J\Pi})^2 + (\tfrac{1}{2}\Gamma_{J\Pi})^2}. \quad (55)$$

This form suggests that one define a cross section $\sigma_c(\alpha ls|J\Pi)$ for the formation of the compound nucleus

$$\sigma_c(\alpha ls|J\Pi) = \pi\lambda^2 \frac{2J+1}{(2I+1)(2i+1)} \frac{\Gamma_{J\Pi}\Gamma(\alpha ls|J\Pi)}{(E-E_{J\Pi})^2 + (\tfrac{1}{2}\Gamma_{J\Pi})^2} \quad (56)$$

and the cross section for compound elastic scattering, σ_{ce}:

$$\sigma_{ce} = \pi\lambda^2 \frac{2J+1}{(2I+1)(2i+1)} \frac{[\Gamma(\alpha ls|J\Pi)]^2}{(E-E_{J\Pi})^2 + (\tfrac{1}{2}\Gamma_{J\Pi})^2}. \quad (57)$$

Note that σ_c depends only upon the initial state and the compound state quantum numbers.

From these definitions it follows that

$$\sigma(\alpha'l's'|\alpha ls|J\Pi) = \sigma_c \frac{\Gamma(\alpha'l's'|J\Pi)}{\Gamma_{J\Pi}}. \quad (58)$$

This result suggests the interpretation that a resonance reaction is a two-step process. The first step is the formation of the compound nucleus σ_c. In the second step the compound nucleus decays. The probability for a particular mode of decay is given by the ratio of its width to the total width, $\Gamma(\alpha'l's'|J\Pi)/\Gamma_{J\Pi}$; the ratio of the cross sections for any two modes of decay is just the ratio of the widths:

$$\frac{\sigma(\alpha'l's'|\alpha l s|J\Pi)}{\sigma(\alpha''l''s''|\alpha l s|J\Pi)} = \frac{\Gamma(\alpha'l's'|J\Pi)}{\Gamma(\alpha''l''s''|J\Pi)}. \tag{59}$$

Note that the ratio is independent of the manner in which the compound nucleus is formed. Hence this ratio should not depend upon the nature of the incident particle on the target nucleus so long as the identical compound nucleus is formed. We may employ the average value (25) to obtain an estimate of the branching ratio (59):

$$\frac{\sigma(\alpha'l's'|\alpha l s|J\Pi)}{\sigma(\alpha''l''s''|\alpha l s|J\Pi)} \approx \frac{T^{(J)}{}_{\alpha'l's'}(E')}{T^{(J)}{}_{\alpha''l''s''}(E'')}, \tag{60}$$

where we have explicitly indicated that each T is a function of the energy of the emergent particle. Equation (60) is an estimate since the fluctuations of Γ away from its average can be large. However, Eq. (60) provides important semiquantitative information which follows from the energy and charge dependence of the transmission factors. At low energies this dependence is indicated by Eq. (23) for neutrons and Eq. (26) for charged particles and is a consequence of the angular momentum and of the Coulomb potential barriers. As the energy becomes more comparable with the barrier height the transmission factors increase very rapidly. This rise is not smooth because the penetration factors $s_l(kr)$ are seriously modified by the single-particle resonance. As the energy increases beyond the barrier height the rise is no longer as rapid and the oscillations less violent as the transmission factors approach unity, their asymptotic value. The barrier height for neutrons can be determined from the equation

$$k_B R = l + \tfrac{1}{2}$$

where k_B is the wave number at the barrier height. The corresponding barrier energy E_B is

$$E_B = 20.66 \frac{(l + \tfrac{1}{2})^2}{R^2}. \quad (E_B \text{ in Mev and } R \text{ in fermis}) \tag{61}$$

For charged particles the corresponding expression is

$$E_B{}^{(z)} = 20.66 \frac{m}{M} \frac{(l + \tfrac{1}{2})^2}{R^2} + 1.44 \frac{Zz}{R}, \quad (E_B \text{ in Mev and } R \text{ in fermis}) \tag{62}$$

where m is the nucleon mass and M the mass of the particle. The behavior of the transmission factors for neutrons is illustrated in Figs. 3 and 4. Tables of transmission factors are listed at the end of Section VI.D.

We can now summarize the dependence of the branching ratio (59) on the energy and charge of the particles which form possible modes of decay of the compound nucleus. If the energy of the emitted neutron is greater than the energy of a charged particle and the latter energy is less than the barrier energy $E_B^{(z)}$, neutron emission is strongly favored.

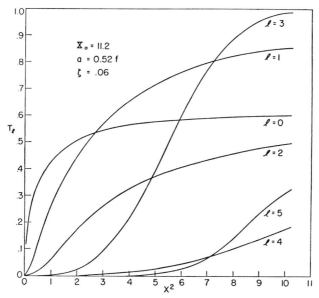

FIG. 3. Transmission factors. These curves are calculated by H. Feshbach, C. E. Porter, and V. F. Weisskopf (to be published) employing a Saxon well, $-V_0(1 + i\zeta)$ $[1 + \exp{(r - R)/a}]$ discussed in Section VI.D. The parameters X_0 and x^2 are $X_0^2 = (2\mu/\hbar^2)V_0R^2$ and $x^2 = (2\mu/\hbar^2)ER^2$. In this study V_0 is 52 Mev ($a = 0.52f$) and $R = (1.15A^{\frac{1}{3}} + 0.4)f$. These curves are for mass number 43. They differ markedly from the black nucleus values.

If the energy of both particles is well above the barrier energy their respective probabilities of emission are comparable. If the energy of the neutral particle is considerably less than the energy of the charged particle, it again becomes possible for their decay widths to be comparable or even for the charged-particle width to exceed the neutron width. A limiting case in which this is obvious is one in which the charged-particle reaction is exoergic, that is energy is released, while the neutron reaction is endoergic, which means that a minimal energy, the threshold energy, must be available before the reaction can proceed. If the available energy is close to or less than this threshold energy, emission of the charged particle will be favored. A familiar example is provided by the α-particle decay of naturally radioactive elements.

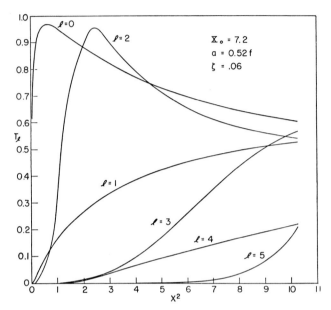

FIG. 4. Transmission factors. See legend for Fig. 3 for a description of the parameters. The target nucleus in this case has a mass number of 197.

Gamma-ray emission may also compete with particle emission because it is usually exoergic, the amount of energy available being of the order of several million electron volts. As a consequence, radiative capture will be the dominant process when resonances occur for sufficiently low energies of the incident particle and when there are no competing endoergic reactions. This requires very low energies indeed when the incident particle is a neutron since the width for re-emission of the neutron with energy unchanged, compound elastic scattering, rises fairly rapidly from zero energy so that it exceeds the gamma-ray width as soon as the neutron energy is of the order of kilovolts, an energy change which is miniscule for the gamma ray and which therefore produces essentially no change in the gamma-ray width. On the other hand when the incident particle is charged, the widths for re-emission are tremendously reduced by the Coulomb barrier, the relative factor for $l = 0$ particles being $\pi\mu e^{-2\pi\mu}$ [Eq. (26)] so that radiative capture can be the dominant compound nuclear process over a considerable range in energy. However if neutron emission [for example, a (p,n) reaction] is possible, it will generally be more probable than γ-emission. When radiative capture is much more important than compound elastic scattering the fluctuations in the elastic scattering coming from the resonance will be considerably damped, for then the parameter Λ in Eq. (38) is considerably less than one.

A similar set of remarks can be made for σ_c the cross section for formation of the compound nucleus, Eq. (56), if two reactions involving the

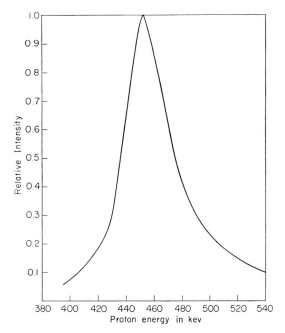

FIG. 5. Thin target excitation curve for $C^{12}(p,\gamma)$. [Taken from W. A. Fowler, C. C. Lauritsen, and T. Lauritsen, Revs. Modern Phys. **20**, 236 (1948).]

same compound nuclear state are compared or if the widths and resonance energies of the compound nuclear states involved are comparable, and the target nuclei similar. We shall not repeat these considerations except to make the obvious comment that the cross section σ_c for incident neutrons is much larger than σ_c for incident charged particles if their energies are comparable and less than $E_B{}^{(z)}$.

We conclude this section with a discussion of the energy dependence, that is, the shape of the $\sigma(\alpha'|\alpha)$ cross section as a function of energy. As can be seen from Eq. (53) if the energy $E_{J\text{II}}$ is sufficiently great, so that $E_{J\text{II}}$ is well above the threshold for the reaction, $\sigma(\alpha'|\alpha)$ is symmetric about the resonance energy (Fig. 5). If the resonance energy is low, for example, the incident particles are thermal neutrons, then the factor λ^2 and the widths Γ may vary over the width of the resonance and the shape of the resonance will now be asymmetric. For example $\sigma(\gamma|\alpha)$, Eq. (54), will be larger on the low-energy side of neutron resonances (see Fig. 6) since $\Gamma(\alpha ls|J\Pi)$ is proportional to λ^{-1}. Negative energy ($E_{J\Pi} < 0$)

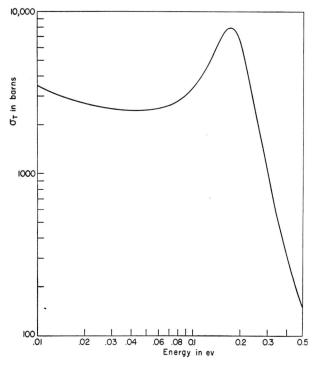

FIG. 6. Total cross section for neutrons bombarding cadmium. The resonance occurs at 0.178 ± 0.002 ev, $\Gamma_\gamma = 113 \pm 5$ mv, $\Gamma_n = 0.65 \pm .02$ mv, J of compound nucleus is 1. The resonance is asymmetrical, because of the variation of the λ^2 factor and Γ_n. The total cross section is approximately equal to the reaction cross section. [Curve taken from D. J. Hughes and R. B. Schwartz, *Neutron Cross-Sections*, 2nd ed. (Brookhaven National Lab., 1958), p. 217.]

resonances also occur and one can observe only the high-energy side of the resonance curve. The shape is also modified if the resonance energy is near the threshold energy E_T for an endoergic process. Suppose we can neglect the energy dependence of Γ and λ^2 over the resonance width. Then the energy dependence of $\sigma(\alpha'|\alpha)$ is

$$\sigma(\alpha'|\alpha) \sim \frac{(E - E_T)^{l'+\frac{1}{2}}}{(E - E_{J\Pi})^2 + (\frac{1}{2}\Gamma_{J\Pi})^2} C_{l'}(\mu').$$

This cross section rises from its zero value at threshold at a rate which increases with increasing l'. The maximum in the cross section is shifted to values of E larger than $E_{J\text{II}}$. Under some special circumstances the maximum may even be absent. It is obviously important to extract the gross-energy dependence of the widths before attempting to determine $E_{J\text{II}}$.

6. Total Cross Section

The $J\text{II}$-component of the total cross section, assuming again that l and s are good quantum numbers, is

$$\sigma_T(\alpha ls|J\text{II}) = \pi\lambda^2 \frac{2J+1}{(2I+1)(2i+1)} \left\{ 4 \sin^2 \delta(\alpha ls|J\text{II}) \right.$$
$$+ \frac{\Gamma(\alpha ls|J\text{II})}{(E-E_{J\text{II}})^2 + (\tfrac{1}{2}\Gamma_{J\text{II}})^2} [\Gamma_{J\text{II}} \cos 2\delta(\alpha ls|J\text{II})$$
$$\left. - 2(E-E_{J\text{II}}) \sin 2\delta(\alpha ls|J\text{II})] \right\}. \quad (63)$$

To obtain the total cross section we must sum over all possible values of l, s and add the potential scattering for the other J's which are non-resonant:

$$\sigma_T(\alpha) = \sum_{ls} \sigma_T(\alpha ls|J\text{II}) + 4\pi\lambda^2 {\sum}' \frac{2J'+1}{(2I+1)(2i+1)} \sin^2 \delta(\alpha l's'|J'\text{II}'). \quad (64)$$

The prime on the second sum indicates the omission of the J and l, s values included in the first sum.

Let us now examine Eq. (63) for $\sigma_T(\alpha ls|J\text{II})$ more closely. The first term inside the braces gives just the potential scattering while the remaining terms are the combined resonance and interference contributions. The maximum value of the cross section occurs at

$$E = E_{J\text{II}} - \tfrac{1}{2}\Gamma_{J\text{II}} \tan \delta(\alpha ls|J\text{II}). \quad (65)$$

The corresponding value of the cross section is

$$[\sigma_T(\alpha ls|J\text{II})]_{\max} = 4\pi\lambda^2 \frac{2J+1}{(2I+1)(2i+1)} \left[\sin^2 \delta(\alpha ls|J\text{II}) \right.$$
$$\left. + \frac{\Gamma(\alpha ls|J\text{II})}{\Gamma_{J\text{II}}} \cos^2 \delta(\alpha ls|J\text{II}) \right]. \quad (66)$$

The minimum value and the energy at which it occurs are

$$[\sigma_T(\alpha ls|J\Pi)]_{\min} = 4\pi\lambda^2 \frac{2J+1}{(2I+1)(2i+1)} \cdot \left[1 - \frac{\Gamma(\alpha ls|J\Pi)}{\Gamma_{J\Pi}}\right] \sin^2 \delta(\alpha ls|J\Pi), \quad (67)$$

$$E = E_{J\Pi} + \tfrac{1}{2}\Gamma_{J\Pi} \cot \delta(\alpha ls|J\Pi).$$

As in the elastic scattering case we see that the minimum will be visible only if the width for elastic scattering is an appreciable fraction of the total width. The conditions under which this is likely to occur have been discussed earlier.

7. Angular Distribution for Reactions

The angular distribution of reaction products when only one compound nuclear state ($J\Pi$) is involved and in the absence of nonresonant processes, is

$$\frac{d\sigma(\alpha'|\alpha)}{d\Omega} = \frac{\lambda^2/4}{(2I+1)(2i+1)[(E-E_{J\Pi})^2 + (\tfrac{1}{2}\Gamma_{J\Pi})^2]}$$

$$\sum (-)^{s'-s} \bar{Z}(l,Jl_2J;sL) \bar{Z}(l_1'Jl_2'J;s'L)$$

$$\cdot P_L(\cos\vartheta) g(\alpha'l_1's'|J\Pi) g(\alpha'l_2's'|J\Pi) g(\alpha l_1 s|J\Pi) g(\alpha l_2 s|J\Pi)$$

$$\cdot \cos[\delta(\alpha'l_1's'|J\Pi) - \delta(\alpha'l_2's'|J\Pi) + \delta(\alpha l_1 s|J\Pi) - \delta(\alpha l_2 s|J\Pi)]. \quad (68)$$

Since l_1 and l_2 represent possible routes between the same initial and final states (this is true for l_1' and l_2' as well), it follows that these partial waves must have the same parity so that

$$l_1 + l_2 \text{ and } l_1' + l_2' \text{ are even.}$$

From the property of the $\bar{Z}$ coefficients that $l_1 + l_2 + L$ must be even, it follows that L is even. Therefore, the angular distribution involves only even Legendre polynomials and is thus *symmetric about 90°*. Note that deviations from symmetry about 90° would occur if nonresonant reaction processes occur. We see that a measurement of the angular distribution would provide information about the relative sign of the g's.

8. Angular Distribution for Elastic Scattering of Neutral Particles

The general expressions for the angular distribution in the form we require in which the cross section is broken up into a potential scattering term, a resonance term, and an interference term are given in Eq. (14) *et seq.* The potential scattering term is obtained directly upon insertion of Eq. (17) for $\mathfrak{J}_p$. Again we assume $\delta(\alpha l s | J \Pi)$ is real. Then

$$\frac{d\sigma_p}{d\Omega} = \frac{\lambda^2}{(2I+1)(2i+1)} \sum [\bar{Z}(l_1 J_1 l_2 J_2; sL)]^2 \sin \delta(\alpha l_1 s | J_1 \Pi_1) \cdot$$
$$\sin \delta(\alpha l_2 s | J_2 \Pi_2) \cos [\delta(\alpha l_1 s | J_1 \Pi_1) - \delta(\alpha l_2 s | J_2 \Pi_2)] P_L(\cos \vartheta). \quad (69)$$

If the phase shifts δ depend only upon l and not on J or s as would be the case if the potential responsible for the potential scattering is central, then the simpler and more familiar expression below can be employed

$$\frac{d\sigma_p}{d\Omega} = \lambda^2 \left| \sum \sin \delta_l e^{i\delta_l} (2l+1) P_l(\cos \vartheta) \right|^2. \quad (70)$$

The resonance term can be easily evaluated for the case of a single isolated resonance considered here. We obtain

$$\frac{d\sigma_R}{d\Omega} = \frac{\lambda^2/4}{[(E-E_{J\Pi})^2 + (\frac{1}{2}\Gamma_{J\Pi})^2](2I+1)(2i+1)}$$
$$\cdot \sum (-)^{s'-s} \bar{Z}(l_1 J l_2 J; sL) \bar{Z}(l_1' J l_2' J; s'L)$$
$$\cdot P_L(\cos \vartheta) g(\alpha l_1' s' | J \Pi) g(\alpha l_2' s' | J \Pi) g(\alpha l_1 s | J \Pi) g(\alpha l_2 s | J \Pi)$$
$$\cdot \cos [\delta(\alpha l_1' s' | J \Pi) + \delta(\alpha l_1 s | J \Pi) - \delta(\alpha l_2' s' | J \Pi) - \delta(\alpha l_2 s | J \Pi)]. \quad (71)$$

Again this angular distribution is symmetric about 90°.

The interference term Eq. (16) is

$$\frac{d\sigma_I}{d\Omega} = -\frac{\lambda^2}{(2I+1)(2i+1)[(E-E_{J\Pi})^2 + \frac{1}{2}(\Gamma_{J\Pi})^2]}$$
$$\sum [\bar{Z}(l_1 J_1 l_2 J; sL)]^2 \Gamma(\alpha l_2 s | J \Pi) \sin \delta(\alpha l_1 s | J_1 \Pi_1)$$
$$\cdot P_L(\cos \vartheta) \bigg[(E - E_{J\Pi}) \cos \{2\delta(\alpha l_2 s | J \Pi) - \delta(\alpha l_1 s | J_1 \Pi_1)\}$$
$$+ \frac{\Gamma_{J\Pi}}{2} \sin \{2\delta(\alpha l_2 s | J \Pi) - \delta(\alpha l_1 s | J_1 \Pi_1)\} \bigg]. \quad (72)$$

In the event that the phase shifts δ depend only on l and not on J and s, the sum over J_1 can be performed. We obtain

$$\frac{d\sigma_I}{d\Omega} = -\frac{\lambda^2(2J+1)}{(2I+1)(2i+1)[(E-E_{J\Pi})^2 + (\frac{1}{2}\Gamma_{J\Pi})^2]}$$
$$\sum_{ll'} (2l+1)(ll'00|L0)^2 \Gamma(\alpha l' | J \Pi) \sin \delta_l \, P_L(\cos \vartheta)$$
$$\cdot [(E-E_{J\Pi}) \cos (2\delta_{l'} - \delta_l) + \tfrac{1}{2}\Gamma_{J\Pi} \sin (2\delta_{l'} - \delta_l)] \quad (73)$$

where

$$\Gamma(\alpha l'|J\Pi) = \sum_s \Gamma(\alpha l's|J\Pi) \tag{74}$$

and where l' is limited by the requirement

$$|\mathbf{l'} + \mathbf{s}| = J. \tag{75}$$

In this case it is possible to cast Eq. (73) in another form which we will need in a later discussion and which is also somewhat easier to use for computation.

$$\frac{d\sigma_I}{d\Omega} = -\lambda \frac{2J+1}{(2I+1)(2i+1)} \sum_l P_l(\cos\vartheta) \, \text{Re} \, \frac{f_p^* e^{2i\delta_l}}{E - E_{J\Pi} + \tfrac{1}{2}i\Gamma_{J\Pi}} \Gamma(\alpha l|J\Pi) \tag{76}$$

where f_p is the potential scattering amplitude:

$$f_p = \frac{\lambda}{2i} \sum_l (2l+1)(2i \sin \delta_l e^{i\delta_l}) P_l(\cos\vartheta). \tag{77}$$

The entire angular distribution may be written in a simpler form when the target nucleus has a zero spin ($I = 0$) and when the incident particle has spin $1/2$ ($s = 1/2$). This simplification is primarily a result of the unique value of $1/2$ for s which in turn has the consequence that the $\mathfrak{I}$ matrix for elastic scattering $\mathfrak{I}(\alpha l's'|\alpha ls|J\Pi)$ is diagonal. We may therefore introduce the abbreviations

$$\begin{aligned} \mathfrak{I}(\alpha ls|\alpha ls|l + \tfrac{1}{2}\Pi) &= \mathfrak{I}(l+) \\ \mathfrak{I}(\alpha ls|\alpha ls|l - \tfrac{1}{2}\Pi) &= \mathfrak{I}(l-). \end{aligned} \tag{78}$$

In terms of these amplitudes the elastic scattering amplitude may be expressed in terms of an operator f which acts upon the spin wave function for the incident particle to yield the scattering amplitude:

$$f = A + B(\boldsymbol{\sigma} \cdot \mathbf{n}), \tag{79}$$

where $\mathbf{n}$ may be expressed in terms of the final momentum (in units of $\hbar$) $\mathbf{k}_f$ and the initial momentum $\mathbf{k}_i$:

$$\mathbf{n} = (\mathbf{k}_f \times \mathbf{k}_i)/|\mathbf{k}_f \times \mathbf{k}_i|. \tag{80}$$

The vector $\mathbf{n}$ is a unit vector perpendicular to the plane of scattering. The functions A and B are:

$$A = \frac{\lambda}{2i} \sum [(l+1)\mathfrak{I}(l+) + l\mathfrak{I}(l-)]P_l(\cos\vartheta) \tag{81}$$

$$B = -\frac{\lambda}{2}\sum [\mathfrak{J}(l\,+) - \mathfrak{J}(l\,-)]P_l^{(1)}(\cos\vartheta) \tag{82}$$

$$P_l^{(1)} = \sin\vartheta\,\frac{d}{d(\cos\vartheta)}\,P_l(\cos\vartheta). \tag{83}$$

For an unpolarized beam (we shall discuss polarization later) the differential cross section is

$$\frac{d\sigma(\alpha|\alpha)}{d\Omega} = |A|^2 + |B|^2. \tag{84}$$

The amplitudes $\mathfrak{J}(l+)$ are given by the sum of the potential scattering amplitude $\mathfrak{J}_p$ [Eq. (18)] and the diagonal element resonance scattering amplitude $\mathfrak{J}_R$ [Eq. (20)]. From the computational point of view Eqs. (79), (80), and (82) form a much simpler sequence than Eqs. (71), (72), and (73), though of course the final result for $d\sigma/d\Omega$ must be identical. In addition the form, Eq. (79), will be most useful for the discussion of polarization phenomena.

9. Angular Distribution for the Elastic Scattering of Charged Particles

This problem requires special consideration because the usual partial wave series for the potential scattering, involving now the long-range Coulomb field, barely converges. This difficulty is overcome by subtracting the known point charge Coulomb scattering amplitude, the remainder being now sufficiently convergent. Assuming the absence of any nonresonant reaction process or channel spin flip the transition matrix is diagonal:

$$\mathfrak{J}_p(\alpha l's'|\alpha ls|J\Pi) = \delta_{ll'}\delta_{ss'}[e^{2i(\delta(\alpha ls|J\Pi)+\sigma_l)} - 1]. \tag{85}$$

The phase σ_l is defined in Eq. (30). The phase shift $[\delta(\alpha ls|J\Pi) + \sigma_l]$ is the consequence of the combined action of the Coulomb and specifically nuclear forces. We may now subtract out the point Coulomb term:

$$\begin{aligned}\mathfrak{J}_p &= \delta_{ll'}\delta_{ss'}[(e^{2i\sigma_l} - 1) + \Delta\mathfrak{J}_p(\alpha ls|J\Pi)] \\ \Delta\mathfrak{J}_p &= e^{2i\sigma_l}[e^{2i\delta(\alpha le|J\Pi)} - 1].\end{aligned} \tag{86}$$

The scattering amplitude coming from the first term in Eq. (86) is just the Coulomb amplitude

$$f_c = \frac{-zZe^2}{2mv^2\sin^2\tfrac{1}{2}\vartheta}\,e^{2i\sigma_0 - i\mu\ln\sin^2\tfrac{1}{2}\vartheta}. \tag{87}$$

The potential scattering angular distribution consists of the Rutherford term obtained from f_c, the term involving $\Delta\mathfrak{J}_p$ bilinearly which is derived by substituting $\Delta\mathfrak{J}_p$ for $\mathfrak{J}$ in Eq. (5) and the interference term which is just the appropriate analog of Eq. (76). We find

$$\frac{d\sigma_p}{d\Omega} = \left(\frac{zZe^2}{2mv^2 \sin^2 \tfrac{1}{2}\vartheta}\right)^2 + \lambda^2 \sum \frac{\bar{Z}^2(l_1 J_1 l_2 J_2; sL)}{(2I+1)(2i+1)}$$
$$P_L(\cos\vartheta) \sin \delta(\alpha l_1 s | J_1 \Pi_1) \sin \delta(\alpha l_2 s | J_2 \Pi_2)$$
$$\cdot \cos [2(\sigma_{l_1} - \sigma_{l_2}) + \delta(\alpha l_1 s | J_1 \Pi_1) - \delta(\alpha l_2 s | J_2 \Pi_2)]$$
$$- 2\lambda \left(\frac{zZe^2}{2mv^2 \sin^2 \tfrac{1}{2}\vartheta}\right) \sum \frac{(2J+1)}{(2I+1)(2i+1)} \cdot$$
$$P_l(\cos\vartheta) \sin \delta(\alpha l s | J\Pi) \cos [2(\sigma_l - \sigma_0) + \mu \ln \sin^2 \tfrac{1}{2}\vartheta + \delta(\alpha l s | J\Pi)]. \quad (88)$$

This form does simplify considerably under two circumstances. First, if δ is a function only of l, that is, the nuclear potential is central, then the scattering amplitude is

$$f_p = f_c + \frac{\lambda}{2i} \sum (2l+1) P_l(\cos\vartheta) e^{2i\sigma_l}(e^{2i\delta_l} - 1)$$

and

$$\frac{d\sigma_p}{d\Omega} = |f_p|^2. \quad (89)$$

Second, if the target nucleus has spin zero and the incident particle is a proton, spin 1/2, then

$$f_p = A_p + B_p \mathbf{\sigma} \cdot \mathbf{n} \qquad \frac{d\sigma_p}{d\Omega} = |A_p|^2 + |B_p|^2$$

where

$$A_p = f_c + \frac{\lambda}{2i} \sum [(l+1)(e^{2i\delta_{l+}} - 1) + l(e^{2i\delta_{l-}} - 1)] e^{2i\sigma_l} P_l(\cos\vartheta) \quad (90)$$

$$B_p = -\frac{\lambda}{2} \sum (e^{2i\delta_{l+}} - e^{2i\delta_{l-}}) e^{2i\sigma_l} P_l^{(1)}. \quad (91)$$

Here the phase shift for the combined action of the nuclear and Coulomb field is $\sigma_l + \delta_{l+}$ for $J = l + 1/2$ and $\sigma_l + \delta_{l-}$ for $J = l - 1/2$. Note that in the evaluation of the cross section, only $\sigma_l - \sigma_0$ appears. This quantity is given by

$$\sigma_l - \sigma_0 = \text{phase } (l + i\mu)(l - 1 + i\mu) \cdots (1 + i\mu). \quad (92)$$

This concludes the description of the potential scattering. We can now go on to include the resonance scattering. The pure resonance term

$d\sigma_R/d\Omega$ is still given by Eq. (71) with the substitution

$$\delta(\alpha ls|J\Pi) \to \sigma_l + \delta(\alpha ls|J\Pi). \tag{93}$$

The interference term will consist of two terms, one arising from the interference with the point charge Coulomb scattering amplitude and the other from the remainder of the potential scattering amplitude. The first term designated by $d\sigma_{Ic}/d\Omega$ is

$$\frac{d\sigma_{Ic}}{d\Omega} = -\lambda \frac{(2J+1)}{(2I+1)(2i+1)} \sum_{ls} P_l(\cos\vartheta) \, \text{Re} \, f_c^* \frac{e^{2i(\delta(\alpha ls|J\Pi)+\sigma_l)}}{E - E_{J\Pi} + \tfrac{1}{2}i\Gamma_{J\Pi}} \Gamma(\alpha ls|J\Pi)$$

$$= \frac{\lambda z Z e^2}{2mv^2 \sin^2 \tfrac{1}{2}\vartheta} \frac{2J+1}{(2I+1)(2i+1)} \sum P_l(\cos\vartheta) \frac{\Gamma(\alpha ls|J\Pi)}{(E - E_{J\Pi})^2 + (\tfrac{1}{2}\Gamma_{J\Pi})^2}$$

$$\cdot [(E - E_{J\Pi}) \cos (2\delta(\alpha ls|J\Pi) + 2(\sigma_l - \sigma_0) + \mu \ln \sin^2 \tfrac{1}{2}\vartheta)$$

$$+ \tfrac{1}{2}\Gamma_{J\Pi} \sin (2\delta(\alpha ls|J\Pi) + 2(\sigma_l - \sigma_0) + \mu \ln \sin^2 \tfrac{1}{2}\vartheta)]. \tag{94}$$

The second term designated, $d\sigma_I'/d\Omega$, is given by

$$\frac{d\sigma_I'}{d\Omega} = -\frac{\lambda^2}{(2I+1)(2i+1)} \sum [\bar{Z}(l_1 J_1 l_2 J; sL)]^2 \Gamma(\alpha l_2 s|J\Pi) \sin \delta(\alpha l_1 s|J_1 \Pi_1)$$

$$\cdot \frac{1}{(E - E_{J\Pi})^2 + (\tfrac{1}{2}\Gamma_{J\Pi})^2} \cdot [(E - E_{J\Pi}) \cos [2\delta(\alpha l_2 s|J\Pi)$$

$$+ 2(\sigma_{l_2} - \sigma_{l_1}) - \delta(\alpha l_1 s|J_1 \Pi_1)]$$

$$+ \tfrac{1}{2}\Gamma_{J\Pi} \sin [2\delta(\alpha l_2 s|J\Pi) + 2(\sigma_{l_2} - \sigma_{l_1}) - \delta(\alpha l_1 s)J_1 \Pi_1)]] P_L(\cos\vartheta). \tag{95}$$

Again if $\delta(\alpha ls|J\Pi)$ depends only upon l, Eq. (95) simplifies and can be shown to be equal to

$$\frac{d\sigma_I'}{d\Omega} = -\lambda \frac{2J+1}{(2I+1)(2i+1)} \sum_l P_l(\cos\vartheta) \Gamma(\alpha l|J\Pi) \, \text{Re} \, \frac{f_p'^* e^{2i(\delta_l - \sigma_l)}}{E - E_{J\Pi} + \tfrac{1}{2}i\Gamma_{J\Pi}} \tag{96}$$

where

$$f_p' = \frac{\lambda}{2i} \sum (2l+1) P_l(\cos\vartheta) e^{2i\sigma_l}(e^{2i\delta_l} - 1). \tag{97}$$

The elastic scattering cross section is then

$$\frac{d\sigma(\alpha|\alpha)}{d\Omega} = \frac{d\sigma_p}{d\Omega} + \frac{d\sigma_{Ic}}{d\Omega} + \frac{d\sigma_I'}{d\Omega} + \frac{d\sigma_R}{d\Omega}. \tag{98}$$

Finally we should mention the special case of spin 0 target nucleus, spin 1/2 incident particle. Then we can write the scattering amplitude operator f in the same form as given by Eq. (79) with the substitution

$$A = A_p + \frac{\lambda}{2i}[(l+1)\mathfrak{I}_R(\alpha l\tfrac{1}{2}|l+\tfrac{1}{2}\Pi) + l\mathfrak{I}_R(\alpha l\tfrac{1}{2}|l-\tfrac{1}{2}\Pi)]P_l(\cos\vartheta)$$

$$B = B_p - \frac{\lambda}{2}[\mathfrak{I}_R(\alpha l\tfrac{1}{2}|l+\tfrac{1}{2}\Pi) - \mathfrak{I}_R(\alpha l\tfrac{1}{2}|l-\tfrac{1}{2}\Pi)]P_l^{(1)}. \tag{99}$$

The elastic scattering cross section is

$$\frac{d\sigma}{d\Omega} = |A|^2 + |B|^2. \tag{100}$$

10. Gamma Rays in Nuclear Reactions

The emission or absorption of gamma rays by nuclei is sensitive to the multipole moment involved. It is, therefore, not convenient to employ the channel spin concept which couples the photon spin and the target (or residual) nucleus spin. It is consequently necessary to transform from the channel spin representation to a representation in multipole moments, and to rearrange Eq. (5) accordingly. This has been done by Simon (*17*). We quote his results in Eqs. (101) and (102).

Particle in and gamma ray out:

$$\frac{d\sigma(N,\gamma)}{d\Omega} = \frac{\lambda^2}{2(2i+1)(2I+1)}$$
$$\sum \mathrm{Re}\,\mathfrak{I}^*(\alpha l_1 s|\alpha L_1'p_1'|J_1\Pi_1)\mathfrak{I}(\alpha l_2 s|\alpha' L_2'p_2'|J_2\Pi_2)$$
$$\cdot (L_1'\,1\,{-}1\,1|l_1'0)(L_2'\,1\,{-}1\,1|l_2'0)\epsilon(l_1',p_1')\epsilon(l_2',p_2')$$
$$\cdot (-)^{s-I'-L+L_1'-L_2'+1}[(2J_1+1)(2J_2+1)]^{\frac{1}{2}}W(J_1L_1'J_2L_2';I'L)$$
$$\cdot \bar{Z}(l_1J_1l_2J_2;sL)\bar{Z}(l_1'L_1'l_2'L_2';1L)P_L(\cos\vartheta). \tag{101}$$

Gamma ray in, particle out:

$$\frac{d\sigma(\gamma,N)}{d\Omega} = \frac{\lambda_\gamma^2}{4(2I+1)} \sum \mathrm{Re}\,\mathfrak{I}^*(\alpha L_1 p|\alpha' l_1' s'|J_1\Pi_1)\mathfrak{I}(\alpha L_2 p|\alpha' l_2' s'|J_2\Pi_2)$$
$$\cdot (L_1\,1\,{-}1\,1|l_10)(L_2\,1\,{-}1\,1|l_20)\epsilon(l_1,p)\epsilon(l_2,p)(-)^{s'-I-L+L_1-L_2+1}$$
$$\cdot [(2J_1+1)(2J_2+1)]^{\frac{1}{2}}W(J_1L_1J_2L_2;IL)\bar{Z}(l_1L_1l_2L_2;1L)\bar{Z}(l_1'J_1l_2'J_2;s'L)$$
$$\cdot P_L(\cos\vartheta). \tag{102}$$

Here L_1, L_1', etc. give the polarities of the gamma rays involved, while p is an index which takes on the value 0 for magnetic radiation and unity for electric radiation. The ϵ symbols in these equations differ from zero when

magnetic radiation: $p = 0$ $\quad \epsilon(l_1,0) = 1$, only if $l_1 = L_1$
electric radiation: $p = 1$ $\quad \epsilon(l_1,1) = 1$, only if $l_1 = L_1 \pm 1$

The Clebsch-Gordan coefficients $(L_1 \, 1 \, -1 \, 1 | l_1 0)$ which appear are

$$(L \, 1 \, -1 \, 1 | L 0) = -\frac{1}{\sqrt{2}},$$
$$(L \, 1 \, -1 \, 1 | L+1, 0) = \left[\frac{L}{2(2L+1)}\right]^{\frac{1}{2}},$$
$$(L \, 1 \, -1 \, 1 | L-1, 0) = \left[\frac{L+1}{2(2L+1)}\right]^{\frac{1}{2}}. \qquad (103)$$

Parity conservation requires that for electric multipole transitions the change in parity of the emitting nucleus be $(-)^L$ while for magnetic multipole transitions it is $(-)^{L+1}$.

For the case of a single resonance these formulae can be written in a somewhat simpler form in which the properties of the gamma ray and final state are separated from the properties of the incident nucleon plus target nucleus.

$$\frac{d\sigma(N,\gamma)}{d\Omega} = \frac{\lambda^2(2J+1)}{2(2i+1)(2I+1)} \frac{1}{(E-E_{J\Pi})^2 + (\frac{1}{2}\Gamma_{J\Pi})^2}$$
$$\sum F_N(L,J)F_\gamma(L,J)P_L(\cos\vartheta) \quad (104)$$

$$F_N(L,J) = \sum \cos\left[\delta(\alpha l_2 s | J\Pi) - \delta(\alpha l_1 s | J\Pi)\right] g(\alpha l_1 s | J\Pi) g(\alpha l_2 s | J\Pi)$$
$$\cdot \bar{Z}(l_1 J l_2 J; sL)(-)^s$$

$$F_\gamma(L,J) = \sum g_\gamma(\alpha' L_1' | J\Pi) g_\gamma(\alpha' L_2' | J\Pi)(L_1' 1 \, -1 \, 1 | l_1' 0)(L_2' 1 \, -1 \, 1 | l_2' 0)$$
$$\cdot \epsilon(l_1', p_1')\epsilon(l_2', p_2') W(JL_1' J L_2'; I'L)\bar{Z}(l_1' L_1' l_2' L_2'; 1L)(-)^{L_1'-L_2'+1-L-I'}.$$

The sum F_N is over l_1, l_2, and s while the sum F_γ is over L_1', L_2', l_1', l_2'. These results are more complicated in appearance than in practice. The sum F_γ usually contains only a few terms since only the lowest possible allowed multipole orders need be considered. For example, when there is no nuclear parity change the magnetic dipole and electric quadrupole might occur simultaneously. Then the pair (L_1', l_1') takes on three separate values $(1,1)$ $(2,1)$ $(2,3)$.

Further simplifications in Eq. (104) can occur for special situations. We list a few examples below:

(a) Target nucleus spin $I = 0$, incident particle a nucleon $i = 1/2$:

$$F_N(L,J) = \Gamma(\alpha l s | J\Pi)\bar{Z}(lJlJ; \tfrac{1}{2}L)(-)^s \qquad (105)$$

where l is $J + 1/2$ or $J - 1/2$ depending on the parity change involved.

(b) Target nucleus spin $I = 0$; incident particle spin $i = 0$

$$F_N(L,J) = \Gamma(\alpha J s(=0)|J\Pi)(2J + 1)(JJ00|L0). \tag{106}$$

(c) Pure multipole radiation $L_1' = L_2' = L_1$

$$F_\gamma(L,J) = \Gamma_\gamma(\alpha'L_1|J\Pi)W(JL_1JL_1;I'L)\sum (L_1\,1\,-1\,1|l_1'0)(L_1\,1\,-1\,1|l_2'0)$$
$$\cdot \bar{Z}(l_1'L_1l_2'L_1;1L). \tag{107}$$

(d) Final state spin $I' = 0$

$$F_\gamma(L,J) = \Gamma_\gamma(\alpha'J|J\Pi)\frac{(-)^{2J+1}}{2J+1}\sum (J\,1\,-1\,1|l_1'0)(J\,1\,-1\,1|l_2'0)$$
$$\cdot \bar{Z}(l_1'Jl_2'J;1L). \tag{108}$$

11. Polarization (18)

Further insight into the properties of compound nuclear levels and the levels of the target and residual nuclei can be obtained by measurement of the polarization of the emergent particles, when these possess a spin. This measurement involves the use of an appropriate analyzer which, for example, can be a scatterer. In that case the asymmetry of the beam scattered by the analyzer for a given angle of scattering is directly related to the measured polarization. A more complicated experiment involves the use of a polarized beam; that is, we require a polarizer. This polarized beam is then scattered by the substance under examination and the polarization of the scattered beam measured by an analyzer.

The simpler experiment, scatterer plus analyzer, in which an unpolarized beam is scattered (we use the term "scatter" here to indicate not only elastic scattering but also any reaction such as inelastic scattering or a transmutation) by the nucleus of interest (called the scatterer) is a direct measure of the interference among amplitudes corresponding to quantum numbers describing the various possible components of the scattered beam. The interfering amplitudes may differ because of their final channel spin, orbital angular momentum, total angular momentum or parity. For an isolated resonance, interference between the resonance and potential scattering amplitudes is always present in the elastic scattering process. In reaction processes, interference with nearby levels (although in this case the resonance is not truly "isolated") or more generally with distant levels as incorporated in the nonresonant "potential" reaction amplitude can occur. Interference is also possible if the final state of the residual nucleus can be excited by at least two differing channels (l_1', s_1' and l_2', s_2').

The second experiment mentioned above is *not* a measure of the interference among different exit channels and thus is a possible experiment even when there is only one exit channel leading to a particular final state. As we shall see for elastic scattering it would yield results which combined with the first experiment and measurement of the angular distribution would permit the complete determination of the phase shifts. Unfortunately this experiment is at the present time not practicable.

A simple and important example is provided by the elastic scattering of spin (1/2) particles; for example, nucleons by spin zero nuclei. The scattering amplitude f as an operator acting on the spin wave function for the incident particle is given in Eq. (79) as

$$f = A + B(\boldsymbol{\sigma} \cdot \mathbf{n}), \tag{79'}$$

where A and B are defined in Eqs. (81) and (82), and $\mathbf{n}$ a unit vector perpendicular to the plane containing the initial and final momenta $\mathbf{k}_i$ and $\mathbf{k}_f$ is defined in Eq. (80). The differential cross section for an unpolarized incident beam is:

$$\frac{d\sigma(\alpha|\alpha)}{d\Omega} = |A|^2 + |B|^2. \tag{84'}$$

Suppose we now follow up the scattering by the target nucleus of interest by a second scattering by another nucleus of spin 0 of known properties, the analyzer. The scattering amplitude for the double scattering will then be

$$f_{\text{SA}} = f_A(\Omega_A) f_S(\Omega_S) \tag{109}$$

where the subscripts A and S refer to analyzer and scatterer and where Ω is a shorthand for the spherical angles of scattering (ϑ, φ). Then the cross section for an initially unpolarized beam is

$$\frac{d\sigma_{\text{SA}}}{d\Omega} = \frac{1}{2} \operatorname{tr} f_{\text{SA}}^+ f_{\text{SA}} = \frac{1}{2} \operatorname{tr} f_S^+ f_A^+ f_A f_S$$

where tr means trace. Then

$$\frac{d\sigma_{\text{SA}}}{d\Omega} = \frac{d\sigma_S}{d\Omega} \frac{d\sigma_A}{d\Omega} + 4(\operatorname{Re} B_S A_S^*)(\operatorname{Re} B_A A_A^*) \mathbf{n}_A \cdot \mathbf{n}_S$$

or

$$\frac{d\sigma_{SA}}{d\Omega} = \frac{d\sigma_S}{d\Omega}\frac{d\sigma_A}{d\Omega}[1 + P_S P_A \mathbf{n}_A \cdot \mathbf{n}_S] \quad (110)$$

where

$$P_S = \frac{2 \text{ Re } B_S A_S^*}{d\sigma_S/d\Omega}. \quad (111)$$

The geometry of the situation is shown in Fig. 7. Since $\mathbf{n}_S \sim (\mathbf{k}_3 \times \mathbf{k}_2)$ we see that the sign of the second term in the brackets in Eq. (110) will, for the direction of scattering indicated by the solid line, be exactly opposite to that for the broken line case. It is this asymmetry which permits the determination of P_S.

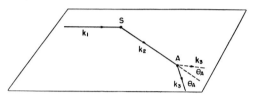

FIG. 7. Double scattering geometry. S is the scatterer and A the analyzer.

We also note that P_S can be interpreted as a measure of the polarization with respect to the direction perpendicular to the plane of scattering; that is, in the direction $\mathbf{n}_S$. This follows from the fact that P_S is the expectation value of $\boldsymbol{\sigma} \cdot \mathbf{n}$ with respect to the amplitude f_S:

$$P_S = \frac{\text{tr } f_S^+ \boldsymbol{\sigma} \cdot \mathbf{n} f_S}{\text{tr } f_S^+ f_S}. \quad (112)$$

A comment follows from the evaluation of P_S in terms of the $\mathfrak{J}$ matrix. Employing Eqs. (81) and (82), we find

$$P_S \frac{d\sigma_S}{d\Omega} = -\frac{\lambda^2}{2}\text{Im}\sum P_l^{(1)} P_{l'}\{(l+1)\mathfrak{J}^*(l+)\mathfrak{J}(l'+) - l\mathfrak{J}^*(l-)\mathfrak{J}(l'-) \\ -(l+1)\mathfrak{J}^*(l-)\mathfrak{J}(l'+) + l\mathfrak{J}^*(l+)\mathfrak{J}(l'-)\}.$$

It is easy to verify that the only nonvanishing contributions to the above sum come from interference terms.

The equivalent results can of course be obtained if a polarized beam is incident upon the scatterer and the asymmetry then measured. In other words in Fig. 7 the first scatterer is a polarizer and the second scatterer, the nucleus under investigation. Then

$$\frac{d\sigma_{PS}}{d\Omega} = \frac{d\sigma_P}{d\Omega}\frac{d\sigma_S}{d\Omega}[1 + P_P P_S \mathbf{n}_P \cdot \mathbf{n}_S]. \quad (113)$$

Measuring the asymmetry yields P_S as in Eq. (110). It is probably not necessary to add that the polarizer need not be a scatterer. A reaction which produces a polarized product is just as good. P_P is, as we have seen

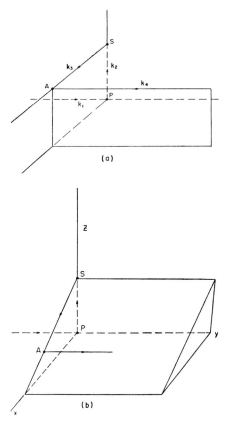

FIG. 8. Triple scattering geometry. P is the polarizer, S the scatterer, and A the analyzer. In (a) $\mathbf{n}_P$, $\mathbf{n}_S$, and $\mathbf{n}_A$ are mutually perpendicular. In (b) $\mathbf{n}_P$ and $\mathbf{n}_S$ are perpendicular but $\mathbf{n}_A$ is not perpendicular to $\mathbf{n}_S$. For definition of the vectors $\mathbf{n}$ see Eq. (80).

from Eq. (112), just the polarization transverse to the plane of motion.

We turn now to the experiment in which a polarized beam is scattered by the scatterer and then scattered by the analyzer as illustrated in Fig. 8 showing a polarizer (P), a scatterer (S), and an analyzer (A). Then

$$\frac{d\sigma_{\text{PSA}}}{d\Omega} = \frac{1}{2} \operatorname{tr} f_A^+ f_S^+ f_P^+ f_P f_S f_A.$$

We obtain

$$\frac{d\sigma_{PSA}}{d\Omega} = \left(\frac{d\sigma}{d\Omega}\right)_P \left(\frac{d\sigma}{d\Omega}\right)_S \left(\frac{d\sigma}{d\Omega}\right)_A \{1 + P_P P_S \mathbf{n}_P \cdot \mathbf{n}_S + P_A P_S (\mathbf{n}_A \cdot \mathbf{n}_S)$$
$$+ P_A P_P (\mathbf{n}_A \cdot \mathbf{n}_P) + P_A P_P [Q_S (\mathbf{n}_P \times \mathbf{n}_S) \cdot \mathbf{n}_A + R_S (\mathbf{n}_P \times \mathbf{n}_S) \cdot (\mathbf{n}_S \times \mathbf{n}_A)]\}, \quad (114)$$

where

$$Q_S = \frac{2 \operatorname{Im} A_S B_S{}^*}{(d\sigma_S/d\Omega)} \qquad R_S = \frac{2|B_S|^2}{d\sigma_S/d\Omega}. \quad (115)$$

It is quite clear that by measuring appropriate asymmetries it is possible to measure Q_S and R_S (see Fig. 8a for Q_S and Fig. 8b for R_S). These together with $d\sigma_S/d\Omega$ and P_S permit the complete determination of the magnitudes and relative phases of A_S and B_S. Note that both Q_S and R_S unlike P_S do not reduce to zero in the absence of interference.

We turn now to the more complicated problems in which the initial and final nuclei spins are not zero and in which the incident or emergent particle may have spin greater than 1/2. The amplitude for the transition from an initial channel spin s, z-component m to a final spin s', m' is (9):

$$f(\alpha' s' m' | \alpha s m) = \lambdabar \sum [\pi(2l+1)]^{\frac{1}{2}} (ls0m|Jm)(l's'\mu'm'|Jm) Y_{l'\mu'}$$
$$\cdot \Im(\alpha' s' l' | \alpha s l | J\Pi), \quad (116)$$

where the sum is over all quantities except those contained in the argument of f. To obtain the cross section for the production of a given quantity $O(\mathbf{i}')$ which is a function only of $\mathbf{i}'$, the emergent particle spin, we take the expectation values of this quantity $O(\mathbf{i}')$ with respect to the final state:

$$\frac{d\sigma(O)}{d\Omega} = \frac{1}{(2i+1)(2I+1)}$$
$$\cdot \sum f^*(\alpha' s_1' m_1' | \alpha s m) f(\alpha' s_2' m_2' | \alpha s m) < s_1' m_1' | O(\mathbf{i}') | s_2' m_2' > \quad (117)$$

or

$$\frac{d\sigma(O)}{d\Omega} = \frac{1}{(2i+1)(2I+1)}$$
$$\cdot \sum f^*(\alpha' s_1' m_1' | \alpha s m) f(\alpha' s_2' m_2' | \alpha s m)(i'I'\nu'\Lambda'|s_1'm_1')(i'I'\nu''\Lambda'|s_2'm_2')$$
$$\cdot <i'\nu'|O(\mathbf{i}')|i'\nu''>. \quad (118)$$

If we let

$$F_{\alpha'\alpha}(sm|i'\nu'\Lambda') = \sum_{s'm'} f(\alpha' s' m' | \alpha s m)(i'I'\nu'\Lambda'|s'm'), \quad (119)$$

then

$$\frac{d\sigma(O)}{d\Omega} = \frac{1}{(2i+1)(2I+1)} \sum <i'\nu'|O(\mathbf{i}')|i'\nu''> F_{\alpha'\alpha}^*(sm|i'\nu'\Lambda')$$
$$F_{\alpha'\alpha}(sm|i'\nu''\Lambda'). \quad (120)$$

In the last equation we have expressed the cross section in a form which directly gives the weight factor for each possible matrix element of $O(\mathbf{i}')$. If this information is not required it is possible (if $i' = 1/2$, Eq. (120) is already in form (121)) to calculate the matrix element of $O(\mathbf{i}')$ in Eq. (117) by making direct use of the transformation properties of O. Suppose we consider a component of O which transforms under rotation like a spherical harmonic of order q, component k. Then one can show (17) that

$$<s_1'm_1'|O_k^{(q)}|s_2'm_2'> = (-)^{I'-k-i'-m_1'}\left[\frac{(2s_1'+1)(2s_2'+1)(2i'+1)}{2q+1}\right]^{\frac{1}{2}}$$
$$W(i's_1'i's_2';I'q) \cdot (s_1's_2' - m_1'm_2'|q-k)$$

so that

$$\frac{d\sigma(O_k^{(q)})}{d\Omega} = \frac{1}{(2i+1)(2I+1)}$$
$$\sum f^*(\alpha's_1'm_1'|\alpha sm)f(\alpha's_2'm_2'|\alpha sm)(-)^{I'-k-i'-m_1'}$$
$$\cdot \left[\frac{(2s_1'+1)(2s_2'+1)(2i'+1)}{2q+1}\right]^{\frac{1}{2}}$$
$$\cdot (s_1's_2' - m_1'm_2'|q-k)W(i's_1'i's_2';I'q). \quad (121)$$

The sum is over s_1', s_2', m_1', m_2', s,m.

We turn now to the special case of polarization. Then

$$O(\mathbf{i}') = \mathbf{i}'$$

and the corresponding $O_k^{(q)}$ is

$$O_{\pm 1}^{(1)} = \mp \frac{i_x' \pm i_y'}{[2i'(i'+1)]^{\frac{1}{2}}} \qquad O_0^{(1)} = \frac{i_z'}{[i'(i'+1)]^{\frac{1}{2}}}.$$

In this case the matrix element $<i'\nu'|\mathbf{i}'|i'\nu''>$ is just the matrix representation of $\mathbf{i}'$. Either expression (120) or (121) is quite suitable for practical evaluation. However, general results are not so readily obtainable and for this purpose it is appropriate to perform the sum over magnetic quantum numbers. This has been done by Simon and Welton (19) and by Simon (17) in terms of the Wigner 9j coefficients. We shall not quote the final formulae, but shall give only the consequent theorems.

The cross section $d\sigma(\mathbf{i})/d\Omega$ is proportional to $\mathbf{n}$

$$\mathbf{n} = \frac{(\mathbf{k}_{\alpha'} \times \mathbf{k}_{\alpha})}{|\mathbf{k}_{\alpha'} \times \mathbf{k}_{\alpha}|}$$

where $\mathbf{k}_{\alpha}$ and $\mathbf{k}_{\alpha'}$ are the momenta of the incident and emergent particles (more precisely in the entrance and exit channel). Polarization results from the interference of different subchannels. If only one exit channel is possible there will be no polarization. There will be no polarization if only S-waves are effective *either* in the entrance or exit channel, and also if only zero channel spin is effective in the exit channel. The cross section can be expanded in terms of the associated Legendre polynominals $P_L^{(1)}$, the maximum value of L being the minimum of $2J$, $2l$, and $2l'$ where J is the maximum total angular momentum entering into the process, l the maximum orbital angular momentum in the entrance channel, and l' the maximum in the exit channel. Clearly $L \geq 1$. These results can be used as a check on calculations employing Eq. (120) or Eq. (121).

12. Overlapping Levels

When the levels of the compound nucleus overlap; that is, the width of the associated resonance is comparable with the spacing between their maxima, the expressions for the transition matrix $\mathfrak{I}$ may require modification. The general expressions for angular distribution, total cross section, polarization, etc. given in terms of $\mathfrak{I}$ remain valid. As we can see directly from these expressions, modification is required only when the overlapping compound nuclear levels have the same total angular momentum J and the same parity Π. In that event Eq. (22) for the resonant part of the scattering must be replaced by the following (6):

$$\mathfrak{I}_R(\alpha'l's'|\alpha ls|J\Pi) = -i e^{i[\delta(\alpha ls|J\Pi)+\delta(\alpha'l's'|J\Pi)]} \sum_i \frac{B_i(\alpha'l's'|\alpha ls|J\Pi)}{E - W_i(J\Pi)}, \quad (122)$$

where the sum is over the possible overlapping levels. The constants B_i and W_i are complex, W_i depending only on J and Π. We have assumed that there is no nonresonant reaction amplitude. This is not just the sum of Breit-Wigner amplitudes since in the latter there is a precise relation between B_i and Im W_i. For example, if we are dealing with isolated elastic scattering resonances B_i is real, and if there are no reactions (pure elastic scattering) B_i is just twice ($-$ Im W_i). Or more generally as can be seen from Eq. (22), B_i is factorable into $g(\alpha ls|J\Pi)$ for the entrance channel and $g(\alpha'l's'|J\Pi)$ for the exit channels. Then $(-\operatorname{Im} W_i)$ is just 1/2 the sum of $[g(\alpha'l's'|J\Pi)]^2$ over all possible $\alpha'l'$, s' including α, l, s. Of course in the limit of no overlap Eq. (122) must reduce to a sum of Breit-Wigner amplitudes.

In the more general situation represented by Eq. (122) we find that these Breit-Wigner relations hold only on the average. For example, the sum of the B_i can be written as a sum of factorable products

$$\sum_i B_i = \sum_\mu \gamma_\mu(\alpha ls|J\Pi)\gamma_\mu(\alpha'l's'|J\Pi). \tag{123}$$

Moreover,

$$\sum_i \operatorname{Im} W_i = -\tfrac{1}{2} \sum |\gamma_\mu(\alpha'l's'|J\Pi)|^2. \tag{124}$$

The usual empirical analysis of overlapping resonances proceeds by making the tentative assumption that the resonance structure can be resolved into a sum of Breit-Wigner amplitudes. Corrections to this ansatz to obtain a better fit can then be made by suitably varying the parameters B_i, W_i away from their Breit-Wigner values. It is worth noting that in most instances these corrections are small.

In the case of pure elastic scattering (no significant contribution from reactions) it is possible to find relations between the resonance parameters which may be useful in the event that the more complicated form Eq. (122) is required. The first of these follows directly from Eqs. (123) and (124),

$$\tfrac{1}{2}\sum B_i = -\sum \operatorname{Im} W_i. \tag{125}$$

It is also possible to show that the first moment of the B_i also has a simple connection to $\operatorname{Im} W_i$:

$$\operatorname{Im} \sum B_i W_i = -\tfrac{1}{2}\left(\sum B_i\right)^2 = -2\left(\sum \operatorname{Im} W_i\right)^2. \tag{126}$$

An important simplification occurs when only two levels overlap. Then

$$\mathfrak{I}_R = -ie^{2i\delta(\alpha ls|J\Pi)} \frac{\Gamma E - S}{(E - E_1)(E - E_2) - i(\Gamma E - S)}$$

The parameters Γ, S, E_1 and E_2 are real.

13. Many Overlapping Levels. Statistical Theory

When many levels overlap, the analysis of the resultant comparatively smooth cross sections in terms of the resonance structure given by Eq. (122) is certainly inconvenient if not impossible. If a sufficiently large number of compound nuclear levels are involved, the resultant transition amplitude and cross section cannot be sensitive to the individual values of the parameters B_i and W_i. Rather they must depend only on appropriate averages of B_i and W_i. For this purpose we need to know the distribution functions for B_i and W_i (or related functions) and the various correlations between B_i's or W_i's. In the absence of such information the "reasonable" assumption of randomness has been made (20) so that the correlation functions were dropped. Then even without knowledge of the distribution functions various relevant averages could be calculated but the fluctuation away from the average could not be evaluated.

However in recent years direct evidence on the nature of the distribution functions, for widths and level spacing have been obtained. This evidence permits the development of a firmer basis for the statistical assumptions we shall need. These experiments have been performed with neutrons of low energy (some tens of electron volts) with the following tentative results which we take from the analysis of Porter and Thomas (21). The distribution of neutron widths is found to follow the following law where $P(\Gamma)d\Gamma$ is the probability that the value of the width will fall between Γ and $\Gamma + d\Gamma$

$$P(\Gamma)\, d\Gamma = \frac{1}{\sqrt{2\pi}} \frac{e^{-x/2}}{\sqrt{x}} dx \qquad x = \frac{\Gamma}{<\Gamma>}. \qquad (127)$$

Here $<\Gamma>$ is the average value of Γ:

$$<\Gamma> = \int_0^\infty \Gamma P(\Gamma)\, d\Gamma.$$

We note that small widths are favored and more important that the dispersion is very large:

$$\delta = \frac{<\Gamma^2> - <\Gamma>^2}{<\Gamma>^2} = 2. \qquad (128)$$

In other words the distribution $P(\Gamma)$ is very broad.

The observed distribution of the spacing of energy levels is consistent with the Wigner distribution (22):

$$P(S)\, dS = \frac{\pi}{2D^2} S e^{-\frac{\pi}{4}\frac{S^2}{D^2}} dS, \qquad (129)$$

where D is the average distance between levels. The dispersion in this case is

$$\delta = \frac{<S^2> - D^2}{D^2} = \frac{4}{\pi} - 1. \qquad (130)$$

These results, Eqs. (127) and (129), by their very form, suggest that there is a random character present in the description of nuclear reactions. In fact it has been shown by Porter and Blumberg (*21*), following a line of attack suggested by Wigner (*22*), that both results follow from the assumption that the matrix elements of the Hamiltonian, whose eigenvalues and eigenvectors are the energy levels and wave functions of the compound nucleus, are random, their distributions being Gaussian.[e] The underlying reasons justifying this hypothesis have yet to be delineated. But the fact that the randomness hypothesis leads directly to distributions (127) and (129) provides a very strong support for its validity. As we shall see it has other far-reaching implications. We shall need almost immediately one important albeit trivial corollary. The distributions of the matrix elements g (Eqs. (23 and 24)) whose square is the width Γ is gaussian:

$$P(g)\,dg = \frac{1}{\sqrt{2\pi\langle g^2\rangle}}\,e^{-\frac{1}{2}\frac{g^2}{\langle g^2\rangle}}. \tag{131}$$

We see that g is positive or negative with equal probability.

The randomness assumption of Wigner and Porter can be employed to simplify the expression for the transition amplitude $\mathfrak{J}(\alpha'l's'|\alpha ls|J\Pi)$ when the contribution to $\mathfrak{J}(\alpha'l's'|\alpha ls|J\Pi)$ comes from many levels. The additional assumption is that so many levels contribute that the cancellations which occur because of randomness of the appropriate matrix element is complete. Then it is possible to show that Eq. (122) reduces to

$$\mathfrak{J}_R(\alpha'l's'|\alpha ls|J\Pi) = -ie^{i[\delta(\alpha ls|J\Pi)+\delta(\alpha'l's'|J\Pi)]}\sum_\lambda \frac{g_\lambda(\alpha'l's'|J\Pi)g_\lambda(\alpha ls|J\Pi)}{E - E_{J\Pi\lambda} + \tfrac{1}{2}i\Gamma_{J\Pi\lambda} + \tfrac{1}{2}i\Gamma'_{J\Pi\lambda}} \tag{132}$$

where

$$\Gamma_{J\Pi\lambda} = \sum g_\lambda{}^2(\alpha'l's'|J\Pi). \tag{133}$$

The sum in Eq. (132) is just a sum of Breit-Wigner type terms over the various compound nucleus levels λ [see Eq. (20)] except for the presence of the Γ' terms in the denominator which modifies the relation between numerator and denominator. However the Γ' terms have an average value of zero; that is,

$$\sum_\lambda \Gamma_{J\Pi\lambda}' = 0. \tag{134}$$

---

[e] So far Porter and Blumberg (*21*) have looked at the situation where the distributions are identical for the diagonal as well as off diagonal elements.

We shall also need the absolute square of the transition amplitude. Employing the randomness assumption for the g-factors in Eq. (132) we obtain

$$|\mathfrak{I}_R(\alpha'l's'|\alpha ls|J\Pi)|^2 = \sum \frac{\Gamma_\lambda(\alpha'l's'|J\Pi)\Gamma_\lambda(\alpha ls|J\Pi)}{(E - E_{J\Pi\lambda})^2 + \frac{1}{4}(\Gamma_{J\Pi\lambda} + \Gamma_{J\Pi\lambda'})^2}.$$

This equation can be simplified by taking into account the usual experimental situation of a large energy spread in the incident beam. It is then appropriate to take an energy average of the above expression. This yields

$$<|\mathfrak{I}_R(\alpha'l's'|\alpha ls|J\Pi)|^2> = \frac{2\pi}{\Delta E}\sum \frac{\Gamma_\lambda(\alpha'l's'|J\Pi)\Gamma_\lambda(\alpha ls|J\Pi)}{\Gamma_{J\Pi\lambda} + \Gamma_{J\Pi\lambda'}}.$$

Here ΔE is the energy spread of the beam. If we now assume that all the Γ_λ are taken from the identical probability distribution Eq. (127), the above equation can be rewritten as follows

$$<|\mathfrak{I}_R(\alpha'l's'|\alpha ls|J\Pi)|^2> = \frac{2\pi}{D_{J\Pi}}\left\langle\frac{\Gamma(\alpha'l's'|J\Pi)\Gamma(\alpha ls|J\Pi)}{\Gamma_{J\Pi} + \Gamma_{J\Pi'}}\right\rangle. \quad (135)$$

The calculation of the average over λ on the right-hand side of this equation is not yet possible because of the absence of any direct information, either experimental or theoretical, on the probability distribution for $\Gamma_{J\Pi'}$. The simplest result is obtained if one assumes that the $\Gamma_{J\Pi'}$ distribution is much broader than that for $\Gamma_{J\Pi}$. Then to a first approximation the fluctuations in the numerator are not correlated with those in the denominator and one obtains approximately for many exit channels

$$\left\langle\frac{\Gamma(\alpha'l's'|J\Pi)\Gamma(\alpha ls|J\Pi)}{\Gamma_{J\Pi} + \Gamma_{J\Pi'}}\right\rangle = \frac{<\Gamma(\alpha'l's'|J\Pi)><\Gamma(\alpha ls|J\Pi)>}{<\Gamma_{J\Pi}>}. \quad (136)$$

At the opposite extreme the distribution for $\Gamma_{J\Pi'}$ is much narrower than that for $\Gamma_{J\Pi}$. Then the values of the numerator and denominator are closely correlated. Equation (136) is then no longer correct. The right-hand side should be multiplied by a correction factor which usually is considerably smaller than unity. The experimental results[d] seem to indicate that the correction factor should be close to unity and thus that Eq. (136) is approximately correct as it stands. It is hardly necessary

---

[d] We refer here to the energy region in which the compound nuclear levels overlap significantly. If they do not, for example, for thermal energies, $\Gamma_{J\Pi'}$ is zero and equality (136) does not hold. However, the average value required can then be calculated with the aid of Eq. (127). The effect of this correction is considerable and has been calculated and applied by Lane and Lynn to low-energy neutron reactions.

to add that further experimental and theoretical studies of this point would be desirable.

For the rest of this discussion we shall assume the correctness of Eq. (136). Recall that the average values of Γ can be expressed in terms of the transmission factors as given by Eq. (25) so that finally[e]

$$<|\mathfrak{J}_R(\alpha'l's'|\alpha ls|J\Pi)|^2> = \frac{T_{\alpha'l's'}{}^{(J)}T_{\alpha ls}{}^{(J)}}{\sum T_{\alpha''l''s''}{}^J}. \quad (137)$$

This is the familiar expression (*23*) in terms of which it becomes possible to make complete and unambiguous predictions for the resonance contribution to reaction cross sections.

We turn to these predictions now. To compare with experiment it is of course necessary to add the nonresonant contribution. Again because of the random character of $\mathfrak{J}_R$ there is no interference between $\mathfrak{J}_p$ and $\mathfrak{J}_R$ and therefore the two contributions, resonant and nonresonant, combine incoherently. In other words, the experimental cross section $d\sigma/d\Omega$ is the sum of the resonant $d\sigma_R/d\Omega$ and nonresonant cross sections $d\sigma_p/d\Omega$ while the interference term $d\sigma_I/d\Omega$ vanishes. The nonresonant cross section is discussed in Section V.B. Employing again the randomness hypothesis and averaging over energy, the cross section $d\sigma_R/d\Omega$ of Eq. (15) reduces to

$$\left\langle \frac{d\sigma_R(\alpha'|\alpha)}{d\Omega} \right\rangle = \frac{\lambda^2}{4} \sum \frac{(-)^{s'-s}}{(2I+1)(2i+1)}$$
$$<|\mathfrak{J}_R(\alpha'l's'|\alpha ls|J\Pi)|^2> \bar{Z}(lJlJ;sL)\bar{Z}(l'Jl'J;s'L) \cdot P_L(\cos\vartheta), \quad (138)$$

where the average $<|\mathfrak{J}_R|^2>$ is given in terms of the transmission factors by Eq. (137). From the properties of the $\bar{Z}$-coefficient we note that L is even, so that the angular distribution is symmetric about 90°. This result is not surprising since asymmetry about 90° can occur only if there are interference terms between states of different parity. Of course, there are no interference terms in Eq. (138). In passing we note that for the same reason, no interference, the statistical model leads to zero polarization.

Finally we give the total cross section for a given reaction

$$<\sigma_R(\alpha'|\alpha)> = \pi\lambda^2 \sum \frac{2J+1}{(2I+1)(2i+1)} \frac{T_{\alpha'l's'}{}^{(J)}T_{\alpha ls}{}^{(J)}}{\sum T_{\alpha''l''s''}{}^{(J)}}. \quad (139)$$

---

[e] When the nonresonant term is appreciable it is necessary to employ the more accurate relation between T and Γ given by Eq. (37) of Section VI.D.

Equations (138) and (139) are identical with those obtained in ref. (*23*) except that in the latter it was assumed that the transmission factors T depended only on α and l. This is correct only if the complex potential model potential is central, containing neither a spin-orbit term which would lead to a J-dependent T, nor a spin-spin force which would produce an s-dependent T. Certainly there is some effect from spin-orbit forces. The existence of a spin-spin force has not yet been established so that it is probably small.

We finally consider the case in which the energy of excitation of the residual nucleus can be so high that the energy levels of that nucleus available for the reaction form an essentially continuous distribution which can be approximately described by a density function. This function $\omega_{I'}(\epsilon)$ depends upon the spin of the levels in the residual nucleus and the energy of excitation ϵ,

$$\epsilon = E - E' + Q_{\alpha'}. \tag{140}$$

In this equation, E is the incident energy, E' the kinetic energy of the final particle, and $Q_{\alpha'}$ the reaction energy. The energy-angle distribution of the emergent particles is given by

$$\left\langle \frac{d^2\sigma_R}{d\Omega\, dE'} \right\rangle = \sum_{I'} \omega_{I'}(E + Q_{\alpha'} - E') \left\langle \frac{d\sigma_R(E'I'\beta'|\alpha)}{d\Omega} \right\rangle. \tag{141}$$

We have made the dependence of σ_R on E' and I' explicit; the remaining possible variables are subsumed by β'.

The well-known result, the isotropy of reaction products when the residual nucleus is highly excited, follows from Eq. (141) when two assumptions are made. The first is that the factors $T_{ls}{}^J$ are independent of s, an approximation which from present evidence seems reasonably precise. The second is that

$$\omega_{I'}(\epsilon) \simeq (2I' + 1)\omega_0(\epsilon). \tag{142}$$

This ansatz was first examined critically by C. Bloch (*24*) who demonstrated its validity as long as I' is not too large. The total cross section per unit energy becomes

$$\left\langle \frac{d\sigma_R}{dE'} \right\rangle = \pi\lambda^2 \sum \frac{(2J+1)(2l'+1)}{(2I+1)(2i+1)} \frac{T_{\alpha'l'}{}^J(E')T_{\alpha l}{}^J(E)\omega_0(E+Q_{\alpha'}-E')}{\int \sum_{\alpha''l''} (2l''+1)T_{\alpha''l''}{}^J(E'')\omega_0(E+Q_{\alpha''}-E'')\, dE''}. \tag{143}$$

It should be recognized that in the denominator we have assumed that (142) holds for *all* exit channels. This is, of course, not necessarily the case but the required changes would complicate Eq. (143) considerably and for that reason we have not indicated them, although in any particular case they are easily inserted.

The Weisskopf form (*20*) for $<d\sigma_R/dE'>$ which is generally employed in the analysis of experimental results requires one further approximation; namely that $T_{\alpha l}{}^J$ is independent of J. Then

$$\left\langle \frac{d\sigma_R}{dE'} \right\rangle \simeq \sigma_{c\alpha}(E) \frac{E'\sigma_{c\alpha'}\omega_0(E + Q_{\alpha'} - E')}{\sum_{\alpha''} \int E''\sigma_{c\alpha''}\omega_0(E + Q_{\alpha''} - E'')\, dE''}, \quad (144)$$

where $\sigma_{c\alpha}(E)$ is the cross section for the formation of the compound nucleus with the target nucleus specified by α:

$$\sigma_{c\alpha} = \pi\lambda^2 \sum (2l + 1)T_{\alpha l}(E). \quad (145)$$

With this result we conclude the description of the major results which flow from the statistical assumption. Some remarks about their relation to experimental results is in order. In the low-energy range where only a few levels of the residual nucleus can be excited, expressions (138) and (139) yield surprisingly good agreement with experiment. The angular distribution is symmetric about 90° and the excitation curves are nicely explained by (139). In fact Eq. (139) can be used to limit the possible angular momenta J' of the residual nucleus which can be involved. The small deviations from these results can usually be explained by the presence of nonresonant direct interactions of which the excitation of collective motions, for example, rotational levels, seem to be particularly important. At the higher energies for which many levels of the target nuclei can be excited, the theory predicts isotropy and an energy distribution given by Eq. (143) or (144). Again the predictions are found to be correct as long as the excitation energy is large enough. In fact for neutrons the major part of the inelastic cross section is isotropic and is estimated correctly. However, transitions to the low lying levels of the residual nucleus are not given correctly. For these cases the nonresonant direct interaction processes are dominant, the angular distribution being not symmetric about 90°. The contribution of the resonant compound nuclear states to these transitions is reduced because these transitions form so small a fraction of the possible modes of decay of these states. In the low-energy range where the only transitions possible are to these states, the resonant contributions usually overwhelm the nonresonant contributions.

14. Density of Levels

We quote here the semiempirical expression obtained by T. D. Newton (*25*) for the spacing $D_0 = \omega_0^{-1}$. (See also reference *26*.)

$$\omega_0^{-1} = D_0 = A^{8/3}(2\bar{\jmath}_N + 1)^{1/2}(2\bar{\jmath}_z + 1)^{1/2}(2U + 3T)^2 e^{8.75 - 2U/T} \quad (146)$$

$$T = \frac{4.1 U^{1/2}}{(\bar{\jmath}_N + \bar{\jmath}_z + 1)^{1/2} A^{1/3}},$$

where D_0 is expressed in electron volts, T and U in millions of electron volts; T is the nuclear temperature while U is the excitation energy ϵ minus a pairing correction. The pairing correction for the various elements is listed in Newton's Table I. The average single-particle angular momenta $\bar{\jmath}_N$ and $\bar{\jmath}_z$ are listed in Newton's Table II. Expression (146) is applicable for small values of E. Its prediction of the level densities as observed by thermal neutrons is at most off by a factor of 3, which considering the large orders of magnitude variation of D_0 constitutes rather good agreement.

REFERENCES

References *1-7a* are general references.

1. J. M. Blatt and V. F. Weisskopf, *Theoretical Nuclear Physics* (John Wiley and Sons, New York, 1952).
2. R. G. Sachs, *Nuclear Theory* (Addison-Wesley Publishing Co., Cambridge, Mass., 1953).
3. A. M. Lane and R. G. Thomas, Revs. Modern Phys. **30,** 257 (1958).
4. Nuclear Reactions I, *Handbuch der Physik*, edited by S. Flügge (J. Springer Verlag, Berlin, 1957), Vol. XL.
5. Nuclear Reactions III, *Handbuch der Physik*, edited by S. Flügge (J. Springer Verlag, Berlin, 1957), Vol. XLII.
6. H. Feshbach, Ann. Phys. (N.Y.) **5,** 357 (1958).
7. P. M. Endt and M. Demeur, eds., *Nuclear Reactions* (North Holland Publishing Co., Amsterdam, 1959).
7a. Since the writing of this chapter the following important general reference has been published: Nuclear Reactions II, Theory, *Handbuch der Physik*, edited by S. Flügge (J. Springer Verlag, Berlin, 1959), Vol. XLI/1.
8. E. P. Wigner and L. Eisenbud, Phys. Rev. **72,** 29 (1947).
9. J. M. Blatt and L. C. Biedenharn, Revs. Modern Phys. **24,** 249 (1952).
10. R. Huby, Proc. Phys. Soc. **A67,** 1103 (1954).
11. M. Jacob and G. C. Wick, Ann. Phys. (N.Y.) **7,** 404 (1959).
12. E. P. Wigner, Phys. Rev. **73,** 1002 (1948).
13. H. Feshbach, C. E. Porter, and V. F. Weisskopf, Phys. Rev. **96,** 448 (1954).
14. M. M. Shapiro, Phys. Rev. **90,** 171 (1953).
15. T. Teichman and E. P. Wigner, Phys. Rev. **87,** 123 (1952).
16. E. Vogt, Resonance Reactions, Theoretical, in *Nuclear Reactions* (ref. 7), Chapter V.

17. A. Simon, Phys. Rev. **92,** 1050 (1953).
18. J. V. Lepore, Phys. Rev. **79,** 137 (1950); L. Wolfenstein, Ann. Rev. Nuclear Sci. **6,** 43 (1956); R. van Wageningen, Some Aspects of the Theory of Elastic Scattering of Spin 0 and Spin 1/2 Particles by Spin 0 Targets, Thesis, University of Groningen (1957).
19. A. Simon and T. A. Welton, Phys. Rev. **90,** 1036 (1953); Errata, Phys. Rev. **93,** 1435 (1954).
20. V. F. Weisskopf and D. H. Ewing, Phys. Rev. **57,** 472, 935 (1940).
21. C. E. Porter and R. G. Thomas, Phys. Rev. **104,** 483 (1956); S. Blumberg and C. E. Porter, *ibid.* **110,** 786L (1958).
22. E. P. Wigner, Proc. Conf. on Neutron Phys. by Time-of-Flight, Gatlinburg, Tennessee, p. 59 (1956); Oak Ridge National Laboratory Report (1956) ORNL-2309.
23. W. Hauser and H. Feshbach, Phys. Rev. **87,** 366 (1952); L. Wolfenstein, *ibid.* **82,** 690 (1951); T. Ericson and V. Strutinsky, Nuclear Phys. **8,** 284 (1958).
24. C. Bloch, Phys. Rev. **93,** 1094 (1954).
25. T. D. Newton, Can. J. Phys. **34,** 804 (1956).
26. J. M. B. Lang and K. J. LeCouteur, Proc. Phys. Soc. **A67,** 586 (1954).

V. B. Direct Interactions

1. Direct Interactions in Inelastic Scattering............................ 670
   BY C. A. LEVINSON
   a. Theoretical Considerations....................................... 672
      (1) Distorted Born Approximation.............................. 674
      (2) Plane Wave Born Approximation............................. 675
      (3) Coupled Equations... 676
      (4) Adiabatic Approximation................................... 676
      (5) The Semiclassical Approach................................ 678
   b. Inelastic Scattering Calculations............................... 680
      (1) Distorted Born Approximation.............................. 680
      (2) Coupled Equations... 682
      (3) The Adiabatic Approximation............................... 682
   c. Experiments with Resolved Final States.......................... 684
   d. Statistical Model versus Direct Interaction..................... 689
   e. Conclusions... 694
   References.. 694

2. The Theory of Stripping and Pickup Reactions........................ 695
   BY M. K. BANERJEE
   a. Dynamics of the Mass Exchange Reactions......................... 699
   b. Cross Section of the Mass Exchange Reaction..................... 702
   c. Angular Distributions in Mass Exchange Reactions................ 709
   d. Dependence of the Cross Section on the Nuclear Wave Function.... 721
   e. Exchange Effects in Stripping Reactions......................... 725
   References.. 731

1. Direct Interactions in Inelastic Scattering
by C. A. LEVINSON

The direct interaction model of inelastic scattering can be considered as an extension of the optical model for elastic scattering. On both physical and formal grounds the relationship is fairly clear. The optical model of elastic scattering treats nuclear matter as a refracting, absorptive medium in which a nucleon propagates with a characteristic wave number and mean free path. The scattering amplitude in the optical model is an average amplitude which smooths out the effects of any resonant structure. The mean free path represents absorption of the elastic wave into inelastic channels or into long-lived compound nuclear states which then decay back into the elastic channel. In the direct interaction model of inelastic scattering the incident wave comes on to the refracting medium and suffers an inelastic collision with some degree of freedom of the medium (that is, target particle, vibrational mode, rotational mode, etc.) after which it propagates out much as in elastic scattering. In a formal derivation (*1*)* the direct interaction amplitude appears as the inelastic amplitude which one computes neglecting the contributions of resonant compound nucleus states. If the resonant states dominate the inelastic process and many such states contribute significantly then the compound statistical model of nuclear reactions (*2*) is a more sensible description to use. In this model the incident nucleon is assumed to amalgamate with the target for an appreciable time (long enough to achieve thermal equilibrium with the target nucleus) before being re-emitted. In the direct interaction picture no such amalgamation takes place and the time of interaction is just that required for the nucleon to pass unimpeded over the target. It takes a 10-Mev nucleon about 10^{-21} sec to cross a nucleus while the time associated with a compound level of energy width 1 ev is $\hbar/1$ ev $\doteq 10^{-15}$ sec. For an incident 10-Mev nucleon a reasonable compound nucleus width is probably much greater than 1 ev. To distinguish direct from compound processes by time methods is, therefore, quite a difficult undertaking. On the other hand the differential cross sections predicted by the two models are different. For almost all experiments above 10 Mev involving discrete resolvable final states the differential cross sections unambiguously point to direct interactions. However, for inelastic processes to unresolved final states (and these are the majority) there is no clear cut indication that one theory is always correct. In this respect it is noted that polarization measurements of the outgoing projectiles would yield important information because nucleons which are "boiled out" of a target are not expected to be polarized while "direct interaction" yields should show polarization.

In considering an inelastic scattering to a discrete final state, a new constraint becomes important, namely, the angular momentum transferred to the target. If one takes this constraint into account in a semiclassical fashion (*3*) (geometrical optics approximation), one finds that the inelastic events must occur only on the locus of a cylinder intersecting the nucleus whose axis is along the momentum transfer direction and whose radius is equal to $\Delta l/Q$, Δl being the angular momentum transfer and Q the magnitude of the momentum transfer. This is a considerably different situation when compared with elastic scattering where scattering takes place throughout the body of the target. It is not surprising that the resulting theoretical differential cross sections for inelastic scattering depend so strongly on Δl.

* The reference list for Section V.B.1 begins on page 694.

At first it was believed that inelastic scattering experiments to discrete final states would easily distinguish between various nuclear models. This has not turned out to be the case. Primarily, this is due to the recently discovered fact (4) that the two main models, shell and collective, are really not so different if one treats the shell model by means of intermediate coupling. Generally, inelastic scattering to discrete final states seems to depend most on the radius of the target, index of refraction, mean free path, and angular momentum transfer. Details of nuclear structure appear to have only a minor effect on the shape of the differential cross section, although the magnitude is probably particularly sensitive to collective admixtures. However, so many details affect the magnitude that it is not feasible at present to use this as a means of investigating nuclear structure.

In this report the theoretical foundations of the direct interaction model will be presented along with the various approximations derived from the exact formulation. Next a brief summary of published work using these approximations will be given followed by a discussion of relevant experiments. Finally, the intriguing question of the role of compound nuclear effects as described by the statistical model will be discussed along with experimental attempts to distinguish compound from direct processes.

a. Theoretical Considerations

It is most convenient to follow Feshbach's (1) unified theory of nuclear reactions.[a] In order to facilitate comparison with his treatment we shall endeavor to use his notation as much as possible.

In this discussion we will ignore the effects of the Pauli principle when the incident particle is identical with particles in the target. Consider a target of A particles described by position, spin, and isotopic spin variables r_k. The set of target energy eigenfunctions ψ_i form a complete set of functions with energies ϵ_i. In terms of them we can expand the wave function $\psi(r_1 \cdots r_A; r_0)$ which describes the scattering of a particle with variables r_0 and initial energy E on the target.

$$\Psi(r_1 \cdots r_A; r_0) = \sum_i \psi_i(r_1 \cdots r_A) u_i(r_0). \tag{1}$$

Let ψ_0 denote the ground state target wave function and normalize ϵ_0 to zero. $u_0(r_0)$ will describe the elastic scattering of the particle and must approach an incoming plane wave plus outgoing spherical waves at large distances. For $i \neq 0$ and $\epsilon_i < E$, $u_i(r_0)$ will describe the scattered

[a] See Sections V.A and VI.D.

particle leaving the target in various excited states ψ_i. $u_i(\mathbf{r}_0)$, therefore, behaves like outgoing spherical waves at large distances. For $\epsilon_i > E$, $u_i(\mathbf{r}_0)$ must be exponentially damped, since otherwise, we would be describing an event where the target was excited by an amount of energy greater than that possessed by the projectile. The Schroedinger equation obeyed by Ψ is

$$H\Psi = E\Psi, \tag{2}$$

where

$$H = H_A(\mathbf{r}_1 \cdots \mathbf{r}_A) + T_0 + V(\mathbf{r}_0, \mathbf{r}_1 \cdots \mathbf{r}_A). \tag{3}$$

H_A is the target Hamiltonian; T_0 is the kinetic energy operator for the variable $\mathbf{r}_0$, and V is the interaction potential between the incoming particle and the target. In general, $V = V(\mathbf{r}_0, \mathbf{r}_1 \cdots \mathbf{r}_A)$, but often the simplifying assumption that V is made up of a sum of two-body forces is made. Then one takes V to be of the form $V = \sum_{k=1}^{A} V(\mathbf{r}_0, \mathbf{r}_k)$. By definition we have

$$H_A(\mathbf{r}_1 \cdots \mathbf{r}_k)\psi_i(\mathbf{r}_1 \cdots \mathbf{r}_A) = \epsilon_i \psi_i(\mathbf{r}_1 \cdots \mathbf{r}_A) \tag{4}$$

with $\epsilon_0 = 0$. If we multiply the Schroedinger equation (2) on the left by $\psi_i^*(\mathbf{r}_1 \cdots \mathbf{r}_A)$ and integrate over the target variables we obtain the set of coupled equations:

$$[T_0 + V_{ii}(\mathbf{r}_0) + \epsilon_i - E]u_i(\mathbf{r}_0) = -\sum_{j \neq i} V_{ij}(\mathbf{r}_0) u_j(\mathbf{r}_0), \tag{5}$$

where

$$V_{ij}(\mathbf{r}_0) \equiv \int \psi_i^*(\mathbf{r}_1 \cdots \mathbf{r}_A) V(\mathbf{r}_0, \mathbf{r}_1 \cdots \mathbf{r}_A) \psi_j(\mathbf{r}_1 \cdots \mathbf{r}_A) d\tau_1 \cdots d\tau_A. \tag{6}$$

These are the basic coupled equations in one variable $\mathbf{r}_0$ which must be solved. The "coupling potential" $V_{ij}(\mathbf{r}_0)$ measures the degree to which states ψ_i and ψ_j mix when the target is perturbed by the presence of the projectile at $\mathbf{r}_0$. We now group the functions $u_i(\mathbf{r}_0)$ into those corresponding to "open" channels and those corresponding to "closed" channels. An open channel is one where $\epsilon_i < E$ in our convention ($\epsilon_0 = 0$) and corresponds to a possible energy conserving process in which the projectile with energy E transfers energy ϵ_i to the target. A closed channel corresponds to $\epsilon_i > E$. Feshbach (1) shows that the closed channel wave functions can be formally eliminated from Eq. (5) and one obtains the result

$$[T_0 + U_{ii}'(\mathbf{r}_0) + \epsilon_i - E]u_i(\mathbf{r}_0) = -\sum_{j \neq i} U_{ij}'(\mathbf{r}_0) u_j(\mathbf{r}_0) \tag{7}$$

where the summation on j goes only over the open channels, and the new "coupling potential" $U_{ij}'(\mathbf{r}_0)$ includes effects of the closed channels and is real. It should be noted that U_{ij}' may vary strongly with the energy E when E is in the neighborhood of a compound nucleus resonance which strongly affects the open channel wave functions. If one removes contributions to U_{ij}' coming from resonating states then it varies slowly with energy. Call this modified potential $U_{ij}(\mathbf{r}_0)$ and remember that it does not include resonance effects. Now, if we substitute U_{ij} for U_{ij}' in Eq. (7), we have a mathematical description of the direct interaction model. These basic equations are, therefore,

$$[T_0 + U_{ii}(\mathbf{r}_0) + \epsilon_i - E]u_i(\mathbf{r}_0) = -\sum_{j \neq i} U_{ij}(\mathbf{r}_0)u_i(\mathbf{r}_0). \tag{8}$$

Let $i = 1$ correspond to the inelastic channel for which we wish the differential cross section. Then setting $i = 1$ in the above equations we can solve for the scattering amplitude f_{10} for transitions from channel 0 to channel 1:

$$f_{10}(\mathbf{k}_1, \mathbf{k}_0) = \sqrt{\frac{k_1}{k_0}} \frac{M}{2\pi\hbar^2} \int V_1^{(-)*}(\mathbf{r}_0) \sum_{j \neq 1} U_{1j}(\mathbf{r}_0)u_j(\mathbf{r}_0) \, d\tau_0, \tag{9}$$

where $V_1^{(-)*}(\mathbf{r}_0)$ is a solution of

$$[T_0 + U_{11}(\mathbf{r}_0) + \epsilon_1 - E]V_1^{(-)}(\mathbf{r}_0) = 0. \tag{10}$$

At large distances $V_1^{(-)}(\mathbf{r}_0)$ consists of an incident plane wave in the direction of $\mathbf{k}_1$ plus converging spherical waves; $\mathbf{k}_1$ is the wave vector of the scattered particle in channel i; M is the nucleon mass. We thus find that the direct interaction amplitude for a channel 0 to channel 1 transition contains a sum of the contributions associated with each of the open channels i, $i \neq 1$. In order to evaluate the direct interaction amplitude as shown in Eq. (9) one must be able to compute the coupling potential $U_{ij}(\mathbf{r}_0)$. In addition the coupled equations, (8) must be solved in order to find $u_j(\mathbf{r}_0)$. These problems are tremendously difficult so approximation methods must be devised. Several approaches will now be discussed.

(1) *Distorted Born Approximation*

For various reasons only channels 0 and 1 may be of importance. Possibly the cross section for exciting channels $i > 1$ may be experimentally small which would make $u_i(\mathbf{r}_0)$ small for $i > 1$. Also the coupling potential U_{1i} to other channels $i > 1$ may be small. In any of these cases the expression for f_{10} reduces to the single term

$$f_{10}(\mathbf{k}_1, \mathbf{k}_0) \doteq -\sqrt{\frac{k_1}{k_0}} \frac{M}{2\pi\hbar^2} \int V_1^{(-)*}(\mathbf{r}_0) U_{10}(\mathbf{r}_0) u_0(\mathbf{r}_0) \, d\tau_0. \tag{11}$$

To lowest order in $V(\mathbf{r}_0, \mathbf{r}_1 \cdots \mathbf{r}_A)$, $U_{10}(\mathbf{r}_0)$ is just $V_{10}(\mathbf{r}_0)$. Corrections due to the closed channels have not been investigated. Similarly $U_{11}(\mathbf{r}_0)$ in this approximation is the average of $V(\mathbf{r}_0, \mathbf{r}_1 \cdots \mathbf{r}_A)$ taken over the excited state $\psi_1(\mathbf{r}_1 \cdots \mathbf{r}_A)$ and has the form of an optical potential. That is, it extends over the nuclear volume and is proportional to the nuclear density. (At least for a short-range two-body nucleon-nucleon interaction.) $u_0(\mathbf{r}_0)$ is usually taken from an optical model calculation of elastic scattering and so is a purely phenomenological quantity. Up till the present it has been assumed that $U_{11}(\mathbf{r}_0)$ is given by the same potential well which acts on $u_0(\mathbf{r}_0)$. This is probably incorrect. In fact, attempts to fit inelastic scattering using the same distorting potential for $V_1^{(-)}(\mathbf{r}_0)$ and $u_0(\mathbf{r}_0)$ have indicated that the necessary optical model needed to fit the inelastic data is not consistent with the one needed to fit elastic scattering. It would certainly be sensible in future calculations to vary the potential representing $U_{11}(\mathbf{r}_0)$ independently and to keep $u_0(\mathbf{r}_0)$ fixed by using the elastic scattering wave functions given by the optical model and experiment.

T. K. Fowler (5) has recently studied the effective distorting potential for $V_1^{(-)}(\mathbf{r}_0)$ using the multiple scattering formalism of Watson and Frances (6). He finds that the real part of this well should be shallower than the real part of the corresponding elastic real well by as much as a factor of $\frac{2}{3}$ in the case of 20-Mev protons on carbon. It should be noted that in his discussion the final projectile wave function is elastically distorted and the initial projectile wave function has the modified distortion. In our discussion the situation is reversed. It can be shown that the two approaches are mathematically equivalent (compare Feshbach's article (1) p. 382, Eqs. (5.46) and (5.48), and see Sections V.A and VI.D).

(2) Plane Wave Born Approximation

One approximation which should certainly be incorrect is the plane wave Born approximation for the functions $V_1^{(-)}(\mathbf{r}_0)$ and $u_0(\mathbf{r}_0)$ in Eq. (11). However, when the integral in Eq. (11) is carried out in the region of the nuclear surface and not extended into the body of the target, then the Born approximation yields an amplitude which is fairly consistent with much experimental data. The physical content of this approximation seems to be that most direct interaction processes occur in the nuclear surface (7) where the incoming wave is not very much affected by the strong optical potential.

If one particular angular momentum transfer λ occurs in the inelastic process then this surface Born approximation yields a differential cross section proportional to $[j_\lambda(QR)]^2$ where R is some average surface distance from the center of the target nucleus, Q is the magnitude of the momentum transfer, and j_λ is a spherical Bessel function of order λ. More generally the plane wave Born approximation with shell model target states gives a differential cross section (when interactions are limited to the nuclear surface) of the form

$$d\sigma/d\Omega = \sum_\lambda A_\lambda [j_\lambda(QR)]^2, \qquad (12)$$

where the A_λ's are constants which depend on the detailed structure of the target and there are three constraints on λ.

(i) $\lambda = |J_i - J_f| - 1,\ |J_i - J_f|,\ \cdots\ |J_i + J_f + 1|$ for a general two-body spin-dependent interaction between projectile and target nucleons. J_i and J_f are initial and final target spins.

(ii) The parity change of the target is $(-1)^\lambda$.

(iii) $\lambda = |l_i - l_f|,\ \cdots\ |l_i + l_f|$ where l_i, l_f are initial and final shell model orbital angular momenta possessed by the excited target nucleon.

(3) *Coupled Equations*

There is reason to believe, in certain situations, that some small set of coupling potentials $U_{ij}(\mathbf{r}_0)$ in Eq. (8) are all equally important. In this case there is no alternative to solving the coupled equations involving these potentials. Consider the case where the ground state of the target nucleus is the lowest member of a rotational band as described by the strong coupling collective model of Bohr and Mottelson (*8*). Then the various members of the band are strongly coupled. Chase, Wilets, and Edmonds (*9*) have considered this case in detail and have solved a set of three coupled equations of the type of Eq. (8). Since they completely ignore other channels, they replace $U_{ij}(\mathbf{r}_0)$ by $V_{ij}(\mathbf{r}_0)$ where $V_{ij}(\mathbf{r}_0)$ is computed from the collective model particle surface interaction. It has recently been pointed out (*4*) that intermediate coupling shell model wave functions are related to each other in much the same way that different members of a rotational band are. In the case of a nuclear ground state described by an intermediate shell model wave function one might anticipate strong coupling to a group of excited states.

(4) *Adiabatic Approximation*

In a direct interaction one excites some mode of excitation in the nucleus. This may simply be a single particle excitation or else something more complicated such as a collective mode. The variables describing the excitation may have time rates of change which are comparable to the projectile velocity as in the case of single particle excitation. On the other hand, as in the case of some collective motions, the variables describing the excitation may vary slowly compared to the projectile coordinates. In this case one can use an "adiabatic" approach to the problem. As an example, consider again the Bohr-Mottelson strong coupling model where the relevant coordinates are the Euler angles of the principal axes of the deformed nucleus. These variables must change slowly compared to the internal motions of the nucleus in order that the internal motions can quickly adjust to the rotating distorted self-consistent field of the nucleus. Hence, it is qualitatively clear that an incident particle with energy comparable to single particle kinetic energies inside a nucleus will have a motion with respect to which the collective motion can be treated adiabatically.

Following Chase (10) we will discuss the adiabatic approximation more quantitatively. Instead of considering all the target variables we will consider only those variables ε which describe some mode of motion which is to be excited in a direct interaction and which will be treated adiabatically. The total Hamiltonian will be taken as:

$$H = T_0 + V(\mathbf{r}_0, \varepsilon) + H_A(\varepsilon); \tag{13}$$

T_0 is the projectile kinetic energy, $V(\mathbf{r}_0, \varepsilon)$ is the interaction potential between projectile and mode of motion described by ε, $H_A(\varepsilon)$ is the target Hamiltonian for motion in the variables ε. It has eigenfunctions $\psi_i(\varepsilon)$. We consider the scattering equation:

$$[T_0 + V(\mathbf{r}_0, \varepsilon)]\omega_{\mathbf{k}_0}(\mathbf{r}_0, \varepsilon) = E\omega_{\mathbf{k}_0}(\mathbf{r}_0, \varepsilon). \tag{14}$$

This corresponds physically to elastic scattering of a projectile on a target with ε fixed at some value. $\mathbf{k}_0$ denotes the incident wave vector and for large values of $\mathbf{r}_0$

$$\omega_{\mathbf{k}_0}(\mathbf{r}_0, \varepsilon) \xrightarrow[r_0 \to \infty]{} e^{i\mathbf{k}_0 \cdot \mathbf{r}_0} + f(\Omega, \varepsilon)\frac{e^{ik_0 r_0}}{r_0} \tag{15}$$

where Ω represents the angles θ, ϕ of $\mathbf{r}_0$. The adiabatic approximation to the solution of the Schroedinger equation

$$H\Psi = E\Psi \tag{16}$$

is taken to be

$$\Psi \doteq \omega_{\mathbf{k}_0}(\mathbf{r}_0, \varepsilon)\psi_0(\varepsilon). \tag{17}$$

Substituting this approximation into the Schroedinger equation we obtain (ϵ_0 normalized to zero again)

$$(H - E)\omega_{\mathbf{k}_0}(\mathbf{r}_0, \varepsilon)\psi_0(\varepsilon) = \left[\text{terms involving }\frac{\partial \omega_{\mathbf{k}_0}(\mathbf{r}_0, \varepsilon)}{\partial \varepsilon}\right]. \tag{18}$$

Chase (10) shows that the terms on the right-hand side of this equation are unimportant when the period of the collective motion is long compared to the time it takes the projectile to cross the nucleus. This criterion is equivalent to

$$\frac{2R}{V}\frac{\omega}{2\pi} = \frac{kR}{2\pi}\frac{\hbar\omega}{T} \ll 1 \qquad (19)$$

where V = projectile velocity, T = kinetic energy of projectile, $\hbar\omega$ = excitation energy of collective mode, R = nuclear radius, $kR \doteq 10$ for 20-Mev alpha particles, $kR \doteq 5$ for 20-Mev protons. So we must have (for 20-Mev or greater projectiles)

$$\hbar\omega/T \ll 1. \qquad (20)$$

The scattering or reaction amplitude is given rigorously by:

$$f_{i0}(\Omega) = -\sqrt{\frac{K_i}{K_0}}\frac{M}{2\pi\hbar^2}\int d\varepsilon\, \phi_i^*(\varepsilon)\int d\tau_0\, e^{-i\mathbf{k}_i\cdot\mathbf{r}_0}V(\mathbf{r}_0, \varepsilon)\Psi(\mathbf{r}_0, \varepsilon), \qquad (21)$$

or in the adiabatic approximation.

$$f_{i0}(\Omega) \doteq -\sqrt{\frac{K_i}{K_0}}\frac{M}{2\pi\hbar^2}\int d\varepsilon\, \phi_i^*(\varepsilon)\left[\int d\tau_0\, e^{-i\mathbf{k}_i\cdot\mathbf{r}_0}V(\mathbf{r}_0, \varepsilon)\omega_{k_0}(\mathbf{r}_0, \varepsilon)\right]\phi_0(\varepsilon). \qquad (22)$$

We note, however, that the quantity in square brackets is simply proportional to the elastic scattering amplitude[b] for fixed values of the variables ε. If we call this elastic amplitude $f^{el}(\Omega, \varepsilon)$, then we can express the general amplitude for transition from the state 0 to the state i as

$$f_{i0}(\Omega) \doteq -\sqrt{\frac{K_i}{K_0}}\int d\varepsilon\, \phi_i^*(\varepsilon)f^{el}(\Omega, \varepsilon)\phi_0(\varepsilon). \qquad (22')$$

Thus the solution for inelastic scattering in the adiabatic approximation is expressed in terms of elastic amplitudes.

(5) The Semiclassical Approach

In this approach the projectile's path through the target medium is described by the methods of classical ray optics. The incident ray is refracted and partially absorbed until it reaches a point where the inelastic event occurs. The inelastically scattered particle must leave the nucleus in such a way that the angular momentum which it possesses differs from the angular momentum carried by the incident particle by an amount Δl where Δl is the spin change of the target. This approximation is based on the W.K.B. approximation in quantum mechanics and improves

[b] Since $k_i \neq k_0$, this is not strictly true but is consistent with an expansion in $\hbar\omega/T$.

as the energy of the projectile increases. Several W.K.B. calculations of inelastic scattering (11) have been carried out at energies of 90 Mev and beyond.

Butler, Austern, and Pearson (3) in their discussion of the semiclassical approach introduce the useful concept of the "active cylinder." Consider a simple case where refractive effects are ignored. Referring to Fig. 1

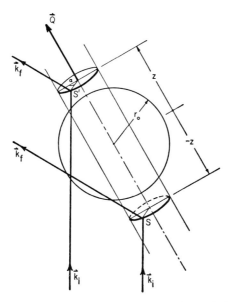

FIG. 1. From Butler, Austern, and Pearson, Phys. Rev. **112**, 1227 (1958). Semiclassical picture of a direct surface reaction under the simplifying approximations that diffraction and absorption effects on the rays can be neglected. The two scattering events depicted occur at opposite ends of the active cylinder; it is the interference between the outgoing rays from such scatterings that gives rise to the interference maxima and minima of the angular distribution.

where $\mathbf{k}_i$ and $\mathbf{k}_f$ are the initial and final wave vectors we have the condition $|(\mathbf{k}_i - \mathbf{k}_f) \times \mathbf{r}| = \Delta l$ for an angular momentum transfer Δl. Now the locus of all points $\mathbf{r}$ which satisfied this condition is a cylinder of radius $\dfrac{\Delta l}{|\mathbf{k}_i - \mathbf{k}_f|}$ whose axis is parallel to the vector $\mathbf{k}_i - \mathbf{k}_f$. As the above authors point out, "it is the interference between the outgoing rays from opposite ends of the active cylinder which gives rise to the interference maxima and minima" which such an approximation yields. Two such interfering rays which scatter at S and S' are shown in Fig. 1. (Of course S and S' must be inside the nucleus for any scattering to occur.) Now, if absorption occurs, then, since the interfering rays will pass through unequal amounts of nuclear matter, the rays will emerge with different magnitude amplitudes and interference effects should tend to wash out. This conclusion contradicts experiment where the most pronounced interference structure is seen for (α,α') scattering and α-particles are the most strongly absorbed of all projectiles presently used. Perhaps a detailed calculation with the semiclassical approach will yield the experimental results (12).

In order to justify the semiclassical approach one must investigate to what extent classical ray optics reproduce a quantum mechanical calculation. McCarthy (13) has verified that above 20 Mev this approximation is good for elastic scattering. He compares the quantum mechanical flux computed in an optical model calculation with the semiclassical rays. At such energies it is probably easier to compute inelastic scattering by means of the distorted Born approximation which includes fewer approximations. The main merit of the semiclassical approach is that it gives one a physical picture which aids one in understanding the effects of distortion and angular momentum constraints.

b. Inelastic Scattering Calculations

(1) *Distorted Born Approximation*

Two examples of distorted Born approximation will be discussed. Levinson and Banerjee (14) have evaluated the amplitude in Eq. (11) for the case $C^{12}(p,p')C^{12*}$, $Q = -4.43$ Mev where E_p is in the range 14–185 Mev. For $U_{10}(\mathbf{r}_0)$ they evaluated the matrix element of a Yukawa two-body operator with a Serber exchange factor $\left(\dfrac{1 + p^x}{2}, \; p^x = \text{space}\right.$ exchange) between shell model wave functions for the C^{12}, 0^+ ground state and the 2^+ excited state. The projectile wave functions $u_0(\mathbf{r}_0)$ and $V_1^{(-)}(\mathbf{r}_0)$ were taken to be optical well solutions where the same well was used for both incoming and outgoing waves. Good fits were obtained to the experimental data when the distorting well was taken to be about $\frac{2}{3}$ as deep as the well needed to describe elastic scattering. As mentioned above this discrepancy was attributed by T. K. Fowler (5) to excitation effects in the nuclear medium caused by the passage of the projectile. Another outstanding discrepancy found was that one needed $U_{10}(\mathbf{r}_0) = 2V_{10}(\mathbf{r}_0)$ in order to fit the magnitude of the cross sections where $V_{10}(\mathbf{r}_0)$ was computed with independent particle shell model wave functions. This same situation arises in inelastic electron scattering on carbon and probably indicates some collective admixtures in the C^{12} wave functions.[c]

[c] R. A. Ferrell, private communication.

N. K. Glendenning (15) has carried out a similar set of distorted wave calculations with one difference. He restricts the inelastic process to the surface of the nucleus. This approximation has been tested by comparing with a calculation without this restriction and is acceptable as far as the shape of the differential cross section is concerned. He obtains fits to the $C^{12}(p,p')$ reaction similar to those of Levinson and Banerjee. He also

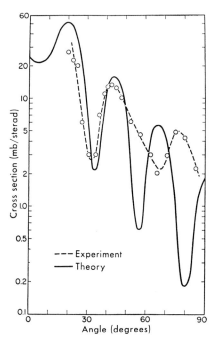

FIG. 2. From N. K. Glendenning, Phys. Rev. **114**, 1297 (1959). Angular distribution of 31.5-Mev alpha particles inelastically scattered from the 2^+ level at 4.43 Mev in C^{12}. Experimental points from H. J. Watters, Phys. Rev. **103**, 1763 (1956).

obtained a fair fit to the $O^{16}(p,p')O^{16*}$, $Q = -6.14$ Mev, $E_p = 19$ Mev process which is a $0^+ \to 3^-$ spin transition. He assumed an independent particle model for the target which predicts this transition to be a $p_{\frac{1}{2}} \to d_{\frac{5}{2}}$ excitation. Next Glendenning computed the differential cross section for exciting the 4.43-Mev 2^+ level in carbon by 31.5-Mev alpha particles. In this case the plane wave surface Born approximation gives a reasonable fit to the data. Glendenning's fit is shown in Fig. 2. The optical model square well parameters used in this calculation were

$$V = -(20 + 14i) \text{ Mev} \quad r < 3.1 \times 10^{-13} \text{ cm.}$$
$$= 0 \quad r > 3.1 \times 10^{-13} \text{ cm.}$$

This calculation shows the presence of diffraction maxima and minima in spite of the qualitative prediction of the semiclassical model that such structure should disappear with strong absorptions present.

(2) *Coupled Equations*

Chase et al. (*9*), as discussed above, have applied the coupled equation approach to the strong coupling collective model. In this approach one can test the distorted wave Born approximation which is a two channel approximation. Chase et al. really performed a three channel calculation (that is, three coupled equations) so they were able to test the effect of a third strongly coupled state on an inelastic calculation. In their calculation they consider three states of spins 0^+ at 0 kev, 2^+ at 50 kev, and 4^+ at 167 kev. In this case for 1-Mev incident neutrons they show that the presence of the 4^+ strongly coupled state changes the total cross section considerably and the shape of the differential cross section to a lesser extent. On the other hand they found that for energies a few hundred kev or more above threshold the adiabatic approximation only introduced errors of 10% or less in the total cross sections and the angular distributions were quite similar to the coupled equation results.

(3) *The Adiabatic Approximation*

Blair (*16*) has shown that a very good approximation for obtaining the elastic amplitude for alpha particle scattering on nuclei is to assume a completely absorbing black nucleus. Using this approximation and following the adiabatic approach he has treated the case where the adiabatic variables discussed above describe either the strong coupling or vibrational collective motion of Bohr and Mottelson. He finds that the resulting inelastic differential cross sections (to terms linear in the deformation) correspond closely to plane wave Born approximation predictions which are experimentally observed in (α,α') processes. This approach is far superior to the plane wave Born approximation (surface interaction) in several respects. First, it is much more realistic in that it takes into account properly the absorptive distortion of the projectile wave functions. Second, the value of R in the Born approximation [compare Eq. (12)] is a parameter which is simply fit to the data while in the adiabatic approach the nuclear radius is given from elastic scattering analysis. Most impressive of all is the fact that the adiabatic theory predicts that the inelastic diffraction pattern will be either in phase or 90° out of phase with the elastic differential cross section: in phase for a change of target parity and out of phase for no change of target parity. This relationship is indeed seen experimentally. Figure 3 shows a fit to the reaction $S^{32}(\alpha,\alpha')S^{32*}$, $Q = -2.24$ Mev, $E_\alpha = 41.7$ Mev which is a $0^+ \to 2^+$ transition. The out of phase structure of the inelastic pattern is clearly seen.

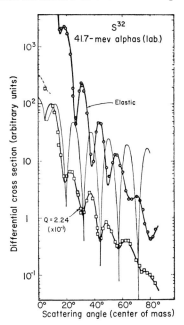

Fig. 3. From John S. Blair, *Inelastic Diffraction Scattering* (16). Angular distribution of alpha particles from S^{32} with incident laboratory energy of 41.7 Mev. The open circles are the elastic cross sections and the open squares are the inelastic cross sections to the 2.24-Mev (2^+) state [multiplied by a factor (10^{-1})] as measured by Robison and Farwell. The solid curve gives the theoretical cross section for a collective quadrupole transition where the radius, R_0, is that deduced from a fit to the elastic angular distribution, 6.19×10^{-13} cm. The dashed curve is the theoretical cross section including Coulomb excitation; this estimate is realistic only for very small angles.

The theory correctly gives the positions of maxima and minima but not the magnitudes for larger angles. The magnitude for small angles is fitted by adjusting the deformation parameter β of the collective model. Independent measurements of β show similar values. Blair's calculation was carried out to first order in β. These results for (α,α') scattering are certainly impressive. It would be interesting to investigate in what manner an adiabatic calculation such as this, carried out to first order in the interaction, differs from a distorted Born approximation calculation. In order to couple in states other then the elastic and final state, higher order interactions are required. Certainly it is easier to compute results with the adiabatic approximation in this case where the elastic amplitude for fixed β can be computed by Fraunhofer optics approximations.

c. Experiments with Resolved Final States

We give here a list of references to nuclear reaction experiments involving resolved final states and beam energies mostly between 10 and 50 Mev. In many of the publications referred to, the direct interaction theory employing the plane wave Born approximation is compared with experiment and striking agreement is found. In Table I, E is the incoming beam energy, and E_x is the excitation energy of the final target state. If $E_x = 0$ this designates that the final target is left in its ground state. This table is certainly not exhaustive but only attempts to give examples concerning most projectiles and targets which have been studied.

A good example to consider is $Li^7(p,p')Li^{7*}$, $E_x = 4.61$. From shell model calculations the excited level is unambiguously predicted to be $\frac{7}{2}^-$ and the ground state is known to be $\frac{3}{2}^-$. This implies that in the Born approximation only $\lambda = 2$ appears; that is, $d\sigma/d\omega \sim |j_2(qR)|^2$. At $E_p = 12$ Mev no sign of $|j_2(qR)|^2$ appears, at 17.5 Mev the dim outlines are present with the diffraction peak just apparent. At 31.8 and 39.6 Mev there is a very pronounced first diffraction peak whose position is quite consistent with the Born approximation using a reasonable radius R. The second peak in $|j_2(qR)|^2$ is not seen. At the lower energies the corrections to the Born approximation are fairly large and tend to wash out the simple $|j_2(qR)|^2$ pattern. The peaks after the first one will appear if the interaction is localized on the surface. However Li^7 is so small that the interaction must occur throughout the nucleus and hence no more than the first diffraction peak is seen. The 48-Mev (α,α') cross section shows no evidence for any but the first diffraction peak which is quite strong.

The $Be^9(\alpha,\alpha')Be^9$, $E_x = 2.43$-Mev reaction starts from a $\frac{3}{2}^-$ state and goes to a state of unspecified spin which is interpreted on the basis of a rotational model by Blair and Henley (17) as being a $\frac{5}{2}^-$ state. The expected $|j_2(qR)|^2$ form of the differential cross section fits nicely and at least four and possibly five diffraction peaks are seen. This indicates a very localized region of interaction. This fact plus the large cross section seen for this reaction and other reactions to other assumed rotational states lead Blair and Henley to favor a rotational model for Be^9. This model consists of a two alpha particle Be^8 "dumbbell" plus an extra neutron. Recently Blair (16) has successfully applied the adiabatic approximation to this process using the radius given by the elastic scattering analysis.

The $Be^9(d,d')Be^{9*}$ reaction at $E_d = 24$ Mev, $E_x = 2.43$ Mev, shows a very striking diffraction pattern which is nicely predicted by the direct interaction theory of these reactions given by Huby and Newns (*18*).

The reaction $C^{12}(p,p')C^{12*}$, $E_x = 4.4$ Mev, is an 0^+–2^+ transition and has been studied for energies of 10 to 185 Mev as well as at some lower energies. The surface Born approximation unambiguously predicts a $|j_2(qR)|^2$ differential cross section. Only at 40 Mev does this shape become apparent. At lower energies the differential cross section is considerably distorted. A distorted wave calculation (*14,15*) has been performed and reasonably good fits obtained to the data. With 40-Mev α-particles a very nice $|j_2(qR)|^2$ pattern is observed showing at least four diffraction maxima.

Sherr and Hornyak (*19*) have investigated the $C^{12}(p,p')C^{12*}(\gamma)C^{12}$ angular correlation between the outgoing proton and the gamma ray emitted when the 4.4-Mev level de-excites. As pointed out by Satchler (*20*) the angular correlation pattern predicted by the Born approximation direct interaction theory should in this case be of the form $\sin^2 2(\theta_\gamma - \theta_q)$ where θ_q is the angle of the recoil nucleus and θ_γ is the angle of the observed gamma ray. Both angles are measured from the beam direction. Distortion effects merely change the pattern so that it is represented by the formula $A + B \sin^2 2(\theta_\gamma - \theta_0)$ in the reaction plane where θ_0 is only slightly different from θ_q. The experiments agree exceedingly well with the theory. The statistical theory of nuclear reactions on the other hand does not predict a correlation pattern related to the recoil direction. Such a pattern arises only in a direct interaction approach.

In the 2s, 1d shell (O^{16}-Ca^{40}) the results are quite similar to those discussed above for the 1p shell. The Ne and Mg (α,α') differential cross sections show amazing fits to $|j_2(qR)|^2$ and the adiabatic approximation, while (p,p') experiments still show considerable distortion effects. The (d,d') reactions show reasonably good correlation with Born approximation direct interaction predictions.

For targets heavier than calcium very few differential cross section reaction experiments to resolved final states have been performed. Level densities are getting large and it is harder to find a nucleus with a resolvable final state when 15-Mev or more protons, for instance, are incident. The 17-Mev (p,p') reaction on Fe^{56} to the 2^+ level at 0.845 Mev shows a typical diffraction pattern. Recently, Kikuchi, Kobayashi, and Matsuda (*21*) have observed the inelastic scattering of 14.1-Mev protons on targets of Ti, Cr, Fe, and Ni to resolved final states. These results more closely resemble $|j_2(qR)|^2$ than any other inelastic proton experiments ever performed. Two or three diffraction maxima are seen in all cases. The effects of distortion seem to be quite small. However, at 8 Mev, the distortion effects become prominent (*22*).

TABLE I. REFERENCES TO NUCLEAR REACTION EXPERIMENTS INVOLVING RESOLVED FINAL STATES AND BEAM ENERGIES MOSTLY BETWEEN 10 AND 50 MEV

Reaction	E (Mev)	E_x (Mev)	Reference
$Li^6(p,p')Li^6$	14.8, 19.4	2.19	1
$Li^6(p,p')Li^6$	40	2.19	2
$Li^6(\alpha,\alpha')Li^6$	31.5	2.19	3
$Li^6(p,He^3)He^4$	15, 18.5	0	4
$Li^6(d,d')Li^6$	15	2.19	5
$Li^7(p,p')Li^7$	12	4.61	6
	17.5	4.61	7
	31.8	4.61, 6.6	8
	39.6	4.61	2
$Li^7(\alpha,\alpha')Li^7$	48	4.61	8
$Li^7(d,d')Li^7$	14.4	0.478, 4.61	9
	15.1	4.61	10
$Li^7(d,t)Li^6$	14.4	0, 2.187	9
$Li^7(d,He^3)He^6$	14.4	0, 1.71	9
$Be^9(p,p')Be^9$	9.9	2.4	11
	12	2.4	12
	31.3	2.45	13
$Be^9(\alpha,\alpha')Be^9$	48	2.45	12
$Be^9(d,d')Be^9$	9	2.43	14
	15.1	2.43	10
	24	2.43	12
$B^{10}(\alpha,\alpha')B^{10}$	43	2.15, 3.58 (5.9 to 6.2)	15
$B^{11}(\alpha,\alpha')B^{11}$	43	2.14, 4.46 (6.81 or 6.76)	15
$B^{10}(\alpha,p)C^{13}$	4.9, 7	0	16
$C^{12}(p,p')C^{12}$	10	4.4	17
	12	4.4	6
	14–19.4	4.4, 7.7, 9.6	18
	40	4.4	19
	185	4.4	20
$C^{12}(\alpha,\alpha')C^{12}$	31.5	4.43, 7.65	3
	43	4.43, 7.65	15
	48	4.43	21
$C^{12}(d,d')C^{12}$	9	4.43	14
	15	4.43	10
	19.1	4.43, 9.61	22
$C^{12}(p,p')C^{12}*(\gamma)C^{12}$ Ang. Correl.	43	4.43	23
$C^{12}(\alpha,p)N^{15}$ and	30, 41.5	0	24
$N^{15}(p,\alpha)C^{12}$	15.5	0	25
$C^{13}(He^3,\alpha)C^{12}$	2, 4.5	0, 4.43	26, 27
$N^{14}(p,p')N^{14}$	9.5	3.9	28
$N^{14}(d,d')N^{14}$	9	3.95	14

TABLE I. REFERENCES TO NUCLEAR REACTION EXPERIMENTS INVOLVING RESOLVED FINAL STATES AND BEAM ENERGIES MOSTLY BETWEEN 10 AND 50 MEV (*Continued*)

Reaction	E (Mev)	E_x (Mev)	Reference
$O^{16}(p,p')O^{16}$	19	6.14, 7.02, 8.87, 9.85, 10.34, 11.08, 11.51, 12.02, 12.53	29
$F^{19}(d,d')F^{19}$	8.9	1.6	30
$F^{19}(p,\alpha)O^{16}$	16	0	31
$F^{19}(\alpha,p)Ne^{22}$	6.00–6.55	1.28	32
$F^{19}(\alpha,p)Ne^{22*}(\gamma)Ne^{22}$ Ang. Correl.	6.4	1.28 to ground	32
$Ne^{20}(\alpha,\alpha')Ne^{20}$	18	1.63, 4.25, 4.97, 5.81, 7.2	33
$Mg^{24}(p,p')Mg^{24}$	18	1.37	34
$Mg^{24}(d,d')Mg^{24}$	8.9	1.368, 4.122, 4.23	35
	15.1	1.37	10
$Mg^{24}(\alpha,\alpha')Mg^{24}$	31.5	1.37, 4.12	3
	42	1.37	36
	48	1.368	21
$Mg^{24}(p,p')Mg^{24*}(\gamma)Mg^{24}$ Ang. Correl.	16.6	1.37	37
$Al^{27}(d,d')Al^{27}$	9	2.23, 2.75, 3.04	38
	15.1	2.23, 2.75	5
$Si^{28}(p,p')Si^{28}$	12	1.78	6
$Si^{28}(d,d')Si^{28}$	8.9	1.78	35
$Al^{27}(\alpha,p)Si^{30}$	30.5	0	39
$P^{31}(\alpha,p)S^{31}$	7.0, 8.1	0	16
	30.5	0	39
$S^{32}(\alpha,\alpha')S^{32}$	43	2.25	15
$A^{40}(\alpha,\alpha')A^{40}$	18	1.46	33
$A^{40}(p,p')A^{40}$	9.51	1.48	40
$Ti(p,p')Ti$	14.1	0.89 and 0.99 (unresolved isotopes)	41
$Cr^{52}(p,p')Cr^{52}$	14.1	1.45	41
$Fe^{56}(p,p')Fe^{56}$	14.1	0.845	41
	17	0.822	42
$Ni(p,p')Ni$	14.1	1.45, 1.33, 1.34 (unresolved isotopes)	41

[1] R. Sherr and W. Hornyak, reported by C. A. Levinson and M. K. Banerjee, Ann. Phys. (N.Y.) **2**, 471 (1957).

[2] S. Chen and N. M. Hintz, Ann. Progr. Report 1957–1958, University of Minnesota Linear Accelerator Laboratory.

[3] H. J. Watters, Phys. Rev. **103**, 1763 (1956).

[4] J. G. Likely and F. P. Brady, Phys. Rev. **104**, 118 (1956).

[5] J. W. Haffner, Phys. Rev. **103**, 1398 (1956).

Table I continued on next page

References to Table I (Continued)

[6] H. E. Conzett, Phys. Rev. **105,** 1324 (1957).

[7] D. R. Maxson and E. Bennett, reported by C. A. Levinson and M. K. Banerjee, Ann. Phys. (N.Y.) **2,** 471 (1957).

[8] R. Silver, "Angular Distribution of (p,p')(p,d) and (α,α') Reactions in Li^7 at Medium Energies," Thesis, University of California at Berkeley.

[9] S. H. Levine, R. S. Bender, and J. N. McGruer, Phys. Rev. **97,** 1249 (1955).

[10] J. W. Haffner, Phys. Rev. **103,** 1398 (1956).

[11] S. W. Rasmussen, Phys. Rev. **103,** 186 (1956).

[12] R. G. Summers-Gill, Phys. Rev. **109,** 1591 (1958).

[13] J. Benveniste, R. G. Finke, and E. A. Martinelli, Phys. Rev. **101,** 655 (1956).

[14] T. S. Green and R. Middleton, Proc. Phys. Soc. **A69,** 28 (1956).

[15] P. C. Roberson, B^{10},B^{11},S^{32}, Thesis, University of Washington (1958); G. Shook, C^{12}, Thesis, University of Washington (1958).

[16] P. von Herrmann and G. F. Pieper, Phys. Rev. **105,** 1556 (1957).

[17] G. E. Fischer, Phys. Rev. **96,** 704 (1954).

[18] R. W. Peele, Phys. Rev. **105,** 1311 (1957).

[19] S. Chen, reported by C. A. Levinson and M. K. Banerjee, Ann. Phys. (N.Y.) **3,** 67 (1958).

[20] H. Tyren and T. A. J. Maris, Nuclear Phys. **3,** 52 (1957).

[21] F. J. Vaughn, "Elastic and Inelastic Scattering of 48 Mev α Particles by Carbon and Mg," Thesis, University of California at Berkeley (1955).

[22] R. G. Freemantle, W. M. Gibson, and J. Rotblat, Phil. Mag. [7] **45,** 1200 (1954).

[23] R. Sherr and W. Hornyak, reported by M. K. Banerjee and C. A. Levinson, Ann. Phys. N.Y. **2,** 499 (1957).

[24] R. Sherr, M. Rickey, G. W. Farwell, *in* Ann. Progr. Report 1957, University of Washington.

[25] D. R. Maxson and E. Bennett, reported by R. Sherr, *in* Proc. Pittsburgh Conf. (1957).

[26] H. D. Holmgren, Phys. Rev. **106,** 100 (1957).

[27] H. D. Holmgren, E. H. Geer, R. L. Johnston, and E. A. Wolicki, Phys. Rev. **106,** 102 (1957).

[28] R. G. Freemantle, D. J. Prowse, and J. Rotblat, Phys. Rev. **96,** 1268 (1954).

[29] W. F. Hornyak and R. Sherr, Phys. Rev. **100,** 1409 (1955).

[30] F. A. El Bedewi, Proc. Phys. Soc. **A69,** 221 (1956).

[31] J. G. Likely and F. P. Brady, Phys. Rev. **104,** 118 (1956).

[32] G. F. Pieper and N. P. Heydenburg, Phys. Rev. **111,** 264 (1958).

[33] L. Seidlitz, E. Bleuler, and D. J. Tendam, Phys. Rev. **110,** 682 (1958).

[34] P. C. Gugelot and P. R. Phillips, Phys. Rev. **101,** 1614 (1956).

[35] S. Hinds, R. Middleton, and G. Parry, Proc. Phys. Soc. **A70,** 900 (1957).

[36] P. C. Gugelot and M. Rickey, Phys. Rev. **101,** 1613 (1956).

[37] H. Yoshiki and R. Sherr, Bull. Am. Phys. Soc. [2] **3,** 200 (1958).

[38] S. Hinds and R. Middleton, Proc. Phys. Soc. **A69,** 347 (1956).

[39] C. E. Hunting and N. S. Wall, Phys. Rev. **108,** 901 (1957).

[40] R. G. Freemantle, D. J. Prowse, A. Hossain, and J. Rotblat, Phys. Rev. **96,** 1270 (1954).

[41] K. Kikuchi, S. Kobayashi, and K. Matsuda, J. Phys. Soc. Japan **14,** 121 (1959); Kikichi, Kobayashi, Matsuda, Nagahara, Oda, Takano, Takeda, and Yamazaki, Institute for Nuclear Study, University of Tokyo, preprint.

[42] G. Schrank, P. C. Gugelot, and I. E. Dayton, Phys. Rev. **96,** 1156 (1954).

d. STATISTICAL MODEL VERSUS DIRECT INTERACTION

So far we have been discussing transitions to discrete resolvable final states. The majority of inelastic events, however, usually involve final target states which are so highly excited as to be too dense to resolve. Often they are even able to decay by further particle emission. The compound statistical model treats these processes quite simply. For example, consider the spectrum of protons emitted in a (p,p') experiment where one observes all outgoing protons for a fixed bombarding energy E_p. Let the final target state be at an energy E_x where the level density is given by $W(E_x)$. Let the emitted proton have energy ϵ. The compound statistical theory of nuclei assumes that this is a result of the decay of a compound nucleus which was formed when the original proton came on to the original target. Let $\sigma_c(\epsilon)$ be the cross section for formation of this compound nucleus by the inverse reaction where a proton of energy ϵ comes on to an excited target state of energy E_x. The theory then gives for the observed proton spectrum for the (p,p') process (23):

$$I(\epsilon) \, d\epsilon = \text{const.} \; \epsilon \sigma_c(\epsilon) W(E_x) \, d\epsilon, \tag{23}$$

where $I(\epsilon)$ is the intensity per unit energy range of emitted protons. The factor $W(E_x)$ is simply a phase space factor which would be present in any theory. Since it is by far the most strongly varying factor with ϵ we see that the form of $I(\epsilon)$ as a function of ϵ is dominated by this factor (the factor comes from a phase space argument plus detailed balancing). Hence a verification that Eq. (23) gives the proper dependence of $I(\epsilon)$ on ϵ really implies that one has chosen a good representation for $W(E_x)$ and not that the entire statistical theory is necessarily correct. The density of states is generally written in two different forms

$$W(E_x) = \text{const.} \; e^{E_z/T} = \text{const.}' \; e^{-\epsilon/T} \tag{24}$$

or

$$W(E_x) = ce^{2\sqrt{aE_x}}. \tag{25}$$

Both forms come from the approximate statistical result (24) that

$$W(E_x) \sim e^{S(E_x)}, \tag{26}$$

where $S(E_x)$ is the entropy of a system (in units of k the Boltzmann constant) with energy E_x. Form (24) results from a Taylor expansion of $S(E_x)$ about $E_{x\max}$ where $E_{x\max}$ is related to E_x and ϵ by the equation $E_x = E_{x\max} - \epsilon$. The expansion is then

$$S(E_{x\max} - \epsilon) = S(E_{x\max}) - \epsilon \left(\frac{\partial S}{\partial E_x}\right)_{E_{x\max}} + \cdots . \tag{27}$$

So T in Eq. (24) is identified with $(\partial S/\partial E_x)^{-1}{}_{E_{x\max}}$ which is a thermodynamical formula for the quantity $k\tau$ of a system at energy $E_{x\max}$ where τ is the temperature. We shall, however, call T the nuclear temperature. If one uses the entropy expression for a Fermi gas of particles then expression (25) results. In this case

$$T(E_{x\max}) = \left(\frac{E_{x\max}}{a}\right)^{\frac{1}{2}}. \tag{28}$$

In principle, then, one can take an experimentally observed $I(\epsilon)$ divide by $\epsilon\sigma_c(\epsilon)$ and plot the logarithm of this against ϵ to give a $1/T$ slope. $\sigma_c(\epsilon)$ should be taken from an optical well calculation of the total reaction cross section. Most calculations to date have not used an optical well with a rounded edge but rather a square well model. It is suspected that the Coulomb barrier effect is overemphasized in this model and that the rounded edge calculation should give a somewhat different and more accurate result. The dependence of nuclear temperature on mass number and excitation can teach us how nuclei are heated up and also something about their structure. For instance, for a Fermi gas it can be shown that $a = \text{const.} \times A$ or $T = \text{const.}\, (E_{x\max}/A)^{\frac{1}{2}}$.

Before one measures a nuclear temperature one would like to believe that the nucleons are really boiling out of a compound system. The statistical nuclear theory says that such nucleons come out of nuclei with a differential cross section which is symmetric about 90° and unpolarized. Up to the present one of the main criteria for the presence of "statistical nucleons" has been the necessary condition of symmetry about 90° in their differential cross section. Polarization measurements of the continuum protons is now feasible and should yield interesting results.

Another interesting prediction of the statistical theory concerns the total cross sections for boiling off various particles in a nuclear reaction. For instance (n,n') processes are much more probable than (n,p) processes, simply because the protons have to come out through a Coulomb barrier. From the standpoint of direct processes where inelastic events often occur near the surface this inhibition of charged-particle emission should be much smaller.

Experiments bearing on this were performed by Paul and Clarke (*25*) at Chalk River in 1952. They bombarded 57 elements with 14.5-Mev neutrons from the $T^3(d,n)\text{He}^4$ reaction and measured activities due to (n,p), (n,α), and $(n,2n)$ reactions. This gave the total cross sections, σ_{ob}, for these processes for a large number of different mass numbers. The statistical theory predicted cross sections σ_{cal} based on the integration of Eq. (23). The level density in Eq. (25) was used where the parameters C and "a" came from other independent experiments. The quantity σ_{ob}/σ_{cal} for the (n,p), (n,α), and $(n,2n)$ processes showed an unexpected behavior. For (n,p) and (n,α) with mass numbers A up to about 50 this ratio fluctuated from 0.1 to 9.0 irregularly. Then for A greater than 50 this ratio increased to values of 100 for $A = 100$ and kept on increasing for A greater than 100. In the case of $(n,2n)$ processes σ_{ob}/σ_{cal} remained between 0.2 and 1.4 for all values of A. The interpretation given by the authors concerning this experiment was that at large values of A where the Coulomb barrier was strongly depressing the yield of charged particles coming from the compound nucleus, some other mechanism was responsible. It was conjectured that this other mechanism was the direct interaction one. This left unsettled of course the question of which final states the direct interaction processes were going to and whether the differential cross sections showed symmetry about 90° or not. It did, however, in a fairly unambiguous way imply that all was not statistical and that when the Coulomb barrier inhibited the statistical processes another process still produced inelastic particles.

Just recently Fulmer and Cohen (26) have studied (p,α) reactions with 23-Mev protons. They measured energy distributions and absolute differential cross sections from 30° to 150°. Their results are consistent with those of Paul and Clarke. For targets of mass number A less than 50 all but the highest energy α-particles were emitted isotropically (Al, Ni, Cu). The high energy α's showed forward peaking characteristic of direct processes. For a Pd target with mass number 106 the forward peaked α's included the highest energy α's and all α's down to 12 Mev. Lower energy α's observed down to 8 Mev were isotropic. At mass number 195 (Pt) all α-particles showed forward peaking and no sign of statistical processes was left. Hence, once again we may conclude that for large A the direct process for charged-particle emission dominates the compound contribution which is becoming vanishingly small due to the large Coulomb barrier. In addition we see that it is the high-energy part of the spectrum, that is, particles leaving low excited targets, that contains the direct interaction particles.

These results are further supported by the work of Eisberg and Igo (27) who found strongly forward peaked differential cross sections for 31-Mev protons in (p,p') processes on Pb, Au, Ta, and Sn for all energies of p' with total cross sections 250 to 290 millibarns. Eisberg and Igo point out that these cross sections are about equal to the area of a diffuse rim of width 1.4×10^{-13} cm and radius equal to the target radius. The de Broglie wavelength of the protons at 30 Mev is of the order of 10^{-13} cm and their mean free path is about 2×10^{-13} cm. Hence, they conclude that in consideration of these magnitudes one can certainly make a case for direct interactions occurring near the surface.

Gugelot's (28) results on (p,p') reactions for $E_p = 18$ Mev again show the same trends. For an Al target ($A = 27$) the most energetic inelastically scattered protons (14–18 Mev) come out forward peaked, but all other p's show an isotropic cross section. For Fe, Ni, and Cu ($A \sim 50$) isotropy sets in for $E_{p'} < 10$ Mev. For $A \sim 100$ isotropy occurs for $E_{p'} \leq 7$ Mev, and finally for Pt ($A = 195$) isotropy never occurs in the inelastic proton spectrum. A very detailed recent investigation of the p' spectrum of 19-Mev incident protons on Rh^{103} for angles between 20° and 90° performed by Blampied and Sherr (29) shows that all inelastic proton differential cross sections down to 5.5 Mev show large forward peaking.

In the case of (α,α') experiments performed by Igo (30) with 40.2-Mev α-particles, all differential cross sections are forward peaked for all $E_{\alpha'}$ and all mass numbers (Al^{27} to Th^{228}).

So far we have been discussing charged-particle reactions and have noted that nonstatistical processes are present and show themselves clearly when the Coulomb barriers become large enough to suppress the compound contributions. In order to learn more about the competition between compound and direct processes we now turn to experiments where neutrons are observed and compound effects are not inhibited.

First let us consider the work of Graves and Rosen (31) who measured the neutron spectra from the interaction of 14-Mev neutrons with various targets by nuclear plate techniques. They studied a mass number range from $A = 12$ to 200 and attempted to fit the neutron spectrum observed to formulae (23) and (24). This gives a mean energy of $2T$ and an energy of T for the peak of the energy distribution. It was found that all spectra in the 0.5- to 5-Mev range could be fit. However, neutrons above 5 Mev did not follow this fit with the same parameters. These high-energy neutrons were produced with cross sections of the order of 100 to 350 millibarns similar to the value found by Eisberg and Igo for their inelastic protons. This cross section is much larger than would be predicted by the statistical curve which fit the low-energy cross section.

The lower energy neutrons from 0.5 to 5 Mev which did show a fit to the statistical formula were produced with approximately ten times this cross section. They yielded temperatures between 0.5 and 1.0 Mev.

Rosen and Stewart (32) measured the differential cross section of the low-energy 0.5- to 4-Mev neutrons from Bismuth and found that they were indeed isotropic while the 4.0- to 12-Mev inelastic neutrons were strongly peaked in the forward direction. Similar work has been done by

Nakada (33) at Livermore who measured $d\sigma/d\omega$ from 30° to 140° for the inelastic neutrons in the energy range 1.5 to 5 Mev when Al, Fe, Sn, Pb, and Bi were bombarded by 14-Mev neutrons. The curves certainly go up in the forward direction and show slight asymmetry about 90°.

What conclusions follow from all this? First, we can certainly rule out statistical processes when the final target state is only excited by a small percentage of the maximum possible energy. The differential cross sections for the high-energy inelastic particles are always forward peaked and much larger in magnitude than predicted by the statistical model and are presumably due to direct interactions. The inelastically emerging particles with lower energies tend to come out with more isotropy. For high Z targets and emerging charged particles, complete isotropy never occurs. In the case of neutrons on bismuth the isotropic particles seem to make up about 90% of the total cross section. For 14-Mev incident neutrons these are in the 0- to 4-Mev range. Now direct interaction differential cross sections are always functions of the momentum transfer in Born approximation. Low-energy particles emerging after a high-energy particle came in could come out in any direction and not change the momentum transfer much. Hence, the direct interaction differential cross sections become more isotropic for the low-energy inelastic particles. (In fact any theory would give smaller angular variations simply because partial waves of smaller angular momentum become more important.) For this reason it might be much easier to obtain unambiguous information with polarization experiments.

It is certainly understandable that direct processes will favor low excited target states since these states are most strongly coupled to the ground state through the operator $V_{io}(\mathbf{r}_0)$ [see Eq. (6)]. The higher excited states have very complex wave functions corresponding to many particles being simultaneously excited, and $V_{io}(\mathbf{r}_0)$ being the matrix element of a sum of two-body operators will be quite small. Physically one can see that only a very few highly excited states of a nucleus can be produced simply by kicking one nucleon or spinning a distorted core.

We have been discussing reactions so far as though we must decide between direct processes and formation of compound nuclei in full temperature equilibrium. We are obviously neglecting situations where some part of the nucleus is involved, and only this part is "heated" before emission takes place. After considering the systematics of 14-Mev neutron reactions Rosen (private communication) concludes that just such processes are occurring, where only some fraction of the nucleons in the nucleus are involved in reactions. This would yield temperatures whose dependence on mass number is characterized by values of A smaller than that of the target.

e. Conclusions

A reasonably consistent picture is beginning to emerge with respect to inelastic scattering processes to discrete excited states. The influence of nuclear distortion on the projectile wave functions can be treated successfully by means of optical wells. If strongly coupled states are present then one must use coupled equations or an adiabatic approximation which includes all orders of the interaction. Characteristic diffraction patterns are most evident when the projectile is strongly absorbed in the target or possesses a high energy and as Blair points out in the case of strong absorption this is consistent with the theory. In the case of higher energies the diffraction pattern is consistent with the W.K.B. approximation. As the effect of absorption becomes smaller, such as in the case of $C^{12}(p,p')C^{12*}$, one finds experimental results with "washed out" diffraction patterns and this also is consistent with distorted wave calculations.

As mentioned in the last section the lowest energy reaction products do seem to "boil out" of the nucleus with some characteristic temperature. However, more work, both theoretical and experimental, remains to be done here.

REFERENCES

1. H. Feshbach, Ann. Phys. (N.Y.) **5**, 357 (1958).
2. H. A. Bethe, Revs. Modern Phys. **9**, 69 (1937); Phys. Rev. **57**, 1125 (1940); D. H. Ewing and V. F. Weisskopf, ibid. **57**, 472, 935 (1940); L. Wolfenstein, ibid. **82**, 690 (1951); W. Hauser and H. Feshbach, ibid. **87**, 366 (1952).
3. S. T. Butler, N. Austern, and C. Pearson, Phys. Rev. **112**, 1227 (1958).
4. J. P. Elliott, Proc. Roy. Soc. **A245**, 128, 562 (1958); M. Redlich, Phys. Rev. **110**, 468 (1958); D. Kurath and L. Pičman, preprint, Nuclear Phys. **10**, 313 (1959).
5. T. K. Fowler, Phys. Rev. Letters **1**, 371 (1958).
6. K. M. Watson, Phys. Rev. **89**, 575 (1953); **105**, 1388 (1956); N. C. Frances and K. M. Watson, ibid. **92**, 291 (1953).
7. N. Austern, S. T. Butler, and H. McManus, Phys. Rev. **92**, 350 (1953).
8. J. Rainwater, Phys. Rev. **79**, 432 (1950); A. Bohr, Kgl. Danske Videnskab. Selskab, Mat.-fys. Medd. **26**, No. 14 (1952); A. Bohr and B. R. Mottelson, ibid. **27**, No. 16 (1953).
9. D. M. Chase, L. Wilets, and A. R. Edmonds, Phys. Rev. **110**, 1080 (1958).
10. D. M. Chase, Phys. Rev. **104**, 838 (1956).
11. P. Benoist, C. Marty, and P. Meyer, Compt. rend. **245**, 1389 (1957); P. Hootan and G. Alcock, in press; E. J. Squires, Nuclear Phys. **6**, 594 (1958).
12. N. Austern, private communication (1959).
13. I. E. McCarthy, Phys. Rev., in press.
14. C. A. Levinson and M. K. Banerjee, Ann. Phys. (N.Y.) **2**, 471 (1957); M. K. Banerjee and C. A. Levinson, ibid. p. 499.
15. N. K. Glendenning, Phys. Rev., in press.
16. J. S. Blair, Phys. Rev. **108**, 827 (1957); **115**, 928 (1959).

17. J. S. Blair and E. Henley, Phys. Rev. **112**, 2029 (1958).
18. R. Huby and H. C. Newns, Phil. Mag. [7] **42**, 1442 (1951).
19. R. Sherr and W. Hornyak, reported by M. K. Banerjee and C. A. Levinson, Ann. Phys. (N.Y.) **2**, 499 (1957).
20. G. R. Satchler, Proc. Phys. Soc. **A68**, 1037 (1955).
21. K. Kikuchi, S. Kobayashi, and K. Matsuda, J. Phys. Soc. Japan **14**, 121 (1959).
22. K. Kikuchi, S. Kobayashi, K. Matsuda, Y. Nagahara, Y. Oda, N. Takano, M. Takeda, and T. Yamazaki, Institute for Nuclear Study, University of Tokyo, preprint.
23. J. M. Blatt and V. F. Weisskopf, *Theoretical Nuclear Physics* (John Wiley and Sons, New York, 1952), Chapter 8.
24. H. A. Bethe, Revs. Modern Phys. **9**, 69 (1937), cf. Eq. (278a) p. 81.
25. E. B. Paul and R. L. Clarke, Can. J. Phys. **31**, 267 (1953).
26. C. B. Fulmer and B. L. Cohen, Phys. Rev. **112**, 1672 (1958).
27. R. M. Eisberg and G. Igo, Phys. Rev. **93**, 1039 (1954).
28. P. C. Gugelot, Phys. Rev. **93**, 425 (1954).
29. W. Blampied and R. Sherr, Bull. Am. Phys. Soc. [2] **2**, 303 (1957).
30. G. Igo, Phys. Rev. **106**, 256 (1957).
31. E. R. Graves and L. Rosen, Phys. Rev. **89**, 343 (1953).
32. L. Rosen and L. Stewart, Phys. Rev. **99**, 1052 (1955).
33. M. P. Nakada, Proc. Intern. Conf. on Neutron Interactions with the Nucleus, Columbia Univ. p. 232 (1957).

2. The Theory of Stripping and Pickup Reactions
by M. K. BANERJEE

Among the reactions of the types to be considered in this section, the (d,p) reaction was the first to be experimentally studied. The first experiments were performed at very low deuteron energies and the angular distributions were isotropic as only S-wave interactions took place. The main interest was in the variation of the reaction cross section with the energy of the incident deuterons. It was also noticed that the (d,p) reaction was more probable than the (d,n) reaction. These features were explained by Oppenheimer and Phillips (*1*).* They pointed out that the proton in the deuteron cannot cross the Coulomb barrier of the target at very low energy and therefore only the neutron can get close to the target and get captured. Such a model, which is radically different from the Bohr compound nucleus theory, was necessary to explain the preference for proton emission which is just the opposite of the prediction of the compound nucleus theory. The full implication of the new model was not clear until experiments were performed at higher energies. At such

* The reference list for Section V.B.2 is on page 731.

energies the (d,p) and (d,n) reactions were found to be equally probable, showing that the Coulomb force plays a secondary role in restricting the penetrability of the proton at low energy. But the most interesting feature of the high-energy experiments was the angular distribution. At high energy the reaction can occur through higher partial waves and the cross section shows marked variation with angle. It was found that in both (d,p) and (d,n) reactions the cross section was peaked either in the forward direction or at some small angle and decreased to very small values at backward angles. These features were explained first by Serber (2) and later more accurately by Butler (3). The explanations were based

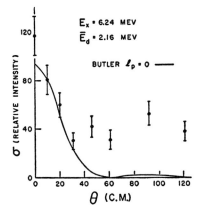

FIG. 1. An example of a (d,n) reaction. [$Al^{27}(d,n)Si^{28}$*(6.24-Mev state).] [A. G. Rubin, Phys. Rev. **108**, 62 (1957).]

on the single-step model of the reaction.[a] When the deuteron gets very close to the target there is a strong interaction between the target and the neutron (or the proton) of the deuteron. As a result the neutron (or the proton) may be stripped off the deuteron and captured by the target, while the proton (or the neutron) continues on. With a model like this it is obvious that the released particle will tend to move in its original direction. Thus it explains the forward peaking of the angular distribution.

Butler showed that the angular distribution is essentially of the form of the square of the spherical Bessel function (5) $j_l(qR)$ of order l; where l is the orbital angular momentum of the state in which the target captures the proton in a (d,n) reaction [or the proton in a (d,p) reaction], q is the change of momentum of the target in the center-of-mass system, and R is a certain radius, slightly larger than the nuclear radius, which may be termed the interaction radius.

[a] For a complete bibliography see "Nuclear Stripping Reactions" (4).

V.B.2. THE THEORY OF STRIPPING AND PICKUP REACTIONS 697

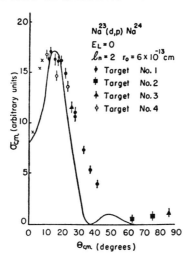

FIG. 2. An example of a (d,p) reaction. [$Na^{23}(d,p)Na^{24}$, ground state.] [W. F. Vogelsang and J. N. McGruer, Phys. Rev. **109**, 1663 (1958).]

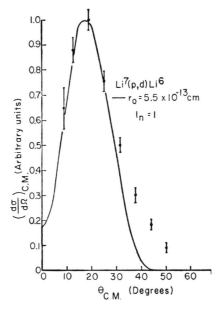

FIG. 3. An example of a (p,d) reaction. [$Li^{7}(p,d)Li^{6}$, ground state.] [J. B. Reynolds and J. G. Standing, Phys. Rev. **101**, 158 (1956).]

It was later found that the angular distribution of the reaction end products in many reactions were of "Butler" type. Examples of such reactions are (p,d), (n,d), (p,α), (α,p), etc. The first two are the reverse of the (d,p) and the (d,n) reactions, respectively, and so it is not surprising that they have the same type of angular distribution. These reverse reactions are called the pickup reactions. It is supposed that the (p,d)

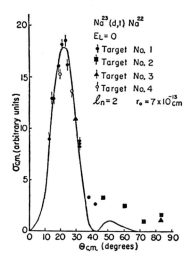

FIG. 4. An example of a (d,t) reaction. [$Na^{23}(d,t)Na^{22}$, ground state.] [W. F. Vogelsang and J. N. McGruer, Phys. Rev. **109**, 1663 (1958).]

reaction occurs through the following mechanism. As the proton approaches the target the strong interaction between the proton and the outermost neutrons of the target comes into play. Thus the proton is able to pick up a neutron from the target and goes on as a deuteron. A similar explanation holds for the (n,d) reaction. In order to explain the angular distribution of the (α,p) reaction it is necessary to consider it as the stripping of a triton from the α-particle. The (p,α) reaction may be

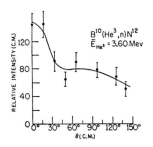

FIG. 5. An example of a (He^3,n) reaction. [$B^{10}(He^3,n)N^{12}$, ground state.] [F. Ajzenberg-Selove, M. L. Bullock, and E. Almqvist, Phys. Rev. **108**, 1284 (1957).]

V.B.2. THE THEORY OF STRIPPING AND PICKUP REACTIONS 699

considered as the pickup of a triton from the target by the proton to form the α-particle.

Some typical angular distributions observed in these processes are shown in Figs. 1 through 8. The general theory of these reactions will be the subject of this article. The theory will be based on the "single step" model mentioned before. In Section V.B.1 the direct interaction model of inelastic scattering has been discussed. The single step model is the direct interaction model for nuclear reactions.

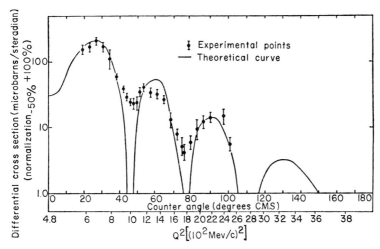

FIG. 6. An example of an (α,p) reaction. [$P^{31}(\alpha,p)S^{34}$, ground state.] [C. E. Hunting and N. S. Wall, Phys. Rev. **108**, 901 (1957).]

a. DYNAMICS OF THE MASS EXCHANGE REACTIONS

The stripping and the pickup reactions may be symbolically represented as

$$(A + B) + C \to A + (B + C). \tag{1}$$

Here A, B, and C may be nucleons or more complex nuclei. $(A + B)$ is a nucleus which has as many neutrons and protons as A and B together. Therefore, $(A + B)$ can be broken up into the components A and B. The symbol $(B + C)$ has a similar explanation. Before the reaction, C is free and A is bound to B. After the reaction, A is free, while C gets bound to B. Therefore, the name "mass exchange reaction" seems appropriate for these reactions.

The symbolic representation (1) may be illustrated with the following examples.

(1) $O^{16}(d,p)O^{17}$: Here A and B stand for the proton and the neutron of the deuteron and C stands for the target O^{16}. $(A + B)$ and $(B + C)$ stand for the deuteron and O^{17}. During the reaction the neutron (B) exchanges its old bond with the proton (A) for the new bond with O^{16} (C).

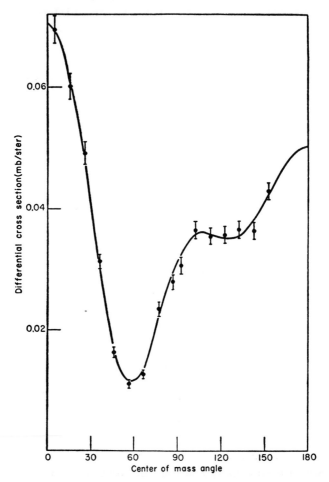

FIG. 7. An example of a (He^3,α) reaction. [$C^{13}(He^3,\alpha)C^{12}$, ground state $E_{He^3} = 2.00$ Mev, $Q_0 = 15.618$ Mev.] [H. D. Holmgren, Phys. Rev. **106**, 100 (1957).]

(2) $F^{19}(p,\alpha)O^{16}$: Here A, B, C, $(A + B)$, and $(B + C)$ stand for O^{16}, H^3, the projectile proton, the target F^{19}, and the α-particle, respectively. During the reaction the triton (B) exchanges its old bond with O^{16} for its new bond with the proton (C).

The mesic units, in which c, $\hbar$, and the mass of the π meson are each equal to unity, will be employed throughout this section.

The same letters A, B, and C will be used to denote the masses of the three particles. The momentum of C before the reaction in the center-of-mass system will be denoted by $\mathbf{K}_i$ and the momentum of A after the reaction in the center-of-mass system will be denoted by $\mathbf{K}_f$.

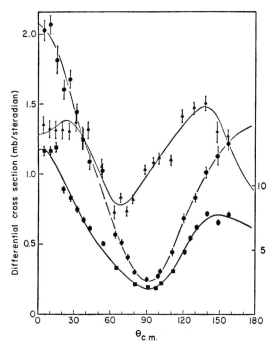

FIG. 8. An example of a (He^3, p) reaction. $[C^{12}(He^3, p)N^{14}.]$ $p_0(\blacktriangle)$ refers to ground state group of protons, $p_1(\bullet)$ to the first excited group, and $p_2(\blacksquare)$ to the second excited state group (for the p_2 group, use the right-hand ordinate scale). [H. D. Holmgren, E. A. Wolicki, and E. H. Geer Illsley, Phys. Rev. **109**, 884 (1958).]

When two particles are bound together their momenta may be divided into two parts; namely, the motion of the center of mass and the internal motion in the bound state. Thus in the initial state, in the center-of-mass system, the particle A has the momentum $-\mathbf{K}_i A/(A + B)$ acquired by sharing the momentum of the center of mass of the nucleus $(A + B)$. In the final state the momentum of A is $\mathbf{K}_f$. Thus the change of momentum is $\mathbf{K}_f + \mathbf{K}_i A/(A + B)$ if we ignore the internal momentum of A. The difference between the final and initial momenta obtained by ignoring the internal momenta will be referred to as the momentum transfer and will be denoted by the letter $\mathbf{q}$.

Thus

$$\begin{aligned} \mathbf{q}_A &= \mathbf{K}_f + \mathbf{K}_i \frac{A}{A+B} \\ \mathbf{q}_B &= -\frac{B}{B+C}\mathbf{K}_f + \frac{B}{A+B}\mathbf{K}_i \\ \mathbf{q}_C &= -\frac{C}{B+C}\mathbf{K}_f - \mathbf{K}_i \end{aligned} \right\}. \qquad (2)$$

It will be seen later that the differential cross section for the mass exchange reaction in the Born approximation can be expressed in terms of the momentum transfers listed above. In Table I the momentum transfers in (d,p), (p,d), (p,α), and (α,p) reactions are listed. In each case the momentum transfers have been expressed in terms of the momenta of the lighter particles in the center-of-mass system; for example, $\mathbf{K}_p$ and $\mathbf{K}_\alpha$ in the case of (p,α) and (α,p) reactions.

b. Cross Section of the Mass Exchange Reaction

In this section a simple theoretical expression for the mass exchange reaction $(A + B) + C \to A + (B + C)$ will be derived under the Born approximation.

The nuclear particles A, B, and C will be regarded as structureless entities. The nucleus $(A + B)$ will be regarded as a compound of the particles A and B. Thus in the $F^{19}(p,\alpha)O^{16}$ reaction, the target F^{19} will be taken to be composed of O^{16} and a triton. It should be understood that this is only a model and is true only for part of the time. F^{19} may look like a mixture of F^{18} and a neutron or like a mixture of O^{18} and a proton with appreciable probabilities. The energy of the nucleus $(A + B)$ at rest will be taken to be equal to $-\epsilon_{AB}$, where ϵ_{AB} is the energy required to break up the nucleus $(A + B)$ into the constituents A and B. In the same scale the energy of the nucleus $(B + C)$ will then be equal to $-\epsilon_{BC}$, where ϵ_{BC} has a similar interpretation. The initial kinetic energy in the center-of-mass system is $K_i^2/2\mu_i$, where

$$\mu_i = \frac{(A+B)C}{A+B+C}$$

is the reduced mass for the motion of $(A + B)$ and C. The final kinetic energy in the center-of-mass system is $K_f^2/2\mu_f$, where $\mu_f = \dfrac{A(B+C)}{A+B+C}$ is the reduced mass for the motion of A and $(B + C)$. The total energy E is then

$$E = \frac{K_i^2}{2\mu_i} - \epsilon_{AB} = \frac{K_f^2}{2\mu_f} - \epsilon_{BC}. \qquad (3)$$

TABLE I. MOMENTUM TRANSFERS. A_T IS THE MASS NUMBER OF THE TARGET NUCLEUS

Reaction	(d,p)	(p,d)	(p,α)	(α,p)
A	p	Final nucleus	Final nucleus	p
B	n	n	H^3	H^3
C	Target	p	p	Target
q_A	$q_p = \mathbf{K}_p - \frac{1}{2}\mathbf{K}_d$	$q_{\text{daughter}} = -\mathbf{K}_d + \frac{A_T - 1}{A_T}\mathbf{K}_p$	$q_{\text{daughter}} = -\mathbf{K}_\alpha + \frac{A-B}{A_T}\mathbf{K}_p$	$q_p = \mathbf{K}_p - \frac{1}{4}\mathbf{K}_\alpha$
q_B	$q_n = -\frac{1}{A_T+1}\mathbf{K}_p - \frac{1}{2}\mathbf{K}_d$	$q_n = \frac{1}{A_T}\mathbf{K}_p + \frac{1}{2}\mathbf{K}_d$	$q_{H^3} = \frac{3}{4}\mathbf{K}_\alpha + \frac{3}{A_T}\mathbf{K}_p$	$q_{H^3} = -\frac{3}{A_T+3}\mathbf{K}_p - \frac{3}{4}\mathbf{K}_\alpha$
q_C	$q_T = \mathbf{K}_d - \frac{A_T}{A_T+1}\mathbf{K}_p$	$q_p = -\mathbf{K}_p + \frac{1}{2}\mathbf{K}_d$	$q_p = \frac{1}{4}\mathbf{K}_d - \mathbf{K}_p$	$q_T = \mathbf{K}_\alpha - \frac{A_T}{A_T+3}\mathbf{K}_p$
$\mathbf{K}_i$	$-\mathbf{K}_d$	$\mathbf{K}_p$	$\mathbf{K}_p$	$-\mathbf{K}_\alpha$
$\mathbf{K}_f$	$\mathbf{K}_p$	$-\mathbf{K}_d$	$-\mathbf{K}_\alpha$	$\mathbf{K}_p$

Therefore the Q-value of the reaction is

$$Q = \frac{K_f^2}{2\mu_f} - \frac{K_i^2}{2\mu_i} = \epsilon_{BC} - \epsilon_{AB}. \tag{4}$$

The letters $\mathbf{r}$, $\mathbf{p}$, and T will be used to denote the coordinate vector, the momentum, and the kinetic energy operators, respectively. The coordinates $\mathbf{r}_A$, $\mathbf{r}_B$, and $\mathbf{r}_C$ of the three particles are not independent variables in the c.m. system as the center of mass is kept fixed. There are two sets of independent variables, which will be used. One consists of $\mathbf{r}_{AB} = \mathbf{r}_A - \mathbf{r}_B$, the separation between the particles A and B, and $\mathbf{r}_i = \mathbf{r}_c - \frac{A\mathbf{r}_A + B\mathbf{r}_B}{A + B}$, the separation between C and the center of mass of $(A + B)$. The other set consists of $\mathbf{r}_{BC} = \mathbf{r}_B - \mathbf{r}_C$, the separation between the particles B and C, and $\mathbf{r}_f = \mathbf{r}_A - \frac{B\mathbf{r}_B + C\mathbf{r}_C}{B + C}$, the separation between A and the center of mass of $(B + C)$. The canonically conjugate momenta are

$$\left. \begin{aligned} \mathbf{p}_{AB} &= \left(\frac{\mathbf{p}_A}{A} - \frac{\mathbf{p}_B}{B}\right)\frac{AB}{A + B} \\ \mathbf{p}_i &= \left(\frac{\mathbf{p}_C}{C} - \frac{\mathbf{p}_A + \mathbf{p}_B}{A + B}\right)\frac{(A + B)C}{A + B + C} \end{aligned} \right\} \tag{5a}$$

$$\left. \begin{aligned} \mathbf{p}_{BC} &= \left(\frac{\mathbf{p}_B}{B} - \frac{\mathbf{p}_C}{C}\right)\frac{BC}{B + C} \\ \mathbf{p}_f &= \left(\frac{\mathbf{p}_A}{A} - \frac{\mathbf{p}_B + \mathbf{p}_C}{B + C}\right)\frac{A(B + C)}{A + B + C} \end{aligned} \right\}. \tag{5b}$$

Let $\mu_{AB} = \frac{AB}{A + B}$ and $\mu_{BC} = \frac{BC}{B + C}$ stand for the reduced masses for the internal motions of the particles A and B, and B and C, respectively. The kinetic energy operator in the c.m. system may be written as

$$T_0 = \frac{p_{AB}^2}{2\mu_{AB}} + \frac{p_i^2}{2\mu_i} = \frac{p_{BC}^2}{2\mu_{BC}} + \frac{p_f^2}{2\mu_f}. \tag{6}$$

For simplicity the nuclear particles will be assumed to be spinless. The nuclear interaction between two particles will be assumed to depend only on the magnitude of the distance between the two particles. Let $\phi_{AB}(\mathbf{r}_{AB})$ and $\phi_{BC}(\mathbf{r}_{BC})$ be the internal radial wave functions of the nuclei $(A + B)$ and $(B + C)$, respectively. These wave functions satisfy the following Schroedinger equations

$$\left. \begin{aligned} \left(\frac{p_{AB}^2}{2\mu_{AB}} + V_{AB} + \epsilon_{AB}\right)\phi_{AB}(\mathbf{r}_{AB}) &= 0 \\ \left(\frac{p_{BC}^2}{2\mu_{BC}} + V_{BC} + \epsilon_{BC}\right)\phi_{BC}(\mathbf{r}_{BC}) &= 0 \end{aligned} \right\}, \tag{7}$$

where V_{AB} and V_{BC} are the nuclear interaction potentials between the pairs (A,B) and (B,C), respectively. Let Ψ describe the state of the three particles under the experimental condition. The Schroedinger equation satisfied by Ψ is

$$(T_0 + V_{AB} + V_{BC} + V_{AC} - E)\Psi = 0, \tag{8}$$

where E is the total energy of the three particles in the c.m. system and is given by Eq. (3) and V_{AC} is the nuclear interaction potential between the pair (A,C). The function $\phi_i(\mathbf{r}_i)$, defined by the integral

$$\phi_i(\mathbf{r}_i) = \int d^3\mathbf{r}_{AB} \phi_{AB}{}^*(\mathbf{r}_{AB}) \Psi(\mathbf{r}_i,\mathbf{r}_{AB}) \tag{9}$$

describes the motion of C with respect to the center of mass of A and B when the last two are bound together to form the nucleus $(A + B)$. It should describe the elastic scattering of C by the nucleus $(A + B)$. The function $\phi_f(\mathbf{r}_f)$, defined by the integral

$$\phi_f(\mathbf{r}_f) = \int d^3\mathbf{r}_{BC} \phi_{BC}{}^*(\mathbf{r}_{BC}) \Psi(\mathbf{r}_f,\mathbf{r}_{BC}) \tag{10}$$

describes the motion of A with respect to the center of mass of B and C when they are bound together to form the nucleus $(B + C)$. If Ψ describes the state of affairs correctly then it must satisfy the boundary condition that for large values of $\mathbf{r}_f$, $\phi(\mathbf{r}_f)$ must look like a purely outgoing wave; that is,

$$\phi_f(\mathbf{r}_f) \xrightarrow[r_f \to \infty]{} g(\theta,\phi) \frac{e^{iK_f r_f}}{\mathbf{K}_f r_f}, \tag{11}$$

where θ, ϕ are the polar angles. The reaction cross section is then given by the expression

$$\sigma(\theta,\phi) = \frac{K_f}{K_i} \frac{\mu_i}{\mu_f} |g(\theta,\phi)|^2. \tag{12}$$

Using Eqs. (3) and (6), Eq. (12) may be written as follows

$$\left(\frac{p_f{}^2}{2\mu_f} - \frac{K_f{}^2}{2\mu_f} + \frac{p_{BC}{}^2}{2\mu_{BC}} + V_{BC} + \epsilon_{BC} \right) \Psi = -(V_{AB} + V_{AC})\Psi. \tag{13}$$

Multiplying each side of the equation on the left by $\phi_{BC}{}^*(\mathbf{r}_{BC})$, integrating over $\mathbf{r}_{BC}$ and using Eqs. (7) and (10), one gets the following equation for $\phi_f(\mathbf{r}_f)$:

$$(\nabla_f{}^2 + K_f{}^2)\phi_f(\mathbf{r}_f) = 2\mu_f \int d^3\mathbf{r}_{BC} \phi_{BC}{}^*(\mathbf{r}_{BC})(V_{AB} + V_{AC})\Psi(\mathbf{r}_f,\mathbf{r}_{BC}). \tag{14}$$

The equation may be solved by the standard method of using the Green's function[b] $-e^{iK_f|\mathbf{r}_f-\mathbf{r}_f'|}/4\pi|\mathbf{r}_f - \mathbf{r}_f'|$ and one gets

$$\phi_f(\mathbf{r}_f) = -\frac{\mu_f}{2\pi} \int \frac{e^{iK_f|\mathbf{r}_f-\mathbf{r}_f'|}}{|\mathbf{r}_f - \mathbf{r}_f'|} \phi_{BC}{}^*(\mathbf{r}_{BC})(V_{AB} + V_{AC})\Psi d^3\mathbf{r}_{BC} d^3\mathbf{r}_f'. \quad (15)$$

It then follows that $g(\theta,\phi)$, to be called henceforth the reaction amplitude is given by the expression

$$g(\theta,\phi) = -\frac{\mu_f}{2\pi} \int d^3\mathbf{r}_f d^3\mathbf{r}_{BC} \phi_{BC}{}^*(\mathbf{r}_{BC}) e^{-i\mathbf{K}_f \cdot \mathbf{r}_f}(V_{AB} + V_{AC})\Psi. \quad (16)$$

The Born approximation may now be introduced by replacing the total wave function Ψ by $\phi_{AB}(\mathbf{r}_{AB})e^{i\mathbf{K}_i \cdot \mathbf{r}_i}$, which describes the state before the reaction. The reaction amplitude, under the Born approximation, is

$$g(\theta,\phi) = -\frac{\mu_f}{2\pi} \int d^3\mathbf{r}_f d^3\mathbf{r}_{BC} \phi_{BC}{}^*(\mathbf{r}_{BC}) e^{-i\mathbf{K}_f \cdot \mathbf{r}_f}(V_{AB} + V_{AC}) e^{i\mathbf{K}_i \cdot \mathbf{r}_i} \phi_{AB}(\mathbf{r}_{AB}). \quad (17)$$

Since the two potentials are assumed to be momentum independent the two plane waves may be multiplied together and the identity

$$\mathbf{K}_i \cdot \mathbf{r}_i - \mathbf{K}_f \cdot \mathbf{r}_f = \mathbf{q}_C \cdot \mathbf{r}_{BC} - \mathbf{q}_A \cdot \mathbf{r}_{AB} \quad (18)$$

may be used to rewrite the reaction amplitude as

$$g(\theta,\phi) = -\frac{\mu_f}{2\pi} \int d^3\mathbf{r}_{AB} d^3\mathbf{r}_{BC} \phi_{BC}{}^*(\mathbf{r}_{BC}) e^{i(\mathbf{q}_C \cdot \mathbf{r}_{BC} - \mathbf{q}_A \cdot \mathbf{r}_{AB})}(V_{AB} + V_{AC}) \phi_{AB}(\mathbf{r}_{AB}) \quad (19)$$

because $d^3\mathbf{r}_f d^3\mathbf{r}_{BC} \equiv d^3\mathbf{r}_{AB} d^3\mathbf{r}_{BC}$.

When the three particles are very close together, that is, in the so-called "inner region" of the configuration space, the strong mutual interactions may cause the wave function Ψ to differ considerably from the approximate form used here. In the compound nucleus theory the compound nuclear eigenfunctions, which belong to the actual Hamiltonian but are subject to some special boundary conditions, are used to describe the wave function Ψ. These eigenfunctions emphasize the strong interaction between the projectile and the target. When Ψ has a good overlap with a single compound nuclear eigenfunction, the reaction exhibits the characteristics of the compound nucleus model. In contrast to this the direct interaction model explains the reaction in terms of a single encounter. The strong correlations between the nucleons of the projectile and those of the target are not expected to play an important

[b] See any standard textbook on quantum mechanics.

role. An approximation of Ψ ignoring such correlation is consistent with the ideas of the direct interaction model. The situation is over-idealized in the Born approximation because the interaction between the target and the projectile is completely ignored. It should be realized that an approximate form of Ψ which ignores the strong correlations may be adequate only in the outer region of the configuration space. Some compound nucleus formation occurs in the inner region. The direct interaction model is valid when the contribution of the compound nuclear part to the reaction amplitude is negligible.

The size of the integral of an interaction potential depends upon the probability of finding the interacting pair within the range of interaction. The pair (A,B) is bound together in the initial stage of the reaction, whereas the pair (A,C) is never bound together. Therefore, the contribution due to the interaction V_{AB} is expected to be larger than that due to the interaction V_{AC}. Although A and C are never bound together, they are both bound to B at different stages of the reaction. Because of this fact the probability of finding A and C close together is larger than what it would be if the particles were entirely free. The probability attains a maximum value when the two nuclei $(A + B)$ and $(B + C)$ are of comparable size. In the usual stripping reaction $(A + B)$ represents the projectile nucleus and $(B + C)$ represents the heavier end product. In the pickup reactions $(A + B)$ represents the target and $(B + C)$ represents the lighter end product. In each case the sizes of the two nuclei are very much dissimilar. As a result, the contribution of V_{AC} is usually very much smaller than that of V_{BC}. Actual calculations have shown that the former is usually less than 5% of the latter. It is convenient to term V_{AC} as the unbound pair interaction and V_{AB} and V_{BC} as the binding interactions.

Neglecting the unbound pair interaction the reaction amplitude may be written as a product of two integrals in the following manner.

$$g(\theta,\phi) = -\frac{\mu_f}{2\pi}\left\{\int d^3\mathbf{r}_{AB} e^{i\mathbf{q}_A\cdot\mathbf{r}_{AB}} V_{AB}(\mathbf{r}_{AB})\phi_{AB}(\mathbf{r}_{AB})\right\}$$
$$\left\{\int d^3\mathbf{r}_{BC} e^{i\mathbf{q}_C\cdot\mathbf{r}_{BC}}\phi_{BC}^*(\mathbf{r}_{BC})\right\}. \quad (20)$$

The integral containing the potential may be simplified with the help of the Schroedinger equations

$$(T_{AB} + V_{AB} + \epsilon_{AB})\phi_{AB}(\mathbf{r}_{AB}) = 0$$
$$(T_{AB} - q_A^2/2\mu_{AB})e^{-i\mathbf{q}_A\cdot\mathbf{r}_{AB}} = 0.$$

Multiplying the first on the left by $e^{-i\mathbf{q}_A\cdot\mathbf{r}_{AB}}$ and the second on the left by

$\phi_{AB}(\mathbf{r}_{AB})$ and subtracting one from the other, one gets, upon integrating over $\mathbf{r}_{AB}$,

$$\int d^3\mathbf{r}_{AB} e^{-i\mathbf{q}_A\cdot\mathbf{r}_{AB}} V_{AB}(\mathbf{r}_{AB})\phi_{AB}(\mathbf{r}_{AB}) = -T\int d^3\mathbf{r}_{AB} e^{-i\mathbf{q}_A\cdot\mathbf{r}_{AB}}\phi_{AB}(\mathbf{r}_{AB}), \quad (21)$$

where
$$T = \epsilon_{AB} + q_A^2/2\mu_{AB}. \quad (22)$$

Therefore,

$$g(\theta,\phi) = T\frac{\mu_f}{2\pi}\left\{\int d^3\mathbf{r}_{AB} e^{-i\mathbf{q}_A\cdot\mathbf{r}_{AB}}\phi_{AB}(\mathbf{r}_{AB})\right\}\left\{\int d^3\mathbf{r}_{BC} e^{i\mathbf{q}_C\cdot\mathbf{r}_{BC}}\phi_{BC}^*(\mathbf{r}_{BC})\right\} \quad (23)$$

and the differential cross section

$$\sigma(\theta,\phi) = \frac{K_f}{K_i}\mu_i\mu_f\left(\frac{1}{2\pi}\right)^2\left|\int d^3\mathbf{r}_{AB} e^{-i\mathbf{q}_A\cdot\mathbf{r}_{AB}}\phi_{AB}(\mathbf{r}_{AB})\right|^2 \left|\int d^3\mathbf{r}_{BC} e^{i\mathbf{q}_C\cdot\mathbf{r}_{BC}}\phi_{BC}^*(\mathbf{r}_{BC})\right|^2. \quad (24)$$

The reaction amplitude for the reverse reaction

$$A + (B + C) \to (A + B) + C$$

is

$$g(\theta,\phi) = -\frac{\mu_i}{2\pi}\left\{\int d^3\mathbf{r}_{BC} e^{-i\mathbf{q}_C\cdot\mathbf{r}_{BC}} V_{BC}(\mathbf{r}_{BC})\phi_{BC}(\mathbf{r}_{BC})\right\}\left\{\int d^3\mathbf{r}_{AB} e^{i\mathbf{q}_A\cdot\mathbf{r}_{AB}}\phi_{AB}^*(\mathbf{r}_{AB})\right\}, \quad (25)$$

where the unbound pair interaction term has been omitted. The potential term may be simplified by repeating the above procedure, and because of the identity

$$T = \epsilon_{AB} + q_A^2/2\mu_{AB} = \epsilon_{BC} + q_C^2/2\mu_{BC}, \quad (26)$$

it follows that

$$g_{\text{Direct}}/\mu_i = g_{\text{Reverse}}/\mu_f, \quad (27)$$

and therefore the reciprocity relation (6)

$$K_f^2\sigma_{\text{Direct}} = K_i^2\sigma_{\text{Reverse}} \quad (28)$$

is satisfied. It is obvious that the reciprocity relation will hold even if the unbound pair interaction is considered.

The two integrals which appear in the expression [Eq. (23)] of the reaction amplitude are the Fourier transforms

$$F_i(\mathbf{q}_A) = \int d^3r\, e^{-i\mathbf{q}_A\cdot\mathbf{r}}\phi_{AB}(\mathbf{r}) \quad (29)$$

of the initial bound state $\phi_{AB}(\mathbf{r})$ with respect to the momentum $\mathbf{q}_A$ and

$$F_f(\mathbf{q}_C) = \int d^3 r\, e^{-i\mathbf{q}_C\cdot\mathbf{r}} \phi_{BC}(\mathbf{r}) \tag{30}$$

of the final bound state $\phi_{BC}(\mathbf{r})$ with respect to the momentum $\mathbf{q}_C$. The reaction amplitude may be written as

$$g(\theta,\phi) \equiv g(\mathbf{q}_A,\mathbf{q}_C) = \frac{\mu_f}{2\pi} T F_i(\mathbf{q}_A) F_f^*(\mathbf{q}_C). \tag{31}$$

The appearance of the Fourier transforms may be understood in terms of the momentum transfers which occur during the reaction. Initially A is bound to B and the net momentum of A is due to the internal motion in the bound state as well as to the motion of the center of mass of $(A + B)$. During the reaction the bond is broken up and A is released into a free state with the final momentum $\mathbf{K}_f$. This final momentum must be exactly equal to the sum of the two types of momentum of A in the initial state. It then follows, from the definition of $\mathbf{q}_A$, that it must be equal to the internal momentum of A at the instant of release. In a similar manner one may see that $\mathbf{q}_C$ must be equal to the internal momentum of C at the instant of its capture by B. The probability that such an event may occur is directly proportional to the probability that one may find the momenta $\mathbf{q}_A$ and $\mathbf{q}_C$ in the initial and the final states, respectively. The square moduli of the two Fourier transforms are just these probabilities.

An inspection of Table I will show that in the usual stripping or pickup reactions one of the two nuclei $(A + B)$ and $(B + C)$ is a light particle, such as a deuteron or an α-particle, while the other is a much heavier particle. The smaller the nucleus the smaller is the uncertainty in the location of one particle relative to the other. As a consequence the uncertainty in the internal momentum will be correspondingly larger. This implies that the Fourier transform of the wave function of a light nucleus will not exhibit any sharp peaking and will tend to be rather flat. Thus the main structure in the angular distribution of a mass exchange reaction arises from the Fourier transform of the heavier nucleus, while the Fourier transform of the lighter nucleus plays the role of a form factor.

c. Angular Distributions in Mass Exchange Reactions

The main features of the angular distributions and their dependence on the angular momenta will be discussed in this section. The assumption that the nuclear particles are spinless will be retained here. Using the symbols $u(r)$ for the bound state radial wave function and l for the internal angular momentum, the nuclear wave functions may be written as

$$\left.\begin{array}{l}\phi_{AB}(r) = u_{AB}(r) y_{m_{AB}}{}^{l_{AB}}(r) \\ \phi_{BC}(r) = u_{BC}(r) y_{m_{BC}}{}^{l_{BC}}(r)\end{array}\right\} \tag{32}$$

where $y_m{}^l$ are the spherical harmonics as defined by Condon and Shortley (8). The plane wave may be expanded as follows

$$e^{i\mathbf{q}\cdot\mathbf{r}} = 4\pi \sum_{\lambda\mu} i^\lambda j_\lambda(qr) y_\mu{}^\lambda(\Omega_r) y_\mu{}^{\lambda*}(\Omega_q) \qquad (33)$$

where $j_\lambda(qr)$ is the spherical Bessel function of order λ and Ω_r and Ω_q stand for the polar angles of $\mathbf{r}$ and $\mathbf{q}$, respectively. Introducing Eqs. (32) and (33) in Eqs. (29) and (30) and using the orthogonality of the spherical harmonics one obtains

$$F_i(\mathbf{q}_A) = 4\pi(-i)^{l_{AB}} y_{m_{AB}}{}^{l_{AB}}(\Omega_{q_A}) R_i(q_A) \qquad (34a)$$

with

$$R_i(q_A) = \int dr \cdot r^2 u_{AB}(r) j_{l_{AB}}(q_A r), \qquad (35a)$$

and

$$F_f(\mathbf{q}_C) = 4\pi(-i)^{l_{BC}} y_{m_{BC}}{}^{l_{BC}}(\Omega_{q_C}) R_f(q_C) \qquad (34b)$$

with

$$R_f(q_C) = \int dr \cdot r^2 u_{BC}(r) j_{l_{BC}}(q_C r). \qquad (35b)$$

Thus the reaction amplitude is

$$g(\theta,\phi) = 8\pi\mu_f T(i)^{l_{BC}-l_{AB}} y_{m_{AB}}{}^{l_{AB}}(\Omega_{q_A}) y_{m_{BC}}{}^{l_{BC}}(\Omega_{q_C}{}^*) R_i(q_A) R_f(q_C). \qquad (36)$$

The square modulus of Eq. (36) gives the reaction amplitude when the initial and the final magnetic quantum numbers are m_{AB} and m_{BC}, respectively. [Since the nuclear particles are spinless the angular momenta of $(A + B)$ and $(B + C)$ are the initial and the final angular momenta.] If these quantum numbers are not measured during the experiment the differential cross section is obtained by averaging over the initial magnetic substates and summing over the final magnetic substates. Using the relation

$$\sum_{m=-l}^{+l} |y_m{}^l|^2 = \frac{[l]}{4\pi}$$

where $[l] \equiv 2l + 1$, one gets

$$\sigma(\theta,\phi) = 4\frac{K_f}{K_i}[l_{BC}]\mu_f\mu_i T^2 R_i{}^2(q_A) R_f{}^2(q_C). \qquad (37)$$

It should be noted that the differential cross section depends on the polar angles θ, ϕ of $\mathbf{K}_f$ through the quantities $q_A = \left|\mathbf{K}_f + \dfrac{A}{A+B}\mathbf{K}_i\right|$ and $q_C = \left|\mathbf{K}_i + \dfrac{C}{B+C}\mathbf{K}_f\right|$. Therefore, it is a function of the angle between $\mathbf{K}_i$ and $\mathbf{K}_f$ alone and not of the azimuthal angle of $\mathbf{K}_f$ about $\mathbf{K}_i$.

The azimuthal symmetry is the result of not measuring the initial and the final magnetic quantum numbers. In such a case the only well-defined vector in the problem is the direction of motion of the projectile and therefore there is an axial symmetry about this vector.

In an actual experiment one usually measures the momenta of the lighter particles before and after the reaction. Thus in a (d,p) experiment the momenta $\mathbf{K}_d$ and $\mathbf{K}_p$ of the deuteron and the proton, respectively, are measured. The relation between $\mathbf{K}_i$ and $\mathbf{K}_f$ and the momenta actually measured in different reactions are listed in Table I. In the case of a (d,p) reaction $\mathbf{K}_i = -\mathbf{K}_d$ and $\mathbf{K}_f = \mathbf{K}_p$. Therefore,

$$q_A \equiv q_p = |\mathbf{K}_p - \tfrac{1}{2}\mathbf{K}_d| \text{ and } q_C \equiv q_{\text{Target}} = \left|\mathbf{K}_d - \mathbf{K}_p \frac{A_T}{A_T + 1}\right|,$$

where A_T is the mass number of the target. The differential cross section is a function of the angle between $\mathbf{K}_d$ and $\mathbf{K}_p$. When the proton is emitted in the forward direction, that is, in the same direction as $\mathbf{K}_d$, the two momentum transfers have their minimum values. Their values increase as the angle of emission of the proton increases and attain their maximum values when the proton is emitted in the backward direction. An inspection of Table I will show that in each case the two relevant momentum transfers are minimum when the lighter end product comes out in the direction of the initial beam and they increase as the angle between the direction of emission of the lighter end product and the initial direction increases. The ranges of variation of the momentum transfers increase as the energy of the incident beam increases.

Apart from the angular dependent factors, the integrals $R_i(q_A)$ and $R_f(q_C)$ are essentially the Fourier transforms. Thus $R_i^2(q_A)$ gives the probability that the internal momentum of A may have the magnitude q_A, when it is in a bound state with B. An idea about the dependence of the Fourier transform on the internal angular momentum may be obtained with the help of a semiclassical model. Consider the bound state of two nuclear particles 1 and 2 with the internal angular momentum l and let $u(r)$ be the radial wave function describing the bound state. To describe the motion of 1 around 2 one may use the semiclassical picture of a circular orbit. The probability of finding the radius r for the orbit is $r^2|u(r)|^2$. If the internal momentum is measured and found to have the value q, the radius of the orbit cannot be specified exactly because of the uncertainty principle. To allow for this uncertainty r may be allowed to have values between l/q and $(l + 1)/q$. Thus the probability of finding the value q is equal to the probability of finding the particle 1 within a shell of inner radius l/q and outer radius $(l + 1)/q$. If the internal angular momentum l is zero then the two radii are 0 and $1/q$. As q increases the outer radius decreases and therefore the probability decreases as well. But the nucleus has a finite size, characterized by a radius R_0. So long as the momentum q is less than $1/R_0$, the sphere, in question, encloses most of the high-density region of the nucleus and only the low-density "tail" of the nucleus is left out. So a slight change in q in this range does not produce any appreciable change in the probability $P(q)$ for finding the particular value of q. Therefore when the maximum value of q is less than $1/R_0$, which is the case for a small nucleus and not too high energy of the incident beam, the probability $P(q)$ changes by a small amount over the whole range of variation of q. This corroborates the remarks made in the previous section concerning the dependence of the nature of the Fourier transform on the size of the nucleus. For a large nucleus with zero internal angular momentum, $P(q)$ will have its maximum value for forward emission and will decrease as the angle of emission increases.

If the internal angular momentum is nonzero the nuclear density distribution is zero at the center and increases to a maximum value at some finite distance and then it falls off. For very small values of q the shell of radii l/q and $(l+1)/q$ lies outside the region of maximum density and the probability of finding particle 1 within the shell is vanishingly small. As q increases the shell approaches the region of high density and $P(q)$ increases. It reaches the maximum value when the shell encloses the peak of the density distribution and as q increases further $P(q)$ falls off. It has been explained before that the probability $P(q) = R^2(q)$, where

$$R(q) = \int dr \cdot r^2 j_l(qr) u(r).$$

Therefore the preceding discussion gives an idea about the variation of the Fourier transform with the momentum for zero and nonzero values of the internal angular momentum.

If the wave functions $u_{AB}(r)$ and $u_{BC}(r)$ are known, the Fourier transform can be obtained by evaluating the integrals (35a) and (35b). However, the exact nuclear wave functions are not known and the best that one can do is to assume a suitable form for the wave functions with the proper angular momenta, which reproduce the sizes of the nuclei correctly. The last two are the most important parameters of the wave function so far as the Fourier transform is concerned.

The discussion will now be specialized to the case of stripping reactions. Since the pickup reactions may be viewed as the reverse of the stripping reactions, a discussion of the latter type is sufficient.

In stripping reactions the nucleus $(A + B)$ is the light projectile which is stripped off and $(B + C)$ is the heavy end product of the reaction. Usually deuteron, triton, and α-particle are employed as projectiles in stripping reactions. In each case the internal angular momentum is zero. Therefore,

$$R_i(q_A) = \int r^2 dr u_{AB}(r) j_0(q_A r)$$
$$= \frac{1}{q_A} \int dr \cdot r \cdot \sin q_A r u_{AB}(r). \quad (38)$$

The usual forms of the light particle wave functions are listed in Table II; and on Figs. 9, 10, and 11 the integral $R_i(q_A)$ has been plotted against

TABLE II. SOME TYPICAL RADIAL WAVE FUNCTIONS

Name	Form
Exponential	$2\alpha^{\frac{3}{2}} e^{-\alpha r}$
Gaussian	$2\pi^{-\frac{1}{4}} b^{-\frac{3}{2}} e^{-r^2/2b^2}$
Hulthen	$\sqrt{\dfrac{2a_1 a_2 (a_1 + a_2)}{(a_1 - a_2)^2}} \dfrac{e^{-a_1 r} - e^{-a_2 r}}{r}$

the appropriate momentum transfer for the different forms of the wave functions.

One could adopt a similar procedure to evaluate $R_f(q_C)$, the Fourier transform of the wave function of the heavier nucleus $(B + C)$. But here a new physical consideration becomes important. It was mentioned in subsection (b) that formation of compound nucleus occurs in the inner region of the configuration space, where all the nuclear particles are within the range of interaction of each other. The direct interaction model is valid when the compound nucleus contribution is not too large, that is, when the contribution from the inner region of the configuration space can be neglected. Since C is the heavy target, small values of r_{BC} imply that all the nuclear particles, except A, are very close together and therefore such a situation corresponds to the inner region of the configuration space. Thus one should leave out small values of r_{BC} from the integral (35b), in order to be consistent in the application of the direct interaction model. But a similar remark cannot be made for r_{AB}, because r_{AB} can be small, yet most of the nucleons, which are contained in C, can be very far away from A and B.

In a more rigorous derivation of the cross section based on the direct interaction model, the initial and the final scattering wave functions appear instead of the plane waves $e^{i\mathbf{K}_i \cdot \mathbf{r}_i}$ and $e^{-i\mathbf{K}_f \cdot \mathbf{r}_f}$ in Eq. (17). Because there is a loss of flux in the elastic channel due to the nonelastic processes the scattering wave functions contain a damping factor which reduces the amplitude of the scattering wave function as it approaches the scattering center. This means that there is very little contribution to the integral from the regions of small values of r_i and r_f. Thus the conclusion

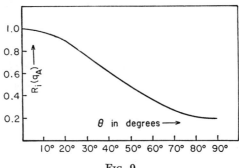

Fig. 9.

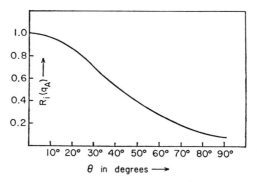

Fig. 10.

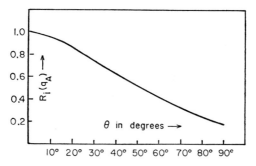

Fig. 11.

Figs. 9, 10, and 11. The integrals $R_i(q) = \int \frac{\sin qr}{q} \phi(r) r\, dr$ have been plotted against the angle θ between $\mathbf{K}_i$ and $\mathbf{K}_f$, where $q = |\mathbf{K}_f - \frac{1}{2}\mathbf{K}_i|$. In this particular case $K_i = 1$ mesic unit and $K_f = 1.2$ mesic units. Figure 9 is for $\phi = \exp(-\beta r)$, with $\beta = 0.68$ mesic unit. Figure 10 is for $\phi = \exp(-r^2/2b^2)$, with $b = 2.05$ mesic units. Figure 11 is for Hulthen wave function with $a_1 = 0.325$ mesic unit and $a_2 = 2.274$ mesic units. The parameters are adjusted so as to give the same probability for finding the neutron and the proton of the deuteron within a distance of 3 mesic units.

is the same as that of the preceding paragraph. Instead of using the complicated scattering wave functions and performing the complete integration, it is a rather attractive approximation to retain the plane waves and use as the lower limit in the integral over r_{BC} a certain "cut-off radius," R_C. The labor, thus saved, outweighs the small error incurred in the approximation.

Another justification of the cut-off procedure follows from the fact that the particle which is captured by the target during a stripping reaction, or which is taken out of the target in a pickup reaction, is usually the one with rather low binding energy. Such a particle spends most of its time outside the nucleus, that is, $u_{BC}(r_{BC})$ is small for small values of r_{BC}. Therefore the central portion of the nucleus $(B + C)$ may be left out of the integral [Eq. (35b)].

The integral $R_f(q_C)$ assumes a rather simple form if the "cut-off radius," R_C, is taken to be larger than the range of interaction of V_{BC}. Dropping the subscripts BC, the equations satisfied by $ru(r)$ and $rj_l(q_C r)$ for $r \geq R_C$ are

$$\left[\frac{d^2}{dr^2} - \frac{l(l+1)}{r^2} - 2\mu\epsilon\right] ru(r) = 0$$

$$\left[\frac{d^2}{dr^2} - \frac{l(l+1)}{r^2} + qc^2\right] rj_l(q_C r) = 0.$$

Multiplying the first equation by $rj_l(q_C r)$ and the second by $ru(r)$, subtracting one from the other, and integrating over r from R_C to infinity, one finds

$$R_f(q_C) = \int_{R_C}^{\infty} dr \cdot ru(r) rj_l(q_C r) = \frac{1}{2\mu T} R_C u(R_C) \left\{ R_C \left[\frac{d}{dr} j_l(q_C r)\right]_{r=R_C} \right.$$
$$\left. - R_C j_l(q_C R_C) \left[\frac{d}{dr} \ln u(r)\right]_{r=R_C} \right\} \quad (39)$$

where T is given by Eq. (26). Introducing the dimensionless quantities

$$\nu = R_C^{\frac{1}{2}} u(R_C) \quad (40)$$

$$\Lambda = R_C \left[\frac{d}{dr} \ln u(r)\right]_{r=R_C} \quad (41)$$

and using the relation

$$r\frac{d}{dr} j_l(qr) = qr j_{l-1}(qr) - (l+1) j_l(qr), \quad (42)$$

$R_f(q_C)$ may be written as

$$R_f(q_C) = \frac{R_C^{-\frac{1}{2}} \nu}{2\mu T} W(l, \Lambda, R_C, q_C) \quad (43)$$

where
$$W(l,\Lambda,R_C,q_C) = q_C R_C j_{l-1}(q_C R_C) - (l + 1 + \Lambda) j_l(q_C R_C). \quad (44)$$

Using this expression for $R_f(q_C)$ in Eq. (37), the cross section becomes

$$\sigma(q_A,q_C) = \frac{K_f}{K_i}[l_{BC}]\frac{\mu_f \mu_i}{\mu^2_{BC}}\frac{\nu^2}{R_C} R_i^2(q_A) W^2(l_{BC},\Lambda,R_C,q_C). \quad (45)$$

It should be remembered that a rather simple model has been used here to describe the structure of the nucleus $(B + C)$, that is, the nucleons of $(B + C)$ are divided into two groups B and C and these two groups are moving around one another in a bound state. Thus, in the case of the reaction $O^{16}(d,p)O^{17}$, B and C are associated with the neutron and O^{16}, respectively, and the model implies that O^{17} looks like a neutron moving about O^{16}. Such a picture holds only part of the time. The nucleus O^{17} may have various other modes of motion with different probabilities. The wave function $u_{BC}(r)$ gives the amplitude of the mode of the nucleus $(B + C)$ in which it appears to be composed of B and C. The square of the dimensionless quantity ν is proportional to the probability that the nucleus $(B + C)$ appears as a compound of the nuclei B and C at the surface of sphere of radius R_C. In keeping with the language of the compound nucleus theory the quantity $2\pi\nu^2/R_C^2\mu_{BC}$ is sometimes called the "width." The value of ν depends on the coupling on the nucleons in the nucleus which determines the probability that the nucleus appears to be composed of B and C. It also depends on the interaction between B and C and the choice of the cut-off radius R_C. The first type of consideration will be dealt with in the next section. For the present, it may be assumed that $(B + C)$ always appears to be composed of B and C. Then a knowledge of the value of $u_{BC}(r)$ is sufficient for the evaluation of ν. The simplest estimate of u_{BC} may be obtained by assuming a square well form for the interaction potential V_{BC} and setting the cut-off radius as equal to the radius of the square well. Then ν is a function of the internal angular momentum, the binding energy, and the range R_C. For a given value of l, ν is a function of $\beta R_C = R_C \sqrt{2\mu_{BC}\epsilon}$ alone. In Fig. 12, ν has been plotted against βR_C for $l = 0, 1$, and 2. A knowledge of ν is required only for the determination of the absolute value of the reaction amplitude. However, the chief success of the direct interaction model lies in predicting correctly the angular distribution. The prediction of absolute values of the cross section cannot be very precise unless the various approximations in the theory are improved upon. Therefore a simple estimate of ν is quite sufficient.

Since $R_i(q_A)$ is the Fourier transform of the smaller nucleus, it serves as the form factor, while the main structure of the angular distribution is given by the function $W(l,\Lambda,R_C,q_C)$. The logarithmic derivative Λ which appears in Eq. (44) may be expressed in terms of βR_C because of the assumption that the cut-off radius R_C is larger than the range of interaction. Outside the range of interaction the bound state wave function of

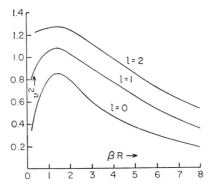

FIG. 12. Plots of the squares of $\nu = R^{\frac{3}{2}}\Psi(R)$ against βR. The wave functions have been generated with a square well of range R. The plots are for $l = 0$, 1, and 2.

angular momentum l and binding energy $\beta^2/2\mu_{BC}$ is $h_l(i\beta r)$, where h_l is the Hankel function of order l. Table III lists the expressions for

TABLE III. $(l + 1 + \Lambda)$ AS FUNCTIONS OF $Z = \beta R_C$

l	$(l + 1 + \Lambda)$
0	$-Z$
1	$-\dfrac{Z^2}{1 + Z}$
2	$-\dfrac{Z^3 + Z^2}{3 + 3Z + Z^2}$
3	$-\dfrac{3Z^2 + 3Z^3 + Z^4}{15 + 15Z + 6Z^2 + Z^3}$

$(l + 1 + \Lambda)$ in terms of βR_C for $l = 0$, 1, 2, and 3. The values of the spherical Bessel functions may be looked up in the excellent tables of these functions published by the National Bureau of Standards, U.S.A. For the special case of $l = 0$, $q_C R_C j_{l-1}(q_C R_C)$ should be replaced by $\cos q_C R_C$. Some typical plots of $W^2(l,\Lambda,R_C,q_C)$ are shown in Fig. 13. These were calculated for $\beta R_C = 3$. These curves exhibit the qualitative features discussed earlier in this section on the basis of a semiclassical picture.

For very large values of β, $l + 1 + \Lambda \simeq -\beta R_C$. Then $W(l,\Lambda,R_C,q_C)$ is also approximately equal to $\beta R_C j_l(q_C R_C)$. The result may be understood in terms of Fig. 14. The scattering wave function which is strongly damped for small values of r is shown as is the wave function of the bound state. Because of the large β, that is, large binding energy, the wave function is strongly damped outside the range of interaction. Thus the

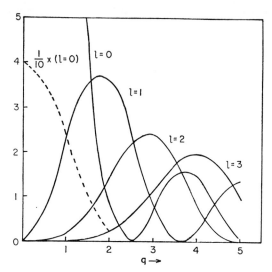

FIG. 13. Plots of the squares of $W(l,\Lambda,R_C,q)$ against q. In each case $\beta R_C = 3$ and Λ has been calculated by generating the wave function with a square well of proper depth and radius equal to R_C. The plots are for $l = 0, 1, 2$ and 3.

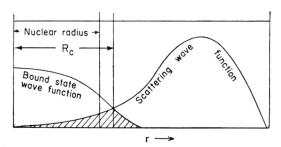

FIG. 14. A schematic plot of the bound state and the scattering radial wave functions against radial distance. The shaded area shows the region of interaction. R_C may be regarded as the mean distance of the region of interaction.

contribution to the integral comes from a fairly localized area in the neighborhood of R_C. Therefore, the integral $R_f(q_C)$ becomes proportional to $u(R_C)j_l(q_C R_C)$. Because of this feature, the mass exchange reactions are sometimes known as "surface reactions."

The localization of the reaction is more marked when the captured (stripping) or the picked up (pickup) particle is not a neutron or a proton but a heavier particle, for example, deuteron, triton, He³, or α-particle. The probability of formation of such particle clusters depends on the probability of finding two or more nucleons in the neighborhood of each other and also on the correlation among the nucleons. The former increases with increasing density of the nucleons. The Exclusion Principle tends to reduce the correlation and the effect becomes larger as the nucleon density increases. Thus the probability of formation of nucleon clusters has a maximum at some value of the density and it turns out that the maximum occurs in the proximity of the nuclear surface where the density is lower than that in the interior of the nucleus.

The radius which appears in the approximation $W(l\Lambda Rq_C) \sim \beta R j_l(q_C R)$ in the "surface reaction" picture may be called the reaction radius. The reaction radius is slightly larger than the cut-off radius, R_C. Both the reaction radius and R_C are usually larger than the "root-mean-square radius" of the nucleus.

The angular distribution of the ground state proton group in the reaction $B^{10}(d,p)B^{11}$, with 7.7-Mev deuterons, will be calculated as an illustration of the methods discussed in this chapter. The experimental data of Evans and Parkinson (7) will be used for comparison. The Q-value of the reaction is 9.28 Mev. The magnitudes of $\mathbf{K}_d = -\mathbf{K}_i$ and $\mathbf{K}_p = \mathbf{K}_f$ are 1.01 mesic units and 1.17 mesic units, respectively. Therefore, $q_p = \sqrt{1.633 - 1.187 \cos \theta}$ mesic units and

$$q_T = \sqrt{2.162 - 2.159 \cos \theta}$$

mesic units. The Hulthen wave function with $a_1 = 0.3248$ mesic unit and $a_2 = 2.2736$ mesic units, will be used for the deuteron wave function. Therefore,

$$R_i(q_p) = \frac{3.60}{(1.739 - 1.187 \cos \theta)(0.802 - 1.187 \cos \theta)}.$$

The value of l must be known in order to calculate the angular distribution. An inspection of the experimental angular distribution shown in Fig. 15 will show that $l = 0$ is not admissible, because the cross section is maximum not at $\theta = 0$ but at some finite value of θ, namely, $\sim 22°$. The surface reaction model may be used to determine the value of l. The solid curves in Fig. 15 are the angular distributions obtained by taking $W(l\Lambda Rq_T) \sim \beta R j_l(q_T R)$ for $l = 1, 2,$ and 3. For each value of l the reaction radius R was chosen so that the first maximum appears at the experimentally observed position. The values required for $l = 1, 2,$ and 3 are approximately 4, 7, and 9.6 mesic units. The radius of B^{11} is about 2.2 mesic units. Therefore the values 2 and 3 for l may be rejected as the corresponding reaction radii are abnormally large.

The form Eq. (44) for $W(l,\Lambda,R_C,q_T)$ may now be used with $l = 1$. The binding energy of the last neutron in B^{11} is 11.51 Mev. Hence $\beta = 1.00$ mesic unit. Taking the value of $(l + 1 + \Lambda)$ from Table III, one gets

$$W(1,\Lambda,R_C,q_T) = q_T R_C j_0(q_T R_C) + \frac{(\beta R_C)^2}{1 + \beta R_C} j_1(q_T R_C). \quad (44')$$

The dashed curve in Fig. 15 shows the angular distribution obtained by

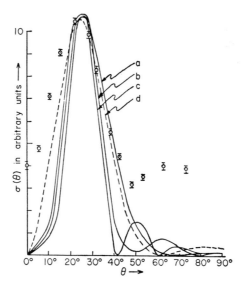

Fig. 15. Angular distribution of the ground state group of protons in the reaction $B^{10}(d,p)B^{11}$ performed with 7.7-Mev deuterons. The experimental points are shown with circles with flags. The theoretical curves a, b, and c are obtained by setting $W(l,\Lambda,R_C,q) \sim j_l(qR_C)$, with $l = 1, 2,$ and 3, respectively. In each case R_C has been adjusted so that the theoretical peak falls on the experimental peak. The dashed curve d is obtained by using the form of Eq. (44') for $W(l,\Lambda,R_C,q)$, with $l = 1$ and $R_C = 3.5$ mesic units.

using Eq. (44') and $R_C = 3.5$ mesic units. The position of the first maximum is in excellent agreement with the experiment.

The preceding calculation shows how the value of l can be determined using the very crude theory for the reaction cross section. The determination of l is based on predicting the location of the first maximum correctly. A study of Fig. 15 will show that even when the location of the first maximum is predicted correctly the details of the angular distribution are not given properly by the simple theory.

One of the reasons for the discrepancy is the use of plane waves instead of the scattering wave functions to describe the motions of the particles before and after the reaction. The cut-off in the radial integral $R_f(q_T)$ simulates the damping of the scattering amplitude. But because of the scattering the momentum transfers q_p and q_T are no longer so sharply defined. The spread in the values of these momenta tends to smooth out the strong maxima and minima predicted by the theory using plane waves. (See note on p. 731.)

d. Dependence of the Cross Section on the Nuclear Wave Function

In deriving the expression (37) for the cross section of the stripping and pickup reactions an extremely simple version was employed for the wave functions of the nuclei $(A + B)$ and $(B + C)$. Besides, the particles A, B, and C were assumed to be spinless. These simplifications will be abandoned in this chapter and the dependence of the reaction cross section on the nuclear wave functions will be studied.

The symbols $|J_A m_A\rangle$, $|J_B m_B\rangle$ and $|J_C m_C\rangle$ will be used to denote the wave functions for the particles A, B, and C, respectively. The letters J and m stand for the total angular momentum and magnetic quantum number, and the subscript indicates to which particle these belong. The symbols $|\alpha J_{AB} M_{AB}\rangle$ and $|\alpha' J_{BC} M_{BC}\rangle$ will denote the wave functions of the nuclei $(A + B)$ and $(B + C)$, respectively. α and α' are additional quantum numbers which may be needed to specify the states of the nuclei uniquely. These wave functions are quite different from the ones given in Eq. (32). In constructing the wave function of, say, $(A + B)$ three angular momenta, namely, J_A, J_B and l_{AB}, have to be coupled together. The convention to be used for coupling the angular momenta will be to couple J_B with the internal angular momentum l_{AB} to obtain the resultant j_{AB} and to couple J_A with j_{AB} to obtain J_{AB}. Such a wave function for $(A + B)$ is

$$|(J_B l_{AB}) j_{AB} J_A J_{AB} M_{AB}\rangle = \sum_{m_A, m_B, m_{AB}} \langle J_B m_B l_{AB} m_{AB} | J_B l_{AB} j_{AB} \mu_{AB}\rangle$$
$$\langle j_{AB} \mu_{AB} J_A m_A | j_{AB} J_A J_{AB} M_{AB}\rangle$$
$$|J_A m_A\rangle |J_B m_B\rangle y_{m_{AB}}{}^{l_{AB}} u_{AB}(r_{AB}). \quad (46a)$$

Similarly, for $(B + C)$ one has

$$|(J_B l_{BC}) j_{BC} J_C J_{BC} M_{BC}\rangle = \sum_{m_B, m_C, m_{BC}} \langle J_B m_B l_{BC} m_{BC} | J_B l_{BC} j_{BC} \mu_{BC}\rangle$$
$$\langle j_{BC} \mu_{BC} J_C m_C | j_{BC} J_C J_{BC} M_{BC}\rangle$$
$$|J_B m_B\rangle |J_C m_C\rangle y_{m_{BC}}{}^{l_{BC}} u_{BC}(r_{BC}), \quad (46b)$$

where the symbol $\langle j_1 m_1 j_2 m_2 | j_1 j_2 j_{12} m_{12}\rangle$ is the vector addition coefficient (8). The wave functions (46) are better than those given in Eq. (32) inasmuch as the spins of A, B, and C have been considered. But they are still not the correct wave functions for the nuclei $(A + B)$ and $(B + C)$. It has been mentioned before that $(A + B)$ may not always appear as a compound of A and B, which is implied in the form of Eqs. (46). However the actual wave function $|\alpha J_{AB} M_{AB}\rangle$ of $(A + B)$ may be expanded in a manner such that $|(J_B l_{AB}) j_{AB} J_A J_{AB} M_{AB}\rangle$ appears in the wave function with an amplitude $\langle (J_B l_{AB}) j_{AB} J_A | \alpha J_{AB}\rangle$. The square of this amplitude gives the probability that the nucleus $(A + B)$ looks like a compound of the particles A and B with the indicated coupling of the angular momenta. The quantity $\langle (J_B l_{AB}) j_{AB} J_A | \alpha J_{AB}\rangle$ is called the "fractional parentage coefficient." The actual wave function $|\alpha' J_{BC} M_{BC}\rangle$ of $(B + C)$ may be similarly expanded such that the wave function $|(J_B l_{BC}) j_{BC} J_C J_{BC} M_{BC}\rangle$ appears in it with the amplitude $\langle (J_B l_{BC}) j_{BC} J_C | \alpha' J_{BC}\rangle$ and the amplitude may be interpreted in a similar manner. Since only these modes of the actual wave functions contribute to the reaction, one may write

$$|\alpha J_{AB} M_{AB}\rangle \sim \langle (J_B l_{AB}) j_{AB} J_A | \alpha J_{AB}\rangle | (J_B l_{AB}) j_{AB} J_A J_{AB} M_{AB}\rangle$$
$$|\alpha' J_{BC} M_{BC}\rangle \sim \langle (J_B l_{BC}) j_{BC} J_C | \alpha' J_{BC}\rangle | (J_B l_{BC}) j_{BC} J_C J_{BC} M_{BC}\rangle. \quad (47)$$

Using such wave functions instead of Eq. (32), the reaction amplitude becomes

$$g(\theta,\phi) = 8\pi\mu_f T \sum_{m_{AB} m_{BC}} y_{m_{AB}}^{l_{AB}}(\Omega_{q_A}{}^*) y_{m_{BC}}^{l_{BC}}(\Omega_{q_C})(i)^{l_{BC}-l_{AB}} R_i(q_A) R_f(q_C)$$
$$\langle (J_B l_{AB}) j_{AB} J_A | \alpha J_{AB}\rangle \langle (J_B l_{BC}) j_{BC} J_C | \alpha' J_{BC}\rangle$$
$$\langle J_B m_B l_{AB} m_{AB} | J_B l_{AB} j_{AB} \mu_{AB}\rangle \langle j_{AB} \mu_{AB} J_A m_A | j_{AB} J_A J_{AB} M_{AB}\rangle$$
$$\langle J_B m_B l_{BC} m_{BC} | J_B l_{BC} j_{BC} \mu_{BC}\rangle \langle j_{BC} \mu_{BC} J_C m_C | j_{BC} J_C J_{BC} M_{BC}\rangle. \quad (48)$$

Limiting the discussion to the case of stripping reactions, when $(A + B)$ stands for a nucleus not heavier than the α-particle, one has $l_{AB} = 0$. Then omitting the subscripts BC in l_{BC} and j_{BC}, the reaction amplitude may be written as

$$g(\theta,\phi) = T 4\sqrt{\pi}\, y_m{}^l(\Omega_{q_C}) i^l R_i(q_A) R_f(q_C)$$
$$\langle (J_B l) j J_C | \alpha' J_{BC}\rangle \langle J_A m_A J_B m_B | J_A J_B J_{AB} M_{AB}\rangle$$
$$\langle J_B m_B l m | J_B l j \mu\rangle \langle j \mu J_C m_C | j J_C J_{BC} M_{BC}\rangle, \quad (49)$$

where the fractional parentage coefficient $\langle (J_B l_{AB}) j_{AB} J_A | \alpha J_{AB}\rangle$ has been set equal to unity. The cross section may be obtained by taking the square modulus of Eq. (49) and averaging and summing over the initial and the final magnetic substates.

$$\sigma(\theta,\phi) = \frac{K_f}{K_i}\frac{\mu_i}{\mu_f}\frac{1}{[J_C][J_{AB}]}\sum_{M_{AB}m_CM_{BC}m_A}|(49)|^2$$

$$= \frac{K_f}{K_i}\frac{\mu_f\mu_i}{\mu_{BC}}\frac{[J_{BC}]}{[J_B][J_C]}|<(J_Bl)jJ_C|\alpha'J_{BC}>|^2\frac{v^2}{R_C}R_i^2(q_A)$$
$$\times W^2(l,\Lambda,R_C,q_C). \quad (50)$$

If J_B and l are not zero the angular momentum j can have values between $|J_B - l|$ and $J_B + l$. If j is not measured in the experiment, the cross section should contain a sum over j. There is no interference between the different j-state contributions.

There are certain selection rules which determine the conditions under which the reaction may proceed. These are mostly based on the conditions of nonvanishing of the vector addition coefficients and of the fractional parentage coefficients.

(1) Conservation of angular momentum.

Since J_{BC} is the vector sum of J_C and j, and j is the vector sum of J_B and l, it follows that

$$|(|J_{BC} - J_C|) - J_B| \leq l \leq J_{BC} + J_B + J_C.$$

Thus, in a (d,p) reaction B, C, and $(B + C)$ stand for the neutron, the target, and the final nucleus, respectively. Denoting the spins of the target and the final nucleus by J_i and J_f, the limits on the values of l become

$$(|J_f - J_i|) - \tfrac{1}{2} \leq l \leq J_i + J_f + \tfrac{1}{2}.$$

(2) Conservation of parity.

The parity of the orbital state l is $(-)^l$. So if the combined parity of B and C is the same as that of $(B + C)$, l must be even. If the two parities are different, l must be odd. In a (d,p) reaction l is even if the parities of the target and the final nucleus are the same. Otherwise l is odd.

(3) Parentage rule.

In order that the fractional parentage coefficient $<(J_Bl)jJ_C|\alpha'J_{BC}>$ may be nonzero it is necessary that the nuclear particles B and C in the specified states $|J_Bm_B>$ and $|J_Cm_C>$ should be able to combine to produce the nucleus $(B + C)$ in the state $|\alpha'J_{BC}M_{BC}>$. It should be mentioned here that if B or C are complex nuclei themselves additional quantum numbers may be necessary to specify their states uniquely. If the states of B, C, and $(B + C)$ satisfy the required condition, then the heavier of the two particles B and C is called the parent of $(B + C)$. Thus, in a (d,p) reaction the target state must be a parent of the state of the final nucleus. The parentage connection is determined by the model assumed to describe the states of the nucleus, that is, L-S coupling or j-j coupling shell model, etc.

(4) Shell model.

The shell model of the nucleus not only determines whether the desired parentage relation exists or not, but also puts a very stringent restriction on the values of l. For the sake of simplicity only the (d,p) reaction will be discussed.

The shell model prescribes the single-particle energy levels. The nuclear states are obtained by filling up these levels in accordance with the Exclusion Principle. Thus, when the target captures a neutron in a (d,p) reaction the neutron must go to one of the levels not occupied by the nucleons of the target. When all the occupied states are specified, the nucleus is said to be in a definite configuration. The parentage connection requires that the configuration of the final state must include all the single-particle states of the target state configuration and that the orbital angular momentum of the additional single-particle state which appears in the former configuration is just equal to the value of l with which the neutron is captured.

The use of the selection rules will now be illustrated with the example of a $B^{10}(d,p)B^{11}$ reaction, where the final nucleus is left in its ground state. The total angular momenta of the ground states of B^{10} and B^{11} are 3^+ and $\frac{3}{2}^-$, where the superscripts indicate the parities of the two states. According to the selection rule on angular momentum, l can have the values 1, 2, 3, 4, and 5. But the parities of the two states are different. So the selection rule on parity restricts the allowed values of l to 1, 3, and 5. According to the j-j coupling shell model the configurations of the two ground states are $(1s_{\frac{1}{2}})^4(1p_{\frac{3}{2}})^6$ and $(1s_{\frac{1}{2}})^4(1p_{\frac{3}{2}})^7$. This implies that the neutron must be captured in the p-state. The actual experimental data, shown in Fig. 15, can be fitted best by taking $l = 1$. Thus one can verify and determine the shell model single particle level spectrum through the study of stripping reactions. This was first pointed out by Bethe and Butler (9) and it has been the main reason for the great interest in experimental and theoretical studies of the stripping and pickup processes.

The reaction $C^{12}(d,p)C^{13}$ provides an interesting illustration of the parentage rule. The second excited state of C^{13} is $\frac{3}{2}^-$, while the ground state of C^{12} is 0^+. Therefore, according to the selection rules on angular momentum and parity it should be possible to reach the second excited state of C^{13} when C^{12} captures a neutron in the p-state. But the configurations of the two states according to the j-j coupling shell model are $(1s_{\frac{1}{2}})^4(1p_{\frac{3}{2}})^7(1p_{\frac{1}{2}})^2$ and $(1s_{\frac{1}{2}})^4(1p_{\frac{3}{2}})^8$. This means that the second excited state of C^{13} does not have the ground state of C^{12} as a parent and according to the parentage rule the reaction should be completely forbidden. However a small cross section for the reaction has been observed. The conclusion is that the j-j coupling assignment is essentially correct.

There is a small admixture of the configuration $(1s_\frac{1}{2})^4(1p_\frac{3}{2})^6(1p_\frac{1}{2})^2$ in the ground state of C^{12}. In fact the amount of admixture of the configuration can be estimated from the reaction cross section. The works of J. B. French and his collaborators (10) may be consulted[e] for a detailed discussion of the methods of determining the fractional parentage coefficients and the amount of configuration mixing from the stripping and pickup cross sections.

e. Exchange Effects in Stripping Reactions

So far a rather special viewpoint has been used to explain the stripping and pickup reactions. Thus in the stripping reaction it is assumed that the lighter end particle is originally a part of the lighter reactant and in the pickup reaction it is assumed that the heavier particle is produced as a result of the lighter one picking up a nuclear particle from the target. Since in all reactions the lighter reactant is used as the projectile, a consequence of the aforementioned picture is that the lighter end product comes out predominantly in the direction of motion of the projectile, that is in the forward direction. In fact a glance at the curves shown in Fig. 12 will make it quite clear that cross sections are invariably very small in the backward direction. In the first stages of the experimental studies of stripping and pickup reactions, the measurements were usually limited to the forward hemisphere. In this region the agreement between the experiment and the Butler curves were quite satisfactory. But when the measurements were extended to the backward hemisphere, numerous examples were found where the cross section in the backward direction was comparable to that in the forward direction. In some cases there was a marked tendency of increase of the cross section in the backward direction.

Owen and Madansky (11) were the first to recognize clearly that the observed rise in the backward direction is due to the exchange effect in stripping reactions. In order to understand the origin of the exchange effect in a simple manner the specific example of the reaction $O^{16}(d,p)O^{17}$ may be considered. In Fig. 16a the state of affairs before the reaction occurs is presented. The neutrons are shown as open circles and the protons as hatched circles. The neutron of the deuteron is marked with a dot in the center of the white circle and the proton of the deuteron is marked with horizontal hatching. The protons in the target nucleus bear vertical hatchings. In order to keep the figures simple the target O^{16} is presented as two neutrons and two protons moving about the carbon nucleus. According to the viewpoint adopted so far the state of affairs after the reaction occurs is that presented in Fig. 16b. It may be noted

[e] See also Section V.G.

that the proton that is emitted is the one originally contained in the deuteron. Since this proton is completely indistinguishable from the protons in the target, the state obtained by exchanging the former with any one of the latter group should also be a possible mode for the reaction. The corresponding situation is represented by Fig. 16c, where the proton that is emitted was originally contained in the target, and the proton of

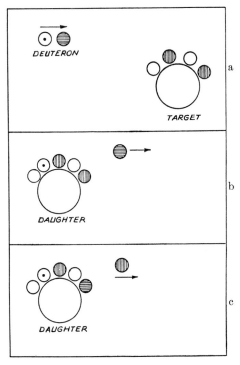

Fig. 16. Schematic representation of the $O^{16}(d,p)O^{17}$ reaction. See the section on exchange effects for explanation.

the deuteron fills the gap thus created. The mechanism shown in Fig. 16b is usually called the "direct process" and the one leading to Fig. 16c is called the "exchange process." In the exchange process the proton comes from the target, while the residual nucleus, to be referred to as the "core," captures the projectile deuteron.

It has been shown in subsection b that the interactions responsible for the direct process are (1) the binding interaction between the neutron and the target nucleons (or that between the neutron and the proton of the deuteron), and (2) the unbound pair interaction between the released proton and the target nucleons. The same interactions are responsible for the exchange process but their roles may be seen a little more clearly if the interactions are regrouped in a new manner. For simplicity of discussion let the neutron and the proton forming the deuteron be called n and p, respectively. The proton with which p is exchanged may be called e. The interactions may now be regrouped as follows:

(1) Interactions of n and p with the core nucleons and (2) interactions of e with n and p.

The first group of interactions lead to the binding of the deuteron with the core in the final state. The second group of interactions is the unbound pair interaction in the exchange case.

The contribution due to the binding interaction is called "heavy particle stripping." The following picture is associated with this process. In a velocity transformed system the target, considered as a compound of the core and the proton e approaches the deuteron. Due to the binding interaction between the deuteron and the core the latter is stripped off the target nucleus and captured by the deuteron, while the proton e is released. The proton e will tend to continue in the original direction of motion of the target, which is the backward direction. Thus the heavy particle stripping will increase the reaction cross section in the backward direction.

The contribution due to the unbound pair interaction, that is, the interaction between the projectile deuteron and the ejected proton e, is termed the "knock-out" process. The idea is that the deuteron comes and directly knocks out e from the target and is itself captured by the core. In subsection b it was mentioned that the contribution of the unbound pair interaction is much smaller than that due to the binding interaction. The remark holds for the two types of exchange processes as well. However, the ratio of the amplitudes for the two processes is not as small as it is in the case of the direct process. The reason is that the wave function describing the bound state of the deuteron and the core has the same extension as that describing the bound state of e with the core. As a result, when both the direct and the heavy-particle strippings are forbidden because of selection rules on angular momentum etc., the knock-out process may give rise to a measurable cross section for the reaction. The picture implied in the knock-out process makes it obvious that the cross section for this process will be peaked in the forward direction.

The mechanism of the exchange processes has been explained with the example of the (d,p) reaction. It is obvious that similar effects will be present in all varieties of stripping and pickup reactions. The subsequent discussions will be limited to the case of the (d,p) reaction alone.

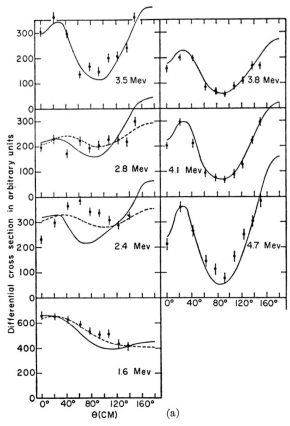

Fig. 17a. Angular distribution of the ground state neutrons from the reaction $B^{11}(d,n)C^{12}$ for bombarding energies ranging from 1.6 to 4.7 Mev. The theoretical curves are obtained by including the heavy-particle stripping term. The full curve is for $R_C = 4.5 \times 10^{-13}$ cm. The dashed curve is for $R_C = 3.8 \times 10^{-13}$ cm. [G. E. Owen and L. Madansky, Phys. Rev. **105**, 1766 (1957).]

It is clear that the heavy-particle stripping falls into the general type of mass exchange reaction represented by the equation

$$(A + B) + C \to A + (B + C).$$

Thus in the case of the reaction $O^{16}(d,p)O^{17}$, the heavy-particle stripping part may be described by associating the symbol A with the proton e, the symbol B with the core and the symbol C with the projectile deuteron. Apart from factors arising from the angular momentum coupling and the symmetries in the nuclear wave functions, the heavy particle stripping amplitude is given by Eq. (36). Written in terms of the momenta $\mathbf{K}_d$

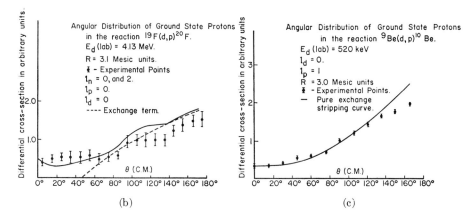

(b)

(c)

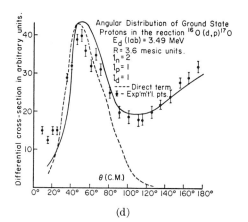

(d)

Fig. 17b. Angular distribution of the ground state protons in the reaction $F^{19}(d,p)F^{20}$. The full curve shows the theoretical prediction including the direct and heavy-particle stripping term. The dashed curve shows the contribution due to the heavy-particle stripping term alone. [M. A. Nagarajan and M. K. Banerjee, Communs. Congr. intern. physique nucléaire, Paris, p. 506 (1959).]

Fig. 17c. Angular distribution of the ground state group of protons in the reaction $Be^9(d,p)Be^{10}$. The experimental data are well explained by considering the heavy-particle stripping term alone. [M. A. Nagarajan and M. K. Banerjee, Communs. Congr. intern. physique nucléaire, Paris, p. 506 (1959).]

Fig. 17d. Angular distribution of ground state protons in the reaction $O^{16}(d,p)O^{17}$. The full curve shows the theoretical calculation including the direct and the heavy-particle stripping terms. The dashed curve shows the contribution due to the direct term alone. [M. A. Nagarajan and M. K. Banerjee, Communs. Congr. intern. physique nucléaire, Paris, p. 506 (1959).]

and $\mathbf{K}_p$ of the projectile deuteron and the end product proton, in the c.m. system, the momentum transfers q_A and q_C are

$$q_A = \left| \mathbf{K}_p + \frac{1}{A_T} \mathbf{K}_d \right|; \quad q_C = \left| \mathbf{K}_d + \frac{2}{A_T + 1} \mathbf{K}_p \right|,$$

where A_T is the mass number of the target nucleus. Thus each is a sum of two vectors, one of which is very small compared to the other. Therefore,

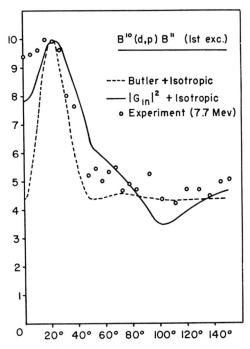

FIG. 18. Comparison of theory and experiment for $B^{11}(d,p)B^{11}$ (first excited state transition) for 7.7-Mev deuterons. For further details see N. T. S. Evans and A. P. French, Phys. Rev. **109**, 1275 (1958).

the variation in the magnitudes of the two q's as the angle between $\mathbf{K}_p$ and $\mathbf{K}_d$ changes from 0° to 180° is very small. As a result the functions $R_i(q_A)$ and $R_f(q_C)$, themselves, change rather slowly with the angle of emission of the proton. This weak variation with angle is a characteristic of the heavy-particle stripping process.

If the "cut-off approximation" is used it should be applied to the evaluation of both $R_i(q_A)$ and $R_f(q_C)$ to obtain the form of Eq. (43) for each.

The most important questions which arise in the estimate of the exchange effect are the relative phases and magnitudes of the direct and the exchange amplitudes. These aspects have been dealt with in detail in the works of Owen and Madansky (*11*), A. P. French (*12*), and Nagarajan and Banerjee (*13*). Some examples of the comparison of the theoretical predictions, including the exchange effects with experimental results are shown in Figs. 17a, b, c, and d.

There are very few cases where the knock-out term makes its presence felt. One example is the $B^{10}(d,p)B^{11*}$ reaction leading to the first excited state of B^{11}. The spins of the ground state of B^{10} and the first excited state of B^{11} are 3^+ and $\frac{1}{2}^-$, respectively. From the description of these states based on the shell model, it is clear that in the direct process the neutron must be captured in a p-state. But then the selection rule (52) on angular momenta is violated and the direct process should therefore be prohibited. If the reaction proceeds through heavy-particle stripping the angular distribution should not show large variation with angle. But the experimental distribution, shown in Fig. 18, shows a marked variation with angle and, in particular, a forward peaking. This strongly suggests that the reaction is a knock-out process. The angular distribution was quite satisfactorily explained on this basis by A. P. French (*12*).

REFERENCES

1. J. R. Oppenheimer and M. Phillips, Phys. Rev. **48**, 500 (1935).
2. R. Serber, Phys. Rev. **72**, 1008 (1947).
3. S. T. Butler, Phys. Rev. **80**, 1095 (1950); Nature **166**, 709 (1950); Proc. Roy. Soc. **A208**, 559 (1951).
4. S. T. Butler, *Nuclear Stripping Reactions* (John Wiley and Sons, New York, 1957).
5. L. I. Schiff, *Quantum Mechanics* (McGraw-Hill, New York, 1949).
6. J. M. Blatt and V. F. Weisskopf, *Theoretical Nuclear Physics* (John Wiley and Sons, New York, 1952), see p. 336.
7. N. T. S. Evans and W. C. Parkinson, Proc. Phys. Soc. **A67**, 684 (1954).
8. E. U. Condon and G. H. Shortley, *Theory of Atomic Spectra* (Cambridge University Press, London and New York, 1935).
9. H. A. Bethe and S. T. Butler, Phys. Rev. **85**, 1045 (1952).
10. J. B. French and B. J. Raz, Phys. Rev. **104**, 1411 (1956); J. B. French and T. Auerbach, *ibid.* **98**, 1276 (1955); J. B. French and A. Fujii, *ibid.* **105**, 652 (1957).
11. G. E. Owen and L. Madansky, Phys. Rev. **105**, 1766 (1957).
12. A. P. French, Phys. Rev. **107**, 1655 (1957); N. T. S. Evans and A. P. French, *ibid.* **109**, 1272 (1958).
13. M. A. Nagarajan and M. K. Banerjee, Communs. Congr. intern. physique nucléaire, Paris, p. 506 (1959).

Note added in proof: The papers of W. Tobocman, Phys. Rev. **115**, 99 (1959), W. Tobocman and G. R. Satchler (to be published), and D. H. Wilkinson, Phil. Mag. [8] **3**, 1185 (1958), deal with the effects of initial and final state interactions on the angular distributions.

V. C. Angular Correlations in Nuclear Spectroscopy

by L. C. BIEDENHARN

1. The Angular-Correlation Process.................................... 736
   a. The Angular-Correlation Process from a Semiclassical View............. 736
   b. Cascades.. 744
   c. Parallel Angular Momenta...................................... 746
   d. Particles with Spin... 746
      (1) Nonrelativistic Case...................................... 747
      (2) Relativistic Case.. 748
   e. Parametrization.. 750
   f. More Involved Correlation Processes.............................. 754
   g. The Density Matrix and Statistical Tensor Formulation................ 760
   h. Coherent and Incoherent Interference............................. 764
   i. Some Concluding Remarks....................................... 769
2. Application of Angular Correlations to Specific Cases.................... 769
   a. Pure Nuclear States... 770
      (1) Direction-Direction Correlation............................. 770
      (2) Polarization-Direction Correlations.......................... 778
   b. Several Intermediate Nuclear States............................. 785
      (1) Direction-Direction Correlation............................. 785
      (2) Triple Correlations....................................... 791
3. Appendix: Summary of Formulas and Notations........................... 800
   a. Intermediate Nuclear States of Sharp Angular Momentum and Parity.... 800
      (1) Direction-Direction Correlations............................ 800
      (2) Polarization-Direction Correlation.......................... 802
      (3) Triple Correlations....................................... 803
   b. Mixed Intermediate States..................................... 804
   References.. 805

Nuclear spectroscopy is concerned with the measurement of the properties of nuclear energy levels, among these properties being the energy of the levels, the angular momentum (or "spin") and parity, various electric and magnetic moments, and the various partial decay probabilities. There is more than this, however, for nuclear spectroscopy also aims at the determination of detailed properties, such as the type of nucleon coupling (LS or jj), and even the discovery of new constants, such as the isotopic spin. The angular correlation of radiations emitted or absorbed in nuclear processes has been one of the principal tools available for the nuclear spectroscopy of *excited* states; mainly for obtaining spins and parities, but by no means restricted to this, as the recent angular-correlation studies on the beta decay interactions show. Such a wide range of application shows, moreover, that angular correlations are properly an aspect of almost every branch of nuclear physics, and a brief survey, such as the present one, can hope to treat only a few topics. We shall confine ourselves to two: the angular correlation (including polarization) of the radiations emitted in cascade, typified by the gamma-gamma correlation; and the angular correlation of the radiations from nuclear reactions—this latter also termed an angular distribution. The important distinction for us between these two topics is that the former proceeds through nuclear states of sharp spin and parity, while the latter generally shows coherent mixing of the intermediate nuclear states. Throughout we shall consider the nuclear states to be *isolated* and not subject to any external perturbing fields. This assumption has two aspects: First, and most important, it is a limitation on the usefulness of the results to be obtained and requires generally that the lifetimes of the intermediate nuclear states be $\lesssim 10^{-10}$ sec; second, it means the exclusion of any discussion of the measurements of electric and magnetic moments in excited states and any corresponding interaction between nuclei and their surrounding electron shells. For both, we refer to the review of Frauenfelder (*1*),* where the subject is treated at length. (See also, Steffen, *2*, and Section IV.C.)

In *atomic* spectroscopy the principal tool for unravelling spectra was the Zeeman effect—which in our present terms we might classify as an angular correlation with an external magnetic field. The magnetic field served to distinguish the (*m*) sublevels of the atomic states both spatially and energetically, and correspondingly the various components of the complete line emitted. Observation of these various components—their number, angular distribution, polarization, and intensity—then led to information on the levels in question. A complete solution to this problem was found in two basic results: (a) The Wigner-Eckart theorem which separated the geometrical and dynamical aspects of the transitions, and (b) the Wigner coefficients which furnished a complete description of the geometrical aspect.

For nuclear spectroscopy, the Zeeman effect is unobservable—even in high fields the splitting is only $\sim 10^{-8}$ ev. Lacking a practical means of identifying the various sublevels energy wise, we must proceed in the next best way to identify them spatially, by unequal populations. This can be done in practice: (a) by alignment or orientation at low temperatures, and (b) by forming the levels from some unidirectional, and perhaps polarized, radiation. Despite these differences in technique, the problem of angular correlations in nuclear spectroscopy is, nonetheless, completely solved by the same two results that were applied to atomic spectroscopy,

* The reference list for Section V.C begins on page 808.

for these results provide the key to *all* calculations involving the conservation of angular momentum. One may proceed, for example, by setting up the matrix elements for the particular sequence of transitions to be considered, and manipulate these in the standard fashion, after separating each matrix element into a Wigner coefficient and a reduced (magnetic quantum number independent) matrix element. Or proceeding in a somewhat more general fashion one may utilize scattering and reaction matrix techniques, as in the Wigner-Eisenbud nuclear reaction theory. In either procedure one is necessarily led to a great many summations over magnetic quantum numbers, corresponding to unobserved or coupled orientations, but—at least in a formal sense!—the desired correlation is obtained directly. From this point of view the introduction of the Racah coefficients appears only as an algebraic convenience, to manipulate and effect various magnetic quantum number sums. It is clear, however, that this point of view is overly restrictive, and indeed obscures the significance of the Racah coefficients. As Blatt has pointed out, what is desired is that we should look carefully at the formulas we have obtained by the above method and ask the question "How can we write these results down at once without ever introducing magnetic quantum numbers?" Section 1 of the present survey attempts to answer just this question and in so doing to make clear the *structure* of the correlation formulas.

We can, in the spirit of quantum mechanics, consider the problem of angular correlations in terms of the operations performed on a system. What are the operations? Typically they are (1) the measurement of the direction of motion of one or more particles of known intrinsic angular momentum and (2) the measurement of the polarization of such a particle (with nonzero intrinsic spin) in a frame of reference whose axis is the direction of motion. The measurement of some process or reaction as a function of these vectors (pseudovectors) will constitute that special class of angular correlations with which we are concerned here. This is rather too general, however, and to make further progress let us assume for the moment that the system in between emissions or absorptions is in a state of sharp total angular momentum, and our measurements consist only of determining directions of motion. We then have the prototype of an angular correlation: the measurements define sharp *linear* momentum states, which, however, are coupled together via sharp *angular* momentum states with conservation of angular momentum. Viewed another way, one may say that angular information contained in a linear momentum measurement is "transmitted" via sharp angular momentum states.

What angular information is contained in a measurement of a direction of motion? The answer—that the orbital angular momentum carried by the radiation is perpendicular to this direction of motion—seems at first sight to be remarkably trivial! Yet this one fact, plus the additional fact that angular momentum is conserved, forms the basis for the entire structure of directional correlations for unpolarized radiations.

Now for classical physics both of these facts are also true, and one might expect therefore that a classical consideration of angular correlation will furnish considerable insight into the structure of angular-correlation theory. There exist many detailed and excellent quantum mechanical discussions of angular-correlation theory with any desired degree of sophistication; it would not be of any particular value for us to attempt to add to this list.[a] Accordingly we shall attempt to show the basic structure of the angular-correlation process, using classical considerations as a guide. These considerations lead in a natural way to a useful interpretation of the Wigner and Racah coefficients, including the 9-j symbol. More importantly these classical considerations focus attention on the essential role of the observed and unobserved directions involved in a correlation. The one drawback to this classical formulation is its limitation to pure states of sharp $|j|$. But even this is not serious, for the generalization of the quantum formulas to processes with coherent mixing is clear from the structure of the correlations obtained.

The present chapter then concerns itself with two disparate subjects: (1) a semiclassical discussion of angular correlations, intended to give insight and an intuitive feeling for the structure of the correlation process itself, and (2) a detailed discussion, with examples, of the application of angular correlations to specific cases. In Section 1 we will make extensive use of Racah and Fano's original work (3). In Section 2, we shall adopt the point of view of Wilkinson's manuscript (4) which has been, even in its unpublished form, of considerable value to experimentalists. Since this latter section is intended to be largely self-contained, and indeed will use results not specifically discussed in Section 1, there will be some unavoidable duplications between 1 and 2.

The author is indebted to many colleagues for discussions on the present chapter, and particularly to Dr. W. T. Sharp. He would also like to thank Messrs. T. Griffy and R. C. Young for their aid in revising the manuscript and examples.

[a] It would also be idle for us to attempt any complete referencing of the work used in this survey, and instead we must refer for this to some of the reviews listed. This is unfair to the authors of original contributions, but quite inevitable. This note is further intended as an apology for the somewhat capricious referencing that will be done.

1. The Angular-Correlation Process

a. The Angular-Correlation Process from a Semiclassical View

We begin this discussion with the simplest case of an angular correlation. Let us consider a beam of spinless particles bombarding a heavy nucleus (so that the center of mass can be considered as fixed on this nucleus throughout the reaction), causing a nuclear reaction in which another beam of spinless particles is observed to emerge. Our observations consist of two things only: a measurement of the two directions of motion $\mathbf{k}_{i(\text{initial})}$ and $\mathbf{k}_{f(\text{final})}$.

For the least complicated case let us further assume that the initial and final nuclei are distinguishable but both have *zero angular momentum*. Under the assumption that the intermediate nuclear state has sharp angular momentum J and parity Π (for example, a well-defined resonance) we have $J = l_i = l_f$, so that but a single entrance (l_i) and exit (l_f) orbital angular momentum enters.

Now let us introduce the basic information we have about the process:

(a) The orbital angular momentum $\mathbf{l}_i$ is perpendicular to $\mathbf{k}_i$. We conclude that $\mathbf{J}(=\mathbf{l}_i)$ *lies with equal probability in the plane perpendicular to* $\mathbf{k}_i$.

(b) The compound system, of angular momentum J, breaks up by emitting a spinless particle along $\mathbf{k}_f$. We may pictorially consider this break-up process to be much like "water spraying off a rotating wheel"; the emission is then azimuthally symmetric around the direction of the angular momentum $\mathbf{J}$. But $\mathbf{l}_f = \mathbf{J}$ must be perpendicular to $\mathbf{k}_f$, and, it follows, *that* $\mathbf{k}_f$ *is equally probable in the plane perpendicular to* $\mathbf{J}$.

These relationships are sketched in Fig. 1, using unit vectors (denoted by a circumflex).

These conditions imposed on the vector $\mathbf{k}_f$ may be summarized by saying that (using $\mathbf{k}_i$ as an axis) *both ϑ and φ have uniform probability*.[b] The (normalized) probability in *solid* angle is then:

$$\frac{dW(\vartheta)}{d\Omega} = (2\pi^2)^{-1}(\sin \vartheta)^{-1}. \tag{1}$$

This classical result (first given by Christy, 5) is of interest in its own right,[c] for it provides the *limiting form* for an angular-correlation process

[b] It will be noted that by this derivation ϑ ranges over 0 to 2π. But the range π to 2π duplicates that for 0 to π with $\varphi \to \varphi + \pi$. Thus the "extra range" merely guarantees that symmetry about $\vartheta = \pi/2$ must obtain.

[c] Applications are discussed by Ericson and Strutinsky (6).

proceeding through a single nuclear state of large angular momentum, involving particles of small spin. We note two characteristic features: the strong forward peaking characteristic of high angular momentum and the symmetry about 90°, resulting from sharp parity.

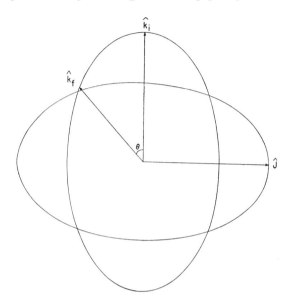

FIG. 1. A nuclear reaction for spinless particles on a spinless target proceeding through a sharp intermediate state of spin J, considered classically. The directions of the initial and final directions of motion $\hat{\mathbf{k}}_i$ and $\hat{\mathbf{k}}_f$, are restricted to be perpendicular to J. The circles show orientations of uniform probability; the circumflex denotes unit vectors.

Aside from this, however, we may regard this simplest example as the paradigm for the general case. To illustrate this, let us again consider this same example but now from a more general point of view.

Any correlation where only two directions of motion are measured must be expressible as a Legendre series in the only angle invariantly defined, the angle between $\mathbf{k}_i$ and $\mathbf{k}_f$. In symbols

$$\frac{dW(\vartheta)}{d\Omega} = \frac{1}{4\pi} \sum_\nu (2\nu + 1) B_\nu P_\nu(\cos \vartheta)$$

$$\cos \vartheta = \hat{\mathbf{k}}_i \cdot \hat{\mathbf{k}}_f. \tag{2}$$

The coefficients B_ν are to be interpreted *as the average value of* $P_\nu(\hat{\mathbf{k}}_i \cdot \hat{\mathbf{k}}_f)$ *evaluated over all possible configurations of the system.* That is:

$$B_\nu = [P_\nu(\hat{\mathbf{k}}_i \cdot \hat{\mathbf{k}}_f)]_{\text{av}}. \tag{3}$$

It is helpful at this point to introduce a method for diagramming the various directional relationships. Let us represent the various vectors by *unit* vectors, all issuing from a common origin. We then give a "spherical mapping" by plotting only the point of intersection of the unit vector with the unit sphere about the common origin. Thus for Fig. 1 we now get the spherical triangle of Fig. 2, the curved lines representing arcs of great

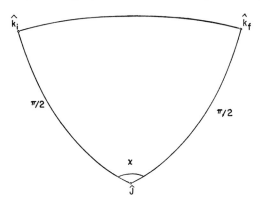

Fig. 2. The situation of Fig. 1 expressed by a spherical mapping. The vertices are the intersections of unit vectors, $\hat{k}_i$, $\hat{k}_f$, $\hat{J}$ (issuing from a common center), with the unit sphere. The arcs are segments of great circles. Note that the angle χ is *uniformly probable*.

circles on the unit sphere. But if we remember now the discussion of Fig. 1, we see that the dihedral angle χ has *uniform probability*, a feature that will be characteristic for all the cases considered below.

Now the addition theorem for the Legendre polynomials,

$$P_\nu(\cos \omega) = P_\nu(\cos \vartheta) P_\nu(\cos \vartheta')$$
$$+ 2 \sum_{m=1}^{\nu} \left[\frac{(\nu-m)!}{(\nu+m)!} \right] P_\nu^m(\cos \vartheta) P_\nu^m(\cos \vartheta') \cos m(\varphi - \varphi'),$$
$$\cos \omega = \cos \vartheta \cos \vartheta' + \sin \vartheta \sin \vartheta' \cos(\varphi - \varphi'), \quad (4)$$

allows us to express P_ν in terms of an arbitrary axis in space. Choosing the vector $\mathbf{J}$ as this axis, we get at once, that

$$[P_\nu(\hat{\mathbf{k}}_i \cdot \hat{\mathbf{k}}_f)]_{\mathrm{av}} = P_\nu(\hat{\mathbf{k}}_i \cdot \hat{\mathbf{J}}) P_\nu(\hat{\mathbf{J}} \cdot \hat{\mathbf{k}}_f) = [P_\nu(\cos \pi/2)]^2. \quad (5)$$

Hence we find for the angular distribution,

$$\frac{dW(\vartheta)}{d\Omega} = \frac{1}{4\pi} \sum_\nu (2\nu+1)[P_\nu(0)]^2 P_\nu(\cos \vartheta).$$

This result is equivalent to, but less perspicuous than, the former distribution $(\sin \vartheta)^{-1}$. Much more important is the fact that this use of the addition theorem proves to be a *general* method.

Let us now add the complication of intrinsic spins, $\mathbf{j}_i$ and $\mathbf{j}_f$ for the initial and final nuclei. By hypothesis, the orientation of these angular momenta are *not* observed, which we equate to the statement that

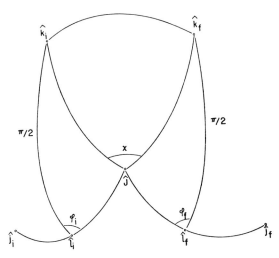

FIG. 3. Vectors for the angular momenta and directions involved in the angular correlation are taken to issue from a common origin and are represented by their intersection with the surface of a unit sphere whose center is the common origin. Since $\mathbf{l}_i$, $\mathbf{j}_i$, and $\mathbf{J}$ have zero sum (that is, they form a triangle) their representative points ($\hat{\mathbf{l}}_i$, $\hat{\mathbf{j}}_i$, $\hat{\mathbf{J}}$ above) lie along a common great circle. The arcs in the diagram schematically indicate segments of great circles; the angles χ, φ_i and φ_f are *uniformly probable*.

$\mathbf{j}_i$ and $\mathbf{j}_f$ *are randomly oriented*, for all configurations that conserve angular momentum.

Consider now the spherical mapping for this case (Fig. 3).

Since $\mathbf{l}_i + \mathbf{j}_i + \mathbf{J} = 0$ (conservation of angular momentum), the three points representing three angular momenta that form a triangle, must lie on a great circle, but *the arc lengths are fixed in value*, proportional to the angles between the vectors.

Just as before the first step is to apply the addition theorem to $[P_\nu(\hat{\mathbf{k}}_i \cdot \hat{\mathbf{k}}_f)]_{av}$ using the angular momentum $\mathbf{J}$ of the compound state as the new axis. The essential point now is to recognize that *the angle χ is once again random*. This is a consequence of our assumption of a sharp compound state: subject only to conservation of angular momentum, the decay of the compound state is to be independent of its formation. In particular, this implies that the break-up process must be azimuthally symmetric around the direction **J**, analogous to the "water off a wheel" picture used earlier. (This point is discussed further on page 741.)

Performing the average over χ yields the result:

$$[P_\nu(\hat{\mathbf{k}}_i \cdot \hat{\mathbf{k}}_f)]_{av} = [P_\nu(\hat{\mathbf{k}}_i \cdot \hat{\mathbf{J}})]_{av} \cdot [P_\nu(\hat{\mathbf{J}} \cdot \hat{\mathbf{k}}_f)]_{av}. \tag{6}$$

The addition theorem is next applied to the spherical triangles $(\hat{\mathbf{l}}_i, \hat{\mathbf{k}}_i, \hat{\mathbf{J}})$, $(\hat{\mathbf{l}}_f, \hat{\mathbf{k}}_f, \hat{\mathbf{J}})$ in Fig. 3. Since $\mathbf{j}_i$ and $\mathbf{j}_f$ are each *random*, the angles φ_i and φ_f are uniformly probable. Using the addition theorem to introduce $\mathbf{l}_i$ as an axis, and averaging over φ_i, one obtains:

$$[P_\nu(\hat{\mathbf{k}}_i \cdot \hat{\mathbf{J}})]_{av} = P_\nu(\hat{\mathbf{k}}_i \cdot \hat{\mathbf{l}}_i) P_\nu(\hat{\mathbf{l}}_i \cdot \hat{\mathbf{J}}), \tag{7}$$

and similarly for the second spherical triangle. The angle between $\mathbf{l}_i$ and **J** is fixed by the requirement that $(\mathbf{l}_i, \mathbf{j}_i, \mathbf{J})$ form a triangle of angular momenta.

The final, classical, angular correlation results at once:

$$\frac{dW(\vartheta)}{d\Omega} = \frac{1}{4\pi} \sum_\nu (2\nu + 1) B_\nu P_\nu(\hat{\mathbf{k}}_i \cdot \hat{\mathbf{k}}_f) \tag{8a}$$

$$(2\nu + 1) B_\nu = [(2\nu + 1)^{\frac{1}{2}} P_\nu(\hat{\mathbf{k}}_i \cdot \hat{\mathbf{l}}_i) P_\nu(\hat{\mathbf{l}}_i \cdot \hat{\mathbf{J}})]$$
$$\times [(2\nu + 1)^{\frac{1}{2}} P_\nu(\hat{\mathbf{k}}_f \cdot \hat{\mathbf{l}}_f) P_\nu(\hat{\mathbf{l}}_f \cdot \hat{\mathbf{J}})]. \tag{8b}$$

This result illustrates the general form of the direction-direction correlation for pure states. We summarize the significant features:

(a) The coefficients for the Legendre series *factor* into two parts, each part determined by the angular momenta associated with the measurement of a given direction of motion. (This factorization property is a general feature of direction-direction correlations proceeding through a sharp intermediate state, even if the radiations are themselves mixed, as will be illustrated explicitly in the following.)

(b) The relevant information comes from two facts, that **l** is perpendicular to **k**, and that the conservation of angular momentum fixes the angles between **l** and **J**.

(c) The symmetry about 90° comes from the fact that the Legendre coefficient introduced by this perpendicularity vanishes for odd ν, that is, $P_\nu(\cos \pi/2) = 0$ unless ν is even. (This symmetry is a general feature reflecting the fact that the intermediate state had sharp parity.)

(d) The question as to whether the individual processes are emissions or absorptions, or the time sequence of the radiations is irrelevant. (This is a general feature of the direction-direction correlation involving sharp nuclear states.)

(e) The fixing of l_i, l_f and J is *dynamical*, and angular correlation thus makes no statement here, aside from conservation of angular momentum magnitude.

The classical discussion of the angular-correlation process of the above paragraphs would be of very limited interest except for the fact that almost all of the concepts essential to deriving this correlation have a precise transcription into a quantum formalism. Let us consider now the same direction-direction correlation as above, except from a quantum mechanical point of view. The first point to note is that the direction of an angular momentum vector is no longer an allowed concept; the maximal description is that of the vector model, where one speaks only of the magnitude of the vector and its projection along some axis. For the angular momentum $\mathbf{l}_i$ carried by a beam of spinless particles traveling along $\mathbf{k}_i$, this has the important consequence that the cylindrical averaging about the beam direction is now *required*, and not, as in the classical case, the result of no information.

The most important consequence of the cylindrical averaging required by quantum mechanics relates, however, to the angle χ, the azimuth angle about the direction of the angular momentum $\mathbf{J}$. This averaging has precisely the same origin as the cylindrical averaging about the beam direction discussed above, for, as we have seen, whenever we have a sharp nuclear state produced by a unidirectional radiation, the angular momentum of the state, $\mathbf{J}$, is cylindrically symmetric about $\mathbf{k}$, the beam direction. But for transitions involving sharp states there is no distinction in the angular distribution between emission and absorption, and since moreover, the relation between the classical vectors $\mathbf{k}$ and $\mathbf{J}$ is symmetric, it follows that this same relation, turned around, states that for the *emission* of radiation by this state, $\mathbf{k}$ *is azimuthally symmetric about the direction of* $\mathbf{J}$. This is just the desired average over χ, which we see is now also required by quantum mechanics.

[The azimuthal averaging of the angle χ is the key to all the manipulations of the present section, and since the justification given in the above paragraph may not seem to be completely compelling, we shall indicate a more formal argument here. The essential point is that the average over χ is but a translation into classical terms of the statement that the density matrix of a state formed by a unidirectional radiation is diagonal along the beam direction. (See discussion on p. 760.) Since the dynamics of the process is not a function of orientation (the scattering matrix expressed as a function of J, Π is independent of M), the assumption of a sharp intermediate state, when combined with the statement that the density matrix is diagonal, implies that the correlation *factors* (see, for example, the discussion of Eq. 54). The result is equivalent, in classical terms, to requiring an averaging over χ.]

The steps performed in the classical calculation are thus compatible with the quantum mechanical calculation, upon noting that the conservation of angular momentum in both cases fixes the angles between the vectors (l_i, j_i, J) and (l_f, j_f, J).

The one major departure from the classical calculation is the quantization of angular momentum, with its attendant introduction of *discrete angles*. We must expect, therefore, to replace the Legendre polynomials of the discrete angles by the appropriate "quantum mechanical angular functions."

There are in principle *three different types* of angle functions that may enter. There are, first, the usual angles between vectors of arbitrary (unquantized) length as, for example, between two linear momenta, $\mathbf{k}_i$ and $\mathbf{k}_f$—the angle functions here are the usual spherical harmonics. Next, one has angle functions between two vectors each of *quantized length*—the appropriate "angle functions" are now the *Racah coefficients*. Finally, there are the functions of angle defined by a vector of quantized length and a vector of unquantized length—these "angle functions" are the *Wigner* (vector-addition) *coefficients*.

For very large angular momenta the distinction between quantized and unquantized length becomes of no importance and we have the classical limiting result that all the angle functions become the usual spherical harmonics.[d]

While the transition from quantum results to classical results is quite direct and involves only the limit $\hbar \to 0$, an attempt to proceed in the opposite direction is hardly straightforward, or even feasible, despite the value of the correspondence principle. In keeping with our purpose of exhibiting the structure of the angular-correlation process rather than detailed proofs, we shall, therefore, simply set down the corresponding quantities as obtained from the many quantum treatments of the problem (see refs. *7–10*). For the addition of angular momenta defined by the triangle, $\mathbf{J} + \mathbf{l}_i + \mathbf{j}_i = 0$, the analog to the Legendre functions of the angle ϑ between $\mathbf{l}_i$ and $\mathbf{J}$ is:

---

[d] It is not our purpose to deal at length with this subject, but it is of interest to note that in approaching this limit the Kramers prescription $l \to (l + \tfrac{1}{2})$ [which comes from $l^2 \to l(l+1)$] serves to define the angle functions to an error of *second* order, $\mathcal{O}(\hbar^2)$. It is considerably more accurate to speak of this "semiclassical" approximation rather than the "classical limit." Note also that if the angular momenta are not sharp, the limit is the more general Jacobi polynomial rather than the spherical harmonics. The classical limit for the Racah and Wigner functions has been discussed in refs. *7–10*.

$$P_\nu(\hat{\mathbf{l}}_i \cdot \hat{\mathbf{J}}) \sim (-)^\nu [(2l_i + 1)(2J + 1)]^{\frac{1}{2}} \cdot W(l_i \nu j_i J; l_i J), \qquad (9)$$

where the angle for the Legendre polynomial on the left is determined by:

$$\cos \vartheta = \hat{\mathbf{l}}_i \cdot \hat{\mathbf{J}} = \frac{j_i(j_i + 1) - l_i(l_i + 1) - J(J + 1)}{2[l_i(l_i + 1)(J)(J + 1)]^{\frac{1}{2}}}. \qquad (10)$$

The coefficient $W(\ldots)$ in Eq. (9) is designated as a Racah coefficient.

The Legendre functions of the 90° angle between $\mathbf{l}_i$ and $\mathbf{k}_i$ are quite different in principle from the Legendre functions above, since this angle does not arise from the addition of angular momenta, but measures instead the orientation of an angular momentum with respect to an unquantized direction. For these we have the Wigner coefficients,

$$P_\nu(\hat{\mathbf{l}} \cdot \hat{\mathbf{k}}) = P_\nu(0) \sim (-)^l \left[\frac{2l + 1}{2\nu + 1} \right]^{\frac{1}{2}} (ll00|\nu 0). \qquad (11)$$

(The notation employed in Eq. 11 is that the Wigner coefficient referring to the addition of angular momenta $\mathbf{a} + \mathbf{b} = \mathbf{c}$, with magnetic quantum numbers $\alpha + \beta = \gamma$, is denoted by $(ab\alpha\beta|c\gamma)$.)

The restriction of ν to even integers that came about in the classical case from the vanishing of $P_\nu(0)$ for odd integers, also obtains for the Wigner coefficients here, which similarly vanish for odd ν.

With these identifications the separate factors in Eq. (8b) assume their quantum mechanical form:

$$(2\nu + 1)^{\frac{1}{2}} P_\nu(\hat{\mathbf{k}}_i \cdot \hat{\mathbf{l}}_i) P_\nu(\hat{\mathbf{l}}_i \cdot \hat{\mathbf{J}}) \sim (2J + 1)^{-\frac{1}{2}} (-)^{j_i - J} \cdot \bar{Z}(lJlJ; j_i \nu) \qquad (12)$$

where the $\bar{Z}$ coefficient has the general definition,[e]

$$\bar{Z}(lJl'J'; j_i \nu) \equiv [(2l + 1)(2l' + 1)(2J + 1)(2J' + 1)]^{\frac{1}{2}} \\ \cdot (ll'00|\nu 0) \cdot W(lJl'J'; j_i \nu). \qquad (13)$$

The angular correlation for a nuclear reaction with spinless particles given by the angular momentum sequence $j_i(l_i)J(l_f)j_f$ thus has the form:

$$\frac{dW(\vartheta)}{d\Omega} = \frac{1}{4\pi} (-)^{j_i - j_f} \sum_\nu A_\nu(i) A_\nu(f) P_\nu(\cos \vartheta), \qquad (14)$$

with

$$A_\nu(i) = \bar{Z}(l_i J l_i J; j_i \nu) \cdot (2J + 1)^{-\frac{1}{2}} \qquad (15a)$$
$$A_\nu(f) = \bar{Z}(l_f J l_f J; j_f \nu) \cdot (2J + 1)^{-\frac{1}{2}}. \qquad (15b)$$

[e] The $\bar{Z}$ coefficients given here differ from the definition of $Z(l_1 J_1 l_2 J_2; sL)$ given in Blatt and Biedenharn (11), by the phase $\exp \frac{1}{2}\pi i(L - l_1 + l_2)$. The necessity for this phase correction in Wigner and Eisenbud's nuclear reaction theory was pointed out by Huby (12).

We will sometimes, in the sequel, designate the left side of Eq. (14) as simply $W(\vartheta)$.

Aside from the normalization $(2J + 1)^{-\frac{1}{2}}$ (which arises from our requirement that $\int dW(\theta) = 1$), we see that the classical result introduces just that combination of terms, the $\bar{Z}$ coefficients, which was found sufficiently useful in the quantum mechanical case to label and tabulate separately. The fact that our classical derivation agrees precisely with the quantum result is meaningful, since the limiting results, Eqs. (9) and (11), are independently derivable. The normalization of the $\bar{Z}$ coefficients is chosen so that each state J has the proper statistical weight, $2J + 1$.

An important consequence of the quantization, which in fact greatly increases the utility of the correlation process as an experimental tool, is that ν no longer ranges from 0 to ∞. There is now an upper limit on ν defined by the rules (obtained from the properties of the Racah coefficients) that:

$$\nu_{\max} \leq \text{minimum } (2J, 2l_i, 2l_f).$$

Parity conservation, expressed by the properties of $(ll00|\nu 0)$, restricts the values of ν to be even.

We note once again that the coefficient of $P_\nu(\cos \vartheta)$ in Eq. (14), factors into two terms, one determined only by the initial link in the correlation ("formation"), the other by the final link ("decay"). This was a consequence of the randomicity of the azimuth angle χ; since this is restricted only by the assumption that the intermediate state is sharp, we see that this *factorization property also holds for mixed radiations*.

b. Cascades

The classical construction used to derive Eqs. (8), and by analogy (14), is a very direct way to construct more complicated correlations. Consider, for example, the correlation where the intermediate state (J) emits an *unobserved* radiation of fixed angular momentum l', going to a state J', which then emits the observed radiation along $\mathbf{k}_f$. For brevity we designate this process by the quantum number sequence: $j_i(l_i)J(l')J'(l_f)j_f$. Figure 4 illustrates the spherical diagram for this more complicated case. One notes now that in addition to the random vectors of Fig. 3, we have the additional randomly oriented vector $\mathbf{l}'$.

Applying the addition theorem to $[P_\nu(\hat{\mathbf{k}}_i \cdot \hat{\mathbf{k}}_f)]_{\mathrm{av}}$ just as before requires now *an extra step caused by the angular momentum* triangle $(\mathbf{JJ'l'})$. That is, after Eq. (6), we must use now:

$$[P_\nu(\hat{\mathbf{k}}_f \cdot \hat{\mathbf{J}})]_{\mathrm{av}} = P_\nu(\hat{\mathbf{J}} \cdot \hat{\mathbf{J}}')[P_\nu(\hat{\mathbf{k}}_f \cdot \hat{\mathbf{J}}')]_{\mathrm{av}}, \qquad (16)$$

and $\hat{J}'$ replaces $\hat{J}$ in (7). The additional, but unobserved, radiation introduces, therefore, another Legendre coefficient. In the quantum analog, one thus simply inserts into Eq. (14) the extra term:

$$[P_\nu(\hat{J} \cdot \hat{J}')]_{QM} \equiv (-)^\nu[(2J + 1)(2J' + 1)]^{\frac{1}{2}} W(J\nu l'J'; JJ'). \quad (17)$$

One can appreciate from this elementary derivation of the cascade correlation how direct and economical the methods of Racah really are! For they operate only with the *observables*, and do not introduce the

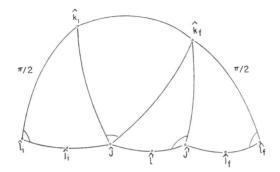

FIG. 4. The spherical mapping for the cascade $j_i(l_i)J(l')J'(l_f)j_f$.

cumbersome machinery of all possible observations (as the "maximal" techniques of magnetic quantum number sums necessarily do). To be sure our derivation above was rather imprecise, but it can be made rigorous without any essential change.

Although it is of slight experimental interest, the situation for a *cascade* of unobserved radiations follows at once.[f] Each unobserved link inserts a factor such as

$$(-)^\nu[(2J + 1)(2J' + 1)]^{\frac{1}{2}} W(J\nu l'J'; JJ') = [P_\nu(\hat{J} \cdot \hat{J}')]_{QM}.$$

Just as for its classical analog, this factor is at most unity. We must not conclude from this fact, however, that in a cascade, say, of three radiations (0, 1, and 2), that the (0,2) correlation is less anisotropic than the (0,1) or (1,2) correlations (counter examples are easily constructed). It should be noted, moreover, that the value of l', the multipolarity of the unobserved radiation, is *not* a restriction on the complexity of the angular correlation.

[f] An interesting example is, however, given by Gove and Litherland, Section II.C.2, pp. 291–293 in Part A.

c. Parallel Angular Momenta

If we suppose that in a cascade an unobserved radiation L_1 links two states j_1, j_2 and that these three angular momenta are related either by $j_1 = L_1 + j_2$, or by $j_2 = L_1 + j_1$ (that is, to say, the angular momenta are "parallel"), then for the classical calculation we see that the extra Legendre coefficient introduced by this radiation is just unity. Thus, in this special case, the unobserved radiation has *no* influence on the correlation whatever. (One might be concerned to distinguish parallel ($\vartheta = 0$) from antiparallel ($\vartheta = \pi$) but for the direction-direction correlation ν is even and the distinction is lost.)

The quantum result, on the other hand, does not show quite this simplicity. The reason is that the quantum equivalent to the P_ν introduced by the unobserved radiation, does not become unity (or even independent of ν) for the case of parallel angular momenta. It is clear then that a given unobserved radiation with parallel angular momenta does indeed generally affect the correlation in the quantum result. The origin of this result is again to be found in the fact that the direction of an angular momentum is not an observable; the triangle formed by "parallel angular momenta" cannot have zero area.

There is, however, one special case where even the quantum result simplifies. This is the situation where *all* of the angular momenta in a given cascade are parallel, which we define as $j_n = L_n + j_{n+1}$ (the case $j_{n+1} = L_n + j_n$ is equivalent, since it corresponds to the same correlation in reverse order). For this situation, as the explicit formulas for the Racah coefficients show by construction, the correlation in *every* case becomes the same as the double correlation for the simplest case, with the radiations having the same angular momenta as the two observed in the cascade, that is, the scheme $j_i(L_i)L_f(L_f)0$, with $j_i = L_i + L_f$.

The simplicity of parallel angular momenta has been often observed (13); a general proof was given by Fano (14).

One notes that, because of the factoring of the correlation process into two links, [the two A_ν in Eq. (14)], a slight extension of Fano's work is possible. Thus if *either* link shows only parallel angular momenta, then the contribution of that link reduces to an equivalent single radiation. In symbols, the cascade link (of n steps), $j_n = j_{n+1} + L_n$ connecting j_0 with j reduces to the single step $j + L_0(L_0)j$. An example of this situation is given in Section 2.

d. Particles with Spin

We have considered so far that the excitation has been produced by *spinless* particles. What changes occur when particles with spin produce the excitation, but the observation is still restricted only to a direction of motion?

In answering this question, it is essential to note that a measurement of a spin direction is *kinematically* connected with the measurement of a direction of motion, according to special relativity (15). Accordingly, the nonobservation of the spin orientation cannot now be equated in general to a *random* orientation. In fact if the particle is massless, there can be but two orientations—along or against the direction of motion. In practice, we distinguish two cases:

(a) Light particles where relativistic effects are essential: photons, leptons.

(b) Massive particles where relativistic effects are generally of little practical importance: nucleons, nuclei.

(1) *Nonrelativistic Case*

For the latter we may assert that the spin orientation is freely orientable with respect to **k**, and that, as a result, an unobserved spin orientation again implies a *random* orientation.

There are now three angular momenta that add to produce the intermediate J; that is, $\mathbf{l} + \mathbf{s} + \mathbf{j}_i + \mathbf{J} = 0$. We may group these angular momenta in six different ways, but only two groupings are generally used. One such is to introduce the *particle angular momentum*, $\mathbf{j}_p = \mathbf{l} + \mathbf{s}$; for the other, one introduces the *channel spin angular momentum*, $\mathbf{j}_s = \mathbf{s} + \mathbf{j}_i$.

Consider first the spherical diagram for the particle angular momentum, j_p. As before, the orbital angular momentum is perpendicular to the beam direction. Moreover, the spin angular momentum is (by hypothesis) random. Thus (see Fig. 5), averaging over the two random vectors (**s** and $\mathbf{j}_i$) we obtain successively, from the addition theorem,

$$[P_\nu(\hat{\mathbf{k}}_i \cdot \hat{\mathbf{J}})]_{\mathrm{av}} = P_\nu(\hat{\mathbf{J}} \cdot \hat{\mathbf{j}}_p)[P_\nu(\hat{\mathbf{k}}_i \cdot \hat{\mathbf{j}}_p)]_{\mathrm{av}} \tag{18}$$

and finally,

$$[P_\nu(\hat{\mathbf{k}}_i \cdot \hat{\mathbf{J}})]_{\mathrm{av}} = P_\nu(\hat{\mathbf{J}} \cdot \hat{\mathbf{j}}_p) P_\nu(\hat{\mathbf{k}}_i \cdot \hat{\mathbf{l}}) P_\nu(\hat{\mathbf{j}}_p \cdot \hat{\mathbf{l}}). \tag{19}$$

We see that the sole effect of the (unobserved) spin is to replace $P_\nu(\hat{\mathbf{l}} \cdot \hat{\mathbf{J}})$ by $P_\nu(\hat{\mathbf{l}} \cdot \hat{\mathbf{j}}_p) P_\nu(\hat{\mathbf{j}}_p \cdot \hat{\mathbf{J}})$, and analogously for the quantum mechanical equivalent.

To summarize then, for the correlation of a nonrelativistic particle with unobserved spin, and total angular momentum j_p, one replaces, in Eq. (15a), the term $A_\nu(i)$ by:

$$A_\nu(j_p) = (2l+1)(2j_p+1)(2J+1)^{\frac{1}{2}}(ll00|\nu 0) W(l\nu s j_p; l j_p) \\ \cdot W(j_p \nu j_i J; j_p J) \cdot (-)^l. \tag{20}$$

This result is simply the quantum transcription of Eq. (19).

The close formal similarity of this case to that of an additional unobserved radiation should be noted.

Let us next consider the *channel spin* case. The channel spin j_s is defined by: $s + j_i + j_s = 0$. Since *both* s and j_i are *random*, we see that

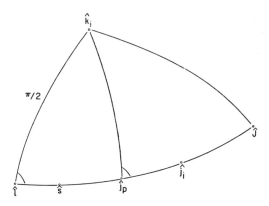

Fig. 5. The spherical mapping for a particle with spin s producing the nuclear state J. If the particle is nonrelativistic, an unobserved spin corresponds to s random. The initial state j_i has random orientation also.

their sum, j_s, is similarly *randomly oriented*. The required average for $P_\nu(\hat{\mathbf{k}}_i \cdot \hat{\mathbf{J}})$ can be read off from the appropriate spherical mapping, and is:

$$[P_\nu(\hat{\mathbf{k}}_i \cdot \hat{\mathbf{J}})]_{av} = P_\nu(\hat{\mathbf{l}} \cdot \hat{\mathbf{k}}_i)P_\nu(\hat{\mathbf{l}} \cdot \hat{\mathbf{J}}). \tag{21}$$

In other words, *the channel spin j_s plays the role of the initial angular momentum j_i in the elementary spinless particle correlation.*

One notes, moreover, that the channel spin result is the simpler of the two cases (j_s versus j_p) in two respects: (1) the j_p case involves an extra Racah coefficient, and—more importantly—(2) the j_s case gives the extreme cases for the angular correlation. This latter point, which is basically the reason for Wigner's introduction of the channel spin in the first place, cannot be discussed without discussing the problem of coherency versus incoherency, and this is given in the concluding section. We emphasize again, however, that the simplicity of the channel spin case is directly the result of j_s being random, which required that *both* j_i and s be random.

(2) *Relativistic Case*

Let us consider next the case for relativistic particles. Because of the kinematical coupling of the spin and direction of motion, any meaningful spin measurement must be carried out in a coordinate system rigidly attached to the direction of motion. For massless particles (with $s \neq 0$), the possible spin orientations are two, along or against the direction of motion.

Now the general situation concerns itself with *two* particles (in general, each with spin) moving either toward or away from each other. The relative orbital angular momentum is measured with respect to their center of momentum. Although both particles may be relativistic [as for example, an electron and neutrino emitted by a heavy (that is, fixed position) nucleus], it is reasonable to consider for the moment, that one of the two particles is heavy, so that its spin may be considered as freely orientable with respect to $\mathbf{k}$. Practically we need consider only leptons (spin $\frac{1}{2}$) and photons (spin 1). For leptons, the possible spin orientations are the maximum allowed in any case so that an unobserved spin is again random in all coordinate systems. Hence, the previous result for a freely orientable spin again applies.

On the other hand, we may consider the case for completely polarized leptons with, say, positive helicity. Then the state $\mathbf{j}_p = 1 + \mathbf{1/2}$ has projection $\frac{1}{2}$ along the direction of motion. The angle ϑ (between the angular momentum and the direction of motion $\mathbf{k}$) is no longer $\pi/2$, as would be the case for spinless particles, but rather has the semiclassical value

$$\vartheta \cong \cos^{-1}\left(\frac{\frac{1}{2}}{j_p + \frac{1}{2}}\right).$$

The average value of $P_\nu(\hat{\mathbf{J}} \cdot \hat{\mathbf{k}}_i)$ is now:

$$[P_\nu(\hat{\mathbf{J}} \cdot \hat{\mathbf{k}}_i)]_{\text{av}} = P_\nu(\hat{\mathbf{j}}_p \cdot \hat{\mathbf{k}}_i) P_\nu(\hat{\mathbf{J}} \cdot \hat{\mathbf{j}}_p). \qquad (22)$$

The quantum analog to the $P_\nu(\cos \vartheta)$ that enters here is

$$P_\nu(\cos \vartheta) \to (-)^{j_p - \frac{1}{2}} \cdot \left[\frac{2j_p + 1}{2\nu + 1}\right] (j_p j_p \tfrac{1}{2} - \tfrac{1}{2}|\nu 0).$$

We note that neither the semiclassical nor the quantum result vanishes for odd ν, since the observation of a helicity is changed by a reflection. (Unless, however, a helicity measurement is also made on the other particle in the correlation the resulting angular correlation still has only ν even.) In the classical *limit*, however (as opposed to the semiclassical result), the angle ϑ becomes $\pi/2$; this indicates also that the odd parity terms are largest for small angular momenta.

If we now average the correlation over the two helicities we have the random case previously discussed. This implies that there must exist a relation (for even ν),

$$P_\nu(\hat{\mathbf{j}}_p \cdot \hat{\mathbf{k}}) = P_\nu(\hat{\mathbf{k}} \cdot \hat{\mathbf{l}}) P_\nu(\hat{\mathbf{l}} \cdot \hat{\mathbf{j}}_p), \tag{23}$$

and its quantum transcription,

$$\left[\frac{2j_p + 1}{2\nu + 1}\right]^{\frac{1}{2}} (j_p j_p \tfrac{1}{2} - \tfrac{1}{2}|\nu 0) = \left[\frac{2l + 1}{2\nu + 1}\right]^{\frac{1}{2}} (ll00|\nu 0) \\ \cdot [(2l + 1)(2j_p + 1)]^{\frac{1}{2}} W(lj_p l j_p; \tfrac{1}{2}\nu), \tag{24}$$

which is indeed the case, as can be verified directly.

The photon situation is very similar to that for leptons. For circularly polarized photons the angle is given by $\cos^{-1}\left(\frac{\pm 1}{J + \tfrac{1}{2}}\right)$, corresponding to right $(+)$ or left $(-)$ circular photons. It follows that

$$P_\nu(\cos\vartheta) \sim (-)^{J-1} \left[\frac{2J + 1}{2\nu + 1}\right]^{\frac{1}{2}} (JJ1 - 1|\nu 0)$$

replaces the term $P_\nu(\cos \pi/2)$ in the correlation for circularly polarized photons. Once again (though it is not true for spin > 1) a *random* mixture of these two polarization states corresponds to a random orientation of the spin vector (1), so that just as for the lepton case, we have a relation (for ν even)

$$P_\nu(\cos\vartheta) = P_\nu(\cos\pi/2) P_\nu(\hat{\mathbf{l}} \cdot \hat{\mathbf{J}}), \tag{25}$$

or the quantum equivalent,

$$\left[\frac{2J + 1}{2\nu + 1}\right]^{\frac{1}{2}} (JJ1 - 1|\nu 0) = \left[\frac{2J + 1}{2\nu + 1}\right]^{\frac{1}{2}} (JJ00|\nu 0) \cdot [2J+1] W(JJJJ; 1\nu). \tag{26}$$

This relation is obtained for magnetic multipoles $(J = L)$; a more complicated relation can be obtained for electric multipoles, but this is unnecessary since (by the duality transformation, $E \leftrightarrow H$), the angular distribution of the two types of multipole must be the same.

Equations (24) and (26) allow one to consider unpolarized leptons and photons in exactly the same way as one considered nonrelativistic particles with spin in the $\mathbf{j}_p$ scheme; Eq. (20). These results are special instances of the general parametrization which we now discuss.

e. Parametrization

If we look at the results obtained thus far on the direction-direction correlation involving pure states, we see that there are two significant features of the coefficients of the Legendre series, Eqs. (3), (15):

(a) The coefficients for the correlation, B_ν, factor into two parts, $A_\nu(i)$ and $A_\nu(f)$, which are each dependent only on the separate transitions, and are independent of each other.

(b) The separate "links," the A_ν, depend upon the type of radiation emitted or absorbed, but this dependence also factors.

These two results, it should be noted, would *not* have been obtained if, instead of the proper angle functions, the $P_\nu(\cos\vartheta)$, we had made an expansion in $(\cos\vartheta)^{2n}$.[g] As we will show in the concluding section on coherent processes, the factorization expressed in (a) is dependent only upon the assumption that the intermediate state is pure, and thus remains true even for mixed radiations.

These two features of the direction-direction correlation are of very great practical value, for one may exploit them to simplify the tabulation of the necessary correlation coefficients. Using (a) we may focus attention on the separate links, and tabulate only the coefficients that may enter any link. Using (b) we may choose a *standard* correlation, the gamma-gamma correlation, and define a multiplicative *particle parameter* to effect the necessary change from the standard correlation.

If the intermediate state is not pure, the same tabulation of correlation coefficients and particle parameters suffices, but now the explicit correlation must be given term by term. This is discussed in the concluding section.

Let us consider now the standard gamma-gamma correlation. Each link, $A_\nu(\gamma)$, has the form

(classical), $\qquad A_\nu(\gamma) \sim (2\nu+1)^{\frac{1}{2}} P_\nu(\hat{\mathbf{L}}\cdot\hat{\mathbf{k}})\cdot P_\nu(\hat{\mathbf{L}}\cdot\hat{\mathbf{J}})$ \hfill (27a)

(quantum), $A_\nu(\gamma) = (2\nu+1)^{\frac{1}{2}} \cdot \left[(-)^{L+1}\left(\frac{2L+1}{2\nu+1}\right)^{\frac{1}{2}}(LL1-1|\nu 0)\right]$

$\qquad\qquad \cdot [((2L+1)(2J+1))^{\frac{1}{2}} \cdot (-)^\nu \cdot W(L\nu j_i J; LJ)]$, \hfill (27b)

where j_i = angular momentum of initial unobserved nuclear state
$\qquad J$ = angular momentum of intermediate nuclear state
$\qquad L$ = multipolarity of gamma ray.

The coefficient tabulated for the standard gamma-gamma correlation is *exactly* this, generalized for radiations of mixed multipolarity. Explicitly this coefficient is:

$$F_\nu(LL'j_iJ) \equiv (-)^{j_i-J-1}[(2L+1)(2L'+1)(2J+1)]^{\frac{1}{2}}(LL'1-1|\nu 0)$$
$$\cdot W(JJLL';\nu j_i), \qquad (28)$$

[g] Alternatively one might express this by saying that the significance of the use of the $P_n(\cos\vartheta)$ instead of $(\cos\vartheta)^n$ is that the P_n have an *addition theorem*, which is the essential step in our earlier discussion.

which reduces to (27b) for $L = L'$ (the different phase in (28) and (27b) results from the rearrangement of the parameters in the Racah coefficient). The normalization is such that $F_0(LLj_iJ)$ is unity. A discussion of the available tabulations of the F_ν coefficients is given in an appendix to this chapter.

This standard correlation requires that the particle parameter for spinless particles be:

$$a_\nu(LL'; \text{spinless}) = (LL'00|\nu0)(LL'1 - 1|\nu0)^{-1}\cos(\xi_L - \xi_{L'})$$
$$= \cos(\xi_L - \xi_{L'}) \cdot \frac{2[L(L+1)L'(L'+1)]^{\frac{1}{2}}}{L(L+1) + L'(L'+1) - \nu(\nu+1)}. \quad (29)$$

For pure states, one sees that this is just the quantum form of the ratio $P_\nu(\cos \pi/2)/P_\nu(\cos \vartheta)$. [The occurrence of the phase dependent term, $\cos(\xi_L - \xi_{L'})$ discussed on p. 775, might be questioned, for this is clearly not geometrical in origin. One might more logically include this with the other dynamical factors that determine the amount of (L,L') mixing. Equation (29) is, however, the more customary form.]

Application and discussion of this result is given in Section 2.

The correlation link in the channel spin scheme, as the previous sections show, is equivalent to the spinless particle situation with j_s replacing j_i [Eq. (21)]. This case is therefore covered by the particle parameter given above. In actual practice the importance of the channel spin scheme in nuclear reactions has led to the separate tabulation of the necessary coefficients, the Z coefficients introduced in Eq. (13), but with a different phase definition (see footnote e).

Finally let us consider the correlation for a particle in the j_p scheme, which as discussed in the preceding sections, will apply both to non-relativistic particles of spin s as well as to leptons. The correlation has the link

$$A_\nu \sim (2\nu + 1)^{\frac{1}{2}} \cdot P_\nu(\hat{\mathbf{k}} \cdot \hat{\mathbf{l}}) \cdot P_\nu(\hat{\mathbf{l}} \cdot \hat{\mathbf{j}}_p) \cdot P_\nu(\hat{\mathbf{j}}_p \cdot \hat{\mathbf{J}})$$
$$= (2l + 1)(2j_p + 1)(2J + 1)^{\frac{1}{2}}(-)^l(ll00|\nu0)$$
$$W(l\nu s j_p; lj_p)W(j_p\nu j_iJ; j_pJ). \quad (30)$$

For the lepton case, Eqs. (23) and (24), this has the alternative form:

$$A_\nu \sim (2\nu + 1)^{\frac{1}{2}} \cdot P_\nu(\hat{\mathbf{k}} \cdot \hat{\mathbf{j}}_p) \cdot P_\nu(\hat{\mathbf{j}}_p \cdot \hat{\mathbf{J}})$$
$$= (2J + 1)^{\frac{1}{2}}(2j_p + 1)(-)^{j_p - \frac{1}{2} + \nu}(j_pj_p\tfrac{1}{2} - \tfrac{1}{2}|\nu0)W(j_p\nu j_iJ; j_pJ). \quad (31)$$

It is clear that one cannot put Eq. (30) into the standard form, but rather one might utilize Eq. (31) to generalize the standard form to half-integer angular momenta, and thus extend the definition of the F_ν. This has not been done too extensively, largely because the j_p correlation is too infrequent to warrant such treatment.[h] One must in this case utilize Eq. (30) and tables of the coefficients that appear there ($\bar{Z}$ and W, see examples in Section 2).

A very striking extension of the concept of particle parameters can be given, as shown by Gardner (*17,18*) and by Lloyd (*13*).[i] To see this extension in its most direct form, let us consider two charged systems (a nucleus and a particle say) interacting via the electromagnetic field. As originally done by Fermi, in 1924, we replace one of the systems (the incident particle) by an equivalent *photon* interaction.[j] Lloyd's parametrization asserts that the observation of the direction of motion of only the incident (or emergent) charged particle, is equivalent—to within a multiplicative factor; that is, the particle parameter a_ν—to the observation of a gamma ray of the same transfer angular momentum connecting the same nuclear states.

There are many important applications of this result, from which we single out two:

(a) The internal conversion angular correlation:

Here an electron in a bound state interacts via the electromagnetic field with its nucleus (in an excited state) and is emitted into an electron state observed by its direction of motion. This direction is correlated with a preceding or succeeding nuclear radiation. The calculation of the a_ν must be done with relativistic wave functions and, so far, only results for the K-shell have been published. The particle parameters, including mixed radiations, are tabulated in ref. *20*.

(b) The Coulomb excitation correlation:

This case is similar to that considered originally by Fermi. A beam of nonrelativistic charged particles (protons, alphas) bombard target nuclei and cause nuclear excitation through the Coulomb coupling. The excited nuclei then radiate gamma rays in returning to their ground state. The correlation is between the direction of motion of the incident particle and the emitted gamma ray. The particle parameter a_ν (with a_0 being the cross section for the process) can be calculated approximately using classical orbits, or more accurately quantum mechanically.

Numerical results for the a_ν have been published for the E2, E1, and

---

[h] A short table has been given by Satchler (*16*).

[i] Lloyd's ideas (expressed in a private communication) were first applied in (*19*) to the internal conversion correlation.

[j] Although Fermi used this as an approximation (and for a fast particle this technique is known as the Weizäcker-Williams approximation), for angular momentum transfers greater than zero, it can be made exact, aside from penetration and radiative reaction.

M1 cases in a quantum treatment (*21,22*). There is an extensive literature on this problem; one should refer to the review of ref. *22* for this. An example of a typical Coulomb excitation is given in Section 2.

It will be recalled that Fermi, in his original formulation of beta decay, took the electromagnetic interaction as a model. It is not surprising then in view of the foregoing discussion that the concept of particle parameters may also be extended to the general beta decay interaction; this leads to a treatment of beta-gamma correlations. The problem of treating beta decay particle parameters is considerably more complicated than the formally similar electromagnetic problems discussed above because of the unknown nature of the interaction constants. Moreover, one has the additional complication of mixed multipoles with similar properties (L,π) that arises from the different interactions $(SVTAP)$. Finally, as a result of Lee and Yang's penetrating analysis, it must now be considered that the general beta decay interaction, to account for the loss of reflection symmetries, may comprise 20 complex coupling constants, with 35 measurable real quantities.

A treatment of the beta decay process with a view toward developing general beta-gamma correlations is too extensive a task to come within the scope of this survey. For the older formulation, which is still largely applicable, we refer to the reviews of refs. *20* and *23*.

An interesting feature of the "new" beta decay is the neutrino helicity. From a formal point of view, this leads to the possibility of polarization correlations in beta decay. For this we refer to the paper of Alder, Stech, and Winther (*24*).

f. More Involved Correlation Processes

When one, in addition to the measurement of two directions of motion, makes other measurements on a cascade or reaction, a more complicated form of the correlation problem is involved. There are two general situations that occur:

(a) One may make additional measurements—spin polarization measurements—on one (or both) of the particles whose directions of motion are measured.

(b) One may measure additional directions of motion in a cascade.

The distinction between these two situations is not in practice as sharp as might at first appear. Since the particles whose spin polarization is measured are typically either leptons or photons, their polarization state is completely described by a single vector (the Poincaré vector), for a fixed direction of motion. This is clearly quite similar to the situation where the additional information is the vector describing another direction of motion, although we must bear in mind the distinction that the polarization vector for relativistic particles is measured with respect to a physically defined frame (the particle's direction of motion).[k]

Let us consider now the measurement of one additional vector, for simplicity taken to be a direction of motion. We take as our prototype the triple correlation of three spinless particles proceeding through sharp states of definite angular momentum; that is, $j_i(l_0)J(l_1)J'(l_2)j_f$. The basic information for the correlation is just as for the simpler direction-direction double correlation: that $\mathbf{1} \cdot \mathbf{k}$ is zero in every case, and that $\mathbf{j}_i$ and $\mathbf{j}_f$ are randomly oriented.

A classical correlation exists for this elementary case, but unfortunately the necessary functions for the classical description are not very familiar. This is immediately clear when one asks: "What angle functions furnish the proper description for three vectors?" For *two* vectors, it is obvious that the $P_\nu(\hat{\mathbf{k}}_1 \cdot \hat{\mathbf{k}}_2)$ must enter, for these are the orthogonal functions of the single polar angle, defined independently of any reference system. For three vectors there are three invariantly defined angles, and the proper orthonormal angle functions hardly look classical at all! The proper functions have been introduced in ref. *20* to describe triple correlations and are:

$$\Lambda\nu_0\nu_1\nu_2(\hat{\mathbf{k}}_0\hat{\mathbf{k}}_1\hat{\mathbf{k}}_2)$$
$$\equiv \sum_{\mu_1\mu_2} (\nu_1\nu_0\mu_2 - \mu_1\mu_1|\nu_2\mu_2)\, Y_{\nu_0}^{\mu_1*}(\hat{\mathbf{k}}_0)\, Y_{\nu_1}^{\mu_2-\mu_1*}(\hat{\mathbf{k}}_1)\, Y_{\nu_2}^{\mu_2}(\hat{\mathbf{k}}_2). \quad (32)$$

Despite the appearance of the Wigner coefficients in this formula, it is nonetheless a classical result, for the Wigner coefficients occur here in the context of a general vector algebra. For example, the Λ_{111} function is just the invariant, antisymmetric, combination of three vectors; that is, $\mathbf{A} \cdot \mathbf{B} \times \mathbf{C}$—(to within an over-all constant). The other Λ similarly represent invariants, though of different orders.

It is useful to put these angle functions into a somewhat different and more symmetrical form. A renormalization (similar to the change from Y_l^0 to P_l) by the factor $[(2\nu_0 + 1)(2\nu_1 + 1)(2\nu_2 + 1)(4\pi)^{-3}]^{\frac{1}{2}}$ is of value. Moreover, it will be observed in Eq. (32) that for $\nu_0 + \nu_1 + \nu_2 =$ odd integer the Λ are not real; the phase, $\exp \frac{1}{2}\pi i(\nu_0 + \nu_1 - \nu_2)$ (correspond-

---

[k] In order to avoid confusion it is well to point out one further distinction. This concerns the fact that the measurement of the Poincaré vector includes the possibility of circular polarization measurements and thus one may obtain odd parity terms (terms that change sign under a coordinate reflection).

The measurement of a "direction of motion" of a particle for a pure state is *not* in actuality the measurement of a direction, but the measurement of a ray since the terms defined by this measurement are unchanged by a reflection ($\mathbf{1} \cdot \mathbf{k} = 0$ implies that $\nu =$ even).

ing to the use of $i^l Y_l^m(\vartheta\varphi)$ as basis functions), corrects this. Finally, one desires to exhibit as far as possible the basic symmetry between the elements making up these angle functions. This step utilizes a symmetrical form of the Wigner coefficients (25). Introducing all these changes, the resulting invariant proper functions of three unit vectors has the definition:

$$P_{\nu_0\nu_1\nu_2}(\hat{\mathbf{k}}_0\hat{\mathbf{k}}_1\hat{\mathbf{k}}_2) \equiv (4\pi)^{3/2}(2\nu_0 + 1)^{-1/2}(2\nu_1 + 1)^{-1/2} \cdot (2\nu_2 + 1)^{-1} \cdot i^{(\nu_0+\nu_1-\nu_2)}$$
$$\cdot \sum_{\alpha\beta\gamma} (-)^\gamma (\nu_0\nu_1\alpha\beta|\nu_2\gamma) Y_{\nu_0}{}^\alpha(\hat{\mathbf{k}}_0) Y_{\nu_1}{}^\beta(\hat{\mathbf{k}}_1) Y_{\nu_2}{}^{-\gamma}(\hat{\mathbf{k}}_2). \quad (33)$$

The $P_{\nu_0\nu_1\nu_2}$ defined by Eq. (33) now have the symmetry properties:

$$P_{\nu_0\nu_1\nu_2}(\hat{\mathbf{k}}_0\hat{\mathbf{k}}_1\hat{\mathbf{k}}_2) = (-)^{\nu_0+\nu_1-\nu_2} P_{\nu_1\nu_0\nu_2}(\hat{\mathbf{k}}_1\hat{\mathbf{k}}_0\hat{\mathbf{k}}_2)$$
$$= (-)^{\nu_1+\nu_2-\nu_0} P_{\nu_0\nu_2\nu_1}(\hat{\mathbf{k}}_0\hat{\mathbf{k}}_2\hat{\mathbf{k}}_1). \quad (34)$$

All other permutations may be reduced to products of these two. The antisymmetry of $\mathbf{A} \cdot \mathbf{B} \times \mathbf{C}$ thus fits the general rule of Eq. (34).

If one of the three indices $(\nu_0\nu_1\nu_2)$, is zero we obtain the usual Legendre function; that is,

$$P_{0\nu_1\nu_2}(\hat{\mathbf{k}}_0\hat{\mathbf{k}}_1\hat{\mathbf{k}}_2) = \delta_{\nu_1}{}^{\nu_2}(2\nu_1 + 1)^{-1/2} P_{\nu_1}(\hat{\mathbf{k}}_1 \cdot \hat{\mathbf{k}}_2).$$

The general form, Eq. (33), is the analog of the addition theorem for the Legendre functions. Let us consider, for example, that the three unit vectors $\hat{\mathbf{k}}_0$, $\hat{\mathbf{k}}_1$, $\hat{\mathbf{k}}_2$ are referred to an arbitrary axis, $\hat{\mathbf{z}}$. If we average over the azimuthal angle of $\hat{\mathbf{k}}_0$ with respect to $\hat{\mathbf{z}}$ we find from Eq. (33) that:

$$[P_{\nu_0\nu_1\nu_2}(\hat{\mathbf{k}}_0\hat{\mathbf{k}}_1\hat{\mathbf{k}}_2)]_{\text{av}} = P_{\nu_0}(\hat{\mathbf{k}}_0 \cdot \hat{\mathbf{z}}) \cdot P_{\nu_0\nu_1\nu_2}(\hat{\mathbf{z}}\hat{\mathbf{k}}_1\hat{\mathbf{k}}_2). \quad (35)$$

A similar result will also apply for averages of $\hat{\mathbf{k}}_1$ or $\hat{\mathbf{k}}_2$ around $\hat{\mathbf{z}}$, utilizing the symmetry under interchange [Eq. (34)].

Like the Legendre functions we obtain a much simpler formula if we specialize one of the vectors to lie along the $\hat{\mathbf{z}}$-axis. In this case we obtain, for $\hat{\mathbf{k}}_0$ along $\hat{\mathbf{z}}$,

$$P_{\nu_0\nu_1\nu_2}(\vartheta_1\vartheta_2\varphi) = (2\nu_2 + 1)^{-1/2} \cdot i^{(\nu_0+\nu_1-\nu_2)}$$
$$\cdot \sum_\beta (\nu_0\nu_1 0\beta|\nu_2\beta) \left[\frac{(\nu_1 - |\beta|)! \, (\nu_2 - |\beta|)!}{(\nu_1 + |\beta|)! \, (\nu_2 + |\beta|)!}\right]^{1/2} P_{\nu_1}{}^{|\beta|}(\cos\vartheta_1) P_{\nu_2}{}^{|\beta|}(\cos\vartheta_2) e^{i\beta\varphi}.$$
$$(36)$$

Here ϑ_1 is the polar angle between $\mathbf{k}_0$ and $\mathbf{k}_1$; ϑ_2 is the polar angle between $\mathbf{k}_0$ and $\mathbf{k}_2$; and φ is the dihedral angle between the $(\mathbf{k}_0, \mathbf{k}_1)$ and $(\mathbf{k}_0, \mathbf{k}_2)$ planes.

The functions defined by Eq. (33) are orthogonal, and normalized such that:

$$\int_0^{2\pi} d\varphi \int_0^{\pi} \sin \vartheta_2 \, d\vartheta_2 \int_0^{\pi} \sin \vartheta_1 \, d\vartheta_1 P_{abc}(\vartheta_1 \vartheta_2 \varphi) P_{\bar{a}\bar{b}\bar{c}}(\vartheta_1 \vartheta_2 \varphi)$$
$$= (4\pi)^3 (2a+1)(2b+1)(2c+1) \delta_a{}^{\bar{a}} \delta_b{}^{\bar{b}} \delta_c{}^{\bar{c}}.$$

For triple correlations, a convenient experimental arrangement observes two of the radiations perpendicular to a third radiation (usually an incident beam direction). Thus $\vartheta_1 = \vartheta_2 = \pi/2$ and the correlation involves the single angle φ. The $P_{\nu_0\nu_1\nu_2}$ for this case appear in Table I.

TABLE I

Triple correlation angle functions, $P_{\nu_0\nu_1\nu_2}(\vartheta_1\vartheta_2\varphi)$, for the case where $\vartheta_1 = \vartheta_2 = \pi/2$

P_{000}	1
P_{022}	$\dfrac{1}{2\sqrt{5}} (3\cos^2\varphi - 1)$
P_{044}	$\dfrac{1}{24}(35\cos^4\varphi - 30\cos^2\varphi + 3)$
P_{202}	$-\dfrac{1}{2\sqrt{5}}$
P_{220}	$-\dfrac{1}{2\sqrt{5}}$
P_{222}	$-\dfrac{1}{\sqrt{70}}(3\cos^2\varphi - 2)$
P_{224}	$-\dfrac{1}{4\sqrt{70}}(5\cos^2\varphi - 1)$
P_{242}	$-\dfrac{1}{4\sqrt{70}}(5\cos^2\varphi - 1)$
P_{244}	$-\dfrac{1}{12}\sqrt{\tfrac{5}{77}}(49\cos^4\varphi - 51\cos^2\varphi + 6)$
P_{404}	$\tfrac{1}{8}$
P_{422}	$\tfrac{1}{4}\sqrt{\tfrac{1}{70}}(2\cos^2\varphi + 1)$
P_{424}	$\tfrac{1}{4}\sqrt{\tfrac{5}{77}}(3\cos^2\varphi - 2)$
P_{440}	$\tfrac{1}{8}$
P_{442}	$\tfrac{1}{4}\sqrt{\tfrac{5}{77}}(3\cos^2\varphi - 2)$
P_{444}	$\tfrac{1}{24}\sqrt{\tfrac{1}{2002}}(490\cos^4\varphi - 545\cos^2\varphi + 109)$

Unfortunately, tabulations of the general $P_{\nu_0\nu_1\nu_2}$ functions do not exist as yet, and the functions must be worked out from the definitions, Eqs. (33) and (36).

After this digression let us return to the subject of the triple correlation. The *classical* triple correlation for the reaction designated by the angular momenta $j_i(l_0)J(l_1)J'(l_2)j_f$ must be expressible by a series in terms of our orthogonal basis functions, that is,

$$W(\hat{\mathbf{k}}_0\hat{\mathbf{k}}_1\hat{\mathbf{k}}_2) = (4\pi)^{-3} \sum_{\nu_0\nu_1\nu_2} (2\nu_0 + 1)(2\nu_1 + 1)(2\nu_2 + 1)$$
$$B_{\nu_0\nu_1\nu_2} P_{\nu_0\nu_1\nu_2}(\hat{\mathbf{k}}_0\hat{\mathbf{k}}_1\hat{\mathbf{k}}_2). \quad (37)$$

This equation is the exact analog of Eq. (2) for the double correlation, and, in a similar fashion, the $B_{\nu_0\nu_1\nu_2}$ are to be interpreted as the average

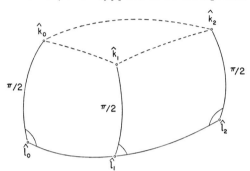

FIG. 6. The spherical mapping for a triple correlation of three spinless particle radiations. The directions of motion of the three particles are $\mathbf{k}_0$, $\mathbf{k}_1$, and $\mathbf{k}_2$. The polar angles labeled $\pi/2$ are known from the information that $\hat{\mathbf{l}} \cdot \hat{\mathbf{k}} = 0$. The initial and final states have zero angular momenta. The reaction proceeds through the intermediate states $|\mathbf{J}| = l_0$ and $|\mathbf{J}'| = l_2$; the three angular momenta l_0, l_1, l_2 form a triangle so that their representative points lie on a great circle.

The angles indicated with an arc are each random.

value of the $P_{\nu_0\nu_1\nu_2}$ evaluated over all possible configurations of the system consistent with our information. That is:

$$B_{\nu_0\nu_1\nu_2} \equiv [P_{\nu_0\nu_1\nu_2}(\hat{\mathbf{k}}_0\hat{\mathbf{k}}_1\hat{\mathbf{k}}_2)]_{av}.$$

To eliminate inessential detail, consider first the special case where $j_i = j_f = 0$, and the sequence of angular momenta is $0(l_0)J(l_1)J'(l_2)0$ with $|\mathbf{J}| = l_0$ and $|\mathbf{J}'| = l_2$. The spherical mapping for this case is given in Fig. 6; the fact that $\mathbf{l}$ is perpendicular to $\mathbf{k}$ is indicated explicitly.

The azimuthal angles about $\hat{\mathbf{l}}_0$, $\hat{\mathbf{l}}_1$, $\hat{\mathbf{l}}_2$ are each *random*. Using the form of the addition theorem given in Eq. (35), it follows that:

$$\begin{aligned} B_{\nu_0\nu_1\nu_2} &= [P_{\nu_0\nu_1\nu_2}(\hat{\mathbf{k}}_0\hat{\mathbf{k}}_1\hat{\mathbf{k}}_2)]_{av} \\ &= P_{\nu_0}(\hat{\mathbf{k}}_0 \cdot \hat{\mathbf{l}}_0)[P_{\nu_0\nu_1\nu_2}(\hat{\mathbf{l}}_0\hat{\mathbf{k}}_1\hat{\mathbf{k}}_2)]_{av} \\ &= P_{\nu_0}(\hat{\mathbf{k}}_0 \cdot \hat{\mathbf{l}}_0)P_{\nu_1}(\hat{\mathbf{k}}_1 \cdot \hat{\mathbf{l}}_1)[P_{\nu_0\nu_1\nu_2}(\hat{\mathbf{l}}_0\hat{\mathbf{l}}_1\hat{\mathbf{k}}_2)]_{av} \\ &= P_{\nu_0}(\hat{\mathbf{k}}_0 \cdot \hat{\mathbf{l}}_0)P_{\nu_1}(\hat{\mathbf{k}}_1 \cdot \hat{\mathbf{l}}_1)P_{\nu_2}(\hat{\mathbf{k}}_2 \cdot \hat{\mathbf{l}}_2)P_{\nu_0\nu_1\nu_2}(\hat{\mathbf{l}}_0\hat{\mathbf{l}}_1\hat{\mathbf{l}}_2). \end{aligned} \quad (38)$$

The structure of this result is apparent, for we have successively referred to $\hat{\mathbf{l}}_0$, $\hat{\mathbf{l}}_1$, and $\hat{\mathbf{l}}_2$ as axes for the $P_{\nu_0\nu_1\nu_2}$ function, and averaged over the azimuth in each case. The one essentially new feature is the appearance of the $P_{\nu_0\nu_1\nu_2}(\hat{\mathbf{l}}_0\hat{\mathbf{l}}_1\hat{\mathbf{l}}_2)$, which, it should be noted, is a function of three angular momentum vectors that form a definite triangle with *fixed* lengths, and *fixed* angles. The three vectors *thus lie in a plane*, and the $P_{\nu_0\nu_1\nu_2}$ function is correspondingly simpler. For example, only two independent angles enter, rather than three. Moreover, the symmetry of the $P_{\nu_0\nu_1\nu_2}$ requires that for $\nu_0 + \nu_1 + \nu_2 =$ odd integer the function vanishes.[1]

It remains only to note now the quantum analog to the classical function $P_{\nu_0\nu_1\nu_2}$ in order to exhibit the quantum mechanical triple correlation result. This function (to within a factor) is Wigner's 9-j coefficient, or as it is also known, Fano's X-coefficient.[m] That is:

$$i^{\nu_2-\nu_1-\nu_0} \cdot [(2J+1)(2J'+1)(2l_1+1)]^{\frac{1}{2}} \cdot X\begin{pmatrix} J l_1 J' \\ J l_1 J' \\ \nu_0 \nu_1 \nu_2 \end{pmatrix} \sim P_{\nu_0\nu_1\nu_2}(\hat{\mathbf{J}}\hat{\mathbf{l}}_1\hat{\mathbf{J}}'). \quad (39)$$

It is interesting to note that this X-coefficient (which is special in that it has two rows the same) vanishes for $\nu_0 + \nu_1 + \nu_2$ an odd integer. This result can be interpreted as the quantum analog to the vanishing of $P_{\nu_0\nu_1\nu_2}(\hat{\mathbf{J}}\hat{\mathbf{l}}_1\hat{\mathbf{J}}')$ for $\nu_0 + \nu_1 + \nu_2$ odd, since the three angular momenta must be coplanar.

Further properties of the X-coefficient will be found in refs. *23, 26, 27 28*, and in the references cited in the list of tables.

It is not difficult to amend this result to take into account initial and final states with nonzero, but unobserved spin. In fact, the required result is readily seen to be exactly the same as discussed for the general initial and final links in the direction-direction correlation. In both cases, it is the cylindrical averaging about $\mathbf{J}$ (or $\mathbf{J}'$) responsible for this simplification.

The result for the triple correlation of three spinless radiations with the angular momentum sequence $j_i(l_0)J(l_1)J'(l_2)j_f$ is then:

Classically:

$$W(\hat{\mathbf{k}}_0\hat{\mathbf{k}}_1\hat{\mathbf{k}}_2) = (4\pi)^{-3} \sum_{\nu_0\nu_1\nu_2} (2\nu_0+1)(2\nu_1+1)(2\nu_2+1)$$
$$\cdot [P_{\nu_0}(\hat{\mathbf{k}}_0 \cdot \hat{\mathbf{l}}_0)P_{\nu_0}(\hat{\mathbf{l}}_0 \cdot \hat{\mathbf{J}})] \cdot [P_{\nu_1}(\hat{\mathbf{k}}_1 \cdot \hat{\mathbf{l}}_1)P_{\nu_0\nu_1\nu_2}(\hat{\mathbf{J}}\hat{\mathbf{l}}_1\hat{\mathbf{J}}')]$$
$$\cdot [P_{\nu_2}(\hat{\mathbf{k}}_2 \cdot \hat{\mathbf{l}}_2)P_{\nu_2}(\hat{\mathbf{l}}_2 \cdot \hat{\mathbf{J}}')] \cdot P_{\nu_0\nu_1\nu_2}(\hat{\mathbf{k}}_0\hat{\mathbf{k}}_1\hat{\mathbf{k}}_2). \quad (40a)$$

[1] This result is a generalization of the familiar result that $\mathbf{A} \cdot \mathbf{B} \times \mathbf{C}$ (that is, P_{111}) vanishes if the three vectors are coplanar. For the general case we need only note that if the three vectors are coplanar, an inversion is equivalent to a rotation by 180° about an axis perpendicular to the plane. The inversion multiplies $P_{\nu_0\nu_1\nu_2}$ by $(-)^{\nu_0+\nu_1+\nu_2}$, while the rotation changes nothing. Thus we conclude that $P_{\nu_0\nu_1\nu_2}$ vanishes if the vectors are coplanar and $\nu_0 + \nu_1 + \nu_2$ is odd.

[m] The proof of Eq. (39) will be published elsewhere.

Quantum mechanically:

$$W(\hat{\mathbf{k}}_0\hat{\mathbf{k}}_1\hat{\mathbf{k}}_2) = (4\pi)^{-3}(2l_1+1)(-)^{j_i-j_f-J+l_1+J'}$$
$$\cdot \sum_{\nu_0\nu_1\nu_2} [(2\nu_0+1)(2\nu_1+1)(2\nu_2+1)]^{\frac{1}{2}} i^{\nu_2-\nu_1-\nu_0}$$
$$\cdot \bar{Z}(l_0Jl_0J;j_i\nu_0)\bar{Z}(l_2J'l_2J';j_f\nu_2)(l_1l_100|\nu_10)X\begin{pmatrix}Jl_1J'\\Jl_1J'\\\nu_0\nu_1\nu_2\end{pmatrix} \quad (40b)$$
$$\cdot P_{\nu_0\nu_1\nu_2}(\hat{\mathbf{k}}_0\hat{\mathbf{k}}_1\hat{\mathbf{k}}_2).$$

(The phase factor in the above result is always ± 1, since $\nu_2 - \nu_1 - \nu_0$ is always an even integer.)

This result can be amended to take into account further complications, such as particles with spin, interference, and the like. Since our purpose in this section is only to illustrate the structure of the triple correlation, we shall defer such cases to the various examples discussed in detail in Section 2. It might be worthwhile to point out, however, that in these more complicated cases it is often quite worthwhile to use a spherical diagram to write out the diagonal term of the desired correlation. Such a diagram, for example, is helpful in deciding the most advantageous coupling scheme for the angular momenta involved, and indicates in a direct way transformations between coupling schemes.

As mentioned at the beginning of this section, the measurement of a spin polarization in addition to two directions of motion is, in a formal sense, an example of a triple correlation. Generally this similarity is somewhat obscured by the fact that (1) the polarization measurement is referred to the particle's direction of motion [which reduces the general $P_{\nu_0\nu_1\nu_2}$ functions to the form of Eq. (36)] and (2) a general spin polarization measurement defines $2s$ vectors (s = spin), so that only for leptons and photons do we get a complete equivalence (description by three vectors).

The problems presented by spin polarization measurements are best treated by the techniques discussed in the next subsection. Applications of the formalism to polarization correlations are given in Section 2.

g. THE DENSITY MATRIX AND STATISTICAL TENSOR FORMULATION

There is another, and very fruitful, way in which we may regard the correlation process, a point of view extensively developed by Fano (*26,29*). In this method we concentrate our attention on the geometric properties of the intermediate state; thus the formation of the state J by a unidirectional radiation with fixed angular momentum is to be considered as giving *information* about the angular properties of the state J (its "orientation," populations, and the like). The subsequent decay of this state then yields an angular distribution that directly results from the angular properties of the state J itself, the process, in fact, being completely symmetric as regards the two radiations. The essential point is, however, that by focusing attention on the general properties of the intermediate state J as derived from partial information, one is led to consider a problem more significant than just the angular correlation process, as we shall see.

The information on the intermediate state produced in an angular correlation is just that discussed several times in the preceeding subsections: conservation of angular momentum $(\mathbf{l} + \mathbf{j}_i + \mathbf{J} = \mathbf{0})$, that $\mathbf{l}$ is perpendicular to $\mathbf{k}$, that $\mathbf{j}_i$ is randomly oriented. One notes that this information includes the statement that the system is *symmetric* about the $\mathbf{k}$ direction. Classically if one wishes to discuss the angular properties of a system symmetric about an axis one would introduce the Legendre polynomials, taking the z-axis to lie along the axis of symmetry. That is:

$$f(\vartheta) = \frac{1}{4\pi} \sum_{\nu} (2\nu + 1)[P_\nu(\hat{\mathbf{k}} \cdot \hat{\mathbf{J}})]_{\text{av}} P_\nu(\hat{\mathbf{k}} \cdot \hat{\mathbf{J}}). \tag{41}$$

We have already determined $[P_\nu(\hat{\mathbf{k}} \cdot \hat{\mathbf{J}})]_{\text{av}}$ under the restrictions imposed by our information.

It is clear that a knowledge of the coefficients $[P_\nu(\hat{\mathbf{k}} \cdot \hat{\mathbf{J}})]_{\text{av}}$ constitutes complete angular information on the distribution $f(\vartheta)$. But $f(\vartheta)$ is just the classical analog of the *density matrix* for the intermediate state, and we may interpret Eq. (41) in a significant way as an *operator expansion of the density matrix for axially symmetric states*. The operators here are the familiar Legendre functions. The coefficients $[P_\nu(\hat{\mathbf{k}} \cdot \hat{\mathbf{J}})]_{\text{av}}$ (the expectation value of the operators) were given the name "tensor parameters" or "statistical tensors" by Fano and the symbol $R_\nu{}^q(JJ)$.

Now let us write down the quantum mechanical analog to Eq. (41). As before we have:

$$[P_\nu(\hat{\mathbf{k}} \cdot \hat{\mathbf{J}})]_{\text{av}} = P_\nu(\hat{\mathbf{l}} \cdot \hat{\mathbf{J}}) P_\nu(\hat{\mathbf{l}} \cdot \hat{\mathbf{k}})$$
$$\sim (-)^l \left[\frac{2l+1}{2\nu+1}\right]^{\frac{1}{2}} (ll00|\nu0)[(2l+1)(2J+1)]^{\frac{1}{2}} \cdot (-)^\nu W(l\nu j_i J; lJ),$$
$$\equiv (4\pi)[(2\nu+1)(2J+1)]^{-\frac{1}{2}} R_\nu{}^0(JJ). \tag{42}$$

But we must also note this time that the operators (the quantum analog to the Legendre functions) measure now the angle between the quantized vector $\mathbf{J}$ and the unquantized vector $\mathbf{k}$; thus it is the Wigner coefficients which are to be introduced. The quantum equivalent to Eq. (41) is therefore:

$$[f(\vartheta)]_{\text{QM}} = \sum_\nu (-)^{J-m} R_\nu{}^0(JJ)(JJm - m|\nu 0),$$
$$\equiv d_{mm}. \qquad (43)$$

The d_{mm} appearing in Eq. (43) are the diagonal elements of the density matrix (diagonal because we have considered states which are cylindrically symmetric).

Equation (43), or rather its more general form for states that are not axially symmetric,

$$d_{mm'} = \sum_{\nu q} (-)^{J-m} R_\nu{}^q(JJ)(JJm - m'|\nu - q), \qquad (44)$$

is Fano's parametrization of the density matrix. [It should be noted that the density matrix in Eq. (44) is normalized such that trace $(\mathbf{d}) = 2J + 1$.] We see that this is no more and no less than an *operator expansion* of the density matrix. One sees, moreover, that the Wigner coefficients in Eq. (44) appear in a new light, *as an explicit matrix formulation of the proper multipole operators*.[n] To exhibit this more clearly let us use the symmetry properties of the Wigner coefficients to rewrite the operator $(JJm - m'|\nu - q)$, that is,

$$(-)^{J-m}(JJm - m'|\nu - q) = (2\nu + 1)^{\frac{1}{2}}(2J + 1)^{-\frac{1}{2}}(J\nu mq|Jm')$$
$$\equiv [(2\nu + 1)/J(J + 1)(2J + 1)]^{\frac{1}{2}}(\mathbf{S}_q{}^{(\nu)})_{mm'}. \qquad (45)$$

To emphasize the multipole operator aspect, we have used $\mathbf{S}_q{}^\nu$ as a symbol for the νth multipole operator (qth component) in accord with the vector spin operators $\mathbf{S}$ in spherical notation $\mathbf{S}_q{}^{(1)}$. These multipole spin operators are orthogonal, upon taking traces (the analog of integrating over angles for the spherical harmonic operators).

Returning to the particular case at hand we note once again that it was the symmetry of our problem about the k-axis which implied the result that *only* $R_\nu{}^0$ differs from zero. Stated in another way, this means that the density matrix of the intermediate state is diagonal (*31*) when we choose our coordinates along $\mathbf{k}$, and—using still another and very convenient description—we may describe the state J by a "population." The population P_m of the mth magnetic sublevel is given by:

$$P_m = d_{mm} = \frac{1}{4\pi} \sum_\nu (2\nu + 1)[J(J + 1)]^{-\frac{1}{2}}[P_\nu(\hat{\mathbf{k}} \cdot \hat{\mathbf{J}})]_{\text{av}}(\mathbf{S}_0{}^{(\nu)})_{mm}. \qquad (46)$$

Many properties of the angular distribution have a direct significance in terms of populations. For example, the restriction that ν be even requires

[n] This is discussed further in (*30*).

that $P_m = P_{-m}$. Similarly, as noted quite early by Christy, populations that vary according to $(Jvm0|Jm)$ define but a single value of ν in the angular distribution.

Turning our attention once again to Eq. (42), we can now give an interpretation of this equation, in terms of "adding information." The information on the initial state, $\mathbf{j}_i$, is that $\mathbf{j}_i$ is random; in the language of statistical tensors this is the statement that

$$R_\nu{}^q(j_i j_i) = \delta_\nu{}^0 \delta_q{}^0 (2j_i + 1)^{\frac{1}{2}}; \qquad (47)$$

that is, "no angular information." The information on the radiation is the statement that it is observed as a plane wave along $\mathbf{k}$, with $\hat{\mathbf{l}} \cdot \hat{\mathbf{k}} = 0$, and known (dynamically) to involve only angular momentum l; in the language of statistical tensors, this becomes the statement:

$$R_\nu{}^q(ll; \text{radiation}) = \delta_q{}^0 (-)^l (4\pi)^{-1} (2l + 1)(ll00|\nu 0). \qquad (48)$$

Thus Eq. (42) takes the form:

Classically:

$$[P_\nu(\hat{\mathbf{k}} \cdot \hat{\mathbf{J}})]_{\text{av}} = P_\nu(\hat{\mathbf{k}} \cdot \hat{\mathbf{l}}) \cdot P_\nu(\hat{\mathbf{l}} \cdot \hat{\mathbf{J}}) \qquad (49\text{a})$$

Quantum mechanically:

$$R_\nu{}^0(JJ) = R_\nu{}^0(ll; \text{radiation})(2J + 1)(-)^\nu \cdot W(l \nu j_i J; lJ). \qquad (49\text{b})$$

We have now obtained in Eq. (49b) another, and more formal, way of regarding the Racah coefficients. The Racah coefficients are the means of coupling two statistical tensors: the statistical tensor of a random state ($\mathbf{j}_i$) with the statistical tensor of a nonrandom state (1) to produce the statistical tensor of the coupled system ($\mathbf{J} = \mathbf{l} + \mathbf{j}_i$).

Generalizing to the case for coherent interference Fano gives,

$$R_\nu{}^q(JJ') = R_\nu{}^q(ll')[(2J + 1)(2J' + 1)]^{\frac{1}{2}} \cdot W(l \nu j_i J'; l'J). \qquad (50)$$

It is very natural at this point to inquire how to couple two nonrandom systems, $\mathbf{j}_0 + \mathbf{j}_1 = \mathbf{J}$. The result gives an additional significance to the X-coefficient:

$$R_{\nu_2}{}^{q_2}(JJ') = \sum_{\substack{\nu_0 q_0 \\ \nu_1 q_1}} [(2J+1)(2J'+1)(2\nu_0+1)(2\nu_1+1)]^{\frac{1}{2}}$$

$$\cdot (\nu_0 \nu_1 q_0 q_1 | \nu_2 q_2) R_{\nu_0}{}^{q_0}(j_0 j_0') R_{\nu_1}{}^{q_1}(j_1 j_1') X \begin{pmatrix} j_0 & j_1 & J \\ j_0' & j_1' & J' \\ \nu_0 & \nu_1 & \nu_2 \end{pmatrix}. \qquad (51)$$

Equation (50) is then a special case of Eq. (51). (Equation (51), except for notational changes, is Eq. (18.14) of ref. *27*.)

Whenever our observation can be expressed in terms of populations, we can conclude that only R_ν^0 enters for coordinates whose z-axis is along the axis of symmetry. An interesting, and important, application of these ideas concerns the angular distribution of radiations from oriented nuclei. The populations of the various sublevels determine the tensor parameters $R_\nu^0(JJ) \sim [P_\nu(\hat{\mathbf{k}} \cdot \hat{\mathbf{J}})]_{\text{av}}$ for the state J, Eq. (46). These in turn can be used to determine the directional correlation (with symmetry axis of the state J as one of the directions) of any subsequent radiation; this employs Eqs. (8) and (15). The angular distribution of radiations from oriented nuclei is completely similar to the usual correlation problem except for the technical difference that the population of the state J (now the initial state) is produced by nuclear orientation, rather than by an initial radiation along a given direction. For example, when the nucleus is polarized by dipole (hyperfine) coupling (32) the population of the mth magnetic sublevel has the formal value (where m_e is the magnetic quantum number of the electron and H is the interaction Hamiltonian),

$$P_m = \sum_{m_e} (mm_e|\exp(-H/kT)|mm_e) \cdot \left\{ \sum_{mm_e} (mm_e|\exp(-H/kT)|mm_e) \right\}^{-1}.$$

(52)

An example of an angular distribution for a radiation emitted by a polarized nucleus is given in Section 2.

One major advantage of the nuclear orientation technique is that the populations thus obtained may have $P_m \neq P_{-m}$. Thus a circular polarization measurement on the subsequent radiation alone may yield odd Legendre coefficients in the distribution. As a result the information obtained from orientation experiments (for example, the celebrated experiment of Wu, *et al.*) has heavily outweighed the disadvantage that the technique is applicable to relatively few nuclei.

h. Coherent and Incoherent Interference

The purpose of the semiclassical procedure of the previous subsections was to display the structure of the correlation process as simply as possible without the confusion of extraneous detail. When we try to extend this classical treatment to coherent interference we are forcing our model too hard, for coherent interference depends upon phases, and in approaching the classical limit the phase oscillates rapidly, yielding the classical motion as the region of stationary phase. (The Coulomb excitation problem shows this very nicely, see refs. *21, 22*.) We cannot expect, therefore, to find anything particularly clarifying by pursuing the classical model further.

Fortunately, the quantum mechanical result for interference is rather easily understood directly. First let us consider incoherent interference.

Such interference occurs, for example, whenever a given angular momentum has several possible values, but is *randomly oriented*. A typical example is the channel spin j_s. The general case for several possible channel spin magnitudes is simply the sum of correlations for each possible j_s, each weighted according to the probability of that value of j_s (see examples in Section 2). The angular distribution produced by this incoherent mixing will have coefficients for the Legendre series which, when normalized, will have magnitudes less than or equal to those for a single channel spin. For this reason the distributions given by the separate angular momenta j_s, which are to be mixed incoherently, are often called the "extreme" angular distributions.

Coherent interference on the other hand results from definite phase relationships. For example, the various orbital angular momenta in a plane wave have definite phase [zero relative phase for a free particle when observed along the direction of motion, using $i^l Y_l^m(\vartheta\varphi)$ for our basis functions]. Thus if several orbital angular momenta can participate in a correlation, we must in general expect coherent interference. If the particle observed in a state of definite linear momentum is not free (not a "plane wave") but is under the influence of Coulomb or nuclear fields say, then the phase relationships between the orbital angular momenta are those of a distorted plane wave and these relative phases (designated as $\xi_l - \xi_{l'}$) will now influence the angular distribution through coherent interference.

It is convenient for practical purposes to distinguish two types of coherent interference that occur in angular correlations. There is first, the coherent interference between various orbital angular momenta for the radiations, mentioned above—designated for brevity as "mixed radiations." Second, there is the coherent interference produced when the cascade or reaction proceeds through several nuclear states of various spins and parities. Nuclear reactions provide the most frequent cases for this latter interference, usually also with mixed radiations. In gamma-gamma correlations the nuclear states are generally sharp; the interference is that of mixed radiations.

It is instructive to consider at this point the relationship of the j_s and j_p schemes for nuclear reactions. Unlike j_s, the j_p scheme shows, in general, *coherent* interference; for with $\mathbf{j}_p + \mathbf{s} + \mathbf{l} = 0$, the randomness of the orientation of **s** is not enough, in general, to keep $\mathbf{j}_p$ and **l** from being coherent.º It is clear that a general treatment of angular correlation for nuclear reactions using the j_p scheme would be appreciably more complicated.

An interesting situation which shows just the opposite behavior occurs for a correlation that proceeds through an unobserved emission of a

º The case of parallel angular momenta is the exception.

particle with spin. For example, consider an inelastic proton excitation leading to a gamma ray, a $(p,p'\gamma)$ process. The angular momenta for this process can be schematized:

$$j_s \binom{l}{l'} J(l_p)J'(L)j_f.$$

In the initial link we have j_s incoherent, and l coherent as before. The final link has been taken to be pure for simplicity, as also for J and J'. The problem concerns the intermediate (unobserved) proton emission. Here the angular momenta are: $\mathbf{l}_p + 1/2 + \mathbf{J} + \mathbf{J}' = 0$. Since now both $\mathbf{l}_p$ and $\mathbf{s}_p$ ($= 1/2$) are randomly oriented, it follows that it is now the j_p scheme which mixes incoherently, and is therefore the simpler. The channel spin for this link, being given by the vector sum of $\mathbf{J}'$ and $\mathbf{s}_p$, is clearly coherent since information on $\mathbf{J}'$ results from the observation of the gamma ray.

A similar situation occurs if a heavy particle (with spin s) is emitted as the first step in a cascade. The angular momenta of this first step are: $j_i + 1 + \mathbf{s} + \mathbf{J} = 0$. Here the "channel spin," $\mathbf{j}_s$, is, strictly speaking, given by $\mathbf{j}_s + \mathbf{s} + \mathbf{J} = 0$, and clearly *this* channel spin is not random. We can avoid this difficulty by using an alternative order for coupling the angular momenta; that is, we introduce: $\mathbf{j}_s' + \mathbf{j}_i + \mathbf{s} = 0$, to define a fictive channel spin, j_s'. This vector is now randomly oriented, and the correlation formulas read then just as before, using this j_s'.

The consideration of coherent processes focuses attention on the fact that the quantum mechanical treatment is bilinear, and each of the "quantum mechanical angle functions"—the Racah and Wigner coefficients—relates to the angular momenta of two (not necessarily different) states. The previous semiclassical treatment because it considered only pure states, could ignore this bilinearity. The one other case, besides sharp states, where we may ignore bilinearity is the random case, where we have incoherent mixing as discussed previously.

The changes necessary in the quantum mechanical angle functions are easily incorporated. In place of the coefficient $(-)^l \left[\dfrac{2l+1}{2\nu+1}\right]^{\frac{1}{2}}$ $(ll00|\nu 0) \sim P_\nu(\hat{\mathbf{l}} \cdot \hat{\mathbf{k}})$, we now have the more general coefficient

$$(-)^l \left[\frac{2l+1}{2\nu+1}\right]^{\frac{1}{2}} (ll'00|\nu 0).^\text{p}$$

[p] The classical limit of this coefficient is: $\left(\dfrac{4\pi}{2\nu+1}\right)^{\frac{1}{2}} \cdot Y_\nu^{l-l'}(\pi/2, 0)$. The perpendicularity is again given by the angle $\pi/2$, but otherwise the classical limit does not appear very useful in suggesting interpretations.

Similarly the Racah coefficient,

$$(-)^\nu[(2l+1)(2J+1)]^{\frac{1}{2}}W(l\nu j_i J; lJ) \sim P_\nu(\hat{\mathbf{l}} \cdot \hat{\mathbf{j}})$$

becomes now the more general coefficient[q]

$$(-)^\nu[(2l+1)(2J+1)]^{\frac{1}{2}}W(l\nu j_i J; l'J'),$$

where we have also made allowance for the fact that the intermediate nuclear state J may show coherent interference. The correlation coefficients, the Z and F coefficients, have already been defined in a manner suitable for the general case, Eqs. (13) and (28).

The structure of the general correlation formula, including coherent interference, is best illustrated by the angular distribution formula for nuclear reactions (33). The differential cross section for the reaction in which a pair of particles designated by α_1 (specifically the target nucleus and incident particle) having channel spin, designated by s_1, react and form another pair of particles, α_2, with channel spin s_2, is given by the formula,

$$d\sigma(\alpha_2 s_2; \alpha_1 s_1) = \frac{\lambda^2_{\alpha_1}}{2s_1+1} \sum_{\nu=0} B_\nu(\alpha_2 s_2; \alpha_1 s_1) P_\nu(\cos\vartheta) d\Omega, \qquad (53\text{a})$$

where

$$B_\nu(\alpha_2 s_2; \alpha_1 s_1) = \frac{(-)^{s_2-s_1}}{4} \sum \bar{Z}(l_{1a}J_a l_{1b}J_b; s_1\nu)\bar{Z}(l_{2a}J_a l_{2b}J_b; s_2\nu)$$
$$\cdot \operatorname{Re}\{(\mathbf{1} - \mathbf{S}^{J_a\Pi_a})^*_{\alpha_2 s_2 l_{2a}; \alpha_1 s_1 l_{1a}} \cdot (\mathbf{1} - \mathbf{S}^{J_b\Pi_b})_{\alpha_2 s_2 l_{2b}; \alpha_1 s_1 l_{1b}}\}. \qquad (53\text{b})$$

(The sum in (53b) is over $J_a, \Pi_a, J_b, \Pi_b, l_{1a}, l_{2a}, l_{1b}, l_{2b}$.)

The typical term of this formula is represented by the angular momentum scheme:

$$s_1 \binom{l_{1a}}{l_{1b}} \begin{matrix} J_a \\ J_b \end{matrix} \binom{l_{2a}}{l_{2b}} s_2.$$

For the first link in the reaction (subscript 1), we have the general coefficient $\bar{Z}(l_{1a}J_a l_{1b}J_b; s_1\nu)$, similarly the second link introduces a general $\bar{Z}$ coefficient. These two links make a contribution to the correlation as weighted by the dynamical factor

$$\operatorname{Re}\{(\mathbf{1} - \mathbf{S}^{J_a\Pi_a})^*_{\alpha_2 s_2 l_{2a}; \alpha_1 s_1 l_{1a}} \cdot (\mathbf{1} - \mathbf{S}^{J_b\Pi_b})_{\alpha_2 s_2 l_{2b}; \alpha_1 s_1 l_{1b}}\},$$

where $\mathbf{S}$ is the scattering matrix. In the examples of Section 2 we approximate the scattering matrix by Breit-Wigner dispersion terms.

An important feature of the correlation process for pure states was

---

[q] The classical limit here (a Jacobi polynomial) is again not very easily interpreted except for pure states.

the fact that the separate links factored. How general is this property? The angular distribution formula, Eq. (53), rather obscures the actual generality of this property, for, as written there, the links are closely tied together through the various elements of the **S** matrix. Now the elements of the **S** matrix are not all independent, for we have two basic restrictions: (1) the unitary condition (conservation of flux) and (2) symmetry (reciprocity). These conditions imply that the S matrix may be written in the form:

$$\mathbf{S}^{J\Pi} = \mathbf{U}^{-1}{}_{J\Pi} \exp(2i\mathbf{\Delta}_{J\Pi})\mathbf{U}_{J\Pi},$$

where **U** is real and orthogonal, and **Δ** is real and diagonal. The $N \times N$ matrix **U** defines N real eigenvectors, $\mathbf{U}_k$, where $k = 1, 2, \ldots, N$; utilizing these one may write the **S** matrix in the following (spectral) form, and obtain

$$B_\nu(\alpha_2 s_2; \alpha_1 s_1) = (-1)^{s_2-s_1} \sum_{J_a\Pi_a} \sum_{J_b\Pi_b} \sum_{k_a=1}^{N} \sum_{k_b=1}^{N}$$
$$\times \sin(\delta_{J_a\Pi_a k_a}) \cdot \sin(\delta_{J_b\Pi_b k_b})$$
$$\times \cos(\delta_{J_a\Pi_a k_a} - \delta_{J_b\Pi_b k_b}) \cdot T, \quad (54a)$$

$$T = \Big[\sum_{l_{1a}l_{2a}} \bar{Z}(l_{1a}J_a l_{1b}J_b; s_1\nu) U(J_a\Pi_a k_a)_{\alpha_1 s_1 l_{1a}} U(J_b\Pi_b k_b)_{\alpha_1 s_1 l_{1b}} \Big]$$
$$\times \Big[\sum_{l_{2a}l_{2b}} \bar{Z}(l_{2a}J_a l_{2b}J_b; s_2\nu) U(J_a\Pi_a k_a)_{\alpha_2 s_2 l_{2a}} U(J_b\Pi_b k_b)_{\alpha_2 s_2 l_{2b}} \Big]. \quad (54b)$$

This formula, although quite complicated in appearance, is actually simpler than Eq. (53), for it has fewer subsidiary conditions on the terms that enter. (The remaining conditions are that the eigenvectors $\mathbf{U}_k$ be orthonormal.) We note the following features of this expression:

(1) The intermediate state occurs bilinearly and each time is characterized by three numbers, J, Π, and k (the latter designating the particular eigenvector of the $N \times N$ scattering matrix).

(2) The two links in the correlation ("entrance" and "exit" channels in nuclear reaction terminology) enter through the T coefficient which factors into two separate parts, and the elements of the T coefficient are all explicitly *real*. For the special case where the intermediate state is *sharp* (that is to say it has a definite value of J, Π and k), one sees that the entire coefficient B_ν factors. This is the typical case for gamma-gamma cascades, and greatly simplifies the discussion. It should be pointed out in connection with cascades, that unlike a reaction, the physical parameters (U,δ) in the two links of Eq. (54) refer now to *different* nuclear states, so that the spectral form is necessary in describing the correlation (see ref. *20*).

The detailed application of these formulas is discussed in Section 2.

i. Some Concluding Remarks

The semiclassical methods presented in the preceding section are of advantage both in furnishing an intuitive feeling for angular correlations, and in providing a mnemonic for writing out complicated correlation formulas with a very direct view of the coupling relations. The method suffers, however, from the drawback that it is unambiguous only for *pure* states. The rule we have used for coherent interference is (a) to write out (symmetrically) the analogous mixed correlation by letting $L, L'; \ldots$ differ in the various Wigner and Racah coefficients and (b) weighting the various correlations by (complex) coefficients (for example, the scattering matrix elements), normalized by $[(2J + 1)(2J' + 1)]^{\frac{1}{2}}$ for each resonant nuclear state. The resulting angular distribution is explicitly real, and a valid parametrization of the most general angular correlation for the process considered. It is, however, *not unambiguous*, for phases such as $(-)^{L+L}$, $(-)^\nu$, ... (necessarily plus or minus only) may be arbitrarily inserted. Such an ambiguity is essential, for it corresponds to the freedom of redefining the meaning to be assigned to the coefficients that weight the various mixed correlations. The origin of this "phase problem" can be found in the fact that the coupling of angular momenta is dependent on order—thus $j + 1 = \mathbf{J}$ differs from $1 + j = \mathbf{J}$, and the corresponding reduced matrix elements (the weighting factors for the mixed correlations) may differ in sign. (The phase $(-)^\nu$ can also enter in a quite different way, through the replacement of emission by absorption, in relating two otherwise similar correlations.)

It is clear that recourse to more fundamental methods is necessary in order to write out mixed correlations with uniquely defined coupling schemes. This has been done in the present chapter for the two most important cases: the general angular distribution formula for nuclear reactions, Eq. (54); and in the gamma-gamma correlation, where the phase problem is explicitly discussed on p. 771,ff. Unfortunately, the literature is not always explicit or correct in treating this problem (cf. the recent discussions by R. Huby, *33a*). This precaution is particularly necessary in interpreting (mixed) triple correlation formulas.

2. Application of Angular Correlations to Specific Cases

It is the purpose of the present section to discuss in detail, by examples, the actual application of the formulas for angular correlation to specific cases. The experimental situations generally fall into two types: correlations proceeding through nuclear states each of definite angular momentum and parity, as is the case with gamma-gamma cascades;

and correlations proceeding through several, coherently interfering, nuclear states, as is generally the case for reactions involving particles.

a. PURE NUCLEAR STATES

(1) *Direction-Direction Correlation*

(a) *Gamma Rays.* The gamma-gamma correlation was the earliest application of angular correlation and still remains the most common. The Lloyd parametrization result allows us, however, to regard the gamma-gamma process even more significantly as the basic, or prototype, angular correlation, from which all other correlations involving the same angular momenta but different particles may be obtained.[r]

(i) *Pure radiations.* Let us consider first the gamma-gamma correlation for gamma rays of definite multipolarity ("pure-pure" case), with the angular momentum scheme $j_1(L_1)J(L_2)j_2$. For this case, it makes no difference whether the physical process is a cascade, or a reaction—for example, the inelastic scattering of the gamma ray, or even (for $j_1 = j_2$) resonance fluorescence, provided only that the restriction to definite angular momenta makes sense physically.

As discussed in Section 1, the correlation breaks into two separate links, one for each observed gamma ray. The correlation is:

$$W(\vartheta) = \sum_\nu A_\nu(1) A_\nu(2) P_\nu(\cos \vartheta), \tag{55}$$

where the coefficients A_ν refer to the separate links. Here $A_\nu(1) = F_\nu(L_1 j_1 J)$ and $A_\nu(2) = F_\nu(L_2 j_2 J)$.

Example. 4(2)2(2)0

$$\begin{aligned}W(\vartheta) &= 1 + F_2(242)F_2(202)P_2(\cos \vartheta) + F_4(242)F_4(202)P_4(\cos \vartheta) \\ &= 1 + (-0.17075)(-0.59761)P_2(\cos \vartheta) + (-0.00848) \\ &\qquad\qquad\qquad\qquad\qquad\qquad\qquad\qquad (-1.06904)P_4(\cos \vartheta) \\ &= 1 + 0.10204 P_2(\cos \vartheta) + 0.00907 P_4(\cos \vartheta). \end{aligned} \tag{56}$$

The tables of F_ν which give a rational fraction for F_ν^2 provide an automatic check for the pure-pure case, in that the final result involves no square roots. Thus in the example we would get:

$$W(\vartheta) = 1 + \tfrac{5}{49}P_2 + \tfrac{4}{441}P_4. \tag{57}$$

[r] The spinless particle case might seem to be more elementary, and thus a better choice for the standard correlation, but the connection in this case between orbital angular momentum and parity eliminates the mixing of angular momentum L with $L+1$ (which can occur for gamma rays) and is thus unsatisfactory.

It should be noted that although F_ν is generally ≤ 1, occasionally [as in this example for $F_4(202)$] it exceeds unity.

(ii) *Mixed transitions.* The situation for a two stage gamma-gamma correlation with mixed radiations is not very much more difficult than the pure-pure case; again we may factor the correlation into two links. The correlation still takes the form of Eq. (55), but now the coefficients A_ν consist of several terms, depending, however, only on the constants of each link.

Consider the mixed-mixed correlation with L_1 and L_1' mixing coherently in the first link, and similarly for L_2, L_2' in the second link; we denote this schematically by $j_1 \begin{pmatrix} L_1 \\ L_1' \end{pmatrix} J \begin{pmatrix} L_2 \\ L_2' \end{pmatrix} j_2$.

The coefficient $A_\nu(1)$ is now given by:

$$A_\nu(1) = F_\nu(L_1 j_1 J) + 2\delta_1 F_\nu(L_1 L_1' j_1 J) + \delta_1^2 F_\nu(L_1' j_1 J), \quad (58)$$

and similarly,

$$A_\nu(2) = F_\nu(L_2 j_2 J) + 2\delta_2 F_\nu(L_2 L_2' j_2 J) + \delta_2^2 F_\nu(L_2' j_2 J). \quad (59)$$

The correlation shows the significant new feature of coherent interference for each mixed radiation, proportional to the nuclear parameters δ—the "mixing coefficients"—and sensitive to their phase (restricted to be $+$ or $-$, only).

The mixed correlation coefficients, $F_\nu(LL'jJ)$, vanish for $\nu = 0$; the total intensity (the coefficient of P_0) is thus normalized to $(1 + \delta_1^2) \times (1 + \delta_2^2)$. [Note that the pure correlation coefficients $F_\nu(LjJ)$ are often written as $F_\nu(LLjJ)$.] In general, the correlation is quite sensitive to the mixing parameters δ. Since these parameters are to be interpreted eventually in terms of a nuclear model it is essential to state explicitly their definition in terms of the nuclear states. As given here, δ^2 is defined as a relative intensity, that is,

$$\delta^2 = \frac{\text{Intensity of radiation } L'}{\text{Intensity of radiation } L}$$

$$= \left(\frac{(j_1|L'|J)}{(j_1|L|J)} \right)^2. \quad (60)$$

Here we have used the ratio of the reduced matrix elements, $(j_1|L|J)$, which are explicitly defined by:

$$<j_1 m_1|T(LM)|Jm> \equiv (j_1|L|J)(JLmM|j_1 m_1) \quad (61)$$

where $T(LM)$ is the appropriate multipole operator effecting the transition.[a] For gamma rays the operators are:

$$T(LM) = \mathbf{j}_{op} \cdot \mathbf{A}^{e,m}(LM)^* \quad (62)$$

[a] Note that this definition differs from that of ref. *20;* compare their footnote 6.

where $\mathbf{j}_{op}$ is the nuclear current operator and the $\mathbf{A}^{e,m}(LM)$ are the electric or magnetic (e,m) standing wave vector potentials (34) normalized to 1 quantum/sec. We have assumed that the wave functions for the nuclear states obey the time reversal law,

$$[\psi(j,m)]_{\text{time reversed}} = (-)^{j-m}\psi(j,-m), \tag{63}$$

and similarly for the vector potentials. With this convention, the reduced matrix elements are explicitly real.

Not only the magnitude, but the sign of δ, can be measured; the definition of δ is taken to be:

$$\delta \equiv \frac{(j_1|L'|J)}{(j_1|L|J)}. \tag{64}$$

Note that this introduces a *standard form for the reduced matrix elements* in which the intermediate state J appears on the right regardless of whether this intermediate state is either the initial or final state for the transition.[t] (It is a useful mnemonic to note that the same convention is adopted in the definition of the F_ν, where the intermediate state is denoted by the last argument.)

Example. $\frac{7}{2}(2)\frac{3}{2}\binom{1}{2}\frac{3}{2}$

$$A_\nu(1) = F_\nu(2\tfrac{7}{2}\tfrac{3}{2}) \tag{65a}$$
$$A_\nu(2) = F_\nu(1\tfrac{3}{2}\tfrac{3}{2}) + 2\delta F_\nu(12\tfrac{3}{2}\tfrac{3}{2}) + \delta^2 F_\nu(2\tfrac{3}{2}\tfrac{3}{2}) \tag{65b}$$
$$W(\vartheta) = (1+\delta^2) + P_2 \cdot [-0.14286][-0.40000 + 2\delta(-0.77460) + \delta^2 \cdot 0]$$
$$= 1 + \delta^2 + P_2 \cdot (0.05714 + \delta 0.22132). \tag{65c}$$

The anisotropy is clearly very sensitive in this example to the param-

---

[t] This convention is not always used; in particular Lloyd uses the convention that the initial state always appears on the right. Since the δ's obey the rule:

$$\frac{(J|L'|j_1)}{(J|L|j_1)} = (-)^{L-L'}\delta, \tag{65}$$

Lloyd's convention would introduce a sign $(-)^{L-L'}$, depending upon whether this radiation was temporally first or second. The convention (of ref. 20) adopted here avoids this dissymmetry.

However, an annoying sign can still occur in special cases, such as the decay of Co^{60} by a $\beta - \gamma - \gamma$ cascade. If the mixing ratio δ for the first gamma ray is determined by the $\beta - \gamma$ correlation, then in the $\gamma_1 - \gamma_2$ cascade the same mixing ratio δ for the first gamma ray now enters with *opposite* sign. The origin of this sign change is found in the fact that different states are intermediate for the two correlations.

The author is indebted to Dr. G. R. Satchler for this example. Quite recently another case has been given by S. Ofer (34a) for the decay of Tb^{156} (5-day).

eter δ. (This example also illustrates the curious "accidental" vanishing of $F_2(2\frac{3}{2}\frac{3}{2})$, which means that any correlation involving a quadrupole transition between two states of spin $\frac{3}{2}$ is isotropic. Thus $\frac{7}{2}(2)\frac{3}{2}(2)\frac{3}{2}$ is isotropic though $\frac{7}{2}(2)\frac{3}{2}(1)\frac{3}{2}$ is not.)

(*iii*) *Cascades with unobserved intermediate gamma rays.* If we consider a cascade of gamma rays, in which only the initial and final gamma ray is observed, the correlation has a fairly simple relation to the standard gamma-gamma case, as discussed in Section 1. The characteristic feature introduced by the unobserved transitions are: (a) For each unobserved gamma ray there is introduced in the coefficient, B_ν, of $P_\nu(\cos\vartheta)$ an additional Racah coefficient, $[(2J_i+1)(2J_{i+1}+1)]^{\frac{1}{2}}(-)^\nu W(J_i\nu L_i J_{i+1}; J_i J_{i+1})$ (where J_i and J_{i+1} are the spins of the two intermediate nuclear states joined by the unobserved gamma radiation, L_i), and (b) the unobserved gammas, if not pure, mix incoherently.

Let us examine the correlation for the scheme

$$j_0 \begin{pmatrix} L_0 \\ L_0' \end{pmatrix} J_1 \begin{pmatrix} L_1 \\ L_1' \end{pmatrix} J_2 \begin{pmatrix} L_2 \\ L_2' \end{pmatrix} j_3;$$

that is, three mixed radiations with the intermediate gamma ray unobserved. The correlation consists of two incoherent parts,

$$W(\vartheta) = W_{L_1}(\vartheta) + \delta_1^2 W_{L_1'}(\vartheta), \tag{66}$$

corresponding to the two L values for the intermediate gamma ray, with an intensity ratio $1:\delta_1^2$ for $L_1:L_1'$. Each part consists of a correlation of the form:

$$W_L(\vartheta) = \sum_\nu A_\nu(0) A_\nu(2) C_\nu(1) P_\nu(\cos\vartheta) \tag{67}$$

where

$$A_\nu(0) = F_\nu(L_0 j_0 J_1) + 2\delta_0 F_\nu(L_0 L_0' j_0 J_1) + \delta_0^2 F_\nu(L_0' j_0 J_1) \tag{67a}$$
$$A_\nu(2) = F_\nu(L_2 j_3 J_2) + 2\delta_2 F_\nu(L_2 L_2' j_3 J_2) + \delta_2^2 F_\nu(L_2' j_3 J_2) \tag{67b}$$

and

$$C_\nu(1) \equiv (-)^\nu [(2J_1+1)(2J_2+1)]^{\frac{1}{2}} W(J_1\nu L_1 J_2; J_1 J_2). \tag{67c}$$

For $W_{L_1'}(\vartheta)$ the sole change is to replace L_1 by L_1' in the coefficient C_ν.

Let us consider as an example the correlation defined by the scheme $6(2)4(1)3(2)1$.

$$W(\vartheta) = 1 + P_2 F_2(264) F_2(213) \cdot [(63)^{\frac{1}{2}} W(4213; 43)]$$
$$\qquad\qquad + P_4 F_4(264) F_4(213) \cdot [(63)^{\frac{1}{2}} W(4413; 43)]$$
$$= 1 + P_2(\cos\vartheta) \cdot (-0.22792)(-0.49487)(0.90469)$$
$$\qquad\qquad + P_4(-0.02980)(-0.44671)(0.68138)$$
$$= 1 + 0.10204 P_2(\cos\vartheta) + 0.00907 P_4(\cos\vartheta). \tag{68}$$

This example illustrates that the multipolarity of the unobserved radiation does not limit the complexity of the distribution, for a term in P_4 occurs even though the unobserved gamma ray had $L = 1$.

There is another interesting feature of this example, for it will be seen that the correlation is precisely the same as that for the example of p. 770, although the nuclear spins involved differ completely. This is an instance of the simplicity of "parallel angular momenta," that is, the situation where $J_{i+1} = L_i + J_i$.

Since the case for mixed radiations so closely parallels that for the mixed gamma-gamma case it is not very useful to give further examples.

(b) *Heavy Particles.* The modifications introduced where heavy particles are involved in a correlation can be summarized in a number of rules, which are illustrated in the examples following. These rules, which may appear somewhat arbitrary, actually have a simple origin as discussed in Section 1. If one bears in mind just what measurements are performed on the system, and, in particular, which elements of the system are random, the rules are quite immediate.

It is convenient to distinguish between spinless particles and particles with spin.

(i) *Spinless particles.* To obtain the correlation for spinless particles, one first considers the basic gamma-ray correlation, involving the same angular momenta as the desired particle correlation.

For every observed spinless particle, one multiplies the coefficient of $P_\nu(\cos \vartheta)$ in the basic correlation by the appropriate particle parameter a_ν.

For pure transitions, the spinless particle parameter is given by:

$$a_\nu(LL) = \frac{2L(L+1)}{2L(L+1) - \nu(\nu+1)}. \tag{69}$$

Example. Radiative capture of α-particles; 1(3)2(2)0.

$$\begin{aligned}W(\vartheta) &= 1 + a_2(33)F_2(312)F_2(202)P_2(\cos \vartheta) \\ &\qquad + a_4(33)F_4(312)F_4(202)P_4(\cos \vartheta) \\ &= 1 + (1.33333)(-0.71714)(-0.59761)P_2(\cos \vartheta) \\ &\qquad + (6.00000)(0.08909)(-1.06904)P_4(\cos \vartheta) \\ &= 1 + 0.57143P_2(\cos \vartheta) - 0.57143P_4(\cos \vartheta).\end{aligned} \tag{70}$$

For mixed transitions, the situation is similar, but a bit more complicated in the case of charged particles. Since mixed transitions are coherent (for an observed direction), the relative phases of the various alpha particle angular momenta enter; these phases because of Coulomb (as well as nuclear) forces are now no longer 0 or π. The charged particle parameters, for mixed angular momenta L and L' are (using Wigner's nuclear reaction theory):

$$a_\nu(LL') = a_\nu(L'L)$$
$$= \cos(\xi_L - \xi_{L'}) \frac{2[L(L+1)L'(L'+1)]^{\frac{1}{2}}}{L(L+1) + L'(L'+1) - \nu(\nu+1)} \quad (71)$$

where,[21]

$$\exp(2i\xi_L) = \exp(2i\sigma_L) \frac{G_L(\eta,kR) - iF_L(\eta,kR)}{G_L(\eta,kR) + iF_L(\eta,kR)} \quad (71a)$$

$$\sigma_L = \arg \Gamma(L + 1 + i\eta) \quad (71b)$$

$$\eta \equiv Z_1 Z_2 e^2/\hbar v = \frac{MZ_1Z_2e^2}{\hbar^2 k} \quad (71c)$$

$$k = (2ME/\hbar^2)^{\frac{1}{2}}. \quad (71d)$$

Example. An inelastic alpha scattering, $(\alpha,\alpha'\gamma)$, where the gamma ray is correlated with the beam direction; $1\binom{1}{3}2(1)1(1)0$.

$$W(\vartheta) = \sum_\nu A_\nu(\alpha) C_\nu(\alpha') A_\nu(\gamma) P_\nu(\cos\vartheta) \quad (72)$$

$$A_\nu(\alpha) = a_\nu(11)F_\nu(112) + 2\delta a_\nu(13)F_\nu(1312) + \delta^2 a_\nu(33)F_\nu(312) \quad (72a)$$
$$C_\nu(\alpha') = (-)^\nu (5 \cdot 3)^{\frac{1}{2}} W(2\nu 11; 21) \quad (72b)$$
$$A_\nu(\gamma) = F_\nu(101) \quad (72c)$$
$$W(\vartheta) = 1 + \delta^2 + [(-2.00000)(0.41833)$$
$$\quad + 2\delta \cos(\xi_1 - \xi_3)(1.22476)(0.23905)$$
$$\quad + \delta^2(1.33333)(-0.71714)](0.59160) \cdot (0.70711)P_2(\cos\vartheta)$$
$$= 1 + \delta^2 + P_2(\cos\vartheta)[-0.35000 + \delta \cos(\xi_1 - \xi_3)(0.24494)$$
$$\quad - \delta^2(0.40000)]. \quad (72d)$$

The phase angle that enters depends upon the target nucleus and the alpha particle energy.[u]

It will be noted in this example that no particle parameter is introduced for the unobserved alpha particle. This is as it should be, since the particle parameter in effect gives information on the orientation of the angular momentum relative to the direction of motion, which for an unobserved direction of motion is randomly oriented regardless of the type particle. If the unobserved alpha had been mixed, the same consideration would show that the correlation is a weighted sum of the correlations for each angular momentum.

[u] The determination of the appropriate phase angles for charged particles is facilitated by the Coulomb function tables of Breit and collaborators (*35*). For large η, asymptotic approximations are useful (see *36*). The tables of C. E. Froberg (*37*) apply to the region of large η. See also the recent article by Breit and Hull (*37a*).

(ii) *Particles with spin.* The modifications introduced by spin are best understood in terms of the angular momentum coupling diagrams of Section 1. We summarize these results in the rules:

(a) If the particle with spin is *observed*, either as a bombarding, or final outgoing radiation, one replaces the factor $F_\nu(LL'jJ)$ for the basic correlation by the factor $F_\nu(LL'j_sJ)a_\nu(LL')$; that is, the channel spin j_s replaces j, and the *spinless* particle parameter is used. For several channel spins, the correlation is the weighted sum of the separate correlations.

(b) If the particle with spin is *not observed*, the factor

$$C_\nu = (-)^\nu[(2J_i + 1)(2J_{i+1} + 1)]^{\frac{1}{2}}W(J_i\nu j_pJ_{i+1}; J_iJ_{i+1}) \qquad (73)$$

replaces the factor

$$C_\nu = (-)^\nu[(2J_i + 1)(2J_{i+1} + 1)]^{\frac{1}{2}}W(J_i\nu LJ_{i+1}; J_iJ_{i+1}), \qquad (74)$$

that is, the particle's total angular momentum j_p replaces L. For several values of j_p, the correlation is a weighted sum of the correlations for separate j_p.

(c) If the particle with spin is observed, but is neither a bombarding nor a final radiation, one replaces j by the "fictitious" channel spin $j_s' = \mathbf{s} + \mathbf{j}_{\text{initial}}$ and proceeds as in (a).

Example. $A(p,\alpha)$ reaction; $\frac{1}{2}(l_p = 1)1(l_\alpha = 1)0$.

There are two channel spins possible; $j_s = 0, 1$. The correlation is therefore the weighted sum of two effective disintegration schemes $0(1)1(1)0$ for $j_s = 0$, and $1(1)1(1)0$ for $j_s = 1$. Thus

$$W(\vartheta) = W_{j_s=0} + \delta^2 W_{j_s=1}, \qquad (75)$$

with δ^2 the ratio of the probabilities of forming the compound state via channel spin 1 relative to channel spin 0. (From an experimental point of view δ^2 is a parameter to be fitted to the data.) Noting that we must introduce a spinless particle parameter for each link we find, for $j_s = 0$,

$$\begin{aligned}W_{j_s=0} &= 1 + [a_2(11)F_2(101)]^2 P_2(\cos\vartheta) \\ &= 1 + 2P_2(\cos\vartheta)\end{aligned} \qquad (76)$$

and similarly for $j_s = 1$,

$$\begin{aligned}W_{j_s=1} &= 1 + [a_2(11)]^2 F_2(111)F_2(101)P_2(\cos\vartheta) \\ &= 1 - P_2(\cos\vartheta).\end{aligned} \qquad (76a)$$

The complete correlation takes the form,

$$W(\vartheta) = (1 + \delta^2) + (2 - \delta^2)P_2(\cos\vartheta). \qquad (76b)$$

Example. Correlation of gamma rays produced by inelastic scattering of protons. The correlation is with respect to the proton beam direction; the scattered proton is not observed; $\frac{1}{2}(l_p = 1)2(l_{p'} = 1)\frac{5}{2}(L = 2)\frac{1}{2}$.

The incident proton is observed, so that we use the channel spins for this stage of the correlation; only $j_s = 1$ enters. The emitted proton is not observed so that it is the total angular momentum j_p that enters incoherently. There are two possibilities, $j_p = \frac{1}{2}, \frac{3}{2}$; the correlation is the weighted sum.

For $j_p = \frac{1}{2}$ we have the scheme: $1(1)2(\frac{1}{2})\frac{5}{2}(2)\frac{1}{2}$ so that

$$W_{j_p=\frac{1}{2}} = \sum_\nu A_\nu(p) C_\nu(p') A_\nu(\gamma) P_\nu(\cos \vartheta) \tag{77}$$

$$A_\nu(p) = a_\nu(11) F_\nu(112) \tag{77a}$$
$$C_\nu(p') = (-)^\nu [(5)(6)]^{\frac{1}{2}} W(2\nu \tfrac{1}{2} \tfrac{5}{2}; 2\tfrac{5}{2}) \tag{77b}$$
$$A_\nu(\gamma) = F_\nu(2\tfrac{1}{2} \tfrac{5}{2}) \tag{77c}$$
$$W_{\frac{1}{2}} = 1 + (-2.00000)(0.41833)(0.89443)(-0.53452) P_2(\cos \vartheta)$$
$$= 1 + 0.40000 P_2(\cos \vartheta). \tag{77d}$$

For $j_p = \frac{3}{2}$, the effective scheme is: $1(1)2(\frac{3}{2})\frac{5}{2}(2)\frac{1}{2}$. The only change is to replace C_ν above by $C_\nu = (-)^\nu (5 \cdot 6)^{\frac{1}{2}} W(2\nu \tfrac{3}{2} \tfrac{5}{2}; 2\tfrac{5}{2})$.

$$W_{j_p=\frac{3}{2}} = 1 + (0.44721)(0.31943) P_2(\cos \vartheta)$$
$$= 1 + 0.14286 P_2(\cos \vartheta). \tag{78}$$

The final correlation is the weighted sum of these two extreme cases:

$$W(\vartheta) = W_{j_p=\frac{1}{2}}(\vartheta) + \delta^2 W_{j_p=\frac{3}{2}}(\vartheta)$$
$$= 1 + \delta^2 + [0.40000 + \delta^2(0.14286)] P_2(\cos \vartheta). \tag{79}$$

Now let us consider a hypothetical situation that will illustrate the use of the "fictitious" channel spin j_s'. Let us take an inelastic gamma process, $(p,p'\gamma)$, in which the incident proton has *zero* orbital angular momentum, and we observe the correlation of the *scattered* proton with the gamma rays. Take the sequence $1(1)\frac{3}{2}(1)\frac{1}{2}$, where the state emitting the proton has spin 1, and is, of course, randomly oriented. The channel spin j_s' is also randomly oriented, and has two possibilities $\frac{1}{2}$ and $\frac{3}{2}$. For the former, we get the scheme: $\frac{1}{2}(1)\frac{3}{2}(1)\frac{1}{2}$ and the correlation:

$$W_{j_s'=\frac{1}{2}}(\vartheta) = \sum_\nu a_\nu(11) F_\nu(1\tfrac{1}{2} \tfrac{3}{2}) F_\nu(1\tfrac{1}{2} \tfrac{3}{2}) P_2(\cos \vartheta)$$
$$= 1 + (-2.00000)(0.50000)(0.50000) P_2(\cos \vartheta)$$
$$= 1 - 0.50000 P_2(\cos \vartheta). \tag{80}$$

For $j_s' = \frac{3}{2}$, we have the scheme $\frac{3}{2}(1)\frac{3}{2}(1)\frac{1}{2}$ and the correlation:

$$W_{j_s'=\frac{3}{2}}(\vartheta) = \sum_\nu a_\nu(11)F_\nu(1\tfrac{3}{2}\,\tfrac{3}{2})F_\nu(1\tfrac{1}{2}\,\tfrac{3}{2})P_2(\cos\vartheta)$$
$$= 1 + (-2.00000)(-0.40000)(0.50000)P_2(\cos\vartheta)$$
$$= 1 + 0.40000 P_2(\cos\vartheta). \tag{81}$$

The final correlation is a weighted sum of these two extreme cases:

$$W(\vartheta) = (1 + \delta^2) + [-0.50000 + \delta^2(0.40000)]P_2(\cos\vartheta). \tag{82}$$

It sometimes happens that but a single value of the channel spin j_s' can enter; for example, in the above case only $j_s' = \frac{3}{2}$ can enter if the gamma emitting state had spin $\frac{5}{2}$. For such a case it is also true that but a single value of j_p may enter. It follows that either procedure must yield the same resulting correlation; the identity of the two methods is another instance of the simplicity of "parallel angular momenta."

(2) *Polarization-Direction Correlations*

The information for nuclear spectroscopy obtained from a direction-direction correlation is not always definitive, and it is very useful to supplement this information by performing more involved correlations, the prime example being the polarization-direction correlation where the polarization (as well as the direction of motion) of one of the particles in the correlation is measured. The most important cases experimentally are two: (a) linear and circular polarization measurements on photons, and (b) circular polarization measurement on leptons. The distinguishing feature of circular polarization measurements is that the terms introduced into the correlation reverse sign under a reflection (ν is odd). Since for pure states the correlation factors into two separate links, we must therefore measure circular polarization terms for *each* link. However, for correlations involving beta decay the fact that the neutrino by its nature has definite helicity means that the measurement of the polarization of the other link in the correlation will now suffice. Parity sensitive experiments on the circular polarization of electrons from beta decay, on the circular polarization of photons in beta-gamma cascades, and even the celebrated Wu experiment are of the form of polarization-direction correlations. We shall not go into this type of experiment further, for although the principles are straightforward, the complications of beta decay theory, typified by the large number of unknown constants and interfering operators, prohibit any brief treatment (*38*, see also *20*, *24*).

Let us consider then circular and linear polarization measurements on photons. What kind of information can one get that prompts such involved experiments? Part of the answer is clear from the foregoing paragraph: parity information from circular polarization. But *linear* polarization measurements depend upon the character (electric or magnetic) of the photon; and thus one may in many cases also obtain parity information from linear polarization measurements. (The limitations on this statement are discussed in the following examples.) Aside from parity information, the fact that polarization-direction correlation yields a different angular distribution from the direction-direction correlation in itself furnishes information. For example, in determining the mixture parameter δ, in a mixed radiation, the direction-direction correlation frequently yields more than one possible answer. The additional information from a polarization-direction correlation is often decisive in these cases (*39*). The much more complicated polarization-polarization correlation, however, offers nothing more than the simpler polarization-direction correlation, with the single exception of circular polarization terms which can be obtained in no other way.

The polarization-direction correlation is an instance of the triple correlation process, and, as such, $P_{\nu_0\nu_1\nu_2}(\hat{k}_0\hat{k}_1\hat{k}_2)$, discussed in Section 1, are the natural angle functions. Here $\hat{k}_0$ and $\hat{k}_2$, say, denote the directions of motion of each of the particles in the correlation. What meaning can one attach to the direction $\hat{k}_1$? The answer depends upon the spin and type of particle whose polarization is measured. For spin $\frac{1}{2}$, the complete description of a pure state (that is, 100% polarized) is furnished by a direction; this can therefore be taken as the vector $\hat{k}_1$, with the caveat that for relativistic particles this vector is measured with respect to the particle's direction of motion (for nonrelativistic particles we may take $\hat{k}_1$ as an arbitrary direction in space). Since $\nu_1 \leq 2s$ we find only $\nu_1 = 0$ and $\nu_1 = 1$ enter for spin $\frac{1}{2}$. The polarization-direction correlation thus consists of two parts: a part involving $P_{\nu_0 0 \nu_2} = (2\nu_2 + 1)^{-\frac{1}{2}} \delta_{\nu_0}{}^{\nu_2} P_{\nu_2}(\hat{k}_0 \cdot \hat{k}_2)$— that is, the polarization independent part—and a part involving $P_{\nu_0 1 \nu_2}$, which is parity sensitive. This is just the situation discussed earlier. For spin 1 particles the complete description of a pure state requires *two* (that is, $2s$) directions in space. For photons, as a result of the fact that photons are completely relativistic, one of these two directions is necessarily along the direction of motion, ($\hat{k}_0$). The remaining direction, measured with respect to the first, is then the Poincaré vector of the usual Stokes polarization description; this direction is to be taken as $\hat{k}_1$. [For $\hat{k}_1$ parallel (antiparallel) to $\hat{k}_0$ we have right (left) circularly polarized photons; $\hat{k}_1$ perpendicular to $\hat{k}_0$, linearly polarized photons; intermediate angles, elliptically polarized photons.] The values of ν_1 range over 0, 1, and $2 = 2s$; thus the correlation will have three types of angular terms, $P_{\nu_0 0 \nu_2}$, $P_{\nu_0 1 \nu_2}$, and $P_{\nu_0 2 \nu_2}$. The first of these occurs in the polarization independent correlation, the usual direction-direction correlation. The second occurs in the circular polarization term, analogous to the $P_{\nu_0 1 \nu_2}$ term in the spin $\frac{1}{2}$ case. The last term is typical of a spin 1 system, and, for photons, is the linear polarization term.

Let us consider now in detail the polarization-direction correlation for photons, proceeding through nuclear states of sharp spin and parity. The angles describing the correlation are: ϑ, the angle between the two directions of motion; β, the polar angle between the photon's direction of motion and its polarization (Poincaré) vector; φ, the azimuthal angle for the photon's polarization vector ($\varphi = 0$ is taken to mean that the Poincaré vector lies in the plane of the directions of motion). The complete correlation is then:

$$W(\vartheta\beta\varphi) = W_0(\vartheta) + W_1(\vartheta\beta) + W_2(\vartheta\beta\varphi). \tag{83}$$

In Eq. (83) the three terms have the meaning:

(a) The polarization independent correlation, $W_0(\vartheta)$;

$$W_0(\vartheta) = \sum_\nu A_\nu(\text{gamma}) A_\nu(\text{radiation 2}) P_\nu(\cos\vartheta).$$

The A_ν have been given in detail previously.

(b) The circular polarization correlation, $W_1(\vartheta\beta)$;

$$W_1(\vartheta\beta) = \sum_\nu D_\nu(\text{gamma}) A_\nu(\text{radiation 2}) P_\nu(\cos\vartheta),$$

$$D_\nu(\gamma) = (\cos\beta) A_\nu(\gamma) \quad \text{with } \nu \text{ an } odd \text{ integer.}$$

Since A_ν (radiation 2) is restricted to even integers for ν (because only a direction of motion is measured by hypothesis), $W_1(\vartheta\beta)$ vanishes identically. [If we also perform a circular polarization measurement on radiation 2, this term will contribute. For example, the circular polarization-circular polarization correlation for two photons would be:

$$W_1(\vartheta\beta_1\beta_2) = (\cos\beta_1)(\cos\beta_2) \sum_{\nu=\text{odd}} A_\nu(1) A_\nu(2) P_\nu(\cos\vartheta).] \tag{84}$$

(c) The linear polarization correlation, $W_2(\vartheta\beta\varphi)$;

$$W_2(\vartheta\beta\varphi) = (\sin\beta) \sum_\nu E_\nu(\gamma) A_\nu(\text{rad. 2}) P_\nu^{|2|}(\cos\vartheta) \cos 2\varphi \tag{85}$$

$$E_\nu(\gamma) = \left\{ (-)^{\sigma_{L_1}} F_\nu(L_1 j_1 J) \left(\frac{2\nu(\nu+1) L_1(L_1+1)}{\nu(\nu+1) - 2L_1(L_1+1)} \right) \right.$$
$$+ 2\delta(-)^{\sigma_{L_1'}} F_\nu(L_1 L_1' j_1 J)(L_1' - L_1)(L_1' + L_1 + 1)$$
$$\left. + \delta^2(-)^{\sigma_{L_1'}} F_\nu(L_1' j_1 J) \frac{2\nu(\nu+1)(L_1')(L_1'+1)}{\nu(\nu+1) - 2L_1'(L_1'+1)} \right\}$$
$$\cdot \left[\frac{(\nu-2)!}{(\nu+2)!} \right]. \tag{85a}$$

Here $\sigma_L = 1$ if the photon, of multipolarity L, is magnetic; $\sigma_L = 0$ if the photon is electric. (The middle term in the brackets is actually symmetric in L_1 and L_1'.) One notes the explicit dependence of the coefficient E_ν, upon the character of the photons. Eq. (85a) is specialized to apply to the physically important case where $L + L'$ is odd, and ν is even.

The azimuthal angle φ is the angle between the plane of the electric vector of the measured radiation and the plane defined by the direction of the measured radiation and the normal to the scattering plane.

The dependence of the formulas above on the order of the two radiations is only formal, and exactly the same formulas result if the photon is temporally the "second" radiation.

In many cases the other radiation in the correlation will also be a gamma ray, and thus one must either distinguish the two gamma rays (by energy measurements), or consider the actually measured correlation to be a weighted average of the two "pure" correlations obtained for polarization measurements on each radiation separately. That is,

$$W_{\text{observed}} = \eta_{12} W(\text{pol } 1 - \text{dir } 2) + \eta_{21} W(\text{pol } 2 - \text{dir } 1), \tag{86}$$

where η_{12} is the over-all efficiency for detection of photon 1 in the polarization sensitive detector, while η_{21} is the over-all efficiency for photons interchanged.

The importance of the relative over-all detector efficiencies η_{12} and η_{21} can be seen as follows. If the efficiencies are the same the polarization sensitive term in W will be absent if

$$\frac{\nu(\nu+1) - 2L_2(L_2+1)}{\nu(\nu+1) - 2L_1(L_1+1)} = \pm \frac{L_2(L_2+1)}{L_1(L_1+1)} \tag{87}$$

for all possible ν; the $(+, -)$ sign refers to the case of radiations of (different, same) character. This actually occurs for

(a) radiations of opposite character and $L_1 = L_2$,
(b) radiations of the same character and $L_1, L_2 = (1,2)$ or $(2,1)$ and
for no other cases.

In both cases there is an *over-all* parity change in the cascade. Therefore, if there is no polarization effect, with equal efficiencies, there must have been an over-all parity change, but the converse does not follow. It should be recognized that when the detector efficiencies are unequal there will also be cases in which the polarization can cancel, or nearly cancel, accidentally. No conclusion should be drawn in such a case and the relative detector efficiencies should be varied.

Example. Let us consider a linear polarization measurement on the pure-mixed gamma correlation of page 772, taking the angular momentum scheme to be $\frac{7}{2}(E2)\frac{3}{2}\begin{pmatrix}M1\\E2\end{pmatrix}\frac{3}{2}$.

(a) Polarization of pure gamma ray measured.

$$W_a = W_0 + \sum_\nu E_\nu(1) A_\nu(2) P_\nu^{|2|}(\cos \vartheta) \cos 2\varphi \tag{88}$$

$$A_\nu(2) = F_\nu(1\tfrac{3}{2}\tfrac{3}{2}) + 2\delta F_\nu(12\tfrac{3}{2}\tfrac{3}{2}) + \delta^2 F_\nu(2\tfrac{3}{2}\tfrac{3}{2})$$
$$= \begin{cases} 1 + \delta^2 & \nu = 0 \\ -0.40000 + 2\delta(-0.77460) & \nu = 2. \end{cases} \tag{88a}$$

$$E_\nu(1) = (-)^\sigma \frac{(\nu - 2)!}{(\nu + 2)!} \frac{2\nu(\nu + 1) \cdot 6}{\nu(\nu + 1) - 12} F_\nu(2\tfrac{7}{2}\tfrac{3}{2})$$
$$= (0.07143) \text{ for } \nu = 2. \tag{88b}$$

The correlation for this arrangement is:

$$W_a(\vartheta) = (1 + \delta^2) + [0.05714 + \delta(0.22132)]$$
$$\cdot [P_2(\cos \vartheta) - \tfrac{1}{2} P_2^{|2|}(\cos \vartheta) \cos 2\varphi]. \tag{88c}$$

Had the first gamma been an M2, the sign of the polarization terms would have been opposite.

(b) Polarization of mixed gamma measured.

$$W_b(\vartheta) = W_0 + \sum_\nu A_\nu(1) E_\nu(2) P_\nu^{|2|}(\cos \vartheta) \cos 2\varphi \tag{89}$$

$$A_\nu(1) = F_\nu(2\tfrac{7}{2}\tfrac{3}{2}) \tag{89a}$$

$$E_\nu(2) = \frac{(\nu - 2)!}{(\nu + 2)!} (-)^1 \frac{2\nu(\nu + 1) \cdot 2}{\nu(\nu + 1) - 4} F_\nu(1\tfrac{3}{2}\tfrac{3}{2})$$
$$+ 2\delta(-)^0 (1) \cdot (4) F_\nu(12\tfrac{3}{2}\tfrac{3}{2})$$
$$+ \delta^2(-)^0 \frac{2\nu(\nu + 1) \cdot 6}{\nu(\nu + 1) - 12} F_\nu(2\tfrac{3}{2}\tfrac{3}{2})$$
$$= 0.20000 - 0.25820\delta \text{ for } \nu = 2 \tag{89b}$$

$$W_b(\vartheta) = 1 + \delta^2 + (0.05714 + 0.22132\delta) P_2(\cos \vartheta)$$
$$+ (-0.02857 + 0.03689\delta) P_2^{|2|}(\cos \vartheta) \cos 2\varphi. \tag{89c}$$

In both cases (a) and (b) the polarization terms are quite sensitive to the value of the mixing parameter δ.

The measured correlation is then the weighted sum of the two cases; that is,

$$W_{\text{measured}} = \eta_{12} W_a(\vartheta) + \eta_{21} W_b(\vartheta), \tag{90}$$

where the η's are the appropriate counter efficiencies.

Example. Coulomb excitation of odd A elements, measuring linear polarization of emitted gamma rays.

The Coulomb excitation process is predominantly E2, and even for odd A target elements where the nuclear transition favors an M1-E2 mixture, the excitation process is, to all practical purposes, pure E2. Thus for the Coulomb excitation-gamma ray correlation we have the effective angular momentum scheme

$$j_i(\text{E2})J\begin{pmatrix}\text{M1}\\\text{E2}\end{pmatrix}j_i,$$

with $j_i = \tfrac{1}{2}, \tfrac{3}{2}, \ldots$.

The polarization-direction correlation measures the polarization of the mixed gamma with respect to the incident charged particle beam. The correlation has the form:

$$W(\vartheta) = W_0(\vartheta) + W_2(\vartheta\varphi) \tag{91}$$

with
$$W_0(\vartheta) = \sum_\nu A_\nu(1)A_\nu(2)P_\nu(\cos\vartheta) \tag{91a}$$

and
$$W_2(\vartheta\varphi) = \sum_\nu A_\nu(1)E_\nu(2)P_\nu^{|2|}(\cos\vartheta)\cos 2\varphi. \tag{91b}$$

The Coulomb excitation "link," $A_\nu(1)$, is given by:

$$A_\nu(1) = a_\nu(\text{E2; Coul. Exc.})F_\nu(2j_iJ). \tag{91c}$$

Here the a_ν are the particle parameters for Coulomb excitation tabulated in (21) and (22).

For the gamma ray "link" we have:

$$A_\nu(2) = F_\nu(1j_iJ) + 2\delta F_\nu(12j_iJ) + \delta^2 F_\nu(2j_iJ) \tag{92}$$

and
$$E_\nu(2) = \left(\frac{(\nu-2)!}{(\nu+2)!}\right)\left\{(-)^1 \frac{2\nu(\nu+1)\cdot 2}{\nu(\nu+1)-4}F_\nu(1j_iJ) \right.$$
$$\left. + 2\delta(-)^0(4)F_\nu(12j_iJ) + \delta^2 \frac{2\nu(\nu+1)\cdot 6}{\nu(\nu+1)-12}F_\nu(2j_iJ)\right\}. \tag{93}$$

A typical example is the Coulomb excitation of Ag^{109}, where $j_i = \tfrac{1}{2}$ and $J = \tfrac{3}{2}$, with a 309 kev mixed E2 + M1 γ-ray emitted. Let us assume a proton energy of 2.70 Mev (center-of-mass); this gives $a_2 = 0.5824$, using the tabulations mentioned above.

The polarization-direction correlation for this case is then:

$$W_0(\vartheta) = 1 + \delta^2 + (0.5824)(-0.5000)[(0.5000) + 2\delta(-0.8660)$$
$$+ \delta^2(-0.5000)]P_2(\cos\vartheta)$$
$$= 1 + \delta^2 + [(-0.1456) + (0.5044)\delta + (0.1456)\delta^2]P_2(\cos\vartheta) \tag{94}$$

$$W_2(\vartheta\varphi) = (-0.2912)[-\tfrac{1}{2}(0.5000) + \tfrac{1}{3}\delta(-0.8660) - \tfrac{1}{2}\delta^2(-0.5000)]$$
$$\cdot P_2^{|2|}(\cos\vartheta)\cdot\cos 2\varphi$$
$$= [(0.0728) + (0.0841)\delta - (0.0728)\delta^2]P_2^{|2|}(\cos\vartheta)\cdot\cos 2\varphi, \tag{94a}$$

and the complete correlation becomes:

$$W(\vartheta,\varphi) = 1 + \delta^2 + [(-0.1456) + (0.5044)\delta + (0.1456)\delta^2]P_2(\cos\vartheta)$$
$$+ [(0.0728) + (0.0841)\delta - (0.0728)\delta^2]P_2^{|2|}(\cos\vartheta) \cdot \cos 2\varphi. \quad (94b)$$

The correlation, W_2, in which linear polarization measurements are made, involves a different function of the mixing parameter δ than the direction-direction correlation and the experiment proves a useful way to resolve the ambiguity which may result from the direction-direction correlation alone (40). The example above is, in fact, taken from this reference.

Our discussion of this experiment has been somewhat oversimplified since in practice thick targets are used, and the above is the thin target result. To compensate for thick targets, we must average the correlation for thin targets over the particle energy spectrum in the thick target, as well as estimate the loss of direction from Rutherford scattering. These corrections (the latter being negligible) are discussed in (40), as well as in (22). Tables of approximate thick target particle parameters are given in this latter reference also.

Example. An interesting example of a polarization-direction correlation is the circular polarization of the gamma rays from the capture of polarized neutrons. Let us assume that the neutron's orbital angular momentum is zero. The neutrons captured by the target nuclei then have total angular momentum $j = \frac{1}{2}$. Now for the state $j = \frac{1}{2}$ a very simple formulation of the density matrix may be given; that is,

$$\mathbf{D} = (1 + \mathbf{P} \cdot \boldsymbol{\sigma}), \quad (95)$$

where $\mathbf{P}$ is the vector giving the polarization direction of the neutrons ($|P|$ is the fractional polarization), and $\boldsymbol{\sigma}$ is the Pauli matrix vector. (We normalize as before, such that $\operatorname{tr} \mathbf{D} = 2j + 1 = 2$.) According to the discussion in Section 1, Fano's statistical tensors, $R_\nu{}^q$, furnish just the *operator* form of the density matrix, that is:

$$\mathbf{D} = [j(j+1)(2j+1)]^{-\frac{1}{2}} \sum_{\nu q} R_\nu{}^{-q}\mathbf{S}_q{}^{(\nu)}(-)^q[2\nu+1]^{\frac{1}{2}}, \quad (96)$$

where $\mathbf{S}_q{}^{(\nu)}$ are the multipole spin operators. Comparing Eqs. (95) and (96), we see that:

$$R_0{}^0 = \sqrt{2}$$
$$R_1{}^q = \sqrt{2} \begin{cases} -2^{-\frac{1}{2}}(P_x + iP_y) & q = 1 \\ P_z & q = 0 \\ 2^{-\frac{1}{2}}(P_x - iP_y) & q = -1. \end{cases} \quad (97)$$

If we choose the **P** direction as the z-axis for our coordinate system, then only $q = 0$ enters, and (assuming 100% polarization, $|P| = 1$) we find that $R_1^0 = \sqrt{2}$. Alternatively we may state that our neutrons have their spin along the z-axis: classically, this implies that

$$[P_\nu(\hat{z} \cdot \hat{j})]_{\mathrm{av}} = 1;$$

quantum mechanically, this is the statement that $\hat{z} \cdot \hat{j} = \frac{1}{2}$ and the quantum analog to P_ν enters—for $\nu = 1$ this is

$$(\tfrac{2}{3})^{\frac{1}{2}}(\tfrac{1}{2}\tfrac{1}{2}\tfrac{1}{2} - \tfrac{1}{2}|10) = [(2\nu + 1)(2j + 1)]^{-\frac{1}{2}}R_\nu^0,$$

which implies $R_1^0 = \sqrt{2}$ as before

This is in complete analogy to the $(ll00|\nu0)$ coefficient introduced by a directional measurement.

Returning to the polarization correlation of the capture gamma rays, we see that the link relating to the capture of the polarized neutron is just,

$$A_\nu(\text{neutron}) = 2(-)^\nu [2J + 1]^{\frac{1}{2}} (\tfrac{1}{2}\tfrac{1}{2}\tfrac{1}{2} - \tfrac{1}{2}|\nu 0) W(\tfrac{1}{2}\nu j_i J; \tfrac{1}{2}J), \quad (98)$$

while the gamma ray link (using the results of the previous section) contributes,

$$A_\nu(\text{gamma}) = F_\nu(L j_f J) \qquad (\nu = \text{even}) \qquad (99)$$
$$D_\nu(\text{gamma}) = (\cos \beta) F_\nu(L j_f J) \qquad (\nu = \text{odd}). \qquad (99\text{a})$$

The final correlation is therefore:

$$W(\vartheta\beta) = 1 - [2(2J + 1)]^{\frac{1}{2}} \cos \beta \, W(\tfrac{1}{2} 1 j_i J; \tfrac{1}{2} J) \cdot F_1(L j_f J) P_1(\cos \vartheta), \quad (100)$$

since A_ν(neutron) differs from zero only for $\nu = 0,1$. Cos ϑ here measures the angle between the polarization vector **P** and the gamma ray; cos β measures the angle between the photon's spin angular momentum and the photon's direction of motion ($\beta = 0$, right circular; $\beta = \pi$, left circular; $\beta = \pi/2$, linearly polarized).

For example, if $j_i = \frac{1}{2}$, $J = 1$, $j_f = 0$ and we measure right circular photons, then:

$$W(\vartheta) = 1 + \cos \vartheta. \quad (101)$$

We have discussed this example of the capture of s-wave polarized neutrons at length, because it shows the relation between the density matrix formulation and Fano's methods in a particularly simple way.

b. Several Intermediate Nuclear States

(1) Direction-Direction Correlation

In practice the restriction to isolated nuclear levels of sharp angular momentum and parity is not always meaningful and the effect of interference of nuclear states on the correlation is essential. A typical situation is a nuclear reaction, where the levels of the excited compound system have widths not negligible compared to the level separation. Although the formulas are easily extended to cover such a situation, it is quite unwieldy to consider that more than a few resonant states are involved in the interference; we shall consider only the case of two interfering levels. Since the general case may be obtained by building up the correlation from pairs of states, this is not really a restriction.

The situation for elastic scattering is rather special, in that the angular distribution arises from the coherent interference of both resonant and nonresonant processes, and the latter is formally equivalent to the consideration of infinitely many nuclear levels. We shall therefore not consider this process either, and refer to the explicit discussion of this case for both charged and neutral particles in reference (11).

Let us then consider the correlation proceeding through two interfering nuclear levels, a and b. It is necessary to distinguish two situations: (i) levels a and b have the *same* spin and parity and (ii) levels a and b differ in spin, or parity, or both. In the first case, there is, in general, coherent interference in both the *total* cross section and in the angular distribution as well. For case (ii), there can be no interference in the total cross section, but a coherent interference pattern for the angular distribution. The angular distribution can depend sensitively on the excitation energy, and it is essential now to determine not only the *shape* of the various distribution patterns that enter, but their (energy-dependent) relative magnitudes as well. The general angular distribution formula discussed in Section 1, (Eq. 35), applies to both cases, and uses the scattering matrix formalism to parametrize the energy dependence. This latter, being dynamical rather than geometrical in origin, requires some more general theory, such as Wigner's theory of nuclear reactions, for quantitative treatment.[v]

---

[v] It is here that the distinction between case (i) and (ii) becomes most important. A suitable approximation for an isolated resonance is furnished by the familiar Breit-Wigner dispersion formula, and, this, in effect, describes the resonance by a rapidly varying phase (Eq. 113, below), with the remaining parameters taken to be slowly varying (with energy). For two levels of differing (J,Π) it suffices simply to consider two such phases. If, however, the levels have the same (J,Π), *three* rapidly varying phases are required (the extra phase, roughly speaking, relates to the "mixing"). The situation where (J,Π) are the same appears to be quite infrequent experimentally, and, moreover, the effect of coherent mixing on the angular correlation is already apparent in the case where (J,Π) differ. We shall therefore confine our attention in the following to the simpler case, and only cite references where the case of two interfering levels of the same (J,Π) are discussed. For the total cross-section the basic formulas have been given by Wigner (41); further discussion appears in (42). An approximate

The general formula for the correlation has been given in Section 1, Eq. (53), in terms of the scattering matrix. If we consider only reactions, so that either α (the label designating the pair of particles emitted or absorbed), or s (the channel spin designation in this section), or both, differ between the entrance and exit channel, then we may write the general result in the form,

$$\frac{d\sigma}{d\Omega} = \frac{\lambda^2 \alpha_1}{2s_1 + 1} \sum_\nu B_\nu P_\nu(\cos \vartheta) \qquad (102a)$$

$$B_\nu = \tfrac{1}{4}(-)^{s_2-s_1} \Sigma \bar{Z}(l_{1a}J_a l_{1b}J_b; s_1\nu) \bar{Z}(l_{2a}J_a l_{2b}J_b; s_2\nu)$$
$$\times \mathrm{Re}\,[S^{J_a \Pi_a *}{}_{\alpha_1 s_1 l_{1a}; \alpha_2 s_2 l_{2a}} S^{J_b \Pi_b}{}_{\alpha_1 s_1 l_{1b} \alpha_2 s_2 l_{2b}}]. \qquad (102b)$$

[The sum in Eq. (102b) is over J_a, Π_a, J_b, Π_b, l_{1a}, l_{1b}, l_{2a}, l_{2b}.]

The general formulas can be simplified greatly if special assumptions are made about the nature of the scattering matrix. We shall assume that there are two resonances in question, of differing spin, or parity, or both. The expression for the scattering matrix, for a definite resonance level of the compound nucleus, with angular momentum J_a and Π_a, has been given by Wigner and Eisenbud (43) in their Eq. (56). In order to get simple expressions, we shall make two additional restrictive assumptions: (1) The channel radius R (equal to the a_s of ref. 43) is independent of the channel spin s, although it may vary for different channels α. (2) The constant matrix $R(\infty)$ in Eq. (46) of Wigner and Eisenbud (43) can be neglected compared to the resonance term. It is advantageous to rewrite their Eq. (56) in an equivalent form. We define the partial widths $\Gamma^{(a)}{}_{\alpha s l}$ in the same way as Wigner and Eisenbud [the $\Gamma_{\lambda s l}$ in their Eq. (55)], but instead of their $\alpha_{\lambda s l}$ we introduce a real quantity $g^{(a)}{}_{\alpha s l}$ defined by

$$g^{(a)}{}_{\alpha s l} = \pm (\Gamma^{(a)}{}_{\alpha s l})^{\tfrac{1}{2}}. \qquad (103)$$

(Here and in the following the superscript (a) is to designate the resonance level with J_a and Π_a.) The ambiguity in sign of the g's does not appear in the formula for the total cross section, but it does affect the angular distribution. In principle a determination, not only of the magnitude of the partial widths $\Gamma^{(a)}{}_{\alpha s l}$, but also of the relative signs of the parameters $g^{(a)}{}_{\alpha s l}$, can be obtained from the angular distribution. Both $g^{(a)}{}_{\alpha s l}$ and $\Gamma^{(a)}{}_{\alpha s l}$ are functions of the channel energy through the usual penetration

---

treatment, that seems to offer some advantages for correlations, has been given by Willard and Biedenharn (42a). An experimental example (two interfering $\tfrac{1}{2}+$ levels in the disintegration of C^{14} by protons) has recently been discussed by Ferguson and Gove (42b).

factor; explicitly,

$$\Gamma^{(a)}{}_{\alpha s l} = 2(\gamma^{(a)}{}_{\alpha s l})^2 P(\eta,l), \tag{104}$$

where
$$P(\eta,l) \equiv \rho[F_l^2(\eta,\rho) + G_l^2(\eta,\rho)]^{-1} \tag{104a}$$

and $(\gamma^{(a)}{}_{\alpha s l})^2$ is the reduced (energy independent) width. [$\eta,\rho = kR$ have been defined in Eqs. (71a–d).]

The phase shifts for the "potential" (hard sphere) scattering, $\xi_{\alpha l}$, have been introduced in Eqs. (71a–d).

We also introduce the notation E_a for the observed resonance energy, that is, in terms of the notation of Wigner and Eisenbud (*43*),

$$E_a = E_\lambda + \Delta_\lambda. \tag{105}$$

In principle, E_a is not a constant but depends on a channel energy through the quantity Δ_λ. In practice the energy variation of Δ_λ can usually be neglected, however, so that E_a can be considered a constant.*

In terms of the notation introduced so far, and making the approximations indicated above, Wigner and Eisenbud's formula for the scattering matrix can be rewritten in the form

$$S^{J_a \Pi_a}{}_{\alpha_2 s_2 l_{2a}; \alpha_1 s_1 l_{1a}} = \exp\left[i(\xi_{\alpha_1 l_{1a}} - \eta_{\alpha_1} \ln 2k_{\alpha_1} R_{\alpha_1})\right] \exp\left[i(\xi_{\alpha_2 l_{2a}} - \eta_{\alpha_2} \ln 2k_{\alpha_2} R_{\alpha_2})\right]$$
$$\times \left[\delta_{\alpha_2 \alpha_1} \delta_{s_2 s_1} \delta_{l_{2a} l_{1a}} + i \frac{g^{(a)}{}_{\alpha_1 s_1 l_{1a}} g^{(a)}{}_{\alpha_2 s_2 l_{2a}}}{E_a - E - i\Gamma_a/2}\right]. \tag{106}$$

In this formula, R_{α_1} is a screening radius for the Coulomb field in channel α_1, and R_{α_2} is a similar quantity for channel α_2. These screening radii drop out of the final formulas. Γ_a is the total width of the level J_a, Π_a, E_a:

$$\Gamma_a = \sum_{\alpha s l} \Gamma^{(a)}{}_{\alpha s l}. \tag{107}$$

Equation (106) is appropriate for the particular value of $J(= J_a)$ and $\Pi(= \Pi_a)$ for which the resonance occurs; a similar equation holds for J_b, Π_b [provided that $(J_a \Pi_a)$ and $(J_b \Pi_b)$ differ].

The angular distribution given by Eq. (102) may be written in the form:

$$W(\vartheta) = W_a(\vartheta) + W_b(\vartheta) + W_{ab}(\vartheta), \tag{108}$$

that is, the sum of two correlations W_a and W_b—just as if the states were pure—and an interference term W_{ab}.

For a pure state, we have the usual simplifications. For W_a we take

---

* The next approximation, a linear dependence of Δ_λ on energy has been considered by Thomas (*44*). Near thresholds even this is unsatisfactory; Δ_λ for this situation has been discussed in (*45*) and (*42*).

the angular momentum scheme: $s_1 \begin{pmatrix} l_{1a} \\ l'_{1a} \end{pmatrix} J_a \begin{pmatrix} l_{2a} \\ l'_{2a} \end{pmatrix} s_2$. The correlation becomes:

$$W_a(\vartheta) = \tfrac{1}{4} I_a (-)^{s_1-s_2} \sum_\nu P_\nu(\cos\vartheta) \cdot [\bar{Z}(l_{1a}J_a l_{1a}J_a; s_1\nu)$$
$$+ 2\delta^{(1a)} \cos\phi_{1a} \bar{Z}(l_{1a}J_a l'_{1a}J_a; s_1\nu) + (\delta^{(1a)})^2 \bar{Z}(l'_{1a}J_a l'_{1a}J_a; s_1\nu)]$$
$$\cdot [\bar{Z}(l_{2a}J_a l_{2a}J_a; s_2\nu) + 2\delta^{(2a)} \cos\phi_{2a} \bar{Z}(l_{2a}J_a l'_{2a}J_a; s_2\nu)$$
$$+ (\delta^{(2a)})^2 \bar{Z}(l'_{2a}J_a l'_{2a}J_a; s_2\nu)]. \quad (109)$$

The "mixing ratios," δ, have the significance here of ratios of (partial widths)$^{\frac{1}{2}}$, that is,

$$\delta^{(1a)} = \frac{g^{(a)}_{\alpha_1 s_1 l'_{1a}}}{g^{(a)}_{\alpha_1 s_1 l_{1a}}} \quad (109a), \qquad \delta^{(2a)} = \frac{g^{(a)}_{\alpha_2 s_2 l'_{2a}}}{g^{(a)}_{\alpha_2 s_2 l_{2a}}}. \quad (109b)$$

The signs, as well as the magnitudes of the δ's, are physical parameters. The phase angle ϕ_{1a} is given by:

$$\phi_{1a} = \xi_{\alpha_1 l_{1a}} - \xi_{\alpha_1 l'_{1a}}. \quad (110)$$

For convenience, the intensity, I_a has been given an explicit designation,

$$I_a \equiv \frac{\Gamma^{(a)}_{\alpha_1 s_1 l_{1a}} \Gamma^{(a)}_{\alpha_2 s_2 l_{2a}}}{(E - E_a)^2 + \tfrac{1}{4}\Gamma_a^2}. \quad (111)$$

An analogous formula holds for the remaining pure term, $W_b(\vartheta)$; the scheme is: $s_1 \begin{pmatrix} l_{1b} \\ l'_{1b} \end{pmatrix} J_b \begin{pmatrix} l_{2b} \\ l'_{2b} \end{pmatrix} s_2$.

These "pure state" correlations, W_a and W_b, are completely similar to the mixed-mixed correlations discussed earlier, with only differences in normalization. Thus the $\bar{Z}$ coefficients incorporate the necessary particle parameters (except for $\cos\phi$) in replacing the F_ν coefficients. (It should be noted that F_ν coefficients for mixed nuclear states are not defined.) The explicit general definitions are given in Eq. (13) for $\bar{Z}$; in Eq. (28) for F_ν.

The interference term in the correlation is quite complicated to write out, since it does not factor and all 16 terms that enter must be written down separately. To facilitate the discussion let us introduce a schematic form for the angular momenta involved:

$$s_1 \begin{matrix} \begin{pmatrix} l_{1a} \\ l'_{1a} \end{pmatrix} J_a \begin{pmatrix} l_{2a} \\ l'_{2a} \end{pmatrix} \\ \begin{pmatrix} l_{1b} \\ l'_{1b} \end{pmatrix} J_b \begin{pmatrix} l_{2b} \\ l'_{2b} \end{pmatrix} \end{matrix} s_2.$$

The two "pure state" correlations given previously, W_a and W_b, both with "mixed-mixed" radiations, each involved only routes through J_a or J_b separately. The interference correlation will involve the 16 possible routes through both J_a and J_b, using but one of the orbital angular momenta from each pair in parentheses in the schematic diagram. Each of the routes is weighted by $(I_a I_b)^{\frac{1}{2}}$, by the appropriate product of four g's, and by $\cos \phi$. The phase angle, ϕ, now contains not only the four hard sphere phases, but also the (energy dependent) phase difference between the Breit-Wigner amplitudes.

The typical term for one of the 16 routes has the form:

$$W_{ab}(\vartheta) = \tfrac{1}{4}(-)^{s_1-s_2}[(E-E_a)^2 + \tfrac{1}{4}\Gamma_a^2]^{-\frac{1}{2}}[(E-E_b)^2 + \tfrac{1}{4}\Gamma_b^2]^{-\frac{1}{2}}$$
$$\cdot g^{(a)}_{\alpha_1 s_1 l_{1a}} g^{(b)}_{\alpha_1 s_1 l_{1b}} g^{(a)}_{\alpha_2 s_2 l_{2a}} g^{(b)}_{\alpha_2 s_2 l_{2b}}$$
$$\cdot \cos[\xi_{\alpha_1 l_{1a}} - \xi_{\alpha_1 l_{1b}} + \xi_{\alpha_2 l_{2a}} - \xi_{\alpha_2 l_{2b}} + \beta_a - \beta_b]$$
$$\cdot \sum_\nu \bar{Z}(l_{1a} J_a l_{1b} J_b; s_1 \nu) \bar{Z}(l_{2a} J_a l_{2b} J_b; s_2 \nu) P_\nu(\cos\theta). \quad (112)$$

Here the phase, β_a, is defined by:

$$\tan \beta_a = 2(E-E_a)/\Gamma_a, \quad (113)$$

and similarly for β_b.

An important feature of the interference distribution is that we may no longer be, as with pure states, restricted to even Legendre coefficients. If the interfering states are of the same parity, only even powers of $\cos \vartheta$ enter; if they are of different parity, only *odd* powers of $\cos \vartheta$ (that is, ν = odd) enter the interference term.

The use of the Breit-Wigner form is but the simplest approximation to the scattering matrix in the vicinity of a resonance. A less extreme form of the one-level approximation retains the assumption of a nonzero $R(\infty)$ matrix; the necessary formulas for the S matrix have been worked out in detail by Thomas (42), see also ref. (46).

A sufficient condition that $R(\infty)$ be neglected in the analysis of resonance cross sections is that $\tfrac{1}{2}\Gamma \ll D$, where D is the level spacing. Thus our use of the Breit-Wigner dispersion term alone is not entirely consistent for the analysis of two interfering resonances; the necessary changes, however, can be incorporated directly into the formulas already discussed.

The modifications in the preceding formulas when a heavy particle is replaced by a gamma ray is straightforward, and need not be explicitly discussed.

Before concluding this section on double correlations involving mixed nuclear states, we should mention "three-stage" correlations with interference; that is, an angular correlation involving an unobserved intermediate radiation. The necessary formulas to treat this process are obtained as a special case of the triple correlation formulas, this latter being the subject of the next section. The essential features of this three-stage correlation—the incoherencies introduced by the unobserved radiation—have all been discussed previously for pure nuclear states.

As an application let us consider a (p,α) process involving two intermediate levels of spin $\frac{5}{2}$ and $\frac{3}{2}$, and opposite parity. Take, for example, the initial state to have spin 0 and the final state to have spin $\frac{1}{2}$. The angular momenta are taken to be,

$$0 \begin{matrix} (2)\frac{5}{2}(3) \\ (1)\frac{3}{2}(2) \end{matrix} \frac{1}{2},$$

so that the orbital angular momenta are assumed to be pure.

The channel spin for the initial state is $\frac{1}{2}$; the effective scheme is then,

$$\frac{1}{2} \begin{matrix} (2)\frac{5}{2}(3) \\ (1)\frac{3}{2}(2) \end{matrix} \frac{1}{2}.$$

The correlations for the "pure" transitions, W_a and W_b, have been discussed in the previous sections. The schemes are:

(a) $\frac{1}{2}(2)\frac{5}{2}(3)\frac{1}{2}$

$$W_a = I_a \sum_\nu P_\nu(\cos\theta) \bar{Z}(2\tfrac{5}{2}2\tfrac{5}{2}; \tfrac{1}{2}\nu) \bar{Z}(3\tfrac{5}{2}3\tfrac{5}{2}; \tfrac{1}{2}\nu)$$
$$= I_a \cdot [6 + \tfrac{48}{7} P_2(\cos\theta) + \tfrac{36}{7} P_4(\cos\theta)]. \quad (114)$$

(b) $\frac{1}{2}(1)\frac{3}{2}(2)\frac{1}{2}$

$$W_b = I_b \sum_\nu P_\nu(\cos\vartheta) \bar{Z}(1\tfrac{3}{2}1\tfrac{3}{2}; \tfrac{1}{2}\nu) \bar{Z}(2\tfrac{3}{2}2\tfrac{3}{2}; \tfrac{1}{2}\nu)$$
$$= I_b[4 + 4P_2(\cos\vartheta)]. \quad (115)$$

The mixed correlation, W_{ab}, is given by:

$$W_{ab} = (-)^0 [I_a I_b]^{\frac{1}{2}} \sum_\nu P_\nu(\cos\vartheta) \bar{Z}(2\tfrac{5}{2}1\tfrac{3}{2}; \tfrac{1}{2}\nu) \cdot \bar{Z}(3\tfrac{5}{2}2\tfrac{3}{2}; \tfrac{1}{2}\nu)$$
$$\cdot \cos[\xi_{\alpha_1 2} - \xi_{\alpha_1 1} + \xi_{\alpha_2 3} - \xi_{\alpha_2 2} + \beta_a - \beta_b]$$
$$= [I_a I_b]^{\frac{1}{2}} \cos[\xi_{\alpha_1 2} - \xi_{\alpha_1 1} + \xi_{\alpha_2 3} - \xi_{\alpha_2 2} + \beta_a - \beta_b]$$
$$\cdot [\tfrac{36}{5} P_1(\cos\vartheta) + \tfrac{24}{5} P_3(\cos\vartheta)]. \quad (116)$$

(2) *Triple Correlations*

Angular correlations concerned with the measurement of three directions of motion are designated as "triple correlations." The incentive for considering such very much more complicated correlations is their potential usefulness in the identification of nuclear states; a usefulness related in part to their complication, which yields, as the other side of the picture, much more information. The experimental situation is not impractical; the typical case concerns a particle induced reaction—for example, a (p,γ,γ) or a (p,p',γ) correlation—and a coincidence measurement of two directions for the reaction products is quite feasible. (Triple correlation measurements in cascades; that is, triple coincidences, would appear less practical.)

The value of triple correlation experiments may be seen by comparison with double correlation experiments. In the latter case, excluding interference, only even order Legendre polynomials occur and commonly only $\nu \leq 4$ is required to fit the experimental data. This provides two significant ratios, which quite often are inadequate to determine spins and matrix elements (mixing ratios, say). By contrast, the triple correlation with angle functions similarly restricted to an order equivalent to $\nu \leq 4$ yields some 15 independent angle functions, implying in general the possibility of some 14 significant ratios. The increased difficulty of the experiments may very well be worth the trouble.

As pointed out earlier, the polarization direction correlation experiments are in a formal sense triple correlation experiments, but there is one distinction which deserves further comment. This concerns the measurement of odd parity terms. For a given link in the correlation, a circular polarization measurement can define odd ν terms. The question then occurs as to whether the triple-direction correlation process itself can define odd ν terms. At first glance this seems to occur; for example, the correlation might involve the term $(\mathbf{\delta} \cdot \mathbf{k}_1)(\mathbf{k}_1 \cdot \mathbf{k}_2 \times \mathbf{k}_3)$, where the term $\mathbf{\delta} \cdot \mathbf{k}_1$ represents a (lepton) circular polarization measurement and the term $\mathbf{k}_1 \cdot \mathbf{k}_2 \times \mathbf{k}_3$ is proportional to the P_{111} function of three vectors. For measurements of directions of motion, with sharp l, this situation nevertheless does *not* occur. Such measurements are really measurements of a ray, and not a direction, since the ν's are even for such cases. But for a circular polarization measurement as well as a direction of motion, ν may be odd. Hence, terms like $(\mathbf{\delta} \cdot \mathbf{k}_1)P_{122}(\hat{\mathbf{k}}_1\hat{\mathbf{k}}_2\hat{\mathbf{k}}_3)$ might seem to occur. Again they do not; the reason this time involves the symmetry properties of the X coefficient for sharp states (see discussion in Section 1); this requires that $\nu_1 + \nu_2 + \nu_3 =$ even. For "odd parity" terms to occur with a single circular polarization measurement and a triple-direction correlation there must be *interference* between states of *different* parity.

Before discussing the formulas for triple correlation it is worthwhile to point up a difficulty connected with the analysis of such correlations, the problem of center-of-mass corrections. In contrast to the double correlation, no point in space may be physically defined for use as a center of coordinates. For the practical case where two of the three radiations are gamma rays, for example (p,γ,γ), this correction is of no importance. Similarly if the third radiation is a gamma ray, the angles describing this radiation refer to the second radiation as an axis, and for sufficiently heavy residual nuclei and/or short half-lives, the de-centering is generally negligible. But for three particle radiations from light nuclei, the corrections for center-of-mass motion may not be negligible. The rather involved calculations for the analysis of the data have been discussed by Newton (47). We shall limit ourselves then to the two cases exemplified by the (p,γ,γ) and (p,p',γ) reactions.

In an effort to simplify the analysis, two experimental arrangements are often used: (a) two of the three radiations are taken to be parallel (or antiparallel), and (b) two of the directions are taken perpendicular to a third direction (generally the beam direction). In both cases only a single angle remains as a variable, and the correlations are correspondingly simpler. The angle functions appropriate to this latter case have been given in Section 1, Table I.

(a) *General Formulas.* Since we are considering the triple correlation process to arise from a nuclear reaction, it is likely that the first two radiations may both be mixed and proceed through interfering nuclear states. The third radiation on the other hand will most likely proceed through a sharp nuclear state. It is this situation to which the following formulas are intended to apply.[x] For the first step in the process one applies the Wigner nuclear reaction theory, just as in the double correlation problem. The natural (that is, the simplest) description of the problem is, however, not quite the channel spin formalism of that problem but, rather the channel spin formalism for the entrance channel, with the total particle angular momentum (j_p) scheme for the exit channel. Both procedures will be given in the following.

(i) *Channel spin case.* The general formula here has been given by Kraus *et al.* (49), and reads,

$$W(\hat{\mathbf{k}}_0\hat{\mathbf{k}}_1\hat{\mathbf{k}}_2) = \tfrac{1}{4}(2s+1)^{-1}\lambda^2_{\alpha_1} \Sigma S^{J_a\Pi_a *}{}_{\alpha_2 s_{2a} l_{2a}; \alpha_1 s l_{1a}} S^{J_b\Pi_b}{}_{\alpha_2 s_{2b} l_{2b}; \alpha_1 s l_{1b}}$$
$$\cdot (-)^{s-J_a-l_{2a}}[(2\bar{J}+1)(2\nu_0+1)(2\nu_1+1)(2\nu_2+1)]^{\tfrac{1}{2}} \cdot \bar{Z}(l_{1a}J_a l_{1b}J_b; s\nu_0)$$
$$\cdot [(2J_a+1)(2J_b+1)(2l_{2a}+1)(2l_{2b}+1)(2s_{2a}+1)(2s_{2b}+1)]^{\tfrac{1}{2}}$$
$$\cdot (l_{2a} l_{2b} 00 | \nu_1 0) W(s_{2a}\nu_2 j_2 \bar{J}; s_{2b}\bar{J}) i^{\nu_0 - \nu_1 - \nu_2} X \begin{pmatrix} J_a l_{2a} s_{2a} \\ J_b l_{2b} s_{2b} \\ \nu_0 \nu_1 \nu_2 \end{pmatrix}$$
$$\cdot A_{\nu_2} \cdot P_{\nu_0 \nu_1 \nu_2}(\hat{\mathbf{k}}_0\hat{\mathbf{k}}_1\hat{\mathbf{k}}_2), \quad (117)$$

[x] Formulas for the general case have been given by Sharp *et al.* (48) and Kraus *et al.* (49).

where,

$$A_{\nu_2} = F_{\nu_2}(L_1 j_f \bar{J}) + 2\delta F_{\nu_2}(L_1 L_2 j_f \bar{J}) + \delta^2 F_{\nu_2}(L_2 j_f \bar{J}), \qquad (117a)$$

and the sum is over s_{2a}, s_{2b}, J_a, Π_a, J_b, Π_b, l_{1a}, l_{1b}, l_{2a}, l_{2b}, and ν_0, ν_1, ν_2. This formula has a recognizable structure: the $\bar{Z}$ coefficient being the usual coefficient for the entrance channel of a particle induced reaction and the A_{ν_2} being the usual correlation function for mixed gamma rays emitted from a pure state; the Wigner coefficient expresses the by now familiar information that $\mathbf{l}_1 \cdot \mathbf{k}_1$ is zero; the two scattering matrix elements furnish the weighting factors for each of the various routes to the state $\bar{J}$. The appearance of the factor $i^{\nu_0-\nu_1-\nu_2}$ in this equation might, at first glance, appear disturbing. Actually, however, a factor of this form is necessary in order that the angular distribution be explicitly real (with real angular functions such as the $P_{\nu_0\nu_1\nu_2}$). Because of the properties of the X-coefficient (interchange of adjacent rows or columns multiplies the coefficient by $(-)^\Sigma$, where Σ is the sum of the nine arguments), it follows that under the interchange $a \leftrightarrow b$, the terms appearing in Eq. (117) are complex conjugated.

We may schematize the angular momenta in the following way:

$$s \begin{pmatrix} l_{1a} \\ l_{1a}' \end{pmatrix} J_a \begin{pmatrix} l_{2a} \\ l_{2a}' \end{pmatrix} \bar{J} \begin{pmatrix} L_1 \\ L_2 \end{pmatrix} j_f,$$
$$ \begin{pmatrix} l_{1b} \\ l_{1b}' \end{pmatrix} J_b \begin{pmatrix} l_{2b} \\ l_{2b}' \end{pmatrix}$$

and note that j_2 is the spin of the intermediate particle (radiation), with s_2 the channel spin for this step, defined by $\mathbf{j}_2 + \bar{\mathbf{J}} + \mathbf{s}_2 = 0$.

(ii) *j_p scheme for the middle radiation.* The preceding formula is to be preferred on the grounds that it employs the scattering matrix in the form considered in the Wigner-Eisenbud nuclear reaction theory. A simpler way to express the angular momentum relationships, however, occurs if we use a scattering matrix that connects entrance channels in the channel spin formalism and exit channels in the j_p formalism. The relation between these two scattering matrix elements is given by the Racah coefficient employed in the sense of recoupling the angular momenta involved, that is,

$$S^{J\Pi}_{\alpha sl, a's'l'} \equiv \sum_{j_p} [(2s'+1)(2j_p+1)]^{\frac{1}{2}} W(l'j_2 J\bar{J}; j_p s') S^{J\Pi}_{\alpha sl, a'j_p l'}. \qquad (118)$$

Upon introducing this definition into the general formula, and using the identity,

$$\sum_{s_1's_2'} (2s_1' + 1)(2s_2' + 1) W(l_1'j_2 J_1 \bar{J}; j_p s_1') W(l_2'j_2 J_2 \bar{J}; j_p' s_2')$$

$$W(s_1'\nu_2 j_2 \bar{J}; s_2' \bar{J}) X \begin{pmatrix} J_1 l_1' s_1' \\ J_2 l_2' s_2' \\ \nu_0 \nu_1 \nu_2 \end{pmatrix}$$

$$= W(l_1'\nu_1 j_2 j_p'; l_2' j_p) \cdot X \begin{pmatrix} J_1 j_p \bar{J} \\ J_2 j_p' \bar{J} \\ \nu_0 \nu_1 \nu_2 \end{pmatrix}, \quad (119)$$

we obtain the general formula for the j_p scheme,

$$W = \tfrac{1}{4}(2s+1)^{-1} \lambdabar^2{}_{\alpha_1} \Sigma [(2\bar{J}+1)(2\nu_0+1)(2\nu_1+1)(2\nu_2+1)]^{\frac{1}{2}}(-)^{s-J_a-l_{2a}}$$

$$\cdot \bar{Z}(l_{1a}J_a l_{1b}J_b; s\nu_0) S^{J_a\Pi_a *}{}_{\alpha_2 j_p l_{2a}; \alpha_1 s l_{1a}} S^{J_b\Pi_b}{}_{\alpha_2 j_p' l_{2b}; \alpha_1 s l_{1b}}$$

$$\cdot \{[(2l_{2a}+1)(2l_{2b}+1)]^{\frac{1}{2}} (l_{2a}l_{2b}00|\nu_1 0) W(l_{2a}\nu_1 j_2 j_p'; l_{2b} j_p)\}$$

$$\cdot [(2j_p'+1)(2j_p+1)(2J_a+1)(2J_b+1)]^{\frac{1}{2}} i^{\nu_0-\nu_1-\nu_2} X \begin{pmatrix} J_a j_p \bar{J} \\ J_b j_p' \bar{J} \\ \nu_0 \nu_1 \nu_2 \end{pmatrix}$$

$$\cdot A_{\nu_2} \cdot P_{\nu_0 \nu_1 \nu_2}(\hat{\mathbf{k}}_0 \hat{\mathbf{k}}_1 \hat{\mathbf{k}}_2). \quad (120)$$

The sum in Eq. (120) is over J_a, Π_a, J_b, Π_b, l_{1a}, l_{1b}, l_{2a}, l_{2b}, j_p, j_p', and ν_0, ν_1, ν_2.

Actually it is easier to obtain Eq. (120) directly from a spherical diagram, that graphically expresses the desired coupling scheme, than to proceed algebraically as we did above.

This formula appears every bit as complicated as the previous result for the channel spin scheme, and the justification for terming it "simpler" is not immediately clear. If we consider the significance of the bracketed term, however, we see that this expresses directly the two facts about the intermediate radiation: (1) that $l_1 \cdot \mathbf{k}_1$ is zero and (2) that the particle has intrinsic spin j_2 which is not observed. This bracketed term is therefore familiar from its occurrence in the double correlation, in contrast to the analogous term in the channel spin formalism. For the middle radiation unobserved, Eq. (120) is clearly simpler.

An example of the utility of the j_p formalism occurs when we consider gamma rays rather than particles in exit channel. The bracketed term then becomes just $-(L_1 L_2 1 - 1|\nu_1 0)$.

Examples. (a) Let us consider a (p,γ,γ) process, for definiteness, say protons, on B^{11} leading to the ground state of C^{12}, taking the angular momentum scheme to be:

$$\tfrac{3}{2}(l_p = 1)2(1)2(2)0.$$

There are two possible channel spins, $j_s = 1$ or 2, with the effective schemes:

(a) $1(1)2(1)2(2)0$

and

(b) 2(1)2(1)2(2)0,

which are added incoherently for the complete correlation.

In both schemes the indices ν are restricted to be: 0 and 2 for ν_0 and ν_1, and for ν_2, 0, 2, and 4. There are six possible angular functions $P\nu_0\nu_1\nu_2$ that may enter, denoted by the indices 000, 022, 202, 220, 222, and 224.

The correlation is then:

$$W = W_{j_s=1} + \delta^2 W_{j_s=2} \tag{121}$$

$$W_{j_s=1} = 15 \cdot \sum_{\nu_0\nu_1\nu_2} a_{\nu_0}(11) F_{\nu_0}(112) F_{\nu_2}(202)(111 - 1|\nu_1 0) i^{\nu_0-\nu_1-\nu_2}$$

$$\cdot [(2\nu_0 + 1)(2\nu_1 + 1)(2\nu_2 + 1)]^{\frac{1}{2}} \cdot X \begin{pmatrix} 2 & 1 & 2 \\ 2 & 1 & 2 \\ \nu_0 & \nu_1 & \nu_2 \end{pmatrix} \cdot P_{\nu_0\nu_1\nu_2}(\hat{\mathbf{k}}_0\hat{\mathbf{k}}_1\hat{\mathbf{k}}_2). \tag{121a}$$

$W_{j_s=2}$ differs from $W_{j_s=1}$ only in that $F_{\nu_0}(122)$ replaces $F_{\nu_0}(112)$; both W's are normalized to unity. Introducing the values of the various coefficients we find:

$$W(\theta_1\theta_2\varphi) = 1 + \tfrac{1}{4}(5)^{\frac{1}{2}} P_{022} \pm \tfrac{1}{4}(5)^{\frac{1}{2}} P_{202} \pm \tfrac{7}{20}(5)^{\frac{1}{2}} P_{220} \mp \tfrac{5}{4}(\tfrac{5}{14})^{\frac{1}{2}} P_{222}$$
$$\mp 3(\tfrac{2}{35})^{\frac{1}{2}} P_{224}. \tag{122}$$

The upper signs apply to $W_{j_s=1}$ the lower to $W_{j_s=2}$. One notes that for the particular channel spin mixture, $\delta^2 = 1$, the correlation becomes *independent* of the beam direction. This mixing ratio is just that which corresponds to $j_p = \tfrac{1}{2}$.

The $P_{\nu_0\nu_1\nu_2}$ have not been tabulated and are rather laborious to obtain. For the special case where the gamma rays are detected perpendicular to the beam, we find the result [using Table I]

$$W\left(\frac{\pi}{2}, \frac{\pi}{2}, \varphi\right) = (3 + 11\delta^2) + 6 \cos^2 \varphi. \tag{123}$$

The distributions corresponding to the separate channel spins are seen to be quite different, that for channel spin 2 being independent of φ.

If one of the two gamma rays is oriented along the beam direction, the angular distribution of the other gamma ray is given by:

$$W(\vartheta_1 0) = 1 + 3(1 + \delta^2)(9 + 5\delta^2)^{-1} \cos^2 \vartheta_1 \tag{124a}$$
$$W(0\vartheta_2) = 1 + 3(3 - \delta^2)(9 + 5\delta^2)^{-1} \cos^2 \vartheta_2$$
$$- 6(1 - \delta^2)(9 + 5\delta^2)^{-1} \cos^4 \vartheta_2. \tag{124b}$$

For the latter correlation one notes the appearance of $\cos^4 \vartheta_2$, despite the fact that the correlation is initiated only by an $l = 1$ proton, (the observation of γ_1 parallel to the beam has thus contributed additional angular information).

If the first gamma ray is unobserved we get the double correlation:

$$W(\vartheta_2) = 1 + \tfrac{1}{4}(1 - \delta^2)(1 + \delta^2)^{-1} P_2(\cos \vartheta_2). \tag{125}$$

Finally, if we do not observe the second gamma ray the correlation is,

$$W(\vartheta_1) = 1 + \tfrac{7}{20}(1 - \delta^2)(1 + \delta^2)^{-1} P_2(\cos \vartheta_1). \tag{126}$$

It will be seen from this example that the triple correlation yields a wide variety of distinct correlations.

We have assumed above that the gamma rays are distinguished by energy measurements (for B^{11} 12 *versus* 4 Mev). If this is not done, the correlation must be suitably averaged using the counter efficiencies for each gamma ray.

(b) Inelastic proton scattering; (p,p',γ). A typical case is the proton bombardment of Fe^{54} leading to a 1.4-Mev E2 gamma ray from the excited residual Fe^{54} nucleus. Let us assume that we are interested in the odd parity states formed in Co^{55*}, that is, we take the incident proton to have $l_p = 1$ and assume that the angular momentum scheme is

$$0^+(l_p = 1)\tfrac{3}{2}^-(l_p' = 1)2^+(2)0.$$

The incident channel spin is fixed to be $\tfrac{1}{2}$; the exit channel spins are $\tfrac{5}{2}$, $\tfrac{3}{2}$, and mix *coherently*. Using the channel spin formalism of the previous section we find,

$$\sigma = \lambda^2 (6\sqrt{5}) \sum_{\nu_0 \nu_1 \nu_2} [(2\nu_0 + 1)(2\nu_1 + 1)(2\nu_2 + 1)]^{\frac{1}{2}} \cdot P_{\nu_0 \nu_1 \nu_2}$$

$$\cdot \bar{Z}(1\tfrac{3}{2}1\tfrac{3}{2}; \tfrac{1}{2}\nu_0) F_{\nu_2}(202) \cdot (1100|\nu_1 0) \cdot i^{\nu_0 - \nu_1 - \nu_2} \Big\{ |S_{\frac{3}{2}}|^2 \cdot 4 \cdot W(\tfrac{3}{2}\nu_2 \tfrac{1}{2} 2; \tfrac{3}{2} 2)$$

$$X \begin{pmatrix} \tfrac{3}{2} & 1 & \tfrac{3}{2} \\ \tfrac{3}{2} & 1 & \tfrac{3}{2} \\ \nu_0 & \nu_1 & \nu_2 \end{pmatrix} + 2 \operatorname{Re}(S_{\frac{3}{2}} S_{\frac{5}{2}}^*) \cdot \sqrt{24}\ W(\tfrac{5}{2}\nu_2 \tfrac{1}{2} 2; \tfrac{3}{2} 2) X \begin{pmatrix} \tfrac{3}{2} & 1 & \tfrac{5}{2} \\ \tfrac{3}{2} & 1 & \tfrac{3}{2} \\ \nu_0 & \nu_1 & \nu_2 \end{pmatrix}$$

$$+ |S_{\frac{5}{2}}|^2 \cdot 6 \cdot W(\tfrac{5}{2}\nu_2 \tfrac{1}{2} 2; \tfrac{5}{2} 2) X \begin{pmatrix} \tfrac{3}{2} & 1 & \tfrac{5}{2} \\ \tfrac{3}{2} & 1 & \tfrac{5}{2} \\ \nu_0 & \nu_1 & \nu_2 \end{pmatrix} \Big\}. \tag{127}$$

One notes that the scattering matrix elements, $S_{\frac{3}{2}}$ and $S_{\frac{5}{2}}$, using the one-level approximation of the previous section, both show the *same* resonance structure, and—if we assume a channel radius independent of the channel spin—differ in phase only by 0 or π. We represent the ratio of the matrix elements by $\delta = g_{\frac{1}{2}}/g_{\frac{3}{2}}$.

Upon introducing the numerical values of the various coefficients we find the result:

$$\sigma = 2\lambda^2 \frac{\Gamma_i \Gamma_{\frac{3}{2}}}{(E-E_0)^2 + (\Gamma/2)^2}$$
$$\cdot \{(1+\delta^2) - \tfrac{1}{5}(2 - 3\delta - 2\delta^2)P_2(\hat{\mathbf{k}}_1 \cdot \hat{\mathbf{k}}_2)$$
$$+ \tfrac{1}{10}(1 - 4\delta + 4\delta^2)P_2(\hat{\mathbf{k}}_0 \cdot \hat{\mathbf{k}}_2) - \tfrac{1}{5}(4 - \delta^2)P_2(\hat{\mathbf{k}}_0 \cdot \hat{\mathbf{k}}_1)$$
$$+ 1/\sqrt{70}\,(14 - \delta - 4\delta^2)P_{222}(\hat{\mathbf{k}}_0\hat{\mathbf{k}}_1\hat{\mathbf{k}}_2)$$
$$- 24/\sqrt{70}\,(\delta^2 + 2\delta)P_{224}(\hat{\mathbf{k}}_0\hat{\mathbf{k}}_1\hat{\mathbf{k}}_2)\}. \quad (128)$$

This result contains correlation information relevant to a wide variety of experimental arrangements. Consider first the arrangement where $\mathbf{k}_1$ and $\mathbf{k}_2$ are perpendicular to $\mathbf{k}_0$, with φ the dihedral angle. Using the results of Table I we find:

$$\sigma = \frac{\lambda^2 \Gamma_i \Gamma_{\frac{3}{2}}}{(E-E_0)^2 + (\Gamma/2)^2} \cdot \frac{3}{10} \cdot \{(13 - 2\delta + 2\delta^2)$$
$$- 4(2 - 3\delta + 2\delta^2)\cos^2\varphi\}. \quad (129)$$

Taking $\mathbf{k}_1$ parallel to $\mathbf{k}_0$ and using the relation,

$$P_{\nu_0\nu_1\nu_2}(\hat{\mathbf{k}}_0\hat{\mathbf{k}}_0\hat{\mathbf{k}}_2) = i^{\nu_0+\nu_1-\nu_2}(2\nu_2+1)^{-\frac{1}{2}} \cdot (\nu_0\nu_1 00|\nu_2 0) \cdot P_{\nu_2}(\hat{\mathbf{k}}_0 \cdot \hat{\mathbf{k}}_2),$$

yields:

$$\sigma(0\vartheta_2) = \frac{2\lambda^2 \Gamma_i \Gamma_{\frac{3}{2}}}{5[(E-E_0)^2 + (\Gamma/2)^2]}$$
$$\cdot \left\{(1+6\delta^2) + \left(\frac{1}{2} + \left(\frac{6}{7}\right)\delta + \left(\frac{24}{7}\right)\delta^2\right)P_2(\cos\vartheta_2)\right.$$
$$\left. - \frac{24}{7}(2\delta + \delta^2)P_4(\cos\vartheta_2)\right\}. \quad (130)$$

One notes again the appearance of a P_4 term in a reaction initiated by p-wave protons.

The remaining special case for the triple correlation puts $\mathbf{k}_2$ parallel to $\mathbf{k}_0$, and measures the correlation of the inelastic proton. This correlation has the form:

$$\sigma(\vartheta_1,0) = \frac{(\tfrac{1}{5})\lambda^2 \Gamma_i \Gamma_{\frac{3}{2}}}{(E-E_0)^2 + (\Gamma/2)^2}$$
$$\cdot \{(11 - 4\delta + 14\delta^2) - 2(2+\delta)^2 P_2(\hat{\mathbf{k}}_0 \cdot \hat{\mathbf{k}}_1)\}. \quad (131)$$

Since the general result of Eq. (128) contains information on all double correlations for the same reaction it is of interest to display these results also. Averaging over $\mathbf{k}_0$ (which, to be sure, is rather impractical experimentally) yields (omitting the resonance factor for brevity):

$$W(\hat{\mathbf{k}}_1 \cdot \hat{\mathbf{k}}_2) = 1 + \delta^2 - \tfrac{1}{5}(2 - 3\delta - 2\delta^2)P_2(\hat{\mathbf{k}}_1 \cdot \hat{\mathbf{k}}_2). \tag{132}$$

Averaging over $\mathbf{k}_2$ yields the (p, p') correlation in which the coherent channel spin interference drops out as it must:

$$W(\hat{\mathbf{k}}_0 \cdot \hat{\mathbf{k}}_1) = 1 + \delta^2 - \tfrac{1}{5}(4 - \delta^2)P_2(\hat{\mathbf{k}}_0 \cdot \hat{\mathbf{k}}_1). \tag{133}$$

The three stage (p,γ) correlation which results from averaging over the direction of the inelastic proton is probably the most accessible of all these correlations experimentally. For this case, we find:

$$W(\hat{\mathbf{k}}_0 \cdot \hat{\mathbf{k}}_2) = (1 + \delta^2) + \tfrac{1}{10}(1 - 4\delta + 4\delta^2)P_2(\hat{\mathbf{k}}_0 \cdot \hat{\mathbf{k}}_2), \tag{134}$$

where it will be noted that coherent interference occurs once again.

We might consider this same (p,p',γ) correlation also from the point of view of the j_p formalism, using Eq. (120). The scattering matrix element that would appear in the resulting correlations is not directly defined in the Wigner-Eisenbud theory but rather is a derived quantity using the relation,

$$S^{J\Pi}{}_{\alpha_2 j_p l_2; \alpha_1 s l_1} = \sum_{s_2} S^{J\Pi}{}_{\alpha_2 s_2 l_2; \alpha_1 s l_1}[(2s_2 + 1)(2j_p + 1)]^{\tfrac{1}{2}} \cdot W(l_2 j_2 J \bar{J}; j_p s_2). \tag{135}$$

Let us, for example, consider that only $j_p = \tfrac{1}{2}$ is effective in the reaction. Introducing numerical values for the coefficients that appear in the above equation we find,

$$S^{\tfrac{3}{2}+}{}_{\alpha_2 j_p=\tfrac{1}{2},1;\alpha_1 \tfrac{1}{2} 1} = S_{\tfrac{3}{2}}(1/\sqrt{5})(-1 + 2\delta). \tag{136}$$

We may utilize this result to obtain directly the correlation in the j_p scheme from the previously calculated correlations, for it shows that we must combine the correlation functions $\sigma_{s_{2a}s_{2b}}$ in the proportion $\tfrac{1}{5}$ for $(\tfrac{3}{2},\tfrac{3}{2})$, $-\tfrac{2}{5}$ for $(\tfrac{3}{2},\tfrac{5}{2})$, $-\tfrac{2}{5}$ for $(\tfrac{5}{2},\tfrac{3}{2})$ and $\tfrac{4}{5}$ for $(\tfrac{5}{2},\tfrac{5}{2})$; numerically, this is equivalent to setting $\delta = -2$ in the correlation function, and dividing by 5. Using the general result, Eq. (128), we thus obtain:

$$W(\hat{\mathbf{k}}_0 \hat{\mathbf{k}}_1 \hat{\mathbf{k}}_2) = 2\lambda^2 |S^{\tfrac{3}{2}+}{}_{\alpha_2 j_p=\tfrac{1}{2},1;\alpha_1 \tfrac{1}{2} 1}|^2 \cdot \{1 + \tfrac{1}{2}P_2(\hat{\mathbf{k}}_0 \cdot \hat{\mathbf{k}}_2)\}. \tag{137}$$

This very simple result occurs because for $j_p = \tfrac{1}{2}$ one cannot define the direction $\mathbf{k}_1$. The fact that the coefficients of the angle functions involving $\mathbf{k}_1$ vanished properly furnishes a useful check upon the general correlation function, which is sufficiently involved to calculate that such checks are valuable.

3. Appendix: Summary of Formulas and Notations

a. Intermediate Nuclear States of Sharp Angular Momentum and Parity

(1) *Direction-Direction Correlations*

The general case involves mixed radiations with the angular momentum sequence: $j_1 \binom{L_1}{L_1'} J \binom{L_2}{L_2'} j_2$. In this sequence, J always denotes the angular momentum of the intermediate nuclear state. The significance of the other angular momenta (j_i, L_i, L_i') depends, however, on the specific application and is given in detail below. For this sequence, the angular correlation is given by ($\cos \vartheta \equiv \hat{\mathbf{k}}_1 \cdot \hat{\mathbf{k}}_2$):

$$W(\vartheta) = \sum_{\nu=0,2,\ldots} [a_\nu(1) A_\nu(1)] \cdot [a_\nu(2) A_\nu(2)] P_\nu(\cos \vartheta), \tag{I}$$

where the coefficients $A_\nu(i)$, ($i = 1, 2$) have the definition:

$$A_\nu(i) \equiv F_\nu(L_i j_i J) + 2\delta_i F_\nu(L_i L_i' j_i J) + \delta_i^2 F_\nu(L_i' j_i J).$$

The F_ν are defined by:

$$F_\nu(LL' j_i J) \equiv (-)^{j_i - J - 1}[(2L + 1)(2L' + 1)(2J + 1)]^{\frac{1}{2}}(LL'1 - 1|\nu 0) \\ W(JJLL'; \nu j_i).$$

Tabulations of the F_ν coefficients are cited on pp. 805–807. (The Wigner coefficients and Racah coefficients involved in F_ν are also tabulated separately; refs. are cited under tables.)

The *mixing parameter*, δ_i, is discussed on p. 772.
The *particle parameters*, $a_\nu(i)$, are discussed below.
The application of Eq. (I) depends upon the type of radiation involved in the separate links (denoted by $i = 1$ or $i = 2$). Because of the special form of Eq. (I), the links can be discussed separately.

(a) *Gamma Rays.* The particle parameter, a_ν, is defined to be unity for gamma rays; the angular momentum j_i denotes the angular momentum of the initial ($i = 1$) or final ($i = 2$) state; and the angular momenta L_i, L_i' are the total angular momenta of the gamma multipole radiations involved. Examples are discussed on pp. 770–772.

(b) *Nuclear Particle Emission*

i. Channel Spin Scheme. The particle parameter for this case (compare discussion p. 774) is given by:

$$a_\nu(L_i L_i') = a_\nu(L_i' L_i) = \cos (\xi_L - \xi_{L'}) \frac{2[L(L + 1)L'(L' + 1)]^{\frac{1}{2}}}{L(L + 1) + L'(L' + 1) - \nu(\nu + 1)}.$$

The phases, ξ_L, which enter for charged particles, are defined by Eqs. 71; tabulations are cited in ref. 35. For uncharged particles,

$$\cos(\xi_L - \xi_{L'}) = 1$$

is a useful approximation, but neglects hard sphere (potential) scattering.

For the channel spin scheme, j_i now denotes the channel spin of the transition; that is, $j_i = |\mathbf{I} + \mathbf{s}|$ where $\mathbf{I}$ is the initial (final) nuclear angular momentum and $\mathbf{s}$ is the angular momentum of the particle absorbed (or emitted) in the transition. L_i and L_i' are the orbital angular momenta of the particle radiation. Examples are discussed on pp. 774–776.

ii. j_p Scheme. Here j_p denotes $|\mathbf{l} + \mathbf{s}|$ where $\mathbf{l}$ and $\mathbf{s}$, are the orbital and spin angular momentum of the particle emitted (or absorbed); L_i and L_i' are to be interpreted as j_p and j_p' — that is, the angular momentum of the radiation (L_i, L_i') is now given by the total angular momentum of the particle (j_p, j_p'), which may be mixed. j_i denotes the angular momentum of the initial (final) nuclear state.

If the spin, $\mathbf{s}$, is half-integral the particle angular momentum is similarly half-integral. The L_i, L_i' that appear in the F_ν are, however, *integral*. (Although a half-integer extension of the definition can be made, it is impractical to discuss as no tables exist.) In this circumstance (compare p. 752) the link $[a_\nu A_\nu]$ itself must be considered as a whole.

$[a_\nu A_\nu(i)] \equiv (-)^{s-j_i+J}(2J+1)^{\frac{1}{2}}\{W(j_pJj_pJ; j_i\nu)\bar{Z}(lj_plj_p; s\nu)$
$+ 2\delta_i W(j_pJj_p'J; j_i\nu)\bar{Z}(lj_plj_p'; s\nu) + \delta_i^2 W(j_p'Jj_p'J; j_i\nu)\bar{Z}(lj_p'lj_p'; s\nu)\}.$

The mixing parameter δ_i is real, and δ_i^2 represents the ratio of the number of j_p' transitions to transitions by j_p. The $\bar{Z}$ coefficient is discussed below. Examples are discussed on pp. 777–778.

(c) *Other Cases*

i. Internal Conversion Electron Correlation. The angular momenta are the same as for the converted gamma ray. The particle parameters are tabulated in ref. *20*.

ii. Coulomb Excitation. The angular momenta are the same as for the (virtual) gamma ray. The particle parameter tabulations are cited in refs. *21* and *22*. An example is discussed on p. 783.

iii. Beta Decay. See refs. *20*, *23*, and *24*.

(d) *Unobserved Intermediate Radiations*. Denote the angular momentum sequence by: $j_1 \begin{pmatrix} L_1 \\ L_1' \end{pmatrix} J(L) \bar{J} \begin{pmatrix} L_2 \\ L_2' \end{pmatrix} j_2$. The radiation connecting the pure intermediate states J and $\bar{J}$ is unobserved in direction, and is taken to have angular momentum L.

The correlation is given by:

$$W(\vartheta) = \sum_{\nu=0,2,\ldots} [a_\nu A_\nu(1)][a_\nu A_\nu(2)][C_\nu]P_\nu(\cos\vartheta). \quad (I')$$

The first link, $[a_\nu A_\nu(1)]$, is given by the sequence $j_1\begin{pmatrix}L_1\\L_1'\end{pmatrix}J$; the second link, $[a_\nu A_\nu(2)]$, by the sequence $\bar{J}\begin{pmatrix}L_2\\L_2'\end{pmatrix}j_2$. These links are handled as before.

The term C_ν is given by:

$$C_\nu = [(2J+1)(2\bar{J}+1)]^{\frac{1}{2}} W(J\nu L\bar{J}; J\bar{J}).$$

If the radiation is a particle of spin s, L denotes the particle angular momentum j_p.

(2) *Polarization-Direction Correlation*

We consider here only measurements of the linear polarization of gamma rays.

The correlation is given by:

$$W(\vartheta\varphi) = W_0(\vartheta) + W_2(\vartheta\varphi),$$

where:
1) $W_0(\vartheta)$ is the polarization independent correlation given by Eq. (I) and,
2) $W_2(\vartheta\varphi)$ is given by:

$$W_2(\vartheta\varphi) = \sum_{\nu=0,2,\ldots} E_\nu(1)[a_\nu A_\nu(2)]P_\nu^{|2|}(\cos\vartheta)\cos 2\varphi. \quad (II)$$

The angle ϑ is defined as before (the angle between the two directions of motion).

The azimuthal angle φ is the angle between the plane of the electric vector of the measured radiation and the plane defined by the direction of the measured radiation and the normal to the scattering plane. For $\varphi = \pi/2$, the electric vector is in the plane of $\mathbf{k}_1$ and $\mathbf{k}_2$.

The coefficient E_ν is given by:

$$E_\nu(1) = \left\{(-)^{\sigma_{L_1}} F_\nu(L_1 j_1 J)\left(\frac{2\nu(\nu+1)L_1(L_1+1)}{\nu(\nu+1)-2L_1(L_1+1)}\right)\right.$$
$$+ 2\delta(-)^{\sigma_{L_1'}} F_\nu(L_1 L_1' j_1 L)(L_1' - L_1)(L_1' + L_1 + 1)$$
$$\left.+ \delta^2(-)^{\sigma_{L_1'}} F_\nu(L_1' j_1 J)\frac{2\nu(\nu+1)(L_1')(L_1'+1)}{\nu(\nu+1)-2L_1'(L_1'+1)}\right\} \cdot \left[\frac{(\nu-2)!}{(\nu+2)!}\right].$$

The notation σ_L is defined as: $\sigma_L = +1$ for magnetic multipoles; $\sigma_L = 0$ for electric multipoles. (This result is specialized to the case where $L + L' + \nu$ is an odd integer.)

The second link, $[a_\nu A_\nu(2)]$, is handled just as in Part 1a of this Appendix. Similarly, unobserved intermediate radiations are treated as discussed in Part 1d. Examples pp. 782–785.

(3) *Triple Correlations*

In analogy to the double correlation we define the "standard triple correlation" in terms of a gamma cascade. Let the sequence of angular momenta be $j_1 \begin{pmatrix} L_1 \\ L_1' \end{pmatrix} J \begin{pmatrix} L_2 \\ L_2' \end{pmatrix} \bar{J} \begin{pmatrix} L_3 \\ L_3' \end{pmatrix} j_2$. Then the correlation has the form:

$$W(\hat{\mathbf{k}}_1\hat{\mathbf{k}}_2\hat{\mathbf{k}}_3) = \sum_{\nu_1\nu_2\nu_3} [a_{\nu_1}(1) A_{\nu_1}(1)] \cdot [a_{\nu_3}(3) A_{\nu_3}(3)]$$
$$\cdot [a_{\nu_2}(2) A_{\nu_1\nu_2\nu_3}(2)] P_{\nu_1\nu_2\nu_3}(\hat{\mathbf{k}}_1\hat{\mathbf{k}}_2\hat{\mathbf{k}}_3). \quad \text{(III)}$$

The initial link—$[a_{\nu_1}(1) A_{\nu_1}(1)]$ defined by the sequence $j_1 \begin{pmatrix} L_1 \\ L_1' \end{pmatrix} J$—
and the final link—$[a_{\nu_3}(3) A_{\nu_3}(3)]$ defined by the sequence $\bar{J} \begin{pmatrix} L_3 \\ L_3' \end{pmatrix} j_2$—are both treated exactly as in Part 1a of this Appendix.

The middle link is given by the term $[a_{\nu_2}(2) A_{\nu_1\nu_2\nu_3}(2)]$. Here $a_{\nu_2}(2)$ is the particle parameter for radiation (2). The term $A_{\nu_1\nu_2\nu_3}(2)$—which is designed to be closely analogous to the A_ν term in the double correlation—is given by:

$$A_{\nu_1\nu_2\nu_3} = F_{\nu_1\nu_2\nu_3}(J; L_2L_2; \bar{J}) + 2\delta F_{\nu_1\nu_2\nu_3}(J; L_2L_2'; \bar{J}) + \delta^2 F_{\nu_1\nu_2\nu_3}(J; L_2'L_2'; \bar{J}),$$

where the coefficients have the explicit definition:

$$F_{\nu_1\nu_2\nu_3}(J; L_2L_2'; \bar{J}) \equiv (-)^{L_2+1} i^{\nu_1+\nu_2+\nu_3} (L_2 L_2' 1 - 1|\nu_2 0)$$
$$[(2\nu_1 + 1)(2\nu_2 + 1)(2\nu_3 + 1)(2L_2 + 1)(2L_2' + 1)(2J + 1)(2\bar{J} + 1)]^{\frac{1}{2}}$$
$$\times \begin{pmatrix} J L_2 \bar{J} \\ J L_2' \bar{J} \\ \nu_1 \nu_2 \nu_3 \end{pmatrix}. \quad \text{(III')}$$

The nine index symbol in Eq. (III') is the Wigner 9-j symbol (or Fano's X-coefficient as it is also called). It should be noted that the $F_{\nu_1\nu_2\nu_3}$ are symmetric in L_2 and L_2'. Since ν_1, ν_2, and ν_3 are all even integers (for transitions between pure states of sharp angular momentum and parity), the phase $i^{\nu_1+\nu_2+\nu_3}$ is ± 1.

Tabulations of this F coefficient do not exist; it is necessary to resort to the separate tabulations of the various functions that enter. (See citations of tables.)

The angle function $P_{\nu_1\nu_2\nu_3}$ is the properly symmetric, normalized, and real function of three directions as discussed on p. 756. There are no tabulations of this (or related) functions. However, Table I gives the functions appropriate to the common experimental arrangement where two of the three directions are perpendicular to the third.

The particular form of the triple correlation given in Eq. (III) above shows the greatest possible simplification; it should be noted that mixed radiations, particles, ... are not appreciably more complicated to consider.

Examples are discussed on pp. 795–799.

b. Mixed Intermediate States

This case is considerably more complicated than that of Part a of this Appendix and two essential changes occur: (1) the relevant coefficients are normalized to give each intermediate state the proper statistical weight (rather than as in Part a defining $F_0 = 1$ for convenience), and (2) a systematic treatment in terms of the scattering matrix is used to properly take into account the mixing ratios and relative phases.

The general result for a particle in-particle out correlation (designated as alternative $a \rightarrow$ alternative b) is the Blatt-Biedenharn formula:

$$\frac{d\sigma_{ab}}{d\Omega} = \frac{\lambda_a^2}{(2i+1)(2I+1)} \sum_{ss'} \tfrac{1}{4}(-)^{s-s'} \sum_{\nu=0} B_\nu(bs'; as) P_\nu(\cos\vartheta)$$

$$B_\nu(bs'; as) = \Sigma[(1 - \mathbf{S}^{J\Pi})_{bs'l_2;asl_1}(1 - \mathbf{S}^{J'\Pi'})^*{}_{bs'l_2';asl_1'}$$
$$\bar{Z}(l_1Jl_1'J'; s\nu)\bar{Z}(l_2Jl_2'J'; s'\nu)]. \quad (IV)$$

The sum is over $l_1l_1'l_2l_2'J\Pi J'\Pi'$.

Here s and s' denote the channel spin of the initial and final systems. ($s = |\mathbf{i} + \mathbf{I}|$, with i, I the angular momenta of the incident particle and target, respectively.) (l_1, l_1') and (l_2, l_2') are the (mixed) initial and final orbital angular momenta, respectively; $J\Pi$ and $J'\Pi'$ are the (mixed) spin and parity of the intermediate states. The notation $(1 - \mathbf{S}^{J\Pi})_{bs'l_2;asl_1}$ indicates that element of the scattering matrix $\mathbf{S}$ (appropriate to angular momentum J and Π) which is involved in the transition from alternative a, channel spin s, initial orbital angular momentum l_1 to the final state, designated by b, s', l_2.

In Eq. (IV) the coefficient $\bar{Z}$ is to be regarded as shorthand for:

$$\bar{Z}(l_1Jl_1'J'; s\nu) = i^{(\nu-l_1+l_1')}Z(l_1Jl_1'J'; s\nu).$$

Tabulations of the Z coefficients are cited below. (The $\bar{Z}$ only differs by ± 1 from the Z coefficients; alternatively one may absorb this change into the elements of the scattering matrix.)

For a gamma ray replacing a particle one replaces the $\bar{Z}$ coefficient by the $\bar{Z}_1$ coefficient, where,

$$\bar{Z}_1(LJL'J'; j_1\nu) = (-)\left[\frac{1 + (-)^{\sigma_L + \sigma_{L'} + L + L' + \nu}}{2}\right]$$
$$\cdot [(2L + 1)(2L' + 1)(2J + 1)(2J' + 1)]^{\frac{1}{2}}(LL'1 - 1|\nu 0)$$
$$W(LJL'J'; j_1\nu).$$

The first factor is either zero or unity, and is related to conserving parity. As before, $\sigma_L = +1$ for magnetic multipole radiation; $\sigma_L = 0$ for electric radiation. Note that j_1 [the angular momentum of the initial (or final) nucleus] replaces the channel spin.

The coefficient Z_1 was introduced by Sharp, Kennedy, Sears, and Hoyle (50), and has been tabulated by these authors. The $\bar{Z}_1$ and Z_1 coefficients are related by:

$$\bar{Z}_1(LJL'J'; j_i\nu) = \exp\frac{\pi i}{2}(L' + \sigma_{L'} - L - \sigma_L + \nu)Z_1(LJL'J'; j_i\nu).$$

The origin of this phase (always ± 1) is exactly the same as for the phase change of the Z's. Just as there, one often prefers to absorb this sign in the definition of the scattering matrix elements.

The F_ν coefficient is related to the $\bar{Z}_1$ coefficient by:

$$(2J + 1)^{\frac{1}{2}}F_\nu(LL'jJ) = (-)^{j-J}\bar{Z}_1(LJL'J; j\nu).$$

Examples applying Eq. (IV) are given on p. 791.

The triple correlation can also be formulated for mixed intermediate states; the relevant formulas are given and discussed on p. 793.

REFERENCES TO TABULATIONS OF THE ANGULAR MOMENTUM FUNCTIONS

Alder, K., Helv. Phys. Acta **25**, 253 (1952). Algebraic table of Racah coefficients, and some Wigner coefficients.

Arima, A., H. Horie, and Y. Tanabe, Progr. Theoret. Phys. (Kyoto) **11**, 143 (1954). Definitions and properties of the X-coefficient.

Biedenharn, L. C., Tables of the Racah Coefficients, Oak Ridge National Laboratory Report ORNL-1098 (1952). Algebraic tables of $W(l_1J_1l_2J_2; sL)$ for $s = \frac{1}{2}, 1, \frac{3}{2}, 2$. Rational fractions tabulated for $W^2(l_1J_1l_2J_2; sL)$ and $Z^2(l_1J_1l_2J_2; sL)$; $s = \frac{1}{2}$ $(\frac{1}{2})3; L = 0, (1)8$.

Biedenharn, L. C., Revised Z Tables of the Racah Coefficients, Oak Ridge National Laboratory Report ORNL-1501 (1953). A reprint of the Z-tables of above reference.

Biedenharn, L. C., J. M. Blatt, and M. E. Rose, Revs. Modern Phys. **24**, 249 (1952). Algebraic tables of $W(l_1J_1l_2J_2; sL)$ for $s = \frac{1}{2}, 1, \frac{3}{2}, 2$.

Biedenharn, L. C., and M. E. Rose, Revs. Modern Phys. **25**, 729 (1953). Four place decimal tables of (a) $F_\nu(LjJ)$; [$\nu = 2,4,6,8$; $L = 1,2,3,4$; $(j,J) = 0, \frac{1}{2}, \ldots \frac{13}{2}$] and (b) $G_\nu(LjJ)$ [equivalent to $F_\nu(LL + 1jJ)$] similar range. Five place tables of the pure and mixed particle parameters for K-shell internal conversion correlations.

Boys, S. F., Phil. Trans. Roy. Soc. **A245**, 95 (1952); M. J. M. Bernal and S. F. Boys, *ibid.* p. 116; S. F. Boys and R. C. Sehni, *ibid*, p. 463 (1954). Properties of $W(abcd; ef)$ function (in different notation). Latter paper gives nine place decimal fraction tables for $e = 0$ $(\frac{1}{2})\frac{9}{2}$.

Condon, E. U., and G. H. Shortley, *Theory of Atomic Spectra* (Cambridge University Press, London and New York, 1935). Algebraic tables of the Wigner coefficients for angular momentum $j = \frac{1}{2}, 1, \frac{3}{2}, 2$.

Falkoff, D. L., G. S. Holladay and R. E. Sells, Can. J. Phys. **30**, 253 (1952). Algebraic table of the Wigner coefficient for $j = 3$.

Ferentz, M., and N. Rosenzweig, Table of F Coefficients, Argonne National Laboratory Report ANL-5234 (1953). Eight place decimal fraction tabulation of $F_\nu(LL'j_iJ)$ for a most extensive range of the parameters. $L \leq L' \leq$ Min $(|J - j_i| + 6, 10)$; $L' = L, L + 1, L + 2, L + 4, L + 6$; $J,j_i = 0(\frac{1}{2})12$; $\nu = 2(2)8$.

Ferguson, A. J., and A. R. Rutledge, Coefficients for Triple Angular Correlation Analysis in Nuclear Bombardment Experiments, Atomic Energy of Canada, Ltd., Reports CRP-615, AECL-420 (1957). Formulas and five place numerical tables. A composite coefficient, $D_{kM}{}^N$, involving a sum over five angular momentum functions (F,Z, two Wigner coefficients, and an X coefficient) is tabulated; a discussion of the use of this coefficient and examples are given; the methods are tailored to the (particle, gamma, gamma) correlation.

Goldstein, M., and C. Kazek, Jr., Table of the Racah and Z Coefficients, Los Alamos Scientific Laboratory Report LAMS-1739 (1954). Eight place decimal fractions for $W(LJLJ; jl)$ and $Z(LJLJ; jl)$; $J, j = 0(\frac{1}{2})\frac{13}{2}$; $L = 0(\frac{1}{2})\frac{17}{2}$; $l = 0(2)16$

Jahn, H. A., Proc. Roy. Soc. **A205**, 192 (1951). Algebraic tables of the Racah coefficient, $W(abcd; ef)$ $e = \frac{1}{2}, 1, \frac{3}{2}, 2$.

Jahn, H. A., and J. Hope, Phys. Rev. **93**, 318 (1954). Properties of the X-Coefficient.

Kennedy, J. M., and M. J. Cliff, Transformation Coefficients between LS and jj coupling, Atomic Energy of Canada, Ltd., Report AECL-224 (1955).

Kennedy, J. M., B. J. Sears, and W. T. Sharp, Tables of the X-Coefficient, Atomic Energy of Canada, Ltd., Reports AECL-106, CRT-569 (1954). Properties of the X-coefficient and rational fraction tables in the notation of Sharp *et al.* (ref. cited below).

$X \begin{pmatrix} abe \\ cdf \\ ghk \end{pmatrix}$ with g, h, k all even integers.

Table 1. $a = c, b = d, e = f$; $a = 1,2$; $b,c = 1(1)5$.
Table 2. $a = 1, c = 2, b = d, e = f$. Same range as above.
Table 3. $a = c, b = d, e = f$; $a = 1,2$; $b,c\frac{1}{2}(1)\frac{9}{2}$; $a = 3,4$; $b\frac{1}{2}(1)\frac{5}{2}$; $c\frac{3}{2}(1)\frac{9}{2}$.
Table 4. $a = 1, c = 2, b = d, e = f$; $b,c\frac{1}{2}(1)\frac{9}{2}$.
Table 5. $a = c, b = \frac{1}{2}, e = f$; $a = 1(1)4$; $d = \frac{3}{2},\frac{5}{2}$; $c = \frac{1}{2}(1)\frac{9}{2}$.

These tables suffice for most triple correlation calculations involving dipole, quadrupole and a few octupole radiations including mixtures.

Matsunobu, H., and H. Takebe, Progr. Theoret. Phys. (Kyoto) **14**, 589 (1955). Tables of the X coefficient; $a = b = \frac{1}{2}$; $e = 0,1$; $c,d = \frac{1}{2}(1)\frac{7}{2}$; $f,k = 0(1)4$; g,h all permitted values.

Melvin, M. A., and N. V. V. J. Swamy, Phys. Rev. **107**, 186 (1957). Algebraic table of the Wigner coefficient for $j = \frac{5}{2}$.

Obi, S., T. Ishidzu, H. Horie, S. Yanagawa, Y. Tanabe, and M. Sato, Tables of the Racah Coefficients, Annals of the Tokyo Astronomical Observatory, Second Series Vol. III, #3 (1953). Rational fraction tables of $W(abcd; ef)$. Part I has all parameters integers $e = 0(1)7$. Parts II and III tabulate three and four half-integer parameters.

Saito, R., and M. Morita, Progr. Theoret. Phys. (Kyoto) **13**, 540 (1955). Table of formulas of Wigner coefficients, $j = \frac{5}{2}$.

Sato, M., Progr. Theoret. Phys. (Kyoto) **13**, 405 (1955). Algebraic tables of $W(abcd; ef)$ for $e = 3, \frac{7}{2}, 4, \frac{9}{2}$.

Sears, B. J., and M. G. Radtke, Algebraic Tables of Clebsch-Gordan Coefficients, Atomic Energy of Canada, Ltd., TPI-75 (1954). $j = \frac{1}{2}(\frac{1}{2})3$.

Sharp, W. T., J. M. Kennedy, B. J. Sears, and M. G. Hoyle, Tables of Coefficients for Angular Distribution Analysis, Atomic Energy of Canada, Ltd., Reports CRT-556, AECL-97 (1953). Besides an excellent summary of the physical applications, this compact report tabulates the Wigner, Racah, Z, and X coefficients as square roots of rational fractions in a concise notation giving the prime factors explicitly. The ranges are: $Z(ljl'j'; sk)$; $s = 0(\frac{1}{2})4$; $l,l' = 0(1)4$; $j,j' = 0(\frac{1}{2})5$ (mostly $j = j'$).

$W(ljl'j; sk)$; parameters as above, and also short table of W for four arguments half-integer.

$X\begin{pmatrix} abc \\ a'bc \\ ghk \end{pmatrix}$; $a,a' = 1,2$; $b,c = 1(\frac{1}{2})5$, g,h,k = even integers ≤ 8.

$(ll'00|k0)$; $l,l' = 0(1)6$; $k = 0(1)12$.
$(ll'1 - 1|k0)$; $l,l' = 0(1)3$; $k = 0(1)6$.

A function Z_1 (alternative to the F_ν coefficient) is also tabulated and discussed.

Simon, A., Numerical Table of the Clebsch-Gordan Coefficient, Oak Ridge National Laboratory Report ORNL-1718 (1954). Ten place decimal fractions for parameters $0(\frac{1}{2})\frac{9}{2}$.

Simon, A., J. H. Vander Sluis, and L. C. Biedenharn, Tables of the Racah Coefficients, Oak Ridge National Laboratory Report ORNL-1679 (1952). Ten place decimal fractions for $W(abcd; ef)$ for $a,c = 0(\frac{1}{2})\frac{15}{2}$; $b,c = 0(\frac{1}{2})\frac{9}{2}$; $e = 0(\frac{1}{2})3$; $f = 0(1)8$.

Smith, K., and J. W. Stevenson, Table of the Wigner 9-j coefficients, Argonne National Laboratory Reports ANL-5776(1957); ANL-5860 Parts I and II (1958). Ten place numerical tables. $a, b, c, d = \frac{1}{2}(1)\frac{7}{2}$; $e, f, g, h, k = 0(1)7$.

GENERAL REFERENCES, SURVEY ARTICLES, AND REVIEWS

Biedenharn, L. C., Blatt, J. M., and Rose, M., Some Properties of the Racah and Associated Coefficients, Revs. Modern Phys. **24**, 249 (1952).

Biedenharn, L. C., and Rose, M. E., Theory of Angular Correlation of Nuclear Radiations, Revs. Modern Phys. **25**, 729 (1953).

Blatt, J. M., and Biedenharn, L. C., The Angular Distribution of Scattering and Reaction Cross Sections, Revs. Modern Phys. **24**, 258 (1952).

Blin-Stoyle, R. J., and Grace, M. A., Oriented Nuclei, in *Handbuch der Physik* (J. Springer Verlag, Berlin, 1957), vol. 42, p. 555.

Coester, F., and Jauch, J. M., Theory of Angular Correlations, Helv. Phys. Acta **26**, 3 (1953).

Condon, E. U., and Shortley, H. G., *Theory of Atomic Spectra* (Cambridge University Press, London and New York, 1935).

Deutsch, M., Angular Correlations in Nuclear Reactions, Repts. Progr. in Phys. (Phys. Soc. London) **14**, 196 (1953).

de Groot, S. R., and Tolhoek, H. A., Theory of Angular Effects of Radiations from Oriented Nuclei, in *Beta- and Gamma-Ray Spectroscopy*, edited by K. Siegbahn, (North-Holland Publishing Co., Amsterdam 1955), pp. 613–623.

Devons, S., and Goldfarb, L. J. B., Angular Correlations in *Handbuch der Physik* (J. Springer Verlag, Berlin, 1957), Vol. 42.

Edmonds, A. R., *Angular Momentum in Quantum Mechanics* (Princeton Univ. Press, Princeton, New Jersey, 1957).

Fano, U., Statistical Matrix Techniques and Their Application to the Directional Correlations of Radiations, Natl. Bur. Standards Report 1214 (1952).

Fano, U., Description of States in Quantum Mechanics by Density Matrix and Operator Techniques, Revs. Modern Phys. **29**, 74 (1957).

Fano, U., and Racah, G., *Irreducible Tensorial Sets* (Academic Press, New York, 1959).

Frauenfelder, H., Angular Correlation of Nuclear Radiation, Ann. Rev. Nuclear Sci. **2** (1953).

Frauenfelder, H., Angular Correlations, in *Beta- and Gamma-Ray Spectroscopy*, edited by K. Siegbahn (North-Holland Publishing Co., Amsterdam, 1955), Chapter 19.

Racah, G., Lectures on Group Theory and Spectroscopy, Princeton University (unpublished, 1951).

Rose, M. E., *Elementary Theory of Angular Momentum* (John Wiley & Sons, New York, 1958).

Schwinger, J., On Angular Momentum, Atomic Energy Commission Report NYO-3071 (1952).

Sharp, W. T., The Quantum Theory of Angular Momentum, Atomic Energy of Canada, Ltd., CRL-43, AECL-465 (1957).

Spiers, J. A., Directional Effects in Radioactivity, Natl. Research Council of Canada Report 1925 (1949).

Steffen, R. M., The Influence of Extranuclear Fields on the Angular Correlation of Nuclear Radiation, Advances in Phys. (Phil. Mag. Suppl.) **4**, 293 (1955).

Wigner, E. P., *Gruppentheorie and ihre Anwendung auf die Quantenmechanik der Atomspektren* (Vieweg, Braunschweig, 1931).

Wigner, E. P., "On the Matrices Which Reduce the Kronecker Products of Simply Reducible Groups," unpublished manuscript (1941).

Wigner, E. P., *Group Theory and Its Application to the Quantum Mechanics of Atomic Spectra*, transl. by J. J. Griffin with additions by the author (Academic Press, New York, 1959).

REFERENCES

1. H. Frauenfelder, in *Beta- and Gamma-Ray Spectroscopy*, edited by K. Siegbahn (North-Holland Publishing Co., Amsterdam, 1955), Chapter 19.
2. R. M. Steffen, Advances in Physics (Phil. Mag. Suppl.) **4**, 293 (1955).
3. G. Racah and U. Fano, unpublished manuscript (1951).
4. D. H. Wilkinson, Illustrations of Angular Correlation Computations Using Racah Coefficient Methods, unpublished manuscript circulated by Brookhaven National Laboratory (1954).
5. R. F. Christy, Proc. Univ. Pittsburgh Conf. on Nuclear Structure, Pittsburgh (1957), p. 421 ff.
6. T. Ericson and V. Strutinsky, Nuclear Phys. **8**, 284 (1958); see also I. Halpern

and V. Strutinsky, Proc. 2nd Intern. Conf. Peaceful Uses Atomic Energy, Paper P/1513 (1958); I. Halpern, Ann. Rev. Nuclear Sci. to be published.
7. G. Racah, Lectures on Group Theory and Spectroscopy, Princeton University (unpublished, 1951).
8. L. C. Biedenharn, J. Math. Phys. **31**, 287 (1953).
9. P. J. Brussaard and H. A. Tolhoek, Physica **23**, 955 (1957).
10. A. R. Edmonds, *Angular Momentum in Quantum Mechanics* (Princeton University Press, Princeton, N.J., 1957); see also E. P. Wigner, Bull. Am. Phys. Soc. [2] **3**, 170 (1958).
11. J. M. Blatt and L. C. Biedenharn, Revs. Modern Phys. **24**, 258 (1952).
12. R. Huby, Proc. Phys. Soc. **67**, 1103 (1954).
13. S. P. Lloyd, Thesis, University of Illinois (1951); Phys. Rev. **85**, 904 (1952); J. A. M. Cox and H. A. Tolhoek, Physica **19**, 671 (1953); J. Weneser and D. Hamilton, Phys. Rev. **92**, 321 (1953); G. R. Satchler, *ibid.* **95**, 1304 (1954).
14. U. Fano, Nuovo cimento **5**, 1358 (1957).
15. E. P. Wigner, Revs. Modern Phys. **29**, 255 (1957).
16. G. R. Satchler, Proc. Phys. Soc. **A66**, 508 (1953).
17. J. W. Gardner, Proc. Phys. Soc. **A62**, 763 (1949).
18. J. W. Gardner, Proc. Phys. Soc. **A64**, 248, 1136 (1951).
19. M. E. Rose, L. C. Biedenharn, and G. B. Arfken, Phys. Rev. **85**, 5 (1952).
20. L. C. Biedenharn and M. E. Rose, Revs. Modern Phys. **25**, 729 (1953).
21. L. C. Biedenharn, M. Goldstein, J. L. McHale, and R. M. Thaler, Phys. Rev. **101**, 662 (1955); **102**, 1567 (1956); **104**, 1643 (1956).
22. K. Alder, A. Bohr, T. Huus, B. R. Mottelson, and A. Winther, Revs. Modern Phys. **28**, 432 (1956).
23. S. Devons and L. J. B. Goldfarb, in *Handbuch der Physik* (J. Springer Verlag, Berlin, 1957), Vol. 42.
24. K. Alder, B. Stech, and A. Winther, Phys. Rev. **107**, 728 (1957).
25. G. Racah, unpublished manuscript (1953); see also E. P. Wigner, "On the Matrices Which Reduce the Kronecker Products of Simply Reducible Groups," unpublished manuscript (1949).
26. U. Fano, Statistical Matrix Techniques and Their Application to the Directional Correlations of Radiations, Natl. Bur. Standards Report 1214, unpublished (1952).
27. U. Fano and G. Racah, *Irreducible Tensorial Sets* (Academic Press, New York, 1959).
28. J. Schwinger, Atomic Energy Commission Report NYO-3071 (1952).
29. U. Fano, Revs. Modern Phys. **29**, 74 (1957).
30. L. C. Biedenharn, Ann. Phys. (N.Y.) **4**, 104 (1958); see also U. Fano, Revs. Modern Phys. **29**, 74 (1957).
31. B. A. Lippman, Phys. Rev. **81**, 162 (1951); O. Halpern and B. A. Lippman, *ibid.* **87**, 1128 (1952).
32. R. J. Blin-Stoyle and M. A. Grace, in *Handbuch der Physik* (J. Springer Verlag, Berlin, 1957), Vol. 42, p. 555.
33. J. M. Blatt and L. C. Biedenharn, Revs. Modern Phys. **24**, 263 (1952), eqs. 4.5 and 4.6.
33a. R. Huby, Proc. Phys. Soc. **72**, 97 (1958); Compt. rend. congr. intern. de phys. nucléaire, p. 875 (1959).
34. J. M. Blatt and V. F. Weisskopf, *Theoretical Nuclear Physics* (John Wiley and Sons, New York, 1952), Appendix B, Part 4, p. 803 ff.

34a. S. Ofer, Phys. Rev. **114**, 870 (1959).
35. I. Bloch, M. H. Hull, Jr., A. A. Broyles, W. G. Bouricius, B. E. Freeman, and G. Breit, Revs. Modern Phys. **23**, 147 (1951); see also M. Abramowitz, Tables of Coulomb Functions, A.M.S. No. 17, Natl. Bur. Standards (1952); H. Feshbach, D. Shapiro, and V. F. Weisskopf, Atomic Energy Commission Report NYO-3077, NDA 15 B-5 (1953); W. T. Sharp, H. E. Gove, and E. B. Paul, Graphs of Coulomb Functions, Report TPI-70, Chalk River, Ontario.
36. M. Abramowitz and P. Rabinowitz, Phys. Rev. **96**, 75 (1954); L. C. Biedenharn, R. L. Gluckstern, M. H. Hull, Jr., and G. Breit, *ibid.* **97**, 542 (1955).
37. C. E. Froberg, Revs. Modern Phys. **27**, 399 (1955).
37a. G. Breit and M. H. Hull, Jr., in *Handbuch der Physik* (J. Springer Verlag, Berlin, 1959), Vol. 41.
38. M. Morita, Progr. Theoret. Phys. (Kyoto) **14**, 27 (1955); I. Hauser, Thesis, State University of Iowa (1956), Nuovo cimento [10] **5**, Suppl. 1, 182 (1957).
39. P. H. Stelson and F. K. McGowan, Phys. Rev. **105**, 1346 (1957).
40. F. K. McGowan and P. H. Stelson, Phys. Rev. **99**, 127 (1955); Oak Ridge National Laboratory Report ORNL-2204 (1956), p. 20; Bull. Am. Phys. Soc. [2] **1**, 164 (1956).
41. E. P. Wigner, Phys. Rev. **70**, 606 (1946).
42. R. G. Thomas, Lectures on the Theory of Nuclear Reactions, University of Mexico, Summer (1956); A. M. Lane and R. G. Thomas, Revs. Modern Phys. **30**, 257 (1958).
42a. H. B. Willard and L. C. Biedenharn, *Proc. Phys. Soc.* **72**, 874 (1958).
42b. A. J. Ferguson and H. E. Gove, Atomic Energy of Canada, Ltd., Report PD-306 (1959).
43. E. P. Wigner and L. Eisenbud, Phys. Rev. **72**, 29 (1947).
44. R. G. Thomas, Phys. Rev. **81**, 148 (1951).
45. F. W. Prosser and L. C. Biedenharn, Phys. Rev. **109**, 413 (1958).
46. G. Breit, in *Handbuch der Physik* (J. Springer Verlag, Berlin, 1959), Vol. 41; see also, H. B. Willard *et al.* in *Fast Neutron Physics*, edited by J. L. Fowler and J. B. Marion (Interscience Publishers, New York, 1959).
47. T. D. Newton, private communication (Chalk River memorandum) (1956).
48. W. T. Sharp, J. M. Kennedy, B. J. Sears, and M. G. Hoyle, Atomic Energy of Canada, Ltd., Report AECL-97 (1954).
49. A. A. Kraus, Jr., J. P. Schiffer, F. W. Prosser, and L. C. Biedenharn, Phys. Rev. **104**, 1667 (1956).
50. W. T. Sharp, J. M. Kennedy, B. J. Sears, and M. G. Hoyle, Atomic Energy of Canada, Ltd., Report CRT-556 (1953).

V. D. Analysis of Beta Decay Data

by M. E. ROSE

1. The Role of Beta Decay in Nuclear Physics............................. 813
2. Outline of the Theory.. 815
3. The Energy Distribution and Angular Correlation: Allowed Transitions..... 819
4. Analysis of the Shapes of Allowed Spectra............................. 820
5. Forbidden Spectra... 821
6. Corrections to the Spectral Shapes.................................... 824
7. ft Values... 825
8. Failure of Parity Conservation.. 829
9. Orbital Capture... 830
   References... 832

In general terms the processes of β-decay are to be described as nuclear transitions between states of equal mass number in which certain light particles are emitted, absorbed, or both. In contrast to γ-ray processes, the β-processes always involve creation or destruction of light fermions (spin $\frac{1}{2}$ particles) one of which is an electron of either positive or negative charge. The necessity of an accompanying light fermion is evident from energy conservation. For the moment we note that conservation of angular momentum and statistics require that there be an even number of fermions involved in the process and in the only practical case one additional particle accompanies the electron. At this point we refer to this particle as the neutrino (ν) and do not attempt to distinguish between particle and antiparticle. Charge conservation requires that this particle be neutral and evidence of spectral shapes, see Section 4, indicates that its mass is extremely small: $m_\nu \leq 1$ kev.

In a simple physical description of the β-process it is possible to think of the β-process as somewhat analogous to photon emission and to introduce phenomenologically an interaction which is responsible for the β-emission. Here the electron-neutrino field plays a role similar to that of the electromagnetic field in photon processes. But one striking difference is the strengths of these two types of coupling. Allowed γ-processes have lifetimes of the order of or less than 10^{-10} sec whereas the fastest β-process of comparable energy has a lifetime of 0.03 sec. Faster β-processes would entail larger energy release and in that case the emission of a nucleon, as soon as this is energetically permissible, would be overwhelmingly more likely.

The weakness of the β-interaction explains why this process, with few exceptions, occurs from the ground states of unstable nuclei. The few exceptions are those in which the slow β-process competes only with highly forbidden γ-transitions from isomeric states.

The possibility of observing the β-processes depends in every case, directly or indirectly, on the emission or absorption of the electron. This is so because the neutrino has apparently no observable interaction with matter except the very weak β-coupling which was responsible for its creation. Therefore, it can produce an inverse β-process which is characterized by the very small cross section[a] of 10^{-43} cm^2.

The types of measurements which are most frequently made, and which are of greatest importance for this discussion are the energy spectra of the electron or positrons emitted in the decay process

$$Z^A \to (Z \pm 1)^A + e^{\mp} + (\bar{\nu}, \nu) \tag{1}$$

where $\bar{\nu}$ and ν are antineutrino and neutrino respectively. Conventionally, $\bar{\nu}$ is associated with β^- emission, ν with β^+ emission. Future reference to the energy spectra will imply a measurement of $N(W)$ where $N(W)dW$ is the number of $\beta^\pm$ with total energy between W and $W + dW$. In general, one is interested only in the shape of the $N(W)$ versus W curve.

Where branching occurs it is often useful to measure branching ratios: that is, ratios of total intensities for competing processes. Actually, two types of branching are known. In one, both nuclei $(Z \pm 1)^A$ are unstable and electron and positron emission compete. Cu^{64} is an example. In the other, more frequently occurring case, a given unstable state can decay to more than one state in the daughter nucleus. More information can be obtained from this second type of branching.

When positron emission is possible it is also always possible to make the transition by orbital capture:

$$Z^A + e^- \to (Z - 1)^A + \nu \tag{2}$$

where the e^- is a bound electron, most often a K electron. The branching ratio for K capture to e^+ emission is designated as K/β^+, see Section 9 below. The observation of orbital capture clearly depends on the emission of X-rays or Auger electrons following the creation of the vacancy in the atomic shell. Here one can also measure the branching ratio representing the competition between K capture and capture from another shell, say the L shell.

---

[a] A cross section of $(11 \pm 4) \times 10^{-44}$ cm^2 for the process $p + \bar{\nu} \to e^+ + n$ (where is an antineutrino) has been measured by Cowan et al. (1).*

* The reference list for Section V.D begins on page 832.

The energetic conditions for the $\beta^\pm$ and capture processes are

$$M_Z{}^A - M^A{}_{Z+1} > 0 \qquad \beta^- \text{ decay} \qquad (3a)$$
$$M_Z{}^A - M^A{}_{Z-1} - 2m > 0 \qquad \beta^+ \text{ decay} \qquad (3b)$$
$$M_Z{}^A - M^A{}_{Z-1} - \epsilon_e > 0. \qquad \text{capture} \qquad (3c)$$

Here $M_Z{}^A$ is the atomic mass for charge and mass numbers Z and A respectively. In each case (Z, A) is the decaying nucleus and when the inequalities are fulfilled the process is energetically possible. The binding energy (in mass units) of the orbital state is ϵ_e and m is the electron rest mass. The actual nuclear energy release is W_0 which is the difference of nuclear masses:

$$W_0 = {}_Z M^A - {}_{Z \pm 1} M^A \qquad \text{for } \beta^\mp \qquad (4)$$

where

$$_Z M^A = M_Z{}^A - Zm, \qquad (4')$$

and for capture the neutrino energy is $W_0 + \epsilon_e$.

1. The Role of Beta Decay in Nuclear Physics

There are at least two reasons for studying the β-decay processes. The first reason is to be found in the need to understand the process itself. It will be seen that the β-process is much more complicated than the electromagnetic phenomena because in the former there are, in principle, twenty coupling constants and ten different types of interaction which must be admitted if the only restrictions imposed on the interaction are those of relativistic invariance. In contrast the emission of γ-rays proceeds by virtue of essentially two real coupling constants: the charge and magnetic dipole moment of the emitting particle.[b] In the past two years our conception of the nature of the β-interaction has undergone a radical revision and in fact, it is only recently that the true nature of the problem has been revealed (2).

The second reason for the interest in β-decay is to be found in its usefulness as a tool in nuclear spectroscopy. This comes about because the nuclear states are characterized by definite angular momentum and parity. The angular momentum conservation rule then restricts the total angular momentum of the leptons. For a given parity change the lepton angular momentum can have a strong influence on the spectrum

[b] While the description of emission of photons by electrons can be said to be understood fairly well this is not true for nuclei where the precise nature of the transition currents is not known. However, this is a complication over and above that discussed in the text and the same lack of knowledge of nuclear structure is, to some extent, an obstacle in the interpretation of β-spectra. The two problems differ insofar as one does not usually attempt to predict the absolute intensity in β-decay.

shape and on the half-life. The actual chain of reasoning is from the spectrum shape to the nuclear spin change $\Delta J = |J_f - J_i|$. However, in order to draw definite conclusions it is often helpful to know the parity change[c] $\Delta \pi$ and some information about branching ratios. Also, where γ-rays follow the β-transition a knowledge of the multipolarity and parity in these γ-transitions is quite often essential in deciphering the decay scheme.

A complete understanding of the application of β-decay data to nuclear spectroscopy rests on the following results of the theory, see Section 2:

(a) β-transitions can be classified into groups characterized by an order of forbiddenness n; $n = 0, 1, 2, \ldots$, are referred to as allowed, first, second, $\ldots$, forbidden. For a given energy release the lifetime generally increases rapidly (by roughly a factor 100 or more) with each increase of n by one unit.

(b) The parity change for order of forbiddenness n is $(-)^n$.

(c) The angular momentum selection rules are expressed as triangular relations. Thus, $\Delta(abc)$ means that a, b, and c form a triangle: $|a - b| \leq c \leq a + b$ and similar inequalities obtained by permuting a, b, c. Then the selection rules for the transition $J_i \to J_f$ are $\Delta(J_fJ_i0)$ and $\Delta(J_fJ_i1)$ for $n = 0$; $\Delta(J_fJ_i0)$, $\Delta(J_fJ_i1)$, $\Delta(J_fJ_i2)$ for $n = 1$; $\Delta(J_fJ_in)$ and $\Delta(J_fJ_in + 1)$ for $n \geq 2$. Note that $\Delta(J_fJ_in - 1)$ for $n = 3$ and $\Delta \pi = -1$ corresponds to a small correction to a first forbidden transition. Similarly $\Delta(J_fJ_in - 1)$ for $n = 2$, $\Delta \pi = 1$ is a small correction to an allowed transition.

(d) The β-transition probability increases rapidly with W_0.

(e) The spectrum shape for $\Delta J = n + 1$ is uniquely associated with this transition but $n = 0$ and $n = 1$, $\Delta J = 0, 1$ usually have the same shape.

(f) Certain exceptions to these rules occur. The exceptions which involve anomalously long lifetimes are the result of either special selection rules or accidental cancellations of matrix elements. As an example of the former we cite the l-forbiddenness transitions (P^{32}) in which a change of orbital angular momentum of a nucleon is involved. Presumably C^{14} and RaE are examples in which fortuitously small matrix elements occur.

With these six rules we can examine two typical cases in which β-decay data are used to draw conclusions about nuclear levels. Case 1a: If in a β-transition the spectrum has the unique shape referred to in rule (e), see Eq. (17), we can not only conclude that $\Delta J = n + 1$ but we can also deduce the value of n, the order of forbiddenness. This is because the shape of the spectrum is also uniquely associated with a given n. Thus, at least relative parity assignments can be made.

[c] We shall write $\Delta \pi = +1$ if the initial and final nuclear parities are the same and $\Delta \pi = -1$ if they are different.

Case 1b: Suppose the emitting state is known to be $\frac{1}{2}^-$ and an allowed shape spectrum is observed in transition β. One concludes for state x that $J = \frac{1}{2}$ or $\frac{3}{2}$ and the parity is $+$ or $-$. A measurement of the half-life and of W_0 can result in a parity determination (Section 7) and for definiteness we assume that x has parity -1. The fact that the ground state transition is not observed despite the larger energy associated with such a transition indicates a large degree of forbiddenness and a large ΔJ. Thus, a tentative assignment is $\frac{5}{2}^-$ for the ground state g corresponding to a

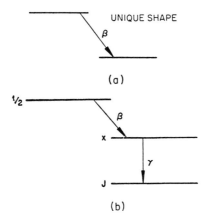

FIG. 1. (a) Simple decay scheme for ground to ground state transition with $\Delta J = n + 1$, $\Delta \pi = (-)^n$. (b) Decay scheme with β-transition to the excited state x followed by γ-transition to the ground state g.

second forbidden transition. To resolve the ambiguity between the choices $\frac{1}{2}^-$ and $\frac{3}{2}^-$ the γ-ray intensity, or better, the internal conversion coefficient of the transition marked γ should be measured. This case is typical of the manner in which several pieces of information are combined to make J and π assignments to excited states. In the example discussed the ground state g would be stable and its spin and parity would be known or amenable to measurement by direct means. In that case an unambiguous assignment of J, π could be made for state x. In other instances assignments can be made from the results of β-γ correlation measurements or the measurement of polarization of a γ-ray following β-emission from oriented nuclei (3).

2. Outline of the Theory

The following brief outline of β-decay theory and the discussion of following sections is presented as a partial clarification of the six rules given in the preceding section. The first step in the theory is the formulation of the β-interaction. This is written in the form of an operator (relativistically invariant) which couples the source (nucleons) and field (leptons). This operator $\mathcal{H}$ will be written first as a sum of two parts

$$\mathcal{H} = \mathcal{H}_- + \mathcal{H}_+ \tag{5}$$

where $\mathcal{H}_\pm$ is responsible for $\beta^\pm$-emission. According to general quantum mechanical rules

$$\mathcal{H}_+ = \mathcal{H}_-^* \tag{5'}$$

where * means hermitian conjugate. Examining only $\mathcal{H}_-$ this in turn will be written as a sum of two parts.

$$\mathcal{H}_- = \mathcal{H}_\beta + \mathcal{H}_\beta' \tag{6}$$

and

$$\mathcal{H}_\beta = g \sum_i (\Psi_f^* O_i \Psi_i)(\psi_e^* O_i \psi_{\bar{\nu}}). \tag{6'}$$

In Eq. (6') O_i is an operator which, in the first scalar product, operates on the nuclear variables and in the second scalar product it is the same type of operator but acts only in the space of the lepton variables. Ψ_f and Ψ_i represent final and initial nuclear states, ψ_e and $\psi_{\bar{\nu}}$ electron and antineutrino states.[d] The lepton scalar product is to be evaluated at the position of the decaying nucleon and a sum over nucleons is implied. The O_i operators can be any operators which preserve relativistic invariance and, aside from a multiplicative constant there are five possible choices:

(a) $O_i = \beta$, a scalar
(b) $O_i = 1, i\boldsymbol{\alpha}$, a polar 4-vector
(c) $O_i = \beta\boldsymbol{\delta}, \beta\boldsymbol{\alpha}$, an antisymmetric 4-tensor
(d) $O_i = \boldsymbol{\delta}, i\gamma_5$, an axial 4-vector
(e) $O_i = \beta\gamma_5$, a pseudoscalar.[e] (7)

The transformation properties listed refer to the extended group of Lorentz transformations. The five choices correspond to the well-known fact that there exists just these five covariants constructed from the four Dirac matrices. In (a) and (e) the sum over i involves one term, in (b) and (d) there are four terms and in (c) there are six terms. Thus for choice (c) we would write

$$\mathcal{H}_\beta \sim (\Psi_f^* \beta\boldsymbol{\delta}\Psi_i) \cdot (\psi_e^* \beta\boldsymbol{\delta}\psi_{\bar{\nu}}) + (\Psi_f^* \beta\boldsymbol{\alpha}\Psi_i) \cdot (\psi_e^* \beta\boldsymbol{\alpha}\psi_{\bar{\nu}}).$$

[d] The assignment of the antineutrino, rather than of the neutrino, to β^--emission is arbitrary as long as the neutrino mass is zero. No experiment so far conceived can distinguish between a zero mass and the upper limit of $m_\nu = 1$ kev.

[e] $\beta\gamma_5$ is antihermitian but $i\beta\gamma_5$, which is hermitian, differs from it by a multiplicative factor: -1 in the contracted interaction term.

Since there is no way of choosing the appropriate form *a priori*, we must admit all five interactions and write

$$\mathcal{H}_\beta = g \sum_X C_X \mathcal{H}_\beta(X) \tag{8}$$

where $\mathcal{H}_\beta(X)$ means the forms of $\mathcal{H}_\beta$ obtained by the five choices in (7), and X stands for S, V, T, A, and P which mean scalar, vector, tensor, axial vector, and pseudoscalar interactions, respectively. The ten coupling constants C_X may each be complex. The gC_X and gC_X' are, therefore, twenty real constants although an over-all phase constant is irrelevant. We can normalize to $\sum_X |C_X|^2 + |C_X'|^2 = 1$.

The second part of $\mathcal{H}_-$ is

$$\mathcal{H}_\beta' = g \sum_X C_X' \mathcal{H}_\beta'(X) \tag{9}$$

where

$$\mathcal{H}_\beta'(X) = \sum_i (\Psi_f{}^* O_i \Psi_i)(\psi_e{}^* O_i \gamma_5 \psi_{\bar{\nu}}). \tag{9'}$$

The effect of the added γ_5 in the lepton scalar product is to change the transformation properties S ↔ P, V ↔ A, T ↔ T. Unless one measures some quantity which is itself a pseudoscalar (for example, $<\mathbf{J}>_{Av} \cdot \mathbf{p}$, that is, the angular distribution of electrons with momentum $\mathbf{p}$ emitted from polarized nuclei with average spin $<\mathbf{J}>_{Av}$) (*4*), the use of $\mathcal{H}_\beta + \mathcal{H}_\beta'$ is an unnecessary complication and the same results,[f] with a renaming of the coupling constants ($C_X \leftrightarrow C_X'$) follow from the use of $\mathcal{H}_\beta$ or $\mathcal{H}_\beta'$. With both $\mathcal{H}_\beta$ and $\mathcal{H}_\beta'$ the results for the intensities are merely additive. We consider only $\mathcal{H}_\beta$ in the following since it suffices for present purposes.

For each of the five interactions one can make an expansion of the scalar products (matrix elements) in the following way: We assume for simplicity that the Coulomb field can be neglected. The $\psi_e \sim \exp(i\mathbf{p} \cdot \mathbf{r})$ and $\psi_{\bar{\nu}} \sim \exp(-i\mathbf{q} \cdot \mathbf{r})$ where $\mathbf{q}$ is the momentum of the antineutrino. Then the lepton scalar product is proportional to

$$e^{-i\mathbf{P} \cdot \mathbf{r}} = \sum_0^\infty \frac{1}{n!} (i\mathbf{P} \cdot \mathbf{r})^n = \sum_0 (2l+1) i^l j_l(Pr) P_l(\cos \Theta). \tag{10}$$

Here $\mathbf{P} = \mathbf{p} + \mathbf{q}$, j_l is a spherical Bessel function (which is replaced by another radial function when the Coulomb field is considered), and P_l a Legendre polynomial with Θ the angle between $\mathbf{P}$ and $\mathbf{r}$. Since r is

[f] It is assumed that $m_\nu = 0$ and that the spin of the neutrino is not observed.

restricted to values ≤ the nuclear radius R and for all β-processes of interest $PR \ll 1$, it follows that in either expansion the successive terms decrease sharply. Using the expansion (10) the term $P_l(\cos \Theta)$ can be coupled to the nuclear operator O_i so that the matrix element in $\mathcal{K}_\beta(X)$ can be written as a sum of the following general type (see also 5):

$$\mathcal{K}_\beta(X) \sim \sum_{lj} (f|T_{jl}|i)(PR)^l \tag{11}$$

where the matrix element $(f|T_{jl}|i)$ has the selection rules

$$\Delta(J_f J_i j), \qquad \Delta\pi = (-)^l$$

and in addition $\Delta(jl1)$ holds: $j = l, l \pm 1$.[g] The largest term is $l = 0$ and $j = 0, 1$ with $\Delta\pi = 1$. These are the allowed transitions. The next term has $l = 1$, $j = 0, 1, 2$, and $\Delta\pi = -1$. These are the first forbidden transitions and $j = 2$ corresponds to the unique type. In general, the latter arise only from the T, A interactions. The second forbidden transitions have $l = 2$ but $j = 2, 3$ only.[g] The terms with $j = 1$ are omitted since for both these and the allowed transitions, $\Delta\pi = 1$ and the second forbidden terms are small corrections to the allowed contributions.[h] The procedure for higher orders should now be clear. We summarize the selection rules in Table I.

TABLE I. SUMMARY OF SELECTION RULES

Interaction	n = 0		n = 1		n = 2	
	$\Delta\pi$	j	$\Delta\pi$	j	$\Delta\pi$	j
S	1	0	−1	1	1	2
V	1	0	−1	1	1	2
T	1	1	−1	0, 1, 2	1	2, 3
A	1	1	−1	0, 1, 2	1	2, 3
P			−1	0	1	1

It is evident that the P interaction, having different parity selection rules ($\Delta\pi = -1$, $j = 0$ for the leading $l = 0$ term) belongs to one order of forbiddenness greater than the other four. The only role of the P interaction of possible practical interest is in $\Delta J = 0$, $\Delta\pi = -1$ (first forbidden transitions). It appears, however, that if the P interaction is important its coupling constant must be extraordinarily large (5). At present there seems to be no definite need for the P interaction. On the other hand,

[g] For the S interaction $j = l$ only. For the P interaction it is necessary to restrict j to the single value $l - 1$, $l \geq 1$; see next paragraph.

[h] For l-forbidden transitions these small corrections would be significant.

the existence of allowed $0 \to 0$ transitions and the unique $\Delta J = n + 1$ transitions makes it certain that at least one of the group S, V and one of the group T, A are necessary.

3. The Energy Distribution and Angular Correlation: Allowed Transitions

Using a system of units[i] for which $m = c = \hbar = 1$ and transition probabilities are expressed in units $mc^2/\hbar = 7.7 \times 10^{20}$ sec^{-1} the transition probability for the $\beta^\pm$-particle to have energy between W and $W + dW$ and for the directions of the β-particle and neutrino (or antineutrino) to make an angle ϑ to $\vartheta + d\vartheta$, is $N_\pm(W) \, dW \sin \vartheta \, d\vartheta$ where

$$N_\pm(W) = \frac{g^2}{4\pi^3} F(\mp Z, W) p W (W_0 - W)^2 \xi \left[1 + \frac{ap}{W} \cos \vartheta \mp \frac{b}{W} \right]. \quad (12)$$

Here

$$\xi = (|C_S|^2 + |C_V|^2)|M_F|^2 + (|C_T|^2 + |C_A|^2)|M_{GT}|^2 \quad (12a)$$
$$\xi a = \tfrac{1}{3}(|C_T|^2 - |C_A|^2)|M_{GT}|^2 - (|C_S|^2 - |C_V|^2)|M_F|^2 \quad (12b)$$
$$\xi b = 2\gamma Re[C_S C_V^* |M_F|^2 + C_T C_A^* |M_{GT}|^2] \quad (12c)$$
$$M_F = (\Psi_f^* \Psi_i); \quad M_{GT} = (\Psi_f^* \sigma \Psi_i). \quad (12d)$$

The matrix elements in (12d) are referred to as Fermi and Gamow-Teller matrix elements respectively. In them the approximation $\beta = -1$ has been made. Throughout our representation is

$$\beta = \begin{pmatrix} 1 & 0 \\ 0 & -1 \end{pmatrix}$$

where each element is a 2 by 2 matrix. When $\mathcal{H}_{\beta'}$ is used each combination of constants $C_X C_Y^*$ is replaced by $C_X' C_Y'^*$ and, with the exception noted in the remarks following Eq. (9'), there are no interference terms in $\mathcal{H}_\beta$, $\mathcal{H}_{\beta'}$.[j]

In Eq. (12) $F(\mp Z, W)$ is the Fermi function. Explicitly

$$F(Z, W) = 2(1 + \gamma)(2pR)^{2(\gamma-1)} e^{\pi y} \frac{|\Gamma(\gamma + iy)|^2}{[\Gamma(2\gamma + 1)]^2} \quad (13)$$

with

$$\gamma = (1 - \alpha^2 Z^2)^{\frac{1}{2}}, \quad y = \alpha Z W / p$$

and $\alpha = e^2/\hbar c = 1/137.03$ is the fine structure constant. The Fermi function is essentially the density of electrons ($\pm$) at the nuclear surface

[i] The units of energy, momentum and length are mc^2, mc, and $\hbar/mc$ respectively.
[j] For a discussion of experiments where the nonconservation of parity in the β interaction becomes germane, see Jackson et al. (6).

normalized so that $F(0, W) = 1$. A complete tabulation of this function is given elsewhere (7). The small values of F for low-energy positrons are a consequence of Coulomb repulsion. In contrast it will be noted that the number of slow electrons can be comparatively large, especially for large values of Z.

The energy-dependent factor $pWq^2 = pW(W_0 - W)^2$ (for zero-mass neutrinos) is simply the volume of momentum space (per unit energy) available to an electron in the energy range specified. The remaining factors in Eq. (12) are:

(i) The angular correlation term (proportional to a) which is observed as an angular correlation between electron and recoil nucleus. In contrast to the first or main term, this angular correlation term distinguishes between the four interactions entering allowed β-decay. For the energy spectrum this term drops out.

(ii) The Fierz interference terms which are present only for mixed interactions. The mixtures occur only between different members of the F and GT interactions (S, V and T, A).

4. Analysis of the Shapes of Allowed Spectra

Evidence to be cited shows that the Fierz interference terms are small if not identically zero. Then with $b = 0$, Eq. (12) shows that

$$[N_\pm(W)/F(\mp Z, W)pW]^{\frac{1}{2}} = K(W_0 - W) \tag{14}$$

where K is independent of energy and the N is total counting rate at W. Since W_0 is often not known,[k] the plot of the left-hand side of Eq. (14) versus W should give a straight line (if $b = 0$) or should in any case extrapolate to $W = W_0$. This type of plot is known as a Kurie (or Fermi) plot and originally it was conceived as a test of the β-decay theory. The fact that very good straight lines are always obtained from the appropriately corrected (see below) and highly precise data shows that no deviations such as would arise from Fierz interference are required within present experimental accuracy. In this connection it is necessary to use very thin sources since distortions arising from finite thickness can simulate a $1/W$ dependence of the spectrum.

Despite the great precision obtainable in spectrum measurements the conclusion drawn from the small Fierz interference is that the F and GT interactions, while predominantly pure, are limited by

$$\frac{ReC_SC_L^*}{|C_S|^2 + |C_L|^2} \approx \frac{ReC_SC_L^*}{|C_L|^2} \lesssim 0.10$$

[k] It is in principle obtainable from the threshold of $n - p$ and $p - n$ reactions.

where C_S and C_L are the smaller and larger, respectively, of the pair C_T, C_A. A similar relation applies for the pair C_S and C_V but the upper limit is considerably larger (8).

The possibility that $b_{GT} \ll 1$ because C_A and C_T are out of phase by $\frac{\pi}{2}$, ($iC_TC_A{}^*$ real) or nearly so is ruled out by the angular correlation experiments in He^6 which depend on a ($M_F = 0$ for the $\Delta J = 1$ transition $He^6 \to Li^6 + e^-$) (9) and here only the moduli $|C_T|$ and $|C_A|$ enter. The results of the experiment are stated to be consistent with $|C_T/C_A| \lesssim 0.01$. The corresponding correlation experiment with $M_{GT} = 0$ would be $C^{10} \to B^{10} + e^+$, $O^{14} \to N^{14} + e^+$ or $Cl^{34} \to S^{34} + e^+$. While such an experiment remains to be done, the angular correlation measurements in A^{35} and Ne^{19} have been carried out and are consistent with $C_S/C_V \approx 0$.[1]

The spectral shape is also important for the question of neutrino mass. If the neutrino mass $m_\nu \neq 0$ the statistical factor would be modified as follows:

$$pW(W_0 - W)^2 \to pW(W_0 - W)[(W_0 - W)^2 - m_\nu^2]^{\frac{1}{2}}. \qquad (15)$$

For W near the endpoint ($W_0 - m_\nu$) the spectrum would have a vertical slope due to the factor $[W_0 - W - m_\nu]^{\frac{1}{2}}$. The spectrum of H^3 with its low endpoint 18.6 kev is most suitable for investigating this question and the results are consistent with $m_\nu \leq 0.002$ (1 kev).[m]

5. Forbidden Spectra

The shapes of forbidden spectra ($n \geq 1$) are usually described in terms of a correction factor. For pure interactions, $X = $ one of S, V, T, A, P, these have been defined by Greuling (12) so that the energy distribution is

$$N_\pm(W) = \frac{g^2}{2\pi^3} C_{nX} F(\mp Z, W) pW(W_0 - W)^2.$$

Thus, for allowed transitions $C_{0X} = \xi$, see Eq. (12). The extension to the general case where all the interactions are operative involves nine additional correction factors for each order n. With a consistent normalization these have been given by Pursey (13).[n] These correction factors are rather

[1] According to Herrmannsfeldt et al. (10), the measurement of the polarization of β-particles emitted by *unpolarized* sources will bear on the question of the nature of the Fermi interaction, but the conclusions at the time of writing are indefinite.

[m] A review of this problem is given by C. S. Wu (11).

[n] See, however, Konopinski (14). The results given by A. Smith (15) are correct except that in the nuclear elements of relativistic operators (for example, βα) it is not permissible to replace β by its nonrelativistic limit.

involved and we do not reproduce them here. However, the following qualitative remarks will be relevant to the use of these formulas.

First, it is necessary to supplement the discussion of Section 2 leading to the order of forbiddenness expansion. If we consider the terms in the T interaction for which $O_i = \beta\sigma_i$ the discussion given there is unchanged. The same remark applies to the operators β, 1, and $\dot{\sigma}$ in the S, V, and A interactions. However, for α, $\beta\alpha$ and γ_5 in the V, T, and A interactions it is necessary to observe first that these have the opposite parity properties as compared to the operators β, 1, $\beta\dot{\sigma}$, $\dot{\sigma}$. Moreover, α, $\beta\alpha$, and γ_5 are operators which are proportional to v/c of the nucleons and, hence, give small matrix elements. The result is that in the expansion (11) the leading term ($l = 0$) gives first forbidden transitions, the $l = 1$ second forbidden transitions, and so on. The selection rules of Table I still apply quite generally. Thus, the operator $\beta\alpha$ combined with $P_0(\cos \Theta)$ gives $j = 1$ (because $\beta\alpha$ is a vector or first rank tensor).

Since first forbidden transitions are far more numerous and of generally greater interest than high order forbidden transitions, the remaining remarks are directed to that case. We first consider the unique transitions ($j = 2$). Since this type of transition is insensitive to the Coulomb field, a fair approximation to the correction factor is obtained from the plane wave expansion of Section 3. Then the correction factor is proportional to

$$<(\mathbf{p} + \mathbf{q})^2>_{Av} = p^2 + q^2 \qquad (16)$$

and the average is over the directions of $\mathbf{p}$ and $\mathbf{q}$. More generally, the nth forbidden correction factor is proportional to $<(\mathbf{p} + \mathbf{q})^{2n}>_{Av}$. When the Coulomb field is included, this becomes

$$C_n(j = n + 1) = \sum_{\nu=0}^{n} \frac{2^{n-2\nu}(2\nu + 1)!}{(2n - 2\nu + 1)!(\nu!)^2} q^{2n-2\nu} L_\nu \qquad (17)$$

where, with $k = \nu + 1$,

$$L_{k-1} = \frac{1}{2p^2 F R^{2k-2}} (g_{-k}^2 + f_k^2) \simeq \left(\frac{2^{k-1}(k-1)!}{(2k-1)!}\right)^2 p^{2k-2} \qquad (18)$$

and g_{-k} is the "large" radial function of an electron with angular momentum $k - \frac{1}{2}$ and parity° $(-)^k$ while f_k is the corresponding function for the same angular momentum and parity $(-)^{k+1}$. Here and in the following $k \geq 1$. The right-hand side is an approximation valid for $\alpha ZR \ll 1$. This approximation is usually excellent for light and medium Z elements and C_n is very nearly symmetrical in p and q for such cases. With the

---

° For a Dirac electron the parity operator is multiplication by β and space inversion ($\mathbf{r} \to -\mathbf{r}$).

approximation represented by the right-hand side of Eq. (18)

$$C_n\ (j = n + 1) = \frac{2^{n-1}}{(2n + 2)!} \frac{1}{pq} [(p + q)^{2n+2} - (p - q)^{2n+2}].$$

It will be noted that the first forbidden unique transitions ($j = 2$, $n = 1$) can contribute to $\Delta J = 1$ and 0 (if $J_i + J_f \geq 2$) but if $\alpha Z R \approx 0.4 Z A^{\frac{1}{3}} \gg W_0 - 1$, which is usually the case, the contribution to $\Delta J < 2$ (or in general $\Delta J < n + 1$) is small. Since there is only one nuclear matrix element with $j = n + 1$ the shape of the spectrum is independent of the values of the nuclear matrix elements.

All other transitions are nonunique in the sense that the shape of the spectrum depends on the ratio of nuclear matrix elements. These can be calculated using some nuclear model but the results are not always reliable. The matrix element ratios can then alternatively be regarded as adjustable parameters and the results obtained by fitting the spectrum can be compared to model calculations. For a $\Delta J = 0$, $\Delta \pi = -1$ transition the operators with $j = 0$ can contribute in all cases. These are $\beta \boldsymbol{\sigma} \cdot \mathbf{r}$ and $\boldsymbol{\sigma} \cdot \mathbf{r}$ in the T and A interactions and $\beta \gamma_5$ and γ_5 in the P and A interactions. Contributions of all these produce interference terms. Also, if $J_i + J_f \geq 1$ the $j = 1$ terms can contribute to $\Delta J = 0$, $\Delta \pi = -1$. The corresponding operators are $\beta \mathbf{r}$ from the S interaction, $\mathbf{r}$ and $\boldsymbol{\alpha}$ from the V interaction, $\beta \boldsymbol{\sigma} \times \mathbf{r}$ and $\beta \boldsymbol{\alpha}$ from T, $\boldsymbol{\sigma} \times \mathbf{r}$ from A. All these can interfere but no interferences arise in the energy spectrum from operators of different j. For $\Delta J = 1$, $\Delta \pi = 1$ all the $j = 1$ operators contribute and interfere. The $j = 0$ operators make no contribution.

For all these cases the correction factors are polynomials in q with (p, Z)-dependent coefficients defined by bilinear combinations of electron wave functions. These combinations are L_ν, M_ν, N_ν for the unmixed case where L_ν is given in Eq. (18) and

$$M_{k-1} = \frac{1}{2p^2 F R^{2k}} (g_k{}^2 + f_{-k}{}^2) \tag{18'}$$

$$N_{k-1} = \frac{1}{2p^2 F R^{2k-1}} (f_{-k} g_{-k} - f_k g_k). \tag{18''}$$

For the mixed case the additional combinations

$$P_{k-1} = \frac{1}{2p^2 F R^{2k-2}} (g_{-k}{}^2 - f_k{}^2) \tag{19}$$

$$Q_{k-1} = \frac{1}{2p^2 F R^{2k}} (g_k{}^2 - f_{-k}{}^2) \tag{19'}$$

$$R_{k-1} = \frac{1}{2p^2 F R^{2k-1}} (f_{-k} g_{-k} + f_k g_k) \tag{19''}$$

occur. Approximations for the Coulomb field valid when $\alpha ZR \ll 1$ are given by Greuling (12). However, corrections due to the departure of the actual field from that of a point nucleus are important for N_ν and M_ν, (also Q_ν and R_ν), especially for $\nu = 0$. The uncorrected values of the six functions $L \ldots R$ for a wide range of parameters on which they depend has been tabulated (16).

A final point of importance is that most of the nonunique first forbidden spectra exhibit the allowed shape. This fact has been emphasized many times in the literature (17) and the circumstance that the shape of the spectrum does not lead directly to a forbiddenness classification should be noted.

6. Corrections to the Spectral Shapes

Before the results of a spectrum measurement can be analyzed a number of corrections must be applied. Corrections for energy resolution (18) are to be applied to the experimental data. Corrections for screening, finite nuclear size, and finite deBroglie wave length may be applied to the theoretical form of the energy distribution.

Screening corrections have been considered by several authors (19). An adequate representation of this effect is obtained by replacing $F(\pm Z, W)$ by

$$F(\pm Z, W \mp V_0) \left[\frac{(W \mp V_0)^2 - 1}{W^2 - 1} \right]^{\frac{1}{2}} \left(\frac{W \mp V_0}{W} \right) \qquad (20)$$

where V_0 is the shift in the potential energy near the nucleus arising from the screening. Using the Thomas-Fermi model this is

$$V_0 \cong 1.13 \alpha^2 Z^{4/3}.$$

For allowed transitions the correction is rarely more than 10 percent and more often of order 1–2 percent. Of course, it is most important for slow β-particles. The corrections are much larger for slow positrons but the transition probability $N(W)$ is then itself exceedingly small. For forbidden transitions the replacement of W by $W \mp V_0$ as in Eq. (20) should be made.

Corrections for the finite size of the nucleus are most important in nonunique forbidden transitions. The numerical values have been given by Rose and Holmes (20).

The corrections for finite de Broglie wave length refer to the use of the exact form for the radial functions rather than those arising from the $\alpha ZR \ll 1$ approximation. These corrections, while usually small, are often pertinent in view of the precision needed in the analysis of β-spectra. The results of reference (16) include this correction.

It is sometimes of interest to consider the effect of another approximation in the theory; namely, the evaluation of the radial functions on the nuclear surface instead of averaging them over the nuclear volume. Since the weight factor in this average is not known, the correction cannot be made with any rigor (21).

7. ft Values[p]

The decay constant of the β-transition is

$$\lambda = \int_1^{W_0} N(W)\, dW = \frac{g^2}{2\pi^3} \xi f(Z, W_0) \qquad (21)$$

where

$$f(Z, W_0) = \int_1^{W_0} pW(W_0 - W)^2 F(Z, W)\, dW. \qquad (21a)$$

Numerical values have been extensively tabulated (22); see Figs. 2 through 6.

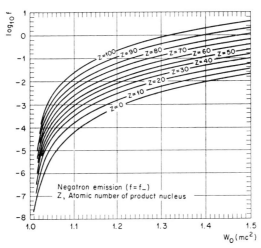

FIG. 2. Values of f for β^- emission, low-energy scale (courtesy Feenberg and Trigg).

It is evident that the parameter governing the intrinsic speed of the transition is not λ since this quantity increases rapidly with W_0. However,

$$ft = \frac{\ln 2}{\lambda} f \qquad (21b)$$

[p] See also Section I.E.1.

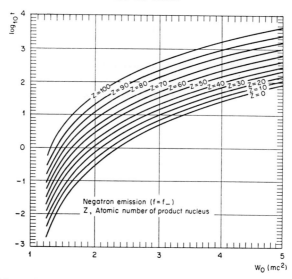

FIG. 3. Values of f for β^- emission, medium-energy scale (courtesy Feenberg and Trigg).

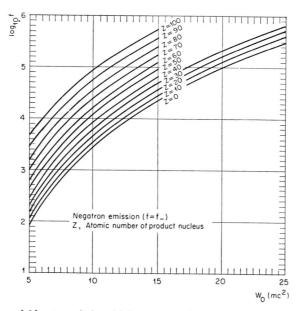

FIG. 4. Values of f for β^- emission, high-energy scale (courtesy Feenberg and Trigg).

where t is the half-life, provides a basis of comparison between different transitions. Usually $\beta \equiv \log_{10} ft$ is the number of interest. The observed values of β for allowed transitions fall into two fairly distinct groups: the super-allowed transitions for which values of β cluster around 3.5 and the normal allowed transitions for which $4.5 \lesssim \beta \lesssim 6.0$.[q] The super-allowed transitions occur between mirror nuclei $N - Z = \pm 1$ and between $N = Z \leftrightarrow N - Z = \pm 2$. First forbidden transitions, which

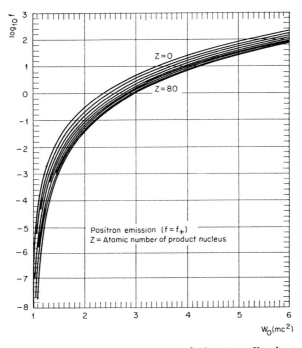

FIG. 5. Values of f for β^+ emission, low-energy scale (courtesy Feenberg and Trigg).

partially overlap the normal allowed ones, generally have $\beta \approx 8$–9 and for second forbidden transitions $\beta \sim 12$–13. Therefore, in many cases a measurement of the ft value permits a forbiddenness and, therefore, a parity change classification.

A further application of the ft values has been made to evaluate the relative strength of Fermi and Gamow-Teller interactions (8). This is based on the fact that

$$ft = \frac{B}{(1 - x)|M_F|^2 + x|M_{GT}|^2} \tag{22}$$

[q] Exceptions already noted are C^{14} ($\beta \approx 9.0$) and P^{32} ($\beta \approx 7.9$).

where x and B are the universal constants

$$x = |C_T|^2 + |C_A|^2 + |C_T'|^2 + |C_A'|^2 = |C_{GT}|^2 = 1 - |C_F|^2 \quad (22a)$$
$$B = \frac{2\pi^3 \ln 2}{g^2} = \frac{43}{g^2}. \quad (22b)$$

Here we have explicitly recognized the terms from the $\mathcal{H}_\beta'$ interaction. Hence, evaluating the matrix elements by some nuclear model one can

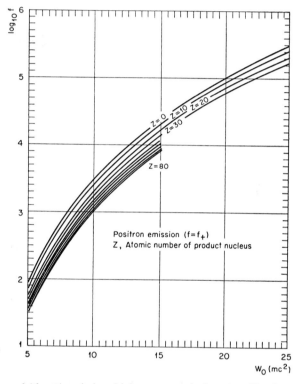

Fig. 6. Values of f for β^+ emission, high energy scale (courtesy Feenberg and Trigg).

draw $B - x$ curves (straight lines) for each transition and these should all intersect at one point. The results of the analysis are

$$B = 2787 \pm 70,$$
$$x = 0.560 \pm 0.012.$$

The errors reflect the spread of the points of intersection. This result for x indicates that the GT interaction is only slightly stronger than the F interaction.

8. Failure of Parity Conservation

The necessity for introducing both interactions, $\mathcal{H}_\beta$ and $\mathcal{H}_\beta'$ in (6), was made evident by the observation of an anisotropy in the angular distribution of β-particles emitted by polarized Co^{60} nuclei (*4*). This result, postulated by Yang and Lee (*2*), implies that the β-interaction does not conserve parity. In effect, a distinction between right and left is made in that the results of a measurement of the β-particle momentum is changed if this vector is reflected in the plane defined by the magnetic field (an axial vector) producing the nuclear polarization.

It is not our purpose to pursue this question in all of its details but it is important for the present considerations to recognize the consequences of this break with the traditional formulation of the theory which was held valid before 1957. A number of other experiments, which have been successfully carried out provide eloquent evidence of parity breakdown and, in addition, are of decisive importance in fixing the interaction. These are summarized herewith:

(a) Electrons emitted in allowed β-decay from unpolarized nuclei exhibit a longitudinal polarization (*23*) equal to $-v/c$ where v is the β-particle velocity. For positrons the polarization is v/c.

(b) The neutrinos emitted in electron capture (see Section 9) are completely polarized (*24*) with a polarization equal to -1.

(c) The electrons emitted in the decay of polarized neutrons exhibit an angular distribution of $1 - 0.09 \cos \vartheta$ where ϑ is the angle of emission with respect to the neutron spin (*25*).

The interpretation of these results is a β-interaction composed of only V and A parts. The results are best represented, according to present data, by

$$C_V = C_V', \quad C_A = C_A' = -1.2 C_V.$$

The breakdown of parity also implies a breakdown of charge conjugation invariance; that is, it is no longer valid to assume in weak interactions, like that discussed here, that the results of an experiment are the same if all particles are replaced by antiparticles. However, with the use of very general arguments, the result that the ratios of all coupling constants are real shows that invariance under combined particle-antiparticle exchange and reflection in a mirror is a valid concept. It should be emphasized that the present data are not very accurate so far as this last conclusion is concerned.

9. Orbital Capture

The process of orbital capture is accompanied by the emission of a monoenergetic neutrino. Since no spectrum can be measured[r] the primary interest centers on branching ratios between competing capture processes to different excited states and to the ground state as well as branching between capture and β^+-emission (to the same daughter state) where the latter is energetically possible.

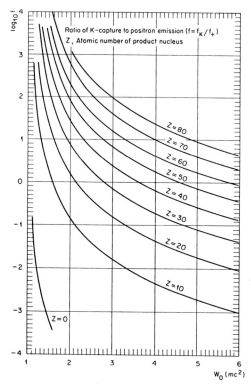

FIG. 7. The K/β^+ branching ratio, low-energy scale (courtesy Feenberg and Trigg).

The decay constant for capture of an $s_{\frac{1}{2}}$ electron (K shell, L_{I} subshell, etc.) is

$$\lambda(s_{\frac{1}{2}}) = \frac{g^2}{4\pi^2} \xi (W_0 + W_s)^2 g_s^2(R) \tag{23}$$

---

[r] Measurements of the weak internal bremsstrahlung spectrum are useful for determining the maximum energy W_0 but do not yield information as to order of forbiddenness. These measurements are, therefore, of secondary importance for purposes of nuclear spectroscopy.

where W_s is the total energy of the bound electron, including rest energy. In terms of the binding energy ϵ_s

$$W_s = 1 - \epsilon_s.$$

In Eq. (23) $g_s(R)$ is the large radial function evaluated at the nuclear

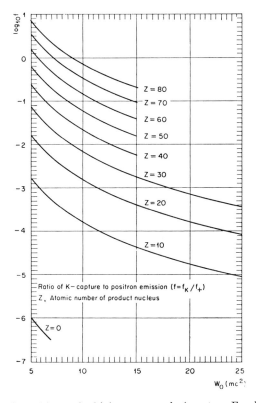

FIG. 8. The K/β^+ branching ratio, high-energy scale (courtesy Feenberg and Trigg).

radius. For a Coulomb field (screening neglected) one obtains

$$g_s^2(R) = \frac{1+\gamma}{2\Gamma(2\gamma+1)} R^{2\gamma-2}(2\alpha Z)^{2\gamma+1}, \qquad K \text{ shell}$$

$$g_s^2(R) = \frac{1}{2}\frac{2\gamma+1}{\Gamma(2\gamma+1)}\frac{1+W_s}{1+2W_s}\frac{R^{2\gamma-2}}{W_s^{2\gamma}}(\alpha Z)^{2\gamma+1}, \qquad L_\text{I} \text{ subshell}$$

$$W_s = \left(\frac{1+\gamma}{2}\right)^{\frac{1}{2}}. \tag{24}$$

Capture from the L_II or L_III subshells is generally much weaker than L_I capture. The only exception occurs in a narrow energy band near the threshold for L_I capture. It should be noted that L_{III} capture occurs only in forbidden transitions and, as stated, is important only in the exceptional case that L_I capture is energetically impossible or nearly so.

In most cases $W_0 \gg W_s$ and then the L_I/K branching ratio (with an obvious change of notation) is given by

$$\lambda(L_I)/\lambda(K) \cong g_{L_I}^2(R)/g_K^2(R). \tag{25}$$

This ratio has been calculated with screening and finite nuclear size corrections (*21*). A deviation of the observed branching ratio from these calculated values occurs only near the K threshold and when this occurs it permits a determination of the energy release W_0:

$$W_0 = (\rho W_K - W_{L_I})/(1 - \rho) \tag{26}$$

where

$$\rho^2 = \frac{g_K^2(R)}{g_{L_I}^2(R)} \frac{\lambda(L_I)}{\lambda(K)}. \qquad (\rho > 0) \tag{26'}$$

The K/β^+ branching ratio for allowed transitions has been given by Feenberg and Trigg (*22*) and the results are shown in Figs. 7 and 8. The λ_+ (or f_+) referred to in these figures is given by Eq. (21) evaluated for positrons. No corrections for screening and finite size have been made but these are always quite small. The branching ratio for forbidden transitions has been discussed by Brysk and Rose (*21*). In general, no information is obtained from the study of such forbidden transitions that would not be available from the analysis of the positron spectrum.

REFERENCES

1. C. L. Cowan, Jr., F. Reines, F. B. Harrison, H. W. Kruse, and A. D. McGuire, Science **124**, 103 (1956), and private communication.
2. See T. D. Lee and C. N. Yang, Phys. Rev. **104**, 254 (1956).
3. See R. J. Blin-Stoyle, M. A. Grace, and H. Halban, in *Beta- and Gamma-Ray Spectroscopy*, edited by K. Siegbahn (North Holland Publishing Co., Amsterdam, 1955), Chapter 19.
4. C. S. Wu, E. Ambler, R. W. Hayward, D. D. Hoppes, and R. P. Hudson, Phys. Rev. **105**, 1413 (1957).
5. M. J. Laubitz, Proc. Phys. Soc. (London) **A69**, 789 (1956).
6. J. D. Jackson, S. B. Treiman, and H. W. Wyld, Jr., Phys. Rev. **106**, 517 (1957).
7. M. E. Rose, in *Beta- and Gamma-Ray Spectroscopy*, edited by K. Siegbahn (North Holland Publishing Co., Amsterdam, 1955), Chapter 9, Appendix II.
8. O. Kofoed-Hansen and A. Winther, Kgl. Danske Videnskab. Selskab **30**, No. 20 (1956).
9. W. B. Hermannsfeldt, R. L. Burman, P. Stähelin, J. S. Allen, and T. H. Braid, Phys. Rev. Letters **1**, 61 (1958); Gatlinburg Conference on Weak Interactions, Bulletin, Am. Phys. Soc., New York meeting (1959). F. Pleasonton, C. H. Johnson and A. H. Snell, Gatlinburg Conference, *ibid*.
10. W. B. Hermannsfeldt, D. R. Maxson, P. Stähelin, and J. S. Allen, Phys. Rev. **107**, 641 (1957).

11. C. S. Wu, in *Beta- and Gamma-Ray Spectroscopy*, edited by K. Siegbahn (North Holland Publishing Co., Amsterdam, 1955), Chapter 11.
12. E. Greuling, Phys. Rev. **61**, 568 (1942).
13. D. L. Pursey, Phil. Mag. [7] **42**, 1193 (1951).
14. E. J. Konopinski, in *Beta- and Gamma-Ray Spectroscopy*, edited by K. Siegbahn (North Holland Publishing Co., Amsterdam, 1955), Chapter 10.
15. A. Smith, Phys. Rev. **82**, 955 (1951).
16. M. E. Rose, C. L. Perry, and N. M. Dismuke, Oak Ridge National Laboratory Report No. 1459 (1953).
17. H. M. Mahmoud and E. J. Konopinski, Phys. Rev. **88**, 1266 (1952).
18. G. E. Owen and H. Primakoff, Phys. Rev. **74**, 1406 (1948).
19. M. E. Rose, Phys. Rev. **49**, 727 (1936); J. Reitz, *ibid.* **77**, 10 (1950). See also C. Longmire and H. Brown, Phys. Rev. **75**, 1102 (1949).
20. M. E. Rose and D. K. Holmes, Phys. Rev. **83**, 190 (1951), and Oak Ridge National Laboratory Report No. 1022 (1951).
21. See H. Brysk and M. E. Rose, Oak Ridge National Laboratory Report No. 1830 (1955); Revs. Modern Phys. **30**, 1169 (1958).
22. E. Feenberg and G. Trigg, Revs. Modern Phys. **22**, 399 (1950).
23. See *Proceedings of the Rehovoth Conference on Nuclear Structure, 1957* (North-Holland Publishing Co., 1958). Session vi, especially pages 376–403 contains a comprehensive account of methods and results.
24. M. Goldhaber, L. Grodzins, and A. W. Sunyar, Phys. Rev. **109**, 1015 (1958).
25. M. T. Burgy, V. E. Krohn, T. B. Novey, G. R. Ringo, and V. L. Telegdi, Phys. Rev. **110**, 1214 (1958).

V. E. Analysis of Internal Conversion Data

by M. E. ROSE

1. Definition of the Process.. 834
2. The Selection Rules. Mixtures of Multipoles............................. 836
3. The Calculation of Internal Conversion Coefficients.................. 838
   a. General Assumptions.. 838
      (1) Perturbation Theory... 838
      (2) Electron Dynamics... 838
   b. Specialized Assumptions: Nuclear Structure and Screening Model........ 839
      (1) Unscreened Coefficients.. 839
      (2) Screening Effect... 839
      (3) Finite Nuclear Size Effect... 839
4. Numerical Values of Internal Conversion Coefficients................. 842
   References... 851

1. Definition of the Process[a]

When a nucleus is in an excited state for which the excitation energy is insufficient for the emission of nuclear particles, the de-excitation will proceed predominantly by either one of two competing mechanisms. Either a γ-ray will be emitted or the nuclear excitation energy will be transferred to one of the orbital electrons resulting in the ejection from the atom of this electron. The latter process is referred to as internal conversion. The branching ratio, giving the number of conversion electrons per second, N_e, to the number of photons per second, N_q, is the internal conversion coefficient α,

$$\alpha = N_e/N_q. \tag{1}$$

Obviously, for radiative transitions α may have any value greater than zero.

There are two exceptions to the de-excitation modes described above. In one the nuclear transition is accompanied by emission of an electron-positron pair (3).* This requires an excitation energy greater than $2mc^2$ (1.02 Mev). The corresponding branching ratio (pairs to photons) is usually about 10^{-4} and rarely exceeds 10^{-3}. The other exception arises

---

[a] The subject of internal conversion has been discussed in many places. The review article by M. E. Rose (1) may be referred to for earlier references. Chapter 5 of "Multipole Fields" (2) gives a summary of the theory.

* The reference list for Section V.E is on page 851.

when both initial and final states have zero angular momentum. Then no electromagnetic radiation can be emitted and only conversion electrons or (with sufficient energy available) pairs can occur. In this case it is meaningless to speak of a conversion coefficient. In any event these two processes, pair emission and radiationless transitions, are usually identifiable and need not be confused with internal conversion. The only case wherein care must be exercised is that one in which, from the energy of the transition and atomic number of the emitter, the conversion coefficient would be expected to be very large and no photons could be observed with experimental techniques available. A distinction between this case (in an even mass emitter) and the $0 \to 0$ radiationless transition may be made on the basis of relative conversion coefficients between different shells or subshells.

It is recognized that the decay constant of the excited state is

$$\lambda = \lambda_0(1 + \alpha)$$

where λ_0 is the decay constant which would characterize a naked nucleus. This λ_0 is the quantity which has to be compared with calculated γ-ray transition probabilities.

The emitted electrons will appear as a line spectrum in a magnetic spectrometer. A given transition can convert in a number of possible shells, assuming energy conservation can be fulfilled. Thus we speak of K, L_I, L_{II}, L_{III}, etc. conversion and also of the conversion coefficients for these various shells and subshells. If the L_I–L_{II} splitting cannot be resolved (small screening separation) one will measure $\alpha(L_I) + \alpha(L_{II})$ where the argument here indicates the initial electronic state. Similarly, for unresolved spin-orbit splitting the L_{II} and L_{III} lines will coalesce and one measures $\alpha(L_{II}) + \alpha(L_{III})$. Where no resolution in the L-shell is possible we obtain the total L-shell conversion with a coefficient given by

$$\alpha(L) = \alpha(L_I) + \alpha(L_{II}) + \alpha(L_{III}). \tag{2}$$

Clearly, one can measure various ratios of conversion coefficients and even where M conversion is weak and L conversion not resolved one can obtain a K/L ratio:

$$K/L = \alpha(K)/\alpha(L). \tag{2'}$$

It is by means of such ratios, which are characteristic of the conversion process, that the possible ambiguity[b] with radiationless transitions can be resolved.

[b] In other cases the $0 \to 0$ transition might be identified by the energy distribution of pairs or by identification of the angular momentum of the initial state from β-decay data and identification of the final state as the ground state in an even-even nucleus. See also Church and Weneser (4).

Similarly, for heavy elements and low-energy transitions, which are incidentally of great interest for nuclear structure considerations, K conversion is often energetically impossible. In that case the ratio of conversion in various L-subshells or the L/M ratio will be of interest.

In addition to the properties of the initial state the conversion coefficients are strongly dependent on the following parameters: k, where kmc^2 is the transition energy; Z, the atomic number; L the angular momentum change; and finally, $\Delta\pi$ the parity change. More will be said about L and $\Delta\pi$ values in the next section but at this juncture it is important to recognize that it is the strong dependence of α on L and $\Delta\pi$ which makes internal conversion measurements so useful as a tool in nuclear spectroscopy.

It will be understood that the conversion coefficient refers to the total number of conversion electrons relative to the total number of gamma rays. However, this does not imply that the measurement of a conversion coefficient requires 4π geometry. The reason for this is found in the fact that when the radiation (electrons or gamma rays) is not observed in directional coincidence with preceding or following radiations or if the emitting nucleus is not oriented, the angular distribution of this radiation is isotropic.

2. The Selection Rules. Mixtures of Multipoles

When the nuclear angular momenta for initial and final states are J_i and J_f the field radiated can have any angular momentum L for which

$$\Delta J = |J_i - J_f| \leq L \leq J_i + J_f. \tag{3}$$

Because the conversion coefficients represent a total intensity and because radiation fields with different L-values are represented by tensors of different rank (2) which cannot interfere in a total intensity, the conversion coefficient is in general a mixture, that is, an incoherent combination of conversion coefficients for fields of pure angular momentum L. Thus,

$$\alpha = \sum_L a_L \alpha_L; \qquad \sum_L a_L = 1 \tag{4}$$

where a_L represents the relative intensities of the γ-rays, of angular momentum L, which are in competition with the conversion electrons. Of course, the L values are limited by Eq. (3).

If we examine the case of γ-ray emission we find, from the properties of Maxwell fields, that if the nuclear parity changes in the transition ($\Delta\pi = -1$) one can have emission of electric multipoles of odd order

(L odd) or magnetic multipoles of even order. If $\Delta\pi = 1$ there can occur only even order electric multipoles and/or odd order magnetic multipoles. The electric multipole for L units of angular momentum[c] is designated as EL and the magnetic multipole as ML. Thus, if $J_i = 1$ and $J_f = 2$ while $\Delta\pi = 1$, the radiated field will be a mixture of M1, E2, and M3. Similarly, the conversion coefficients for a given shell will be

$$\alpha = a_1\alpha(\text{M1}) + a_2\alpha(\text{E2}) + a_3\alpha(\text{M3}) \tag{5}$$

with $a_1 + a_2 + a_3 = 1$. For a given type of multipole (M or E) the relative intensity for multipoles with L and $L + 2$ units of angular momentum is

$$a_{L+2}/a_L \sim (R/\lambda)^4 \ll 1$$

where R is the nuclear radius and λ is the wavelength of the radiation. As an example, for $A = 200$ and $k = 1$ (511 kev),

$$(R/\lambda)^4 = [1.2 \times 10^{-13} A^{\frac{1}{3}}/2\pi(\hbar/mc)]^4 = 3.2 \times 10^{-9}.$$

Also in many cases but not in all cases, a_{L+1}/a_L will be small. Therefore, one can restrict the mixture (4) to two multipoles at most, $L = \Delta J$ and $L = \Delta J + 1$, and in some cases to the single value $L = L_{\min} = \Delta J$. Which situation pertains is a question to be decided by comparison of the observations and the theoretical values of α. Most mixtures are of the M1 + E2 type. Of course, where either J_i or $J_f = 0$, $L = \Delta J$ only—without approximation.

The parity selection rule is summarized by the statement that for $\Delta\pi = (-)^{\Delta J}$ the lowest order multipole ($L = \Delta J$) is electric and the second multipole, if present, is magnetic while for $\Delta\pi = (-)^{\Delta J+1}$ the lowest and next multipoles are magnetic and electric, respectively. In the transition $J \to J$ the lowest order (radiative) multipole is $L = 1$ and for $\Delta\pi = 1, -1$ the radiation is M1, E1 respectively. For $J \geq 1$ the occurrence of quadrupole radiation (E2, M2) is possible. In the case $\Delta J = 0$, $\Delta\pi = 1$ it is also possible to have an admixture of E0 conversion electrons.[b] There are no M0 conversion electrons in any case, since there is no M0 field.

The actual order of reasoning is from the comparison of measured and calculated α to the ΔJ and $\Delta\pi$ values.

Where a mixture is present the mixture parameters a_L, which are purely nuclear parameters, are clearly the same for all shells and subshells. By measuring absolute conversion coefficients as well as conversion ratios a check on the relevant mixture ratio a_L/a_{L+1} can be obtained. Thus

[c] The unit of angular momentum throughout is $\hbar$.

from the K/L ratio in an $M1 + E2$ mixture

$$\frac{1 - a_2}{a_2} = \frac{a_1}{a_2} = \frac{(K/L)\alpha(L; \text{E2}) - \alpha(K; \text{E2})}{\alpha(K; \text{M1}) - (K/L)\alpha(L; \text{M1})} \qquad (6)$$

from which a_2 can be deduced and compared with

$$a_2 = \frac{\alpha(K) - \alpha(K; \text{M1})}{\alpha(K; \text{E2}) - \alpha(K; \text{M1})}. \qquad (6')$$

To simplify the notation we have suppressed the dependence of a_L on $\Delta\pi$. In Eqs. (6) and (6'), $\alpha(K)$ and K/L are measured quantities while the other conversion coefficients would be taken from tables of calculated values, see Section 4 below.

3. The Calculation of Internal Conversion Coefficients

The procedure for calculating internal conversion coefficients is not a simple one and it is beyond the scope of this treatment to describe the theory in detail. Nevertheless, it is important to know the assumptions which underlie the calculations. This discussion will be restricted to practical calculations which are in use at the time of writing.

The assumptions made are of two kinds. The first refers to rather general assumptions which are characteristic of all the calculations. The second are those detailed assumptions which distinguish the various calculations which have been made.

a. General Assumptions

(1) *Perturbation Theory*

One uses the lowest perturbation theory in quantum electrodynamics for which a nonvanishing result is obtained. This means second order perturbation theory for N_e since the initial and final states have no quanta and there is a set of intermediate states with one quantum present (5).

(2) *Electron Dynamics*

The Dirac electron theory is used and therefore a strict solution of the relativistic many body problem is out of the question. This point refers to the influence of other electrons on the converted electron and this effect is referred to as screening. The methods of handling the screening problem are discussed below.

b. Specialized Assumptions: Nuclear Structure and Screening Model

(1) *Unscreened Coefficients*

In the simplest form the nucleus is treated as a point charge. If screening is not included the electron wave functions are well-known Coulomb field Dirac waves (6) and the problem, which is primarily one of calculating radial integrals, can be carried forward analytically to the point where all quadratures are expressed explicitly in terms of calculable functions. This will be referred to as the unscreened, point nucleus conversion coefficients.

(2) *Screening Effect*

Keeping the model of a point nucleus, the screening effect of the other electrons, in both initial and final states, can be included by replacing the Coulomb potential $-\alpha Z/r$ by a screened potential $-(\alpha Z/R)\Phi(r)$. In one type of screening calculation Φ is the Thomas-Fermi-Dirac screening function (7). Somewhat elaborately, exchange terms can be included (8) but this does not change the results appreciably. It is found that screening reduces the K conversion coefficients for a point nucleus by a factor f where f is 0.90–1.0 over the range of k and Z values defined by

$$0.3 \leq k \leq 2.0$$
$$0 < Z \leq 96.$$

Here and in the following, all references to numerical results refer to EL and ML, $L = 1, 2, 3, 4, 5$. For the L-shell f is much less, representing a larger screening effect, as is to be expected. While f is a function of the subshell, of k, Z, and of the radiation type, it is a rather insensitive function of these parameters and in general $0.6 \lesssim f \lesssim 0.8$. Empirical evidence indicates that the screening factor f for the M shell is about 0.6 or 0.7 for heavy elements.

(3) *Finite Nuclear Size Effect*

The coefficients calculated with these assumptions are referred to as screened, point nucleus coefficients. The third refinement is the inclusion of the finite size of the nucleus. The calculations now become somewhat more involved and a whole new set of assumptions must be introduced. To explain these it is convenient to separate the effects of the finite size of the nucleus into two parts (9). In all cases screening is included as discussed in Section b,2 above.

Nonstructure Dependent Coefficients. When the nucleus is treated as a point charge, the internal conversion coefficient is the ratio of two quantities N_e and N_q both of which depend on nuclear structure in that in both (the square of the modulus of) a nuclear matrix element depending on nuclear wave functions and the nuclear transition current appears. But the same matrix element occurs in both N_e and N_q, assuming that only a single L-value is involved. Therefore, aside from the possible appearance of a ratio of mixture parameters a_L/a_{L+1}, the coefficient is independent of nuclear structure. So far no systematic attempts to calculate this mixture ratio from nuclear models has been made. Hence, this ratio is treated as an adjustable parameter, chosen to fit the data.

When the nucleus is no longer treated as a point the conversion coefficient appears in the form

$$\alpha = \Sigma \beta_0 |1 + \beta_1 \mathfrak{R}|^2 \tag{7}$$

where the sum is over final electron states. In Eq. (7) β_0 is the contribution to the conversion coefficient from each final state calculated as in Section b,2 but with electron wave functions in the field of a charge distribution of finite extent. In these terms there is no real structure dependence inasmuch as the nuclear matrix elements cancel out in forming the ratio N_e/N_q. There is a trivial kind of structure dependence in that it is necessary to make an assumption about the nuclear charge density. But the results are insensitive to the details of this assumption as long as the choice of charge density is reasonable (see Sliv and Band, *8*). Also in Eq. (7) β_1 is a quantity which, like β_0, depends only on electron wave functions, and β_1 is generally very small (of order 10^{-2}) (*9*). $\mathfrak{R}$ is a ratio of nuclear matrix elements to which we turn our attention in the following paragraph. If $\mathfrak{R}$ is not large compared to unity, so that $\beta_1 \mathfrak{R}$ is neglected, the conversion coefficients are referred to as finite size coefficients but without structure dependence. We refer to such an influence of the nuclear size as a static effect (*10*). In a rough way of speaking, this corresponds to no penetration of the electron inside the nucleus. As long as $\mathfrak{R}$ is not large compared with unity it would appear to make little difference whether one assumes electron penetration inside the nucleus or not. The assumption that the nuclear currents are confined to the nuclear surface (Sliv and Band, *8*) ($\mathfrak{R} = 1$ essentially) gives basically the same result as $\mathfrak{R} = 0$. The latter assumption is used in the calculation described below, see Section 4. In general $\alpha' = \Sigma \beta_0$ is appreciably smaller than the point nucleus coefficients and this is especially true for M1 transitions where the reduction due to this effect of finite size can be as large as 50%. For other multipoles the finite size coefficients are usually within 10% of the point nucleus values.

Structure Dependent Coefficients. Since $\mathfrak{R}$ is inversely proportional to the matrix element whose square modulus gives N_q it is clear that one may expect abnormally large values of $\mathfrak{R}$ for retarded transitions; that is, those cases where λ_0 is much smaller than "normal."[d] Consequently, these transitions should be associated with a departure of the observed conversion coefficients from the calculated values wherein $\mathfrak{R}$ is neglected, or $\beta_1\mathfrak{R}$ is treated as small in comparison with unity (*11*). Thus, with the realization of this effect, the role of internal conversion is radically altered. First the measurement of internal conversion coefficients had been regarded as a method of obtaining nuclear data in a manner independent of assumptions about nuclear models. Now it becomes a tool for investigating special features of nuclear structure which are manifested in abnormally slow transitions. Note that the *order of magnitude* of the conversion coefficient will, in almost every case, be unchanged although the change in the coefficients arising from the nuclear structure effects may be as large as a factor 2–3.[e] The very large change in λ_0 therefore entails a correspondingly large change in λ, or in the mean life of the excited state.

A number of fairly detailed discussions of this nuclear structure effect, which may be properly regarded as a dynamic effect (*10*), have appeared in the literature. Church and Weneser (*11*) first pointed out the importance of the effect for M1 transitions. The possibility of anomalous E1 coefficients has been discussed by Nilsson (*13*) and by Nilsson and Rasmussen (*14*). It may be emphasized that this part of the conversion coefficient, depending as it does on a nuclear model, should be separated in the calculations from the static part, which depends essentially on electron dynamics only. Since these two parts add coherently it does not suffice to merely list conversion coefficients. Instead for each of the $2j_i + 1$ final states[f] one needs a (complex) matrix element. So far these have been tabulated only for the K-shell in the case of an unscreened point nucleus (M. E. Rose, *15*). Green and Rose (*9*) have shown how these matrix elements can be combined with the computed K-conversion coefficients, in which screening and static nuclear size effects are included, to obtain the corrected matrix elements in an approximation sufficiently accurate for most purposes.

---

[d] Presumably, abnormally slow transitions for nuclei near magic numbers are connected with the l-forbiddenness selection rule which, in the single particle model, would rule out $s_{\frac{1}{2}} \leftrightarrow d_{\frac{3}{2}}$ and so forth. For distorted nuclei large retardation is associated with so-called K-hindering. On this view other transitions would be "normal."

[e] An anomaly of at least this magnitude seems to occur in the 482 kev transition in Ta^{181} (see Reiner, *12*).

[f] j_i is the angular momentum of the initial electronic state.

Finally, it is important to realize that large dynamic structure effects are not to be expected in all cases. They are possibly associated with K-hindering (see above) and should appear only in some heavy nuclei. The proper procedure would then involve a comparison of measured and computed conversion coefficients where the latter include only static nuclear size effects.

4. Numerical Values of Internal Conversion Coefficients

In this section we shall present a tabulation of conversion coefficients. Because of space limitations this tabulation represents only a comparatively small part of the numerical information now available. Therefore we shall discuss the other tables in existence at the time of writing.

First, Tables I and II given here present K-shell and total L-shell conversion coefficients, respectively, for $Z = 25$ to 95 in steps of 10 and for 10 multipoles (EL and ML with $L = 1, 2, 3, 4,$ and 5). The electric conversion coefficients are denoted by α_L and the magnetic by β_L. Values of k from 0.05 to 2.0 are included where, of course, the lowest k must be above the threshold. The threshold is the binding energy of the electron in the model used: This is screening with a Thomas-Fermi-Dirac screening function (no exchange) and the static effect of nuclear size represented by a constant nuclear charge density with $R = 1.2 \times 10^{-13} A^{\frac{1}{3}}$ cm. The static nuclear size effect is based on the "no penetration" model ($\mathfrak{R} = 0$).

The L-shell coefficients were obtained by adding the L-subshell contributions. The separate L-subshell values, together with the K-shell coefficients for all Z values from 25 to 95 are given in a comprehensive tabulation, appearing elsewhere (15). This tabulation also includes M-shell coefficients for a point nucleus and for no screening. The M-subshell ratios should be relatively unaffected by these assumptions. The same reference includes K-shell matrix elements for the purpose of introducing dynamic nuclear size effects.

The K-shell conversion coefficients with screening (exchange included) and static nuclear size effects (surface current model) incorporated have been given by Sliv and Band (8). These tables give coefficients for the same 10 multipoles for all Z values between 33 and 98 inclusive. The energy range in this tabulation starts somewhat higher ($k \geq 0.15$) and goes to higher values ($k \leq 5.0$).

The influence of the static effect of finite nuclear size for the K-shell can be estimated by comparing the coefficients in references (15) or (8) with those calculated without screening (which makes a small change in the K-shell) and for a point nucleus (16). For the L-shell the comparison may be made with screened, point nucleus values (17).

TABLE I. K-Shell Conversion Coefficients

k	α_1	α_2	α_3	α_4	α_5	β_1	β_2	β_3	β_4	β_5
					K Shell, $Z = 25$					
0.05	2.10 (0)	6.53 (1)	1.59 (3)	3.86 (4)	9.22 (5)	1.21 (0)	4.28 (1)	1.29 (3)	3.62 (4)	1.05 (6)
0.10	2.70 (−1)	5.20 (0)	8.19 (1)	1.26 (3)	1.92 (4)	1.68 (−1)	3.13 (0)	5.50 (1)	9.47 (2)	1.66 (4)
0.15	7.76 (−2)	1.11 (0)	1.31 (1)	1.48 (2)	1.66 (3)	5.52 (−2)	7.18 (−1)	8.90 (0)	1.11 (2)	1.40 (3)
0.20	3.19 (−2)	3.67 (−1)	3.47 (0)	3.16 (1)	2.83 (2)	2.56 (−2)	2.58 (−1)	2.54 (0)	2.49 (1)	2.47 (2)
0.40	3.86 (−3)	2.56 (−2)	1.45 (−1)	7.73 (−1)	4.09 (0)	4.31 (−3)	2.50 (−2)	1.40 (−1)	7.84 (−1)	4.40 (0)
0.60	1.18 (−3)	5.72 (−3)	2.41 (−2)	9.66 (−2)	3.82 (−1)	1.62 (−3)	7.00 (−3)	2.89 (−2)	1.19 (−1)	4.91 (−1)
0.80	5.31 (−4)	2.10 (−3)	7.26 (−3)	2.40 (−2)	7.83 (−2)	8.30 (−4)	2.97 (−3)	1.01 (−2)	3.40 (−2)	1.15 (−1)
1.0	2.98 (−4)	1.01 (−3)	3.04 (−3)	8.75 (−3)	2.48 (−2)	5.04 (−4)	1.58 (−3)	4.66 (−3)	1.36 (−2)	3.95 (−2)
1.5	1.16 (−4)	3.05 (−4)	7.28 (−4)	1.67 (−3)	3.79 (−3)	2.11 (−4)	5.31 (−4)	1.25 (−3)	2.90 (−3)	6.63 (−3)
2.0	6.48 (−5)	1.47 (−4)	3.05 (−4)	6.11 (−4)	1.21 (−3)	1.19 (−4)	2.61 (−4)	5.37 (−4)	1.08 (−3)	2.14 (−3)
					K Shell, $Z = 35$					
0.05	3.78 (0)	7.40 (1)	1.04 (3)	1.88 (4)	1.85 (5)	5.10 (0)	1.75 (2)	3.52 (3)	6.18 (4)	1.08 (6)
0.10	5.39 (−1)	8.25 (0)	1.00 (2)	1.17 (3)	1.30 (4)	6.72 (−1)	1.23 (1)	1.81 (2)	2.59 (3)	3.72 (4)
0.15	1.66 (−1)	1.97 (0)	1.92 (1)	1.79 (2)	1.66 (3)	2.13 (−1)	2.75 (0)	3.02 (1)	3.34 (2)	3.67 (3)
0.20	7.03 (−2)	6.97 (−1)	5.65 (0)	4.44 (1)	3.39 (2)	9.61 (−2)	9.53 (−1)	8.68 (0)	7.77 (1)	7.13 (2)
0.40	9.08 (−3)	5.48 (−2)	2.80 (−1)	1.36 (0)	6.58 (0)	1.52 (−2)	8.82 (−2)	4.70 (−1)	2.40 (0)	1.35 (1)
0.60	2.89 (−3)	1.31 (−2)	5.13 (−2)	1.92 (−1)	7.12 (−1)	5.53 (−3)	2.41 (−2)	9.64 (−2)	3.73 (−1)	1.55 (0)
0.80	1.34 (−3)	4.98 (−3)	1.64 (−2)	5.14 (−2)	1.59 (−1)	2.77 (−3)	1.00 (−2)	3.34 (−2)	1.11 (−1)	3.60 (−1)
1.0	7.56 (−4)	2.46 (−3)	7.12 (−3)	1.97 (−2)	5.35 (−2)	1.65 (−3)	5.23 (−3)	1.53 (−2)	4.39 (−2)	1.24 (−1)
1.5	3.03 (−4)	7.86 (−4)	1.83 (−3)	4.13 (−3)	9.13 (−3)	6.57 (−4)	1.73 (−3)	4.07 (−3)	9.26 (−3)	2.10 (−2)
2.0	1.68 (−4)	3.87 (−4)	7.99 (−4)	1.59 (−3)	3.08 (−3)	3.56 (−4)	8.28 (−4)	1.71 (−3)	3.43 (−3)	6.74 (−3)

Table I continued on next page

TABLE I. K-SHELL CONVERSION COEFFICIENTS (Continued)

k	α_1	α_2	α_3	α_4	α_5	β_1	β_2	β_3	β_4	β_5
					K Shell, $Z = 45$					
0.05	5.36 (0)	3.41 (1)	8.92 (1)	2.00 (2)	4.61 (2)	1.59 (1)	5.35 (2)	3.54 (3)	1.25 (4)	3.72 (4)
0.10	8.38 (−1)	9.06 (0)	7.46 (1)	5.80 (2)	4.62 (3)	2.06 (0)	3.57 (1)	4.01 (2)	4.09 (3)	4.15 (4)
0.15	2.73 (−1)	2.57 (0)	1.94 (1)	1.40 (2)	1.01 (3)	6.42 (−1)	7.86 (0)	7.44 (1)	6.77 (2)	6.09 (3)
0.20	1.20 (−1)	9.82 (−1)	6.55 (0)	4.21 (1)	2.69 (2)	2.85 (−1)	2.74 (0)	2.20 (1)	1.71 (2)	1.34 (3)
0.40	1.65 (−2)	8.84 (−2)	4.10 (−1)	1.77 (0)	7.67 (0)	4.31 (−2)	2.42 (−1)	1.22 (0)	6.24 (0)	3.03 (1)
0.60	5.44 (−3)	2.26 (−2)	8.26 (−2)	2.85 (−1)	9.82 (−1)	1.51 (−2)	6.45 (−2)	2.49 (−1)	9.67 (−1)	3.63 (0)
0.80	2.57 (−3)	9.02 (−3)	2.77 (−2)	8.29 (−2)	2.43 (−1)	7.34 (−3)	2.64 (−2)	8.53 (−2)	2.71 (−1)	8.62 (−1)
1.0	1.49 (−3)	4.61 (−3)	1.27 (−2)	3.37 (−2)	8.85 (−2)	4.29 (−3)	1.37 (−2)	3.89 (−2)	1.08 (−1)	3.00 (−1)
1.5	6.04 (−4)	1.56 (−3)	3.56 (−3)	7.87 (−3)	1.71 (−2)	1.67 (−3)	4.42 (−3)	1.03 (−2)	2.31 (−2)	5.12 (−2)
2.0	3.41 (−4)	7.89 (−4)	1.62 (−3)	3.20 (−3)	6.17 (−3)	8.78 (−4)	2.09 (−3)	4.31 (−3)	8.51 (−3)	1.65 (−2)
					K Shell, $Z = 55$					
0.10	1.14 (0)	7.09 (0)	2.88 (1)	1.83 (2)	9.22 (2)	5.44 (0)	8.78 (1)	6.05 (2)	3.23 (3)	1.63 (4)
0.15	3.93 (−1)	2.65 (0)	1.38 (1)	6.91 (1)	3.51 (2)	1.68 (0)	1.93 (1)	1.42 (2)	9.41 (2)	6.17 (3)
0.20	1.78 (−1)	1.13 (0)	5.82 (0)	2.89 (1)	1.45 (2)	7.38 (−1)	6.73 (0)	4.49 (1)	2.86 (2)	1.81 (3)
0.40	2.62 (−2)	1.21 (−1)	4.83 (−1)	1.87 (0)	7.27 (0)	1.08 (−1)	5.83 (−1)	2.69 (0)	1.22 (1)	5.55 (1)
0.60	8.94 (−3)	3.35 (−2)	1.11 (−1)	3.59 (−1)	1.16 (0)	3.68 (−2)	1.53 (−1)	5.58 (−1)	2.00 (0)	7.17 (0)
0.80	4.33 (−3)	1.41 (−2)	4.09 (−2)	1.16 (−1)	3.27 (−1)	1.76 (−2)	6.22 (−2)	1.92 (−1)	5.81 (−1)	1.76 (0)
1.0	2.56 (−3)	7.51 (−3)	1.98 (−2)	5.09 (−2)	1.29 (−1)	1.01 (−2)	3.19 (−2)	8.74 (−2)	2.33 (−1)	6.22 (−1)
1.5	1.06 (−3)	2.70 (−3)	6.09 (−3)	1.32 (−2)	2.84 (−2)	3.78 (−3)	1.01 (−2)	2.29 (−2)	5.02 (−2)	1.09 (−1)
2.0	6.04 (−4)	1.41 (−3)	2.92 (−3)	5.72 (−3)	1.09 (−2)	1.94 (−3)	4.71 (−3)	9.57 (−3)	1.86 (−2)	3.55 (−2)

TABLE I. K-Shell Conversion Coefficients (Continued)

k	α_1	α_2	α_3	α_4	α_5	β_1	β_2	β_3	β_4	β_5
					K Shell, $Z = 65$					
0.15	5.09 (−1)	2.06 (0)	5.89 (0)	1.58 (1)	4.38 (1)	4.11 (0)	4.25 (1)	2.01 (2)	7.79 (2)	2.82 (3)
0.20	2.38 (−1)	1.07 (0)	3.84 (0)	1.34 (1)	4.82 (1)	1.79 (0)	1.49 (1)	7.58 (1)	3.52 (2)	1.60 (3)
0.40	3.83 (−2)	1.46 (−1)	5.01 (−1)	1.69 (0)	5.87 (0)	2.55 (−1)	1.29 (0)	5.25 (0)	2.09 (1)	8.35 (1)
0.60	1.36 (−2)	4.47 (−2)	1.36 (−1)	4.08 (−1)	1.24 (0)	8.53 (−2)	3.38 (−1)	1.13 (0)	3.72 (0)	1.23 (1)
0.80	6.67 (−3)	2.01 (−2)	5.51 (−2)	1.49 (−1)	4.06 (−1)	4.00 (−2)	1.36 (−1)	3.94 (−1)	1.12 (0)	3.17 (0)
1.0	4.01 (−3)	1.12 (−2)	2.86 (−2)	7.12 (−2)	1.77 (−1)	2.27 (−2)	6.93 (−2)	1.80 (−1)	4.58 (−1)	1.16 (0)
1.5	1.71 (−3)	4.32 (−3)	9.70 (−3)	2.09 (−2)	4.43 (−2)	8.23 (−3)	2.17 (−2)	4.77 (−2)	1.01 (−1)	2.09 (−1)
2.0	9.85 (−4)	2.35 (−3)	4.88 (−3)	9.57 (−3)	1.82 (−2)	4.10 (−3)	1.00 (−2)	1.98 (−2)	3.76 (−2)	6.95 (−2)
					K Shell, $Z = 75$					
0.15	5.85 (−1)	9.38 (−1)	6.37 (−1)	2.69 (−1)	1.28 (−1)	9.59 (0)	8.05 (1)	1.72 (2)	1.62 (2)	9.17 (1)
0.20	2.96 (−1)	7.93 (−1)	1.61 (0)	3.13 (0)	6.55 (0)	4.19 (0)	3.02 (1)	9.98 (1)	2.70 (2)	6.72 (2)
0.40	5.12 (−2)	1.61 (−1)	4.67 (−1)	1.40 (0)	4.39 (0)	5.93 (−1)	2.66 (0)	9.36 (0)	3.13 (1)	1.06 (2)
0.60	1.88 (−2)	5.52 (−2)	1.55 (−1)	4.38 (−1)	1.27 (0)	1.94 (−1)	7.08 (−1)	2.11 (0)	6.22 (0)	1.84 (1)
0.80	9.69 (−3)	2.69 (−2)	7.10 (−2)	1.86 (−1)	4.92 (−1)	9.00 (−2)	2.85 (−1)	7.54 (−1)	1.97 (0)	5.14 (0)
1.0	5.94 (−3)	1.58 (−2)	3.97 (−2)	9.73 (−2)	2.36 (−1)	5.01 (−2)	1.45 (−1)	3.51 (−1)	8.29 (−1)	1.96 (0)
1.5	2.61 (−3)	6.63 (−3)	1.50 (−2)	3.22 (−2)	6.73 (−2)	1.78 (−2)	4.52 (−2)	9.40 (−2)	1.89 (−1)	3.75 (−1)
2.0	1.52 (−3)	3.75 (−3)	7.94 (−3)	1.56 (−2)	2.94 (−2)	8.64 (−3)	2.07 (−2)	3.95 (−2)	7.18 (−2)	1.29 (−1)

Table I continued on next page

TABLE I. K-SHELL CONVERSION COEFFICIENTS (Continued)

k	α_1	α_2	α_3	α_4	α_5	β_1	β_2	β_3	β_4	β_5
K Shell, $Z = 85$										
0.20	3.58 (−1)	3.75 (−1)	1.76 (−1)	7.08 (−2)	2.70 (−2)	9.75 (0)	5.97 (1)	8.84 (1)	5.33 (1)	2.40 (1)
0.40	6.48 (−2)	1.58 (−1)	3.88 (−1)	9.94 (−1)	2.71 (0)	1.35 (0)	5.39 (0)	1.50 (1)	3.88 (1)	1.01 (2)
0.60	2.51 (−2)	6.54 (−2)	1.73 (−1)	4.62 (−1)	1.26 (0)	4.47 (−1)	1.44 (0)	3.69 (0)	9.42 (0)	2.39 (1)
0.80	1.34 (−2)	3.51 (−2)	9.10 (−2)	2.31 (−1)	5.87 (−1)	2.04 (−1)	5.84 (−1)	1.38 (0)	3.21 (0)	7.52 (0)
1.0	8.43 (−3)	2.22 (−2)	5.52 (−2)	1.33 (−1)	3.13 (−1)	1.13 (−1)	2.98 (−1)	6.52 (−1)	1.41 (0)	3.04 (0)
1.5	3.84 (−3)	1.01 (−2)	2.32 (−2)	4.93 (−2)	1.01 (−1)	3.89 (−2)	9.30 (−2)	1.80 (−1)	3.39 (−1)	6.33 (−1)
2.0	2.28 (−3)	5.93 (−3)	1.29 (−2)	2.51 (−2)	4.69 (−2)	1.85 (−2)	4.25 (−2)	7.66 (−2)	1.32 (−1)	2.26 (−1)
K Shell, $Z = 95$										
0.40	7.79 (−2)	1.43 (−1)	2.86 (−1)	5.77 (−1)	1.17 (0)	3.25 (0)	1.05 (1)	2.07 (1)	3.69 (1)	6.48 (1)
0.60	3.22 (−2)	7.65 (−2)	1.92 (−1)	4.72 (−1)	1.03 (0)	1.05 (0)	2.86 (0)	6.05 (0)	1.25 (1)	2.59 (1)
0.80	1.78 (−2)	4.65 (−2)	1.19 (−1)	2.87 (−1)	6.84 (−1)	4.78 (−1)	1.18 (0)	2.39 (0)	4.82 (0)	9.78 (0)
1.0	1.15 (−2)	3.14 (−2)	7.85 (−2)	1.80 (−1)	3.99 (−1)	2.62 (−1)	6.08 (−1)	1.18 (0)	2.24 (0)	4.32 (0)
1.5	5.50 (−3)	1.55 (−2)	3.64 (−2)	7.55 (−2)	1.49 (−1)	8.85 (−2)	1.91 (−1)	3.38 (−1)	5.84 (−1)	1.01 (0)
2.0	3.34 (−3)	9.52 (−3)	2.11 (−2)	4.06 (−2)	7.33 (−2)	4.12 (−2)	8.79 (−2)	1.47 (−1)	2.37 (−1)	3.79 (−1)

TABLE II. L Shell Conversion Coefficients

k	α_1	α_2	α_3	α_4	α_5	β_1	β_2	β_3	β_4	β_5
\multicolumn{11}{c}{L Shell, $Z = 25$}										
0.05	1.93 (−1)	1.09 (1)	8.14 (2)	5.43 (4)	3.16 (6)	1.12 (−1)	5.67 (0)	3.00 (2)	1.61 (4)	8.46 (5)
0.10	2.38 (−2)	5.85 (−1)	1.63 (1)	4.83 (2)	1.39 (4)	1.54 (−2)	3.51 (−1)	8.33 (0)	2.07 (2)	5.24 (3)
0.15	6.81 (−3)	1.12 (−1)	1.88 (0)	3.41 (1)	6.29 (2)	4.98 (−3)	7.45 (−2)	1.15 (0)	1.86 (1)	3.08 (2)
0.20	2.80 (−3)	3.53 (−2)	4.29 (−1)	5.56 (0)	7.46 (1)	2.30 (−3)	2.58 (−2)	2.98 (−1)	3.59 (0)	4.44 (1)
0.40	3.34 (−4)	2.33 (−3)	1.45 (−2)	9.12 (−2)	5.91 (−1)	3.81 (−4)	2.33 (−3)	1.42 (−2)	8.82 (−2)	5.61 (−1)
0.60	1.02 (−4)	5.09 (−4)	2.27 (−3)	1.00 (−2)	4.48 (−2)	1.42 (−4)	6.33 (−4)	2.76 (−3)	1.22 (−2)	5.46 (−2)
0.80	4.63 (−5)	1.87 (−4)	6.70 (−4)	3.36 (−3)	8.37 (−3)	7.28 (−5)	2.67 (−4)	9.41 (−4)	3.33 (−3)	1.19 (−2)
1.0	2.57 (−5)	8.87 (−5)	2.76 (−4)	8.29 (−4)	2.50 (−3)	4.37 (−5)	1.40 (−4)	4.26 (−4)	1.29 (−3)	3.92 (−3)
1.5	1.01 (−5)	2.69 (−5)	6.49 (−5)	1.53 (−4)	3.57 (−4)	1.84 (−5)	4.69 (−5)	1.12 (−4)	2.65 (−4)	6.24 (−4)
2.0	5.56 (−6)	1.28 (−5)	2.68 (−5)	5.46 (−5)	1.10 (−4)	1.02 (−5)	2.27 (−5)	4.75 (−5)	9.69 (−5)	1.96 (−4)
\multicolumn{11}{c}{L Shell, $Z = 35$}										
0.05	4.43 (−1)	3.98 (1)	4.08 (3)	2.76 (5)	1.53 (7)	5.57 (−1)	3.52 (1)	2.30 (3)	1.37 (5)	7.34 (6)
0.10	5.75 (−2)	1.73 (0)	6.77 (1)	2.32 (3)	6.89 (4)	7.32 (−2)	1.91 (0)	5.15 (1)	1.41 (3)	3.79 (4)
0.15	1.70 (−2)	3.08 (−1)	6.83 (0)	1.49 (2)	3.01 (3)	2.30 (−2)	3.77 (−1)	6.38 (0)	1.12 (2)	1.99 (3)
0.20	7.19 (−3)	9.47 (−2)	1.42 (0)	2.23 (1)	3.36 (2)	1.03 (−2)	1.25 (−1)	1.55 (0)	2.01 (1)	2.66 (2)
0.40	9.13 (−4)	6.21 (−3)	4.19 (−2)	2.99 (−1)	2.20 (0)	1.60 (−3)	1.01 (−2)	6.40 (−2)	4.18 (−1)	2.81 (0)
0.60	2.88 (−4)	1.40 (−3)	6.46 (−3)	3.08 (−2)	1.52 (−1)	5.70 (−4)	2.63 (−3)	1.18 (−2)	5.39 (−2)	2.52 (−1)
0.80	1.33 (−4)	5.21 (−4)	1.91 (−3)	7.08 (−3)	2.69 (−2)	2.83 (−4)	1.07 (−3)	3.86 (−3)	1.40 (−2)	5.18 (−2)
1.0	7.57 (−5)	2.56 (−4)	7.98 (−4)	2.49 (−3)	7.91 (−3)	1.68 (−4)	5.53 (−4)	1.71 (−3)	5.29 (−3)	1.65 (−2)
1.5	2.97 (−5)	7.95 (−5)	1.94 (−4)	4.67 (−4)	1.12 (−3)	6.71 (−5)	1.78 (−4)	4.33 (−4)	1.04 (−3)	2.49 (−3)
2.0	1.67 (−5)	3.88 (−5)	8.27 (−5)	1.71 (−4)	3.52 (−4)	3.64 (−5)	8.47 (−5)	1.80 (−4)	3.71 (−4)	7.60 (−4)

Table II continued on next page

TABLE II. L SHELL CONVERSION COEFFICIENTS (Continued)

k	α_1	α_2	α_3	α_4	α_5	β_1	β_2	β_3	β_4	β_5
					L Shell, $Z = 45$					
0.05	8.32 (−1)	1.23 (2)	1.43 (4)	8.83 (5)	4.36 (7)	2.02 (0)	1.56 (2)	1.23 (4)	7.71 (5)	4.02 (7)
0.10	1.09 (−1)	4.63 (0)	2.28 (2)	7.91 (3)	2.21 (5)	2.61 (−1)	7.77 (0)	2.37 (2)	6.91 (3)	1.91 (5)
0.15	3.31 (−2)	7.41 (−1)	2.13 (1)	5.02 (2)	9.89 (3)	8.06 (−2)	1.45 (0)	2.69 (1)	5.03 (2)	9.36 (3)
0.20	1.43 (−2)	2.17 (−1)	4.18 (0)	7.27 (1)	1.12 (3)	3.56 (−2)	4.62 (−1)	6.15 (0)	8.45 (1)	1.16 (3)
0.40	1.91 (−3)	1.34 (−2)	1.05 (−1)	8.67 (−1)	6.93 (0)	5.25 (−3)	3.45 (−2)	2.26 (−1)	1.54 (0)	1.08 (1)
0.60	6.24 (−4)	3.04 (−3)	1.54 (−2)	8.22 (−2)	4.44 (−1)	1.82 (−3)	8.61 (−3)	3.94 (−2)	1.86 (−1)	9.01 (−1)
0.80	2.93 (−4)	1.15 (−3)	4.47 (−3)	1.81 (−2)	7.54 (−2)	8.80 (−4)	3.41 (−3)	1.25 (−2)	4.66 (−2)	1.77 (−1)
1.0	1.70 (−4)	5.73 (−4)	1.87 (−3)	6.23 (−3)	2.14 (−2)	5.12 (−4)	1.73 (−3)	5.41 (−3)	1.71 (−2)	5.47 (−2)
1.5	6.79 (−5)	1.84 (−4)	4.62 (−4)	1.16 (−3)	2.93 (−3)	1.96 (−4)	5.35 (−4)	1.32 (−3)	3.21 (−3)	7.82 (−3)
2.0	3.83 (−5)	9.15 (−5)	2.00 (−4)	4.28 (−4)	9.11 (−4)	1.03 (−4)	2.49 (−4)	5.36 (−4)	1.12 (−3)	2.32 (−3)
					L Shell, $Z = 55$					
0.05	1.29 (0)	3.54 (2)	4.17 (4)	2.27 (6)	9.35 (7)	5.18 (0)	5.83 (2)	5.30 (4)	3.38 (6)	1.67 (8)
0.10	1.67 (−1)	1.14 (1)	6.21 (2)	2.04 (4)	5.27 (5)	6.56 (−1)	2.46 (1)	8.69 (2)	2.64 (4)	7.19 (5)
0.15	5.21 (−2)	1.68 (0)	5.72 (1)	1.33 (3)	2.48 (4)	2.01 (−1)	4.40 (0)	9.08 (1)	1.78 (3)	3.31 (4)
0.20	2.29 (−2)	4.63 (−1)	1.10 (1)	1.95 (2)	2.87 (3)	8.82 (−2)	1.37 (0)	1.97 (1)	2.85 (2)	4.00 (3)
0.40	3.23 (−3)	2.55 (−2)	2.48 (−1)	2.24 (0)	1.83 (1)	1.21 (−2)	9.56 (−2)	6.53 (−1)	4.62 (0)	3.35 (1)
0.60	1.09 (−3)	5.65 (−3)	3.37 (−2)	2.03 (−1)	1.17 (0)	4.31 (−3)	2.31 (−2)	1.09 (−1)	5.29 (−1)	2.64 (0)
0.80	5.26 (−4)	2.14 (−3)	9.42 (−3)	4.29 (−2)	1.93 (−1)	2.04 (−3)	8.91 (−3)	3.35 (−2)	1.28 (−1)	5.00 (−1)
1.0	3.05 (−4)	1.06 (−3)	3.85 (−3)	1.43 (−2)	5.30 (−2)	1.17 (−3)	4.41 (−3)	1.42 (−2)	4.56 (−2)	1.50 (−1)
1.5	1.25 (−4)	3.49 (−4)	9.40 (−4)	2.53 (−3)	6.84 (−3)	4.36 (−4)	1.33 (−3)	3.37 (−3)	8.25 (−3)	2.06 (−2)
2.0	7.06 (−5)	1.76 (−4)	4.06 (−4)	9.14 (−4)	2.05 (−3)	2.23 (−4)	6.05 (−4)	1.34 (−3)	2.80 (−3)	5.92 (−3)

TABLE II. L SHELL CONVERSION COEFFICIENTS (Continued)

k	α_1	α_2	α_3	α_4	α_5	β_1	β_2	β_3	β_4	β_5
\multicolumn{11}{c}{L Shell, $Z = 65$}
0.05	1.72 (0)	8.51 (2)	9.63 (4)	4.43 (6)	1.57 (8)	1.16 (1)	1.73 (3)	2.00 (5)	1.25 (7)	5.56 (8)
0.10	2.45 (−1)	2.69 (1)	1.51 (3)	4.48 (4)	1.06 (6)	1.75 (0)	7.43 (1)	2.91 (3)	9.00 (4)	2.33 (6)
0.15	7.81 (−2)	3.85 (0)	1.40 (2)	3.05 (3)	5.36 (4)	5.34 (−1)	1.28 (1)	2.84 (2)	5.73 (3)	1.04 (5)
0.20	3.48 (−2)	1.02 (0)	2.66 (1)	4.60 (2)	6.45 (3)	2.34 (−1)	3.91 (0)	5.93 (1)	8.75 (2)	1.23 (4)
0.40	5.18 (−3)	5.06 (−2)	5.81 (−1)	5.46 (0)	4.44 (1)	3.30 (−2)	2.55 (−1)	1.81 (0)	1.31 (1)	9.52 (1)
0.60	1.81 (−3)	1.08 (−2)	7.61 (−2)	4.92 (−1)	2.88 (0)	1.10 (−2)	6.11 (−2)	2.91 (−1)	1.43 (0)	7.23 (0)
0.80	8.94 (−4)	4.02 (−3)	2.07 (−2)	1.02 (−1)	4.74 (−1)	5.17 (−3)	2.32 (−2)	8.74 (−2)	3.37 (−1)	1.33 (0)
1.0	5.35 (−4)	2.02 (−3)	8.28 (−3)	3.35 (−2)	1.30 (−1)	2.92 (−3)	1.14 (−2)	3.64 (−2)	1.18 (−1)	3.93 (−1)
1.5	2.23 (−4)	6.67 (−4)	1.97 (−3)	5.67 (−3)	1.63 (−2)	1.05 (−3)	3.34 (−3)	8.39 (−3)	2.08 (−2)	5.21 (−2)
2.0	1.27 (−4)	3.39 (−4)	8.40 (−4)	2.02 (−3)	4.71 (−3)	5.26 (−4)	1.50 (−3)	3.26 (−3)	6.93 (−3)	1.47 (−2)
\multicolumn{11}{c}{L Shell, $Z = 75$}
0.05	2.12 (0)	1.85 (3)	2.00 (5)	7.51 (6)	1.96 (8)	2.76 (1)	4.67 (3)	6.72 (5)	4.29 (7)	1.58 (9)
0.10	3.38 (−1)	6.22 (1)	3.39 (3)	9.00 (4)	1.84 (6)	4.65 (0)	2.15 (2)	9.05 (3)	2.81 (5)	6.85 (6)
0.15	1.10 (−1)	8.77 (0)	3.17 (2)	6.39 (3)	1.01 (5)	1.41 (0)	3.63 (1)	8.46 (2)	1.70 (4)	3.00 (5)
0.20	5.02 (−2)	2.27 (0)	6.12 (1)	9.87 (2)	1.29 (4)	6.12 (−1)	1.08 (1)	1.71 (2)	2.52 (3)	3.45 (4)
0.40	7.88 (−3)	1.05 (−1)	1.35 (0)	1.25 (1)	9.73 (1)	8.55 (−2)	6.86 (−1)	4.82 (0)	3.47 (1)	2.52 (2)
0.60	2.85 (−3)	2.14 (−2)	1.75 (−1)	1.16 (0)	6.61 (0)	2.83 (−2)	1.58 (−1)	7.47 (−1)	3.67 (0)	1.86 (1)
0.80	1.45 (−3)	7.87 (−3)	4.69 (−2)	2.43 (−1)	1.11 (0)	1.32 (−2)	5.91 (−2)	2.21 (−1)	8.49 (−1)	3.38 (0)
1.0	8.81 (−4)	3.90 (−3)	1.85 (−2)	7.89 (−2)	3.06 (−1)	7.36 (−3)	2.87 (−2)	9.08 (−2)	2.94 (−1)	9.76 (−1)
1.5	3.75 (−4)	1.28 (−3)	4.23 (−3)	1.32 (−2)	3.83 (−2)	2.63 (−3)	8.28 (−3)	2.05 (−2)	5.06 (−2)	1.27 (−1)
2.0	2.17 (−4)	6.51 (−4)	1.78 (−3)	4.52 (−3)	1.09 (−2)	1.29 (−3)	3.66 (−3)	6.86 (−3)	1.67 (−2)	3.52 (−2)

Table II continued on next page

TABLE II. L SHELL CONVERSION COEFFICIENTS (Continued)

k	α_1	α_2	α_3	α_4	α_5	β_1	β_2	β_3	β_4	β_5
\multicolumn{11}{c}{L Shell $Z = 85$}
0.05	2.06 (0)	3.92 (3)	3.85 (5)	1.03 (7)	1.76 (8)	8.20 (4)	1.32 (4)	2.14 (6)	1.25 (8)	3.62 (9)
0.10	4.36 (−1)	1.41 (2)	7.17 (3)	1.59 (5)	2.79 (6)	1.25 (1)	6.19 (2)	2.69 (4)	7.91 (5)	1.78 (7)
0.15	1.47 (−1)	1.96 (1)	6.85 (2)	1.22 (4)	1.74 (5)	3.83 (0)	1.01 (2)	2.40 (3)	4.67 (4)	7.78 (5)
0.20	6.77 (−2)	5.12 (0)	1.35 (2)	1.98 (3)	2.37 (4)	1.67 (0)	2.95 (1)	4.68 (2)	6.79 (3)	8.86 (4)
0.40	1.13 (−2)	2.32 (−1)	3.12 (0)	2.76 (1)	2.04 (2)	2.33 (−1)	1.82 (0)	1.24 (1)	8.81 (1)	6.26 (2)
0.60	4.30 (−3)	4.60 (−2)	4.12 (−1)	2.67 (0)	1.47 (1)	7.65 (−2)	4.11 (−1)	1.88 (0)	9.12 (0)	4.53 (1)
0.80	2.25 (−3)	1.66 (−2)	1.10 (−1)	5.73 (−1)	2.56 (0)	3.54 (−2)	1.53 (−1)	5.49 (−1)	2.08 (0)	8.15 (0)
1.0	1.41 (−3)	8.10 (−3)	4.31 (−2)	1.89 (−1)	7.30 (−1)	1.98 (−2)	7.35 (−2)	2.25 (−1)	7.14 (−1)	2.35 (0)
1.5	6.22 (−4)	2.57 (−3)	9.61 (−3)	3.13 (−2)	9.27 (−2)	6.99 (−3)	2.10 (−2)	5.04 (−2)	1.22 (−1)	3.02 (−1)
2.0	3.69 (−4)	1.29 (−3)	3.91 (−3)	1.05 (−2)	2.61 (−2)	3.41 (−3)	9.18 (−3)	1.93 (−2)	4.00 (−2)	8.40 (−2)
\multicolumn{11}{c}{L Shell $Z = 95$}
0.05	2.44 (0)	8.63 (3)	7.43 (5)	1.09 (7)	8.42 (7)	2.70 (2)	3.85 (4)	6.13 (6)	3.33 (8)	7.00 (9)
0.10	5.10 (−1)	3.23 (2)	1.41 (4)	2.56 (5)	3.51 (6)	4.08 (1)	1.81 (3)	7.56 (4)	2.14 (6)	4.16 (7)
0.15	1.80 (−1)	4.60 (1)	1.42 (3)	2.18 (4)	2.62 (5)	1.25 (1)	2.93 (2)	6.47 (3)	1.21 (5)	1.83 (6)
0.20	8.64 (−2)	1.21 (1)	2.89 (2)	3.77 (3)	3.91 (4)	5.42 (0)	8.47 (1)	1.24 (3)	1.70 (4)	2.11 (5)
0.40	1.63 (−2)	5.31 (−1)	6.89 (0)	5.67 (1)	3.88 (2)	7.63 (−1)	5.14 (0)	3.22 (1)	2.18 (2)	1.49 (3)
0.60	6.60 (−3)	1.14 (−1)	1.01 (0)	6.17 (0)	3.20 (1)	2.52 (−1)	1.15 (0)	4.86 (0)	2.23 (1)	1.07 (2)
0.80	3.62 (−3)	4.08 (−2)	2.75 (−1)	1.38 (0)	5.91 (0)	1.16 (−1)	4.25 (−1)	1.41 (0)	5.09 (0)	1.92 (1)
1.0	2.32 (−3)	1.97 (−2)	1.09 (−1)	4.64 (−1)	1.73 (0)	6.52 (−2)	2.04 (−1)	5.80 (−1)	1.76 (0)	5.55 (0)
1.5	1.09 (−3)	6.17 (−3)	2.42 (−2)	7.84 (−2)	2.28 (−1)	2.31 (−2)	5.83 (−2)	1.30 (−1)	3.02 (−1)	7.27 (−1)
2.0	6.63 (−4)	3.14 (−3)	9.72 (−3)	2.63 (−2)	6.47 (−2)	1.13 (−2)	2.56 (−2)	5.00 (−2)	9.98 (−2)	2.04 (−1)

REFERENCES

1. M. E. Rose, in *Beta and Gamma Ray Spectroscopy*, edited by K. Siegbahn, North-Holland Publishing Co., Amsterdam, 1955, Chapter 9.
2. See M. E. Rose, *Multipole Fields* (John Wiley and Sons, New York, 1955).
3. M. E. Rose, Phys. Rev. **76**, 678 (1949); **78**, 184 (1950).
4. E. Church and J. Weneser, Phys. Rev. **103**, 1035 (1956).
5. N. Tralli and G. H. Goertzel, Phys. Rev. **83**, 399 (1951).
6. M. E. Rose, Phys. Rev. **51**, 484 (1937).
7. K. Umeda, J. Fac. Sci. Hokkaido Imp. Univ., Ser. II **3**, 171 (1942).
8. J. Reitz, Phys. Rev. **77**, 10 (1950); see also L. A. Sliv and I. M. Band, issued in U.S.A. as report 57 ICC Kl.
9. T. A. Green and M. E. Rose, Oak Ridge National Laboratory Report ORNL-2395 and Phys. Rev., **110**, 105 (1958).
10. M. E. Rose, Proc. Rehovoth Conf. on Nuclear Structure (North-Holland Publishing Co., Amsterdam, 1957).
11. E. Church and J. Weneser, Phys. Rev. **104**, 1382 (1956).
12. A. S. Reiner, Proc. Rehovoth Conf. on Nuclear Structure (North-Holland Publishing Co., Amsterdam, 1957).
13. S. G. Nilsson, University of California Radiation Laboratory Report UCRL-3803.
14. S. G. Nilsson and J. O. Rasmussen, University of California Radiation Laboratory Report UCRL-3889.
15. M. E. Rose, *Internal Conversion Coefficients* (North-Holland Publishing Co., Amsterdam, 1958). Much of the material contained in this reference has been circulated by private communications.
16. M. E. Rose, G. H. Goertzel, B. I. Spinrad, J. Harr, and P. Strong, Phys. Rev. **83**, 79 (1951).
17. M. E. Rose, privately circulated tables and Appendix IV of "Beta and Gamma Ray Spectroscopy" (*1*).

V. F. Analysis of Gamma Decay Data

by D. H. WILKINSON

1. Modes of De-excitation... 852
2. General Selection Rules for Electromagnetic Transitions.................. 853
3. Internal Conversion... 857
4. Nomenclature... 857
5. Isotopic Spin Selection Rules.. 858
6. Units of Transition Strength... 858
7. The Independent-Particle Model.. 863
8. Methods of Observation.. 865
9. Empirical Data: the Light Elements..................................... 867
10. Empirical Data: the Heavy Elements................................... 872
11. Special Selection Rules... 879
12. Branching Ratios... 882
13. Sum Rules... 882
    Appendix: Fractional Parentage and the Parentage Coefficients.......... 884
    References 888

In this chapter we discuss the factors influencing the emission of gamma rays from nuclei as far as possible in terms of the physics of the situation. We also quote the relevant theoretical results but make no attempt to derive them. The discussion is to a large degree determined by the availability in the current literature of accounts likely to appeal to experimenters. It is not therefore supposed necessarily to reflect the importance of the various topics. For example, we give considerable discussion of the independent-particle model and explicit instructions on how to carry out simple calculations with it and to evaluate the fractional parentage coefficients because this is not published elsewhere in detail. On the other hand the collective models are referred to relatively sketchily because they have been the subject of many detailed accounts.

As far as possible references are given to review and summary articles and, in particular, experimental references are always made to previous compilations.

1. Modes of De-excitation

If a nucleus is energetically able to emit a heavy particle (say a nucleon) it usually does so very rapidly and the intrinsically slower mode of de-excitation by the emission of electromagnetic radiation is a weak competitor. The reason is that whereas the nucleon "pre-exists" within the nucleus, the nuclear charges are coupled to the electromagnetic field via the fine structure constant $e^2/\hbar c = \frac{1}{137}$. The effect of this may be seen very easily from a semiclassical viewpoint.

The intrinsic probability per unit time of heavy particle emission from a nuclear orbit is of the order v/R where v is a speed characteristic of the nucleon motion and R is a dimension characteristic of the nuclear size. In the correspondence-principle limit, the power radiated by a charge moving in the same way ($\omega \sim v/R$ and acceleration $\sim v^2/R$) is of order $(e^2/c^3)(v^2/R)^2$, and since in this limit the photon energy $\hbar\omega = \hbar v/R$, the probability per unit time of photon emission is of order $(e^2/\hbar c)(v/c)^2$ (v/R). This rate is smaller than that for heavy particle emission by at least the order of the fine structure constant. In practice, of course, both heavy particle and photon emission are complicated by a host of other factors but it is the occurrence of $e^2/\hbar c$ in all electromagnetic processes which make them intrinsically slow. This slowness is just an accident of nature, a consequence of the actual values that e, $\hbar$, and c have happened to take (compare the analogous emission of π-mesons in nucleon collisions which is very strong because there the comparable coupling constant is of order unity or greater).

But if an excited nucleus is stable against heavy particle emission; or if that emission is slowed down because the energy available is very small; or there is a formidable quantum mechanical barrier, Coulomb or centrifugal, to be penetrated; or if it is inhibited by some selection rule; then we shall encounter de-excitation by an electromagnetic process.[a]

2. General Selection Rules for Electromagnetic Transitions

a. Single Photon Transitions

The wave length of photons from nuclear processes is, as a rule, large compared with the dimensions of the nucleus. Thus λ for a photon of 1 Mev is 2.0×10^{-11} cm while the radius of the uranium nucleus is about 7.5×10^{-13} cm. This means that to a good first approximation we can ignore the variation of the electric field $\mathcal{E}$ of the radiation across the nucleus and write the interaction energy simply as $\mathcal{E}ez$ (for a single proton). The transitions, calculated using time-dependent perturbation

---

[a] Very occasionally the intrinsically very much weaker process of beta decay may compete with an electromagnetic transition if the latter is slow because the energy is low and the spin change is high (see Section 6a). For example, the positron emission from the first excited (0+) state of Al^{26} at 0.22 Mev has a half-life of 6.7 sec. The lifetime for the electromagnetic de-excitation to the (5+) ground state is presumably extremely long.

theory to first order, are therefore chiefly governed by matrix elements of the form: $\int \psi_i z \psi_f d\tau$, where ψ_i and ψ_f are wave functions of the initial and final states between which the transition is made. When the wave functions (for a single spinless charge) are separated into radial and angular parts, it is seen that the integration over the θ variable vanishes unless $l_i = l_f \pm 1$, that is, unless the angular momentum changes by one unit and there is a change of parity (1).*

In making these remarks we have ignored the importance of the intrinsic spin of the nucleons, and the fact that there may be many of them coupled together. When these factors are considered, the rule that the parity must change remains true while the rule on the change of angular momentum is replaced by the requirement that the vectorial change should be unity, namely, that a state of spin J may combine only with spins $J - 1, J, J + 1$ (2). This is electric dipole radiation.

If this rule is not obeyed the radiation will not, of course, be zero because in writing the perturbation as $\mathcal{E}ez$ we have wrongly assumed the field to be constant over the nucleus. The correction terms may be written in the form $\cdots + (z/\lambda)^L + \cdots$ with the associated matrix elements:

$$\cdots \int \psi_i z^L \psi_f \, d\tau \cdots .$$

Successive terms in this series of correction terms decrease by the factor λ/R. Since the intensity of the radiation is determined by the square of the matrix element the contributions from these successive terms are roughly as $(R/\lambda)^{2L}$. If we proceed as before we find that the parity change associated with each term is $(-)^L$ [b] and that the spin change (vectorial) is L,[c] which is therefore the angular momentum transported by the

---

* The reference list for Section V.F begins on page 888.

[b] The parity rule is almost self evident. We are concerned with matrix elements of the form $\int \psi_i z^L \psi_f \, d\tau$ which will vanish if the integrand is an odd power of $\cos \theta$. If the parities of ψ_i and ψ_f are the same L must be even to make the whole integrand even; if ψ_i and ψ_f have different parities L must then be odd to give the matrix element a finite value.

[c] Photons in the free state have an intrinsic angular momentum of unity which, however, must point along, or against, their direction of propagation (3). This polarization is a manifestation of the transverse nature of the electromagnetic field in the far or wave zone. It is interesting to note that this transversality is never complete and that E or H, or both, must have a radial component in all regions of space. This is because the expression for the total angular momentum transmitted by the electromagnetic field (never zero as we shall see) can be written in a form depending on $\mathbf{r} \cdot \mathbf{H}$ and $\mathbf{r} \cdot \mathbf{E}$ (see p. 803 in Blatt and Weisskopf, 2). In addition to their intrinsic spin, photons may be thought of as having an orbital angular momentum and this enables spin changes greater than unity to take place. Such orbital angular momentum has its axis at right angles to the direction of propagation of the photon and so can

photon itself. Since R/λ is a small number we can usually ignore all terms except the first nonzero one. (Only alternate terms contribute to any given transition because of the requirement on the parity.) Electric 2^L-pole radiation is due to the Lth term.

We have been speaking of the radiation as due to the coupling between the charges of the particles in the nucleus and the electromagnetic field. We can similarly write down the coupling between the electromagnetic field and the distribution of magnetization in the nucleus due both to the circulation of charges and to the intrinsic magnetic moments. This coupling leads to the definition of magnetic radiation which splits up into multipoles just as the electric radiation, with the same spin rules and angular momentum transport by the photon (see p. 584 in Blatt and Weisskopf, 2). However, the 2^L-pole magnetic radiation is now associated with the parity change $(-)^{L+1}$.

Since the current density which gives rise to the magnetic radiation is smaller than the charge density by the factor v/c, where v is the speed of the charges, we must expect *a priori* that magnetic radiation will be weaker than electric radiation of the same multipolarity by a factor, in the nucleus, of order $(v/c)^2 \sim 0.05$.

If now the initial and final nuclear spins are J_i and J_f with $J_i \neq J_f$ we should expect the dominant electric radiation to be $2^{|J_i-J_f|}$-pole if the parity change is $(-)^{|J_i-J_f|}$ and otherwise $2^{|J_i-J_f|+1}$-pole. If $J_i = J_f$ then we expect to find electric dipole radiation if there is a change of parity, and electric quadrupole radiation if there is not. Correspondingly the dominant magnetic radiation is $2^{|J_i-J_f|}$-pole if the parity change is $(-)^{|J_i-J_f|+1}$. With $J_i = J_f$ we have magnetic dipole radiation if the parity does not change and magnetic quadrupole if it does.

Now if $2^{|J_i-J_f|}$-pole electric radiation (or dipole if $J_i = J_f$) is allowed by the parities of the states we should expect this to overwhelm the magnetic radiation which will be $2^{|J_i-J_f|+1}$-pole (or quadrupole if $J_i = J_f$) and, hence, weaker—both on account of the extra power of $(R/\lambda)^2$ and of the $(v/c)^2$ factor. Such a transition, in which electric radiation of the lowest multipole order allowed by the spins is also allowed by the parities is sometimes called a "parity-favored" transition. If, however, the parities do not allow such electric radiation then the order of the magnetic radiation will be less than that of the electric. The factor $(R/\lambda)^2$ now discourages the electric radiation relative to the magnetic while the

---

never cancel the intrinsic spin which is along the direction of propagation. For this reason a single photon cannot remove zero (vectorial) angular momentum and a one-photon transition between two states of $J = 0$ is accordingly *rigorously* forbidden. There is of course no objection to a $J \to J$ transition if $J \neq 0$ because we can then, for example, arrange that $\mathbf{J} = \mathbf{J} + \mathbf{1}$.

factor $(v/c)^2$ discourages the magnetic as before. In this "parity-unfavored" situation it is no longer clear whether we should expect the electric or the magnetic transition to predominate.

There are a few obvious exceptions to these general remarks:

1. If we have J_i or $J_f = 0$ then only electric or only magnetic radiation can occur (2^J-pole in either case), the former if the parity change is $(-)^J$, the latter if it is $(-)^{J+1}$.

2. If we have $J_i = J_f = \frac{1}{2}$ then there can be only dipole radiation, electric if the parity changes and magnetic if it does not.

3. If $J_i = J_f = 0$ then *no radiation whatever can occur by single-photon emission* such as we are discussing now: it is *rigorously* forbidden. There are no monopole transitions by a single photon.

b. Zero-Zero Transitions

We have already examined and explained the absence of single-photon emission in $J = 0 \rightarrow J = 0$ transitions as being due to the longitudinal polarization of the unit spin of the photon. We may more physically say that single-photon emission cannot take place from a system unless electromagnetic disturbances are produced outside that system, that is, unless there is a whip crack in a line of force. If both initial and final states have $J = 0$ then the transition between them would not be detected from the outside because both are spherically symmetrical and so can differ, at most, in the radial distribution of their charge. This difference, however, has no observable consequences outside the system, both states behaving like identical point charges. There is, therefore, no changing electric field produced at large distances. The different states differ inside the nucleus itself, however, because the initial and final radial distributions will, in general, be different, and so electromagnetic transitions of some sort are clearly possible even though single-photon emission is not.

It is, for example, clear that the simultaneous emission of two photons is possible since we can then arrange a cancellation of the spins whether or not there is a change of parity involved. Alternatively, if no change of parity is involved between the states of spin zero, it is possible that the energy of de-excitation be removed by the emission of a single orbital (internal conversion) electron. If there is a change of parity then we will have two internal conversion electrons, or one internal conversion electron and one photon. If the energy difference between the two states of $J = 0$ is greater than $2mc^2$ then simultaneous emission of a negative and positive electron (pair emission) is possible (*4*). This takes place readily if there is no change of parity. Pair emission with a change of parity is not rigorously forbidden but is a higher order effect. In fact the only known zero-zero transitions are between states of even parity. For example, 0+ states of O^{16}, Ca^{40}, and Zr^{90} at 6.05, 3.46, and 1.75 Mev, respectively, decay to the ground states by pair emission; 0+ states of Ge^{72} at 0.68 Mev and of Po^{214} at 1.41 Mev decay by the emission of an orbital electron (5).

3. Internal Conversion

We are usually concerned with normal electromagnetic transitions for which single-photon emission is possible. An inevitable parallel process is internal conversion in which an orbital electron emerges and carries away the energy of de-excitation. This process is very nearly, but not entirely, a competitive process to photon emission, and to first order does not diminish the rate of photon emission but rather provides an additional mechanism of decay, which therefore accelerates the total decay rate. A negligible fraction of internal conversion (of order $e^2/\hbar c$) may be thought of as the emergence from the nucleus of a real photon with subsequent photoelectric conversion in an electron shell of the same atom (see p. 20 in Burhop, 4). Accordingly if the total internal conversion coefficient, the ratio of electron to photon emissions, is α_T, the actual lifetime τ_{obs} is related to the partial lifetime τ_γ for photon emission alone by, very nearly:

$$\tau_\gamma = \tau_{obs}(1 + \alpha_T).$$

It is with the partial lifetime τ_γ and with the associated radiative width, $\Gamma_\gamma = \hbar/\tau_\gamma$, that we are concerned in this chapter but it must be remembered that a measured lifetime may have to be corrected for internal conversions before we interpret it in terms of the probable multipolarity. α_T is strongly dependent upon the energy of the transition E_γ, upon the multipolarity and parity change, and upon the nuclear charge Z (6,7).

Internal conversion by pair creation also takes place where energetically possible but this process is never very probable (except for $0 \rightarrow 0$ transitions) (6,7).

4. Nomenclature

The multipolarity L is the vectorial spin change between initial and final states, $\mathbf{J}_i = \mathbf{J}_f + \mathbf{L}$ and is the total angular momentum taken away by the photon. The radiation is then specified by stating L, and whether it is electric or magnetic in character (which with L implies whether or not there is a parity change). A concise notation is E1, M1, E2, M2, and so forth, in which E or M stands for electric or magnetic and the number is L. Radiation of given multipolarity and parity change is radiation of a given *type*.

5. Isotopic Spin Selection Rules

In addition to the general selection rules that were discussed in Section 2.a there are selection rules on the isotopic spins T of the nuclear states *(8)*. The general rule is that we must have $\Delta T = 0, \pm 1$.[d] This has never found an application in practice if only because no states differing by 2 or more in isotopic spin have yet been identified in any nucleus.

A special rule exists for E1 transitions in self-conjugate nuclei (nuclei for which $Z = N$) *(8)*. It is $\Delta T = \pm 1$; the transitions must change the isotopic spin.[e] Since the isotopic spin is not a good quantum number, if only because of the relaxation of charge-independence by the Coulomb forces *(9)* this rule is not rigorous but it is effective and will be discussed later.

For a discussion of these and other isotopic spin selection rules see the review by Burcham *(10)* and Section V.H.

6. Units of Transition Strength

a. THE WEISSKOPF UNITS

The strengths of radiative transitions are usually expressed in terms of their mean lifetime τ_γ (after correction for internal conversion if this is significant) or their radiative width Γ_γ. These are related by $\Gamma_\gamma \tau_\gamma = \hbar$. It is useful to remember that a radiative width of 1 ev is the same as a mean life of 6.6×10^{-16} sec.

Another and often more useful way of expressing a transition strength is in terms of the strength of a transition of the same energy and type calculated according to some simple model for a nucleus of the same size. It is then immediately apparent if the transition is surprisingly fast or surprisingly slow. It is obviously desirable that the model should be a general one and also bear some relation to nature.

The model most generally used is the extreme one-particle model. For the electric transitions we suppose that a single proton moves in an

[d] This rule limiting ΔT to 1 reflects the fact that in a radiative transition only one nucleon ($t = \frac{1}{2}$) may change its quantum numbers.

[e] That $T = 0 \rightarrow T = 0$ transitions are forbidden for E1 radiation is evident because an E1 moment involves the separation of neutrons from protons but $T = 0$ means that neutrons and protons are exactly equivalent in the specification of the state. If this is so for both initial and final states no moment can be induced. The fuller rule is not evident.

orbit of angular momentum L with total spin $J_i = L + \frac{1}{2}$ within some featureless, chargeless, spinless, infinitely massive potential well, and then makes a transition to its final state of zero orbital angular momentum so that $J_f = \frac{1}{2}$. In order to calculate the speed of the EL or electric 2^L-pole transition we must know the radial wave functions in the initial and final states. These are both taken conventionally as simple rectangles of radial extension R which we identify with the radius of the nucleus, namely, the wave function is taken as constant throughout the nucleus. In this way we avoid the need to specify the form of the potential within which the proton moves, the binding energies and so on (*11*).

The estimates of transition probability thus arrived at are called the Weisskopf units and are:[f]

$$\tau_\gamma^{-1}(EL) = \frac{2(L+1)}{L[(2L+1)!!]^2}\left(\frac{3}{L+3}\right)^2 \frac{e^2}{\hbar c}\left(\frac{\omega R}{c}\right)^{2L} \omega \ \sec^{-1}. \quad (1)$$

If we now measure E_γ in million electron volts and R in fermis (10^{-13} cm) this takes the numerical value:

$$\frac{4.4(L+1)}{L[(2L+1)!!]^2}\left(\frac{3}{L+3}\right)^2 \left(\frac{E_\gamma}{197}\right)^{2L+1} R^{2L} \times 10^{21} \ \sec^{-1}. \quad (2)$$

Finally for complete definiteness we need a further convention about R. We know that, to a reasonable approximation:

$$R = r_0 A^{\frac{1}{3}}.$$

For r_0 we use the value 1.2 fermis although it seems that the value appropriate to the light elements should be somewhat greater (*12*). We then finally find, expressing the results in terms of the radiative width Γ_γ in electron volts with E_γ in million electron volts:

$$\Gamma_{\gamma W}(E1) = 6.8 \times 10^{-2} A^{\frac{2}{3}} E_\gamma^3 \ \text{ev}, \quad (3)$$
$$\Gamma_{\gamma W}(E2) = 4.9 \times 10^{-8} A^{\frac{4}{3}} E_\gamma^5 \ \text{ev}, \quad (4)$$
$$\Gamma_{\gamma W}(E3) = 2.3 \times 10^{-14} A^2 E_\gamma^7 \ \text{ev}, \quad (5)$$
$$\Gamma_{\gamma W}(E4) = 6.8 \times 10^{-21} A^{\frac{8}{3}} E_\gamma^9 \ \text{ev}, \quad (6)$$
$$\Gamma_{\gamma W}(E5) = 1.6 \times 10^{-27} A^{\frac{10}{3}} E_\gamma^{11} \ \text{ev}. \quad (7)$$

The calculation of the Weisskopf units for the magnetic transitions is not quite so straightforward because of the complication of the intrinsic magnetic moments of the nucleons. We have already noted that we should *a priori* expect the magnetic matrix elements to be smaller than the corresponding electric matrix elements by a factor of order v/c. If we

[f] $(2L+1)!! = 1 \cdot 3 \cdot 5 \cdots 2L+1$.

apply the uncertainty principle to estimate this factor, we should set it at about $\hbar/McR$ where M is the nucleon mass. However, the intrinsic magnetic moment of a nucleon is about 2 nuclear Bohr magnetons and to allow for the importance of this we may crudely add the intrinsic contribution to that of the orbit. Conventionally we then relate the squares of the matrix elements for magnetic and electric transitions by $10(\hbar/McR)^2$. This leads as before to:

$$\tau_\gamma^{-1}(ML) = \frac{20(L+1)}{L[(2L+1)!!]^2}\left(\frac{3}{L+3}\right)^2 \frac{e^2}{\hbar c}\left(\frac{\hbar}{McR}\right)^2 \left(\frac{\omega R}{c}\right)^{2L} \omega \text{ sec}^{-1}$$

$$= \frac{1.9(L+1)}{L[(2L+1)!!]^2}\left(\frac{3}{L+3}\right)^2 \left(\frac{E_\gamma}{197}\right)^{2L+1} R^{2L-2} 10^{21} \text{ sec}^{-1} \quad (8)$$

$$\Gamma_{\gamma W}(M1) = 2.1 \times 10^{-2} E_\gamma^3 \text{ ev}, \quad (9)$$
$$\Gamma_{\gamma W}(M2) = 1.5 \times 10^{-8} A^{\frac{2}{3}} E_\gamma^5 \text{ ev}, \quad (10)$$
$$\Gamma_{\gamma W}(M3) = 6.8 \times 10^{-15} A^{\frac{4}{3}} E_\gamma^7 \text{ ev}, \quad (11)$$
$$\Gamma_{\gamma W}(M4) = 2.1 \times 10^{-21} A^2 E_\gamma^9 \text{ ev}, \quad (12)$$
$$\Gamma_{\gamma W}(M5) = 4.9 \times 10^{-28} A^{\frac{8}{3}} E_\gamma^{11} \text{ ev}. \quad (13)$$

These Weisskopf estimates can be read directly off the nomogram (Fig. 1) for E1–5 and M1–5 radiation. The nomogram also contains a figure which shows the effect of changing r_0 in the range 1.0–1.7 fermis.

If our observed radiative width is now Γ_γ we say that the transition has a strength of $|M|^2 = \Gamma_\gamma/\Gamma_{\gamma W}$ Weisskopf units. $|M|^2$ is some measure of the square of the matrix element of the actual transition relative to that of the extreme one-particle transition.

b. THE MOSZKOWSKI UNITS

The factor $10(\hbar/McR)^2$ used to relate magnetic to electric transitions in the Weisskopf estimates is, of course, very crude. In particular it should depend on L, because in general the orbital magnetic moment of the radiating particle will depend on L. Moszkowski's (13) units replace the factor 10 above by:

$$\left[\frac{L+3}{L+2}\left(\mu_p L - \frac{L}{L+1}\right)\right]^2$$

in the evaluation of the magnetic transition probabilities. $\mu_P = 2.79$, is the proton's intrinsic magnetic moment in nuclear magnetons. If this is done the above expressions for the speeds are increased by factors of 0.93, 3.8, 9.1, 16, and 22 for M1, M2, M3, M4, and M5 transitions, respectively. We shall use throughout the simple Weisskopf estimates but bear in mind this slightly refined version and refer to it occasionally.

c. The Energy Limitation

It is clear that this use of wave functions which are uniform throughout the nucleus imposes some limitation on the energy of a transition for which it is sensible to consider the Weisskopf units at all. If the transition energy is very high, the nucleon in the final state (consider absorption) will have a short wave length, and so the approximation of a uniform wave function is very bad. The radial overlap integral will be very low and the Weisskopf estimate would be a gross overestimate even if the states were as simple as assumed in the model. In fact, for transitions of high enough energy the Weisskopf units violate the sum rules—see Section 13. The physical reason for the fall off of transition strength at high energies is that the nucleon momentum is great in the upper state but the photon itself brings in little momentum. This high nucleon momentum must come therefore from the wave function of the lower state and this is unlikely. By such consideration of the need for a good overlap between the *momentum* wave functions of the nucleons in the initial and final states we may make a semiquantitative estimate of the energy E_{limit} above which the $\Gamma_{\gamma W}$, which ignore this difficulty, must be treated very cautiously. We find

$$E_{\text{limit}} \sim \frac{\hbar}{R} \sqrt{\frac{2V}{M^*}} \sim 80 \ A^{-\frac{1}{3}} \ \text{Mev} \tag{14}$$

where $M^* \sim 0.6M$ is the effective nucleon mass *(14)* and V is the depth of the nuclear potential. This energy is quite high and is above the region of discrete levels. The Weisskopf units may therefore be used for discrete states without this concern.

d. Statistical Weight Factors

Some authors have occasionally inserted "statistical weight factors" into the experimental, or Weisskopf, estimates of radiative width before quoting the strength in single-particle units. This practice is most strongly to be deprecated and can only lead to confusion. It is, of course, true that the actual spins involved influence the transition probability but in the absence of a detailed model which accounts for these spins there is no generally valid statistical factor that can be defined. If one is testing a particular model then the spins take their places naturally in the theoretical estimate as will be seen shortly. But in this case the Weisskopf estimates themselves would not be used as the basis for comparison but rather a correctly computed single-particle unit using more realistic wave functions, and a particle jumping not between $l = L$ and $l = 0$ but between whatever orbits are suggested by the model.

An exception to the remarks about statistical weights is made if we wish to compare transitions in two different nuclei but between states which a model regards as of the same origin in the two cases, and which have a different order in the two nuclei. Suppose the transition is $J_a \to J_b$ with speed $|M_{ab}|^2$ in the first nucleus and $J_b \to J_a$ with $|M_{ba}|^2$ in the second. We then expect by the principle of detailed balance:

$$\frac{|M_{ab}|^2}{|M_{ba}|^2} = \frac{2J_b + 1}{2J_a + 1}.$$

It is still unwise, however, to insert any such factor as $2J + 1$ in quoting the $|M|^2$ values as such.

It is sometimes useful to know the speed of a strict one-particle electric transition as on the Weisskopf model, using the same assumption about rectangular wave functions, but where the initial and final orbits are l_i and l_f rather than L and 0 and where the initial and final spins are j_i and j_f rather than $L + \tfrac{1}{2}$ and $\tfrac{1}{2}$. Such a transition has:

$$|M|^2 = (2l_f + 1)[C_{000}^{l_i l_f L}]^2 [U(\tfrac{1}{2} j_i l_f L; l_i j_f)]^2.$$

Here the C are the Clebsch-Gordan coefficients (15–17) and the U are related to the Racah coefficients W (17) by:

$$U(abcd;\, ef) = [(2e + 1)(2f + 1)]^{\frac{1}{2}} W(abcd;\, ef).$$

It is now instructive to investigate the effect of a spin-flip on a one-particle electric transition. For simplicity consider $l_i = l_f + L$ so that $j_i = l_f + L \pm \tfrac{1}{2}$ and $j_f = l_f \pm \tfrac{1}{2}$. Three EL transitions between these four states are possible and one of them, $l_f + L - \tfrac{1}{2} \to l_f + \tfrac{1}{2}$ involves a (classical) flipping of the intrinsic spin. We find, using the above expression:

$$\frac{|M_{l_f+L+\frac{1}{2} \to l_f+\frac{1}{2}}|^2}{|M_{l_f+L-\frac{1}{2} \to l_f-\frac{1}{2}}|^2} = \frac{(l_f + L)(2l_f + 1)}{l_f(2l_f + 2L + 1)}.$$

So the intensities of the two transitions that do not flip the intrinsic spin are comparable and become equal in the classical limit of $l_f \to \infty$. Also:

$$\frac{|M_{l_f+L-\frac{1}{2} \to l_f+\frac{1}{2}}|^2}{|M_{l_f+L-\frac{1}{2} \to l_f-\frac{1}{2}}|^2} = \frac{L}{l_f(2l_f + 2L + 1)}.$$

So the intensity of the spin-flip transition is much less than that of the others, and tends to zero in the classical limit. These results are qualitatively to be expected since the intrinsic spin takes no part in the mechanism of electric radiation.

A complete discussion of statistical weight factors for use in clearly specified circumstances is available (18).

This leads us naturally to a discussion of the more detailed predictions about radiative transitions which are made by the independent-particle model of the nucleus.

7. The Independent-Particle Model

a. THE EQUIVALENT-PARTICLE FORMULATION

The Weisskopf estimates are derived from the extreme one-particle model in which the angular momentum of the nucleus is due solely to the last particle. In practice many particles are involved in the specification of most nuclear states and this has profound effects on the theoretical prediction.

The theoretical estimate proceeds as follows: First of all decide between which orbits l_i and l_f the radiating particle jumps. This $l_i \rightarrow l_f$ jump takes place in the presence of $n - 1$ equivalent nucleons in the l-orbit. The spins of the initial and final states are J_i and J_f, respectively, so the transition is:

$$l^{n-1}l_i(\alpha_i T_i S_i L_i, J_i) \rightarrow l^{n-1}l_f(\alpha_f T_f S_f L_f, J_f)$$

in LS coupling, or:

$$j^{n-1}j_i(\alpha_i T_i J_i) \rightarrow j^{n-1}j_f(\alpha_f T_f J_f)$$

in jj coupling. Here α_i, α_f are any residual quantum numbers necessary to complete the specification of the initial and final states.

We now express the initial and final states in terms of the fractional parentage expansion (see Appendix) over the parent states of l^{n-1} such as $(\alpha_p T_p S_p L_p, J_p)$ in LS coupling.

For the electric multipoles we are now able to write down the predicted speed Γ_γ of the shell-model transition in terms of $\Gamma_{\gamma sp}$ the radiative width calculated for a *spinless* particle of protonic charge jumping between orbits l_i and l_f in an *infinitely massive uncharged* potential well. We have *(19)* LS coupling:

$$\Gamma_\gamma = \Gamma_{\gamma sp} \times \left[n \sum_p U(Ll_f L_i L_p; l_i L_f) U(LL_f J_i S_i; L_i J_f) \right.$$

$$\left. \times <\alpha_p|\} \alpha_i> <\alpha_p|\} \alpha_f> (\tfrac{1}{2}\delta_{T_i T_f} - \mathfrak{Z}_p) \right]^2.$$

The $<|\}>$ are the fractional parentage coefficients (see Appendix).

$$\mathfrak{Z}_p = (-)^{2T_p - T_i - T_f} \frac{\sqrt{3}}{2} U(1\tfrac{1}{2}T_i T_p; \tfrac{1}{2}T_f) C^{1T_f T_i}_{0 M_T M_T}. \qquad [M_T = \tfrac{1}{2}(N - Z)]$$

The radiation will be zero unless we conform to the special selection rules for electric radiation in LS coupling—see Section 11a. We must also conform to the isotopic spin selection rules—see Section 5 and Chapter V.H jj coupling:

$$\Gamma_\gamma = \Gamma_{\gamma sp} \times \left[n \sum_p U(Lj_f J_i J_p; j_i J_f) U(Ll_f j_i s; l_i j_f) \right.$$
$$\left. \times \langle \alpha_p | \} \alpha_i \rangle \langle \alpha_p | \} \alpha_f \rangle (\tfrac{1}{2} \delta_{T_i T_f} - \mathfrak{z}_p) \right]^2. \quad (s = \tfrac{1}{2})$$

These expressions for Γ_γ contain many factors by which the $\Gamma_{\gamma sp}$ are multiplied. These factors can be very small and occasionally zero. We must therefore expect that the Γ_γ will show a very wide spread, and some very small values can be expected. On the other hand we see that Γ_γ contains n, the number of particles concerned in the specification of the state. This factor tends to increase Γ_γ above $\Gamma_{\gamma sp}$ and it is therefore possible, though extremely rare, to find theoretical transitions which are faster than the single-particle value. This last factor, one of enhancement, may be thought of as due to the spatial correlations imposed among the n particles by the Pauli principle.

A factor that is included implicitly in these expressions when applied to E1 transitions is the so-called "effective charge factor" which it is sometimes useful to remember even when the crude Weisskopf estimates are being used—for example, Section 10e below. This arises because we must at the least admit that a nucleus is of mass A and bears a charge Z and is not infinitely massive and neutral (apart from the radiating proton) as the model assumes. It is easily seen that the other $Z - 1$ protons radiate in opposition to the "radiating" proton and that their acceleration is $A - 1$ times less. The total radiation is then of order, for E1 transitions,

$$(1 - Z/A)^2 = (N/A)^2 \sim \tfrac{1}{4}$$

of that given by the simple model. For a jumping neutron the radiation comes entirely from the charges on the rest of the nucleus and so the radiation is of order $(Z/A)^2 \sim \tfrac{1}{4}$.

Electric radiation of higher order depends on correspondingly higher orders of the displacement and so for a jumping proton we expect the use of a charged potential well to introduce a factor $(1 - Z/A^L)^2 \sim 1$ and for a jumping neutron $(Z/A^L)^2 \sim$ zero. This naive argument is in fact incorrect and it can be shown that (20) owing to a center-of-mass effect the correct factors for electric multipoles other than dipole, for a harmonic coscillator potential, are exactly unity and identically zero for a jumping proton and for a jumping neutron, respectively.

The corresponding expressions for magnetic transitions are slightly more complicated because both space and spin contributions to the matrix element are significant (19).

b. THE SINGLE-PARTICLE WIDTHS

Typical expressions for $\Gamma_{\gamma sp}$ are (19):

$$\Gamma_{\gamma sp}(E1) = 2.48 \times 10^{25} \times E_\gamma^3 \left[\frac{<l_i 0 | r \cos \theta | l_f 0>}{C_{000}{}^{1 l_f l_i}} \right]^2 \text{ electron volts} \quad (15)$$

$$\Gamma_{\gamma sp}(E2) = 7.94 \times 10^{44} \times E_\gamma^5 \left[\frac{<l_i 0 | r^2 (3 \cos^2 \theta - 1) | l_f 0>}{C_{000}{}^{2 l_f l_i}} \right]^2$$
$$\text{electron volts.} \quad (16)$$

Here E_γ is measured in million electron volts and r in centimeters.

Expressions unlikely to be in error by more than 50% in the $1p$-shell are (19):

$$\Gamma(E1)_{\gamma sp} = 0.80 E_\gamma^3 \text{ electron volts for } l_i = 0; \text{ namely, } 2s \to 1p$$
$$\text{transitions} \quad (17)$$
$$= 0.99 E_\gamma^3 \text{ electron volts for } l_i = 2; \text{ namely, } 1d \to 1p \text{ transitions}$$
$$\Gamma(E2)_{\gamma sp} = 1.05 \times 10^{-5} E_\gamma^5 \text{ electron volts for } l_i = 1; \text{ namely,}$$
$$\text{transitions within the } 1p\text{-shell.} \quad (18)$$

Note that the $\Gamma_{\gamma sp}$ are not the $\Gamma_{\gamma W}$ but should be evaluated for realistic wave functions appropriate to the actual orbits between which the transitions are made. They are, of course, close to the $\Gamma_{\gamma W}$.

8. Methods of Observation

We have now established and discussed our units and the sort of deviation from them that we should expect real transitions to display even if the independent-particle model were fully valid. We must now examine the empirical data expressed in terms of these units.

Before we do this however we will make a distinction, which is chiefly operational but possibly also fundamental, between transitions in light and heavy nuclei. The operational distinction is implied by the uncertainty relationship: $\Gamma_\gamma \tau_\gamma = \hbar$ which expresses the alternative experimental approaches we may make to the problem of electromagnetic transitions. If the transition is slow then we can measure τ_γ, the lifetime, directly; if it is fast then it is the radiative width Γ_γ which reveals itself, in the absolute yield of a reaction of radiative capture or indirectly in a branching ratio against heavy-particle emission of known probability.

These methods do not yet overlap. The shortest directly measured lifetimes are about 10^{-11} sec which is about 10^{-4} ev. To measure a radiative width as small as 10^{-3} ev is a *tour de force*. Some special techniques bridge the gap from both sides: Doppler shift work from the side of lifetime measurements, and resonance fluorescence from the side of radiative widths, for example. But such special methods are of limited applicability and for most practical purposes there exists a gap of several orders of magnitude between the two types of measurement.

In the light nuclei (say $A < 40$) we study and identify individual levels formed by the capture of bombarding particles. These levels are always at an excitation of several million electron volts. The emission of dipole and sometimes quadrupole radiation can then be measured. Resonance fluorescence is also proving a useful tool. The higher multipoles have so far not been detected in capture radiation. On the other hand the lifetimes of even the lowest states of the light nuclei are usually too short for direct measurement, although a few are accessible to Doppler techniques and it is occasionally possible to establish by these techniques that certain multipolarities can be reasonably excluded (see Section 9).

In the heavy nuclei the study of individual levels formed by particle bombardment is barely beginning for cases where the spins and parities of both initial and final states are known. But here the lowest states frequently have a measurable lifetime. Alternatively they can be excited by Coulomb excitation, whose yield conveys the same information as the measurement of a lifetime (or often a partial lifetime if mixed M1 + E2 transitions are possible since it picks out the E2 component[g]). Since the transitions of lower energy and high multipolarity have the longer lifetime it tends to be about them that the most information is available.

In the light elements we are well informed about dipole transitions of high energy (some millions of electron volts) while in the heavy elements such information is largely lacking and we rather know about transitions of high multipolarity and low energy (some tens or hundreds of thousands of electron volts).

Another reason for keeping separate the transitions of light nuclei is that there [at any rate in the $1p$-shell finishing at O^{16} and at the very beginning of the next, $(2s,1d)$, shell] we can do rather better than simply compare the measured transitions with the Weisskopf estimates. The situation has been well enough studied theoretically to enable many transitions to be identified with those between particular theoretical states of the full independent-particle model in intermediate coupling

---

[g] Note that the actual lifetime say of a $\frac{5}{2}+ \to \frac{3}{2}+$ transition may be much less than deduced from the $\frac{3}{2}+ \to \frac{5}{2}+$ Coulomb excitation because of the added M1 component of the free de-excitation.

(*21,22*). A comparison of the experimental and computed radiative widths is then very interesting. The correspondence is usually good for dipole transitions.

Even if such detailed comparison cannot be made we can still compare the total distributions of $|M|^2$ found in practice with those predicted by the full independent-particle model for states within the same general range of excitation. As we have remarked earlier we should expect the theoretical distributions to be broad.

9. Empirical Data: the Light Elements

a. ELECTRIC DIPOLE TRANSITIONS

Figure 2 shows the experimental distribution of $|M|^2$ for E1 transitions in the light elements ($A < 20$) as the histogram (*23,24*). The full line is

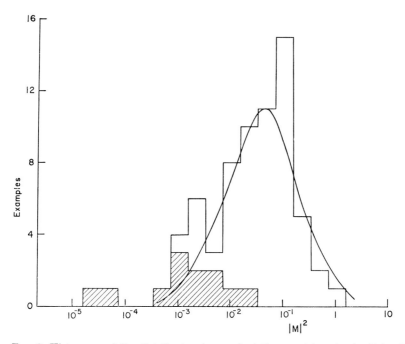

FIG. 2. Histogram of the distribution in speed of E1 transitions in the light elements ($A \leq 20$). $|M|^2 = \Gamma_\gamma/\Gamma_{\gamma W}$ where the $\Gamma_{\gamma W}$ are the Weisskopf units defined in the text (Section 6a). The shaded histogram represents transitions that violate the isotopic spin selection rule. The full curve is the prediction for the distribution of transition strengths made by the full independent-particle model in intermediate coupling.

the prediction of the independent-particle model in intermediate coupling[h] using the range 2–6 for the intermediate coupling parameter a/K (25). The shaded histogram shows the E1 transitions which violate the isotopic spin selection rule—see Section 5. The first comment is that the experimental and theoretical distributions are very similar in form. Both show a symmetrical grouping of transition strengths on the logarithmic scale of $|M|^2$. The theoretical group centers on $|M|^2 \sim 0.045$ and the experimental on $|M|^2 \sim 0.055$. We have used $r_0 = 1.2$ fermis in writing the numerical value of the Weisskopf unit.[i] As we remarked earlier this is rather too small for the light nuclei. Had we used $r_0 = 1.35$ fermis which is closer to the experimental mark for the light elements (12) the experimental distribution would have centered on $|M|^2 \sim 0.044$. We infer from this remarkably close correspondence of experiment and theory that the independent-particle model is indeed a rather good approximation to reality for at least the greater part of the individual nucleon wave functions for these light elements and also that the mixing of configurations between *major* shells cannot yet be very significant.[j]

From a practical point of view, if we have no *a priori* knowledge of special circumstances our figure for the most likely $|M|^2$ value for an E1 transition in the light elements is 0.055 and, as we can see from Fig. 2 it is probable (6 to 1 on) that the speed of any allowed transitions will fall within a factor of 7 or so of this. There is some evidence that the transitions of lower energy tend to have larger $|M|^2$ values but it is slight (24).

b. The Isotopic Spin Selection Rule on E1 Transitions[k]

We can see from Fig. 2 that the E1 transitions which violate the isotopic spin selection rule (8) are indeed rather slower than those which conform to it or for which it has no relevance. It is equally obvious that the separation between allowed and forbidden transitions is not complete and that the $|M|^2$ value cannot be made an infallible guide to the change of isotopic spin. However if $|M|^2 > 0.015$ then there is a 10:1 chance that there is an isotopic spin change if the nucleus is of $T_z = 0$. Similarly if $|M|^2 < 1.5 \times 10^{-3}$ there is a 10:1 chance that no isotopic spin change is involved.

It is possible to make some slightly more specific remarks about these forbidden E1 transitions. We may write the wave function of a state

---

[h] See Chapter VI.B.
[i] We ignore here the difference between $\Gamma_{\gamma W}$ and $\Gamma_{\gamma sp}$.
[j] For the present purpose the $(2s,1d)$ shell is one, of course, and the subshells $1p_{\frac{3}{2}}$ and $1p_{\frac{1}{2}}$ are also completely mixed already by the intermediate coupling approach.
[k] See also Chapter V.H.

of chief isotopic spin T as (9):

$$\psi = \psi(T) + \sum_{T'} \alpha_T(T')\psi(T').$$

If the E1 transition from this state is forbidden for the T-component but allowed to a T'-component, then we should expect the transition to take place roughly $\alpha_T(T')^{-2}$ times slower than an allowed transition. (This assumes that the lower state to which the transition takes place is pure;

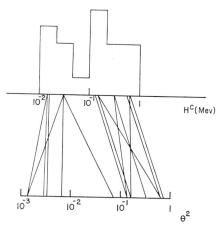

FIG. 3. Distribution of the matrix elements $H^c{}_{TT'}$ of the Coulomb perturbation for the light nuclei. The correlation between H^c and θ^2, the reduced width in single-particle units for emission of the last nucleon to the ground state of $A - 1$, is also shown.

if it too is significantly mixed in isotopic spin it will make a similar contribution to the speed of the transition.) So to guess the speed of an isotopic spin forbidden transition we must guess the values of $\alpha_T(T')$. A recipe for this has been evolved (24).

The $\alpha_T(T')$ themselves depend on the $H^c{}_{TT'}$, the matrix elements of the Coulomb perturbation between states of T and T'. We may very roughly write:

$$\alpha_T(T') \sim H^c{}_{TT'}/\Delta_{TT'}$$

where $\Delta_{TT'}$ is taken as the separation between the state of T and the nearest state of T' if the state T is not immersed in states T' or 3 times the mean spacing between states if T' states surround T.

There appears to be some correlation between H^c, so computed, and θ^2, the reduced width in units of $\hbar^2/MR$, for nucleon emission to the ground state of $A - 1$ from state T. This is shown in Fig. 3. From this we can see that if θ^2 is between 0.1 and 1 then $H^c \sim 0.25$ Mev; if $\theta^2 < 0.05$, $H^c \sim 0.025$ Mev.

To estimate the speed of this forbidden transition we take H^c from Fig. 3 if θ^2 is known. If θ^2 is not known we must take $H^c = 0.1$ Mev. We then take $\Delta_{TT'}$ from the level scheme, hence $\alpha_T(T')$ and so

$$|M|^2 = \alpha_T(T')^2 \times 0.055$$

using the most likely figure for an allowed transition.

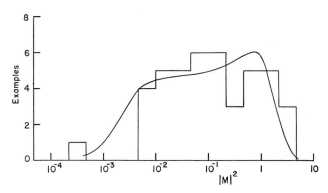

FIG. 4. The experimental (histogram) and theoretical (curve) distributions of $|M|^2$ for M1 transitions in the light nuclei.

There are many uncertainties along this chain of estimates and we may well finish up with an error of one or even two orders of magnitude in the transition speed.

c. MAGNETIC DIPOLE TRANSITIONS

Similarly Fig. 4 shows the experimental and theoretical $|M|^2$ distributions for the M1 transitions in the light elements ($A < 20$) (23,24). Again there is a remarkably close correspondence which is the more convincing because the form of the distributions has changed markedly from that for E1 transitions (Fig. 2). This reinforces our conclusion that the independent-particle model is a good description of the larger part of the wave function.

The empirical recommendation here is that the most likely *a priori* value of $|M|^2$ is about 0.15 which guess is probably accurate to a factor of 20 or so either way.

G. Morpurgo (unpublished) has recently pointed out that M1 transitions which do not change the isotopic spin in self-conjugate nuclei, while they are not forbidden, are expected to be about 100 times slower than normal (compare the E1 selection rule, Section 5). There is good experimental support for this. We therefore estimate $|M|^2 \sim 10^{-3}$ for such transitions.

d. COMPARISON BETWEEN E1 AND M1 TRANSITIONS

The question often arises as to whether we can tell apart dipole transitions with and without change of parity in the light elements. The answer is that, since $|M|^2 = 0.055$ for E1 transitions and $|M|^2 = 0.15$ for M1 transitions are only a factor of about 7 apart in absolute speed

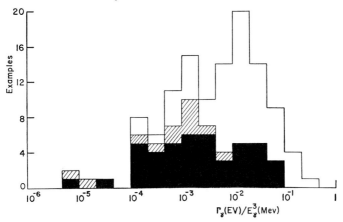

FIG. 5. Distributions of reduced transition speeds [$\Gamma_\gamma(\text{ev})/E_\gamma^3$ (Mev)] for dipole transitions in the light elements. The plain histogram is the ordinary E1 transitions, the shaded histogram is the E1 transitions that violate the isotopic spin rule, and the solid histogram is the M1 transitions.

(say for O^{16}) and since both E1 and M1 distributions of speeds are wide (Figs. 2 and 4), we cannot. The situation is represented pictorially in Fig. 5 which shows the distribution of Γ_γ/E_γ^3, namely, the reduced transition strength, assuming a dipole transition. It is seen that the two distributions completely intermingle. However, if this quantity (when Γ_γ is measured in electron volts and E_γ in millions of electron volts) is greater than 0.04 then the transition is very probably E1 while if it is less than 4×10^{-4} then we probably have an M1 transition or an E1 transition which violates the isotopic spin selection rule.

e. ELECTRIC QUADRUPOLE TRANSITIONS

Not many examples of electric quadrupole transitions are known in the light elements. The situation is summarized in Fig. 6.[1] This figure also shows the prediction of the independent-particle model. The comparison with Figs. 2 and 4 is very striking and shows at once that the shell

---

[1] Some of the techniques used for extracting these data from experiment are very rough (24) and some of these points are correspondingly unreliable; that at $|M|^2 \sim 700$ is clearly wrong, see Section 13.

model's account of the E2 transitions is totally inadequate, in contrast to the excellent account given for the E1 and M1 transitions. It seems that the most likely value for $|M|^2$ is about 5 with an uncertainty of roughly a factor of 5 either way (although much wider fluctuations are obviously to be found). The theoretical prediction is about $|M|^2 \sim 0.15$. We defer a discussion of this situation until a later section, and now merely remark that because of this apparent enhancement of the E2 transitions, we frequently cannot reliably separate E2 from M1 transitions in the light elements on the basis of their speed alone, particularly if the energy is high. As an illustration of this, consider a hypothetical

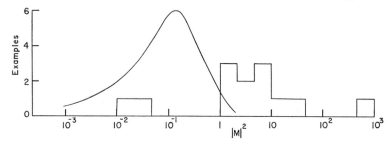

FIG. 6. The experimental (histogram) and theoretical (curve) distributions of $|M|^2$ for E2 transitions in the light nuclei.

transition of 15 Mev in O^{16} with the radiative width $\Gamma_\gamma = 6$ ev. If we regard this as an M1 transition we should have $|M|^2 = 0.09$ which, as we can see from Fig. 4 is perfectly reasonable. If on the other hand we interpret it as an E2 transition we have $|M|^2 = 4$ which Fig. 6 shows to be also quite possible.

In general the feasibility of a given interpretation in the light elements may be judged with the aid of Figs. 2, 4, and 6.

f. Higher Multipole Transitions

Data on transitions of higher order are almost nonexistent in the light elements. We should comment in passing that the only E3 transition of known strength—that from the 3⁻ second excited state of O^{16} at 6.14 Mev has $|M|^2 \sim 30$ and so clearly demonstrates that E3 transitions as well as E2 can be enhanced.

10. Empirical Data: the Heavy Elements

a. The General Picture

Above $A \sim 40$ we enter the second region discussed above, namely that in which data on transitions generally of rather high multipole

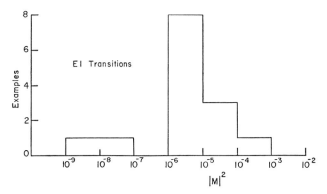

FIG. 7. Distribution of $|M|^2$ for E1 transitions in heavy nuclei.

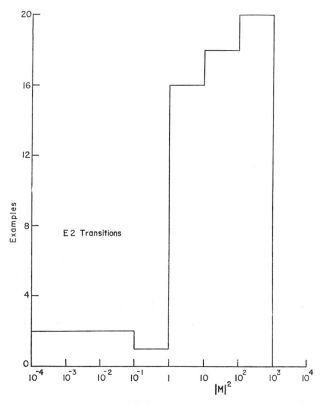

FIG. 8 Distribution of $|M|^2$ for E2 transitions in heavy nuclei.

order become abundant from studies of isomeric lifetimes. The intervening region $20 < A < 40$ is an interesting one which, however, has not been studied in a systematic fashion as yet so we shall not discuss it in detail.

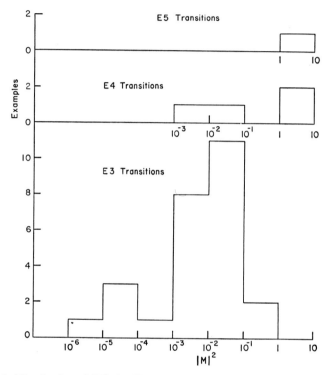

FIG. 9. Distribution of $|M|^2$ for E3, E4, and E5 transitions in heavy nuclei.

The data on absolute radiative widths in the heavy elements are presented[m] in Figs. 7 through 10 for the various multipoles. Here we have no independent-particle model prediction with which to compare these data in detail but it is clear that, frequently, the speed is much below the Weisskopf estimate[n]—further below than we should expect from our

---

[m] The data represented here are taken chiefly from the compilations of Goldhaber and his associates (26). No attempt has been made to bring them completely up to date since these distributions are still representative. New data have been used for the E1 and E2 transitions where material advances in knowledge have been made.

[n] The very striking and unexplained close grouping of the M4 transitions is at $|M|^2 > 1$. It may be that here the slightly more refined Moszkowski unit, 16 times bigger than the Weisskopf unit (see Section 6b), is the more realistic measure. However, in view of our remarks on the further diminution of transition probabilities by

earlier considerations and with a greater spread. From this we must probably conclude that, in contrast with the situation among the light elements, the bulk of the individual wave functions of the states involved are not simple independent-particle states but involve considerable and essential interconfigurational mixing.

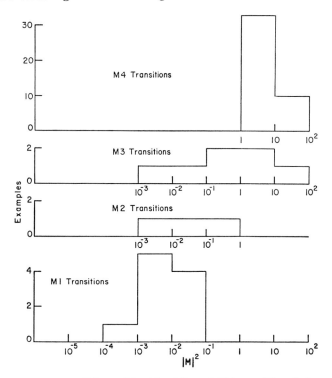

FIG. 10. Distribution of $|M|^2$ for M1, M2, M3, and M4 transitions in heavy nuclei.

Owing to the low energy of most of these isomeric transitions it is frequently, but not always, possible to distinguish the type of a transition on the basis of its lifetime. Reference to the appropriate figures will permit a judgment of ambiguity to be made in any particular case.

b. SYSTEMATIC VARIATIONS OF $|M^2|$

Further evidence of systematic behavior within the transitions of a given type can be sought by displaying the reduced transition speeds as

---

the several other effects of the full independent-particle model (see Section 7a), it may be that here we are seeing some collective enhancement as in the E2 transitions about to be discussed.

functions of neutron or proton number. When this is done no striking correlations are perceptible except for the E2 transitions where the behavior is so regular as immediately to demand an interpretation in terms of some especially simple behavior of nuclear matter. This is displayed in Fig. 11. Already in Fig. 8 the E2 had been singled out as

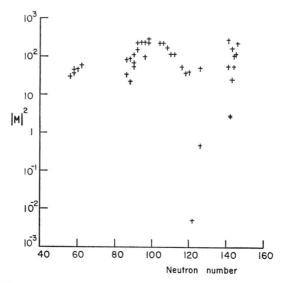

Fig. 11. $|M|^2$ as a function of neutron number for E2 transitions in the heavy nuclei. Below a neutron number of 126 we have displayed data from even-even nuclei only.

being the only electric transitions which were (usually) *faster* than the Weisskopf estimate.

c. The Collective Model

We are clearly driven by these very large values of $|M|^2$ for E2 transitions to postulate some collective form of motion of the protons in which they are correlated in their behavior in a semiclassical way. This is certainly difficult to describe in independent-particle model terms though, in principle, a description can be made through the language of configuration interaction. The interpretation of this collective effect as the rotational motion of permanently deformed nuclei has been extensively discussed (*27*). The systematic trend of $|M|^2$ with neutron number displayed in Fig. 11 is then a reflection of a systematic increase in the deformation of the nucleus as we move away, in either direction, from the doubly closed shell at Pb^{208} (82 protons and 126 neutrons) where the nucleus is rather accurately spherical.

These fast E2 transitions above $N \sim 90$ are between the lowest members of sequences of rotational states of even parity and spins, in even-even nuclei, $J = 0, 2, 4, \ldots$. The energy ratio between the excitations of the second and first excited states (rather accurately 3.3) suggests the rotational nature of the states. These data demonstrate the existence of two regions of permanent deformation one below and one above Pb^{208}. We shall note later a third region in the lighter elements. Below $N \sim 90$ this ratio of excitation energies drops abruptly from 3 to 2 but the E2 transitions remain fast, suggesting another form (vibrational) of collective motion (*28*). We shall return to this later.

We must remember that these systematics of Fig. 11 apply only to *particular* transitions in these nuclei and that other E2 transitions, between states not so simply related in the rotational or vibrational scheme, would not fit in at all. It is only because measuring techniques so far employed (lifetime measurements and Coulomb excitation) have tended by their nature to pick out the transitions between closely related rotational states that this simple pattern emerges. The systematics cannot be applied indiscriminately to any E2 transition in this region. Of course the model carries its own selection rules (see Section 11b) and these may be taken as a guide to probable behavior when the initial and final states with which we are concerned can be identified with states of the model.

We may now look back at the E2 transitions in the light nuclei (Fig. 6). Here too we found many $|M|^2$ values which were greater than unity. This we can now interpret as evidence for some collective motion even within and near the $1p$-shell. For certain cases which have been studied in detail it is possible to construct a self-consistent picture of this process (*29,30*, and Elliott and Flowers, *22*).

d. Slow E1 Transitions

It is also instructive to contrast Fig. 7 which shows the distribution of E1 transition strengths in the heavy nuclei with Fig. 2 which gives the corresponding picture for the light elements. The very strong inhibition of the transitions in the heavy elements as compared with the fast transitions in the light elements suggests the operation of some selection rule. Some of these E1 transitions in heavy elements take place between states which may be associated on the collective model with "pear shaped" deformations (*31*) and states of the normal sequence of opposite parity associated with ellipsoidal deformations. A gross change in the form of the nuclear deformation is thus involved. This change of deformation clearly involves the readjustment of several particles from the independent-particle viewpoint, and so the transition is highly forbidden. More generally we may say that in strongly deformed nuclei other selection rules must be satisfied in addition to the usual ones on spin and parity (see Section 11b).

e. Transitions from Closely Spaced Levels

Strong configurational mixing of the kind to which we have just referred for states of low excitation is not expected on the basis of the

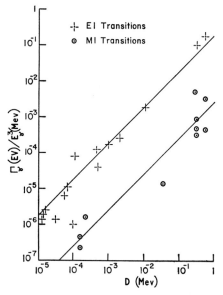

Fig. 12. Reduced transition speeds $[\Gamma_\gamma(\text{ev})/E_\gamma^3 \text{ (Mev)}]$ as a function of level spacing D in the region of the emitting states for E1 and M1 transitions.

shell-model itself. If the excitation is high, however, and states become crowded closely together then we should expect configuration mixing to take place and the transitions to be correspondingly discouraged. We do in fact see this happening for the closely spaced states formed by the capture of thermal neutrons. In this circumstance we are led to suppose, for any model, that radiative widths will be of the form (see p. 644 in

Blatt and Weisskopf, 2)

$$\Gamma_\gamma = \Gamma_\gamma \text{ (pure) } D/D_s;$$

where Γ_γ (pure) stands for the order of the width expected on the basis of the pure one-particle model, D is the experimental level spacing, and D_s is some level spacing characteristic of the separation of the major shells (presumably $\sim$10 Mev).

In order to investigate this prediction we may take very roughly for Γ_γ (pure) the value $\frac{1}{4}\Gamma_{\gamma W}$ (allowing for the "effective charge" factor, see Section 7a) for E1 and $\Gamma_{\gamma W}$ for M1 transitions. A logarithmic plot of $\Gamma_\gamma/\Gamma_\gamma$ (pure) *versus* D should now be a straight line of unit slope from which we find D_s. Figure 12 shows plots of Γ_γ (ev)$/E_\gamma^3$ (Mev) for E1 and M1 radiation following slow neutron capture using data from throughout the periodic table (*32*). It is seen that the relation in fact holds good even though D changes by a factor of more than 10^4. The values of D_s associated with the lines drawn through the points are 3 Mev for the E1 transitions and 10 Mev for the M1 transitions. These are both of the expected order. The data of Fig. 12 run through the region in which, according to Fig. 7, E1 transitions of low energy are peculiarly inhibited. It is clear that no such *unexpected* inhibition applies to the transitions of high energy.

This stresses the fact that it is dangerous to generalize from bodies of data such as are presented in Figs. 7 and 8 since the nature of current experimental techniques may have resulted in the selection of particular and nonrepresentative classes of nuclear transitions.

11. Special Selection Rules

We must always conform to the general selection rules on spin, parity, and isotopic spin (see Sections 2 and 5). But in addition any detailed model such as the shell model or collective model will carry its own special selection rules on radiative transitions.

a. The Shell Model

In the shell model generally[o] we can have the phenomenon of l-forbiddenness. For electric transitions we must have $|l_i - l_f| \leq L$ and for magnetic transitions $|l_i - l_f| \leq L - 1$ (assuming always that the parity change is correct). Here l_i and l_f have their meanings of Section 7a. Thus

[o] I am indebted to Dr. A. M. Lane for a very helpful discussion of the shell-model selection rules.

in M1 transitions no particle may change its orbit. In particular no M1 transition is possible which involves a closed shell. If the transition is not *l*-forbidden it will be zero because of orthogonality of the wave functions. These rules do not appear to be very potent in practice, and for practical purposes of estimating transition strengths they can be disregarded, except perhaps in the light nuclei where explicit calculations for realistic configurations can sometimes be made.

If we work in LS coupling we have the rule that $\Delta S = 0$ in an electric transition. Also in an electric $T = 0 \to T = 0$ transition we must not change the spatial symmetry or partition (listed as the square bracketed quantity $[XY \ldots]$). In an M1 transition neither L nor S may change by more than one and they may not both change. For M1 transitions all $\Delta T = 0$ transitions are also forbidden between states that are not degenerate in energy in the LS limit.

If we are in jj coupling we can have j-forbiddenness. We must have $|j_i - j_f| \leq L$ and in general the jumping nucleon must of itself be able to give radiation of the type we are interested in. For example the E3 transition $[(g_{\frac{9}{2}})^3]_{J=\frac{7}{2}} \to [(g_{\frac{9}{2}})^2 p_{\frac{1}{2}}]_{J=\frac{1}{2}}$ is forbidden because $|j_i - j_f| = 4$. The radiation must then be M4. This rule may have a measure of validity in practice (*26*).

Since in practice we have intermediate coupling and, in any case, can only approximately assign quantum numbers to individual nucleons these rules cannot be strict.

There is no clear experimental distinction in strengths between transitions which according to the simple shell model involve a radiating proton or a radiating neutron even though the naive estimates of speed may differ by many orders of magnitude for neutron and proton.[p]

b. The Collective Model

The collective model in the strong coupling limit of permanent deformations also gives clear selection rules. The most important concerns the quantum number K which is the projection of the total angular momentum J onto the nuclear symmetry axis (*33*). The selection rule is $\Delta K \leq L$. This rule in general seems to have some considerable validity. An extreme and astonishing example appears to be afforded by the E1 transition between the 9− and 8+ states in Hf^{180} at 1.143 and 1.086 Mev, respectively. This transition, if the identification with states of the model is correct, has $\Delta K = 9$ and $|M|^2 \sim 10^{-15}$ (*5*).

[p] Compare the remark in Section 7a.

In such deformed nuclei the individual j-values of the nucleons are no longer good quantum numbers but, in the strong coupling limit, their projections on the nuclear symmetry axis are. We can define Ω, the sum of such projections. Usually $\Omega = K$ for normal rotations so we need not consider selection rules on Ω as distinct from K. Certain types of vibrational distortion can give $\Omega \neq K$ but this does not seem to have been detected experimentally so far.

Even though a transition is not K-forbidden it may still be strongly inhibited by some other selection rule. Consider a deformed odd A nucleus which makes an E1 transition satisfying $\Delta\Omega = 0, \pm 1$, the Ω being here due to the last nucleon. Now each Ω belongs to a particular j of which it is the projection and of which it is merely a magnetic substate in the spherical limit (the ordinary shell model). And to each j belongs a particular l. We must now have $\Delta j = 0, \pm 1$ (compare j-forbiddenness in the ordinary shell model) and also $\Delta l = \pm 1$ (compare l-forbiddenness) (*34*). Violation of this rule explains many of the strongly inhibited E1 transitions in odd A heavy nuclei (Section 10d and Fig. 7). Alternative representations of this situation are also available (*34*).

Again we must note that these rules on K etc. can only be approximate because they represent the situation in an idealized extreme coupling scheme.

Another interesting selection rule of the collective model concerns the behavior of vibrational states. Over a considerable range of neutron number (40–80) even-even nuclei frequently show the even parity spin sequence 0-2-2 with an energy ratio between the excitation of the second and first excited states of about 2 (*28*). It is tempting to associate this situation with the excitation of one and then two quadrupole vibrational quanta. We should then, in first order, expect the spin sequence 0-2-(0, 2, 4), the second state being a degenerate triplet. Quite frequently two members of the expected triplet are found close together but only in rare and ill-substantiated cases does the full triplet reveal itself. If this account were correct we should expect the $2 \to 2$ transition from the second to first excited states to be predominantly E2 since the vibrational model gives no M1 radiation; we should also expect it to be much stronger than the $2 \to 0$ branch to the ground state because the latter would be a 2-quantum jump. Both these expectations are borne out in practice. The collective nature of these transitions is emphasized by their speed which is great ($|M|^2 \gg 1$, see Fig. 11).

There is growing evidence that a third region of permanent deformation and strong coupling exists in the neighbourhood of $A = 25$ reaching perhaps as low as F^{19}. Gamma-ray transitions here show good evidence of the operation of the K selection rule although the situation is not clear-cut (35).

12. Branching Ratios

a. THE SHELL MODEL

The branching ratios predicted by the extreme one-particle model using the Weisskopf approximation to the radial wave functions can be calculated using the Weisskopf units and the $(2l_f + 1)C^2U^2$ expression quoted in Section 6d. This estimate however is inaccurate in practice except in extremely simple cases because of the rough wave functions and also because, as seen in Section 7a, the other particles usually needed to specify the state are enormously important in modifying the single-particle transition probabilities. So in general the independent-particle model makes no simple prediction other than the gross dependence on the energy of the transition. In jj coupling, however, it is useful to work out the $(2l_f + 1)C^2U^2$ factor for the jumping nucleon *by itself*. If this factor is small it is likely that the actual transition will be slow.

b. THE COLLECTIVE MODEL

The collective model makes definite predictions of branching ratios in simple cases (33). Consider competing transitions of multipole order L from a state of J_i, K_i to final states J_f, K_f all of which final states are members of the same rotational family. Then the intrinsic strengths of the competing transitions (after removal of E_γ^{2L+1} from the widths) stand in the ratios:

$$\frac{J_i K_i \to J_f K_f}{J_i K_i \to J_{f'} K_f} = \left[\frac{C^{J_i L J_f}_{K_i K_f - K_i K_f} + b(-)^{J_f+K_f} C^{J_i L J_f}_{K_i, -K_f - K_i, -K_f}}{C^{J_i L J_{f'}}_{K_i K_f - K_i K_f} + b(-)^{J_{f'}+K_f} C^{J_i L J_{f'}}_{K_i, -K_f - K_i, -K_f}} \right]^2.$$

The quantity b is the same for all states of the K_f band and can be calculated from the appropriate wave functions (36). Sometimes $b = 0$; for example, when $L < K_i + K_f$, when either K_i or K_f is zero, and for E2 transitions with all states (initial and final) within the same rotational sequence of $K = \frac{1}{2}$.

13. Sum Rules

It is often useful to enquire whether a given transition can reasonably be of a particular type. We have seen that transitions enhanced above the single-particle speed are possible but there must clearly be a limit to such enhancement.

Many sum rules can be written down to describe the cross section for photon absorption weighted by specified powers of the photon energy and integrated over all energy. These sum rules can therefore be used to derive upper limits on the radiative widths of single levels.

For electric transitions of multipolarity L we have:

$$\Gamma_\gamma \leq 6 \frac{(L+1)}{L(2L+3)[(2L+1)!!]^2} \frac{e^2}{\hbar c} \left(\frac{\omega R}{c}\right)^{2L} ZE_\gamma = \frac{(L+3)^2}{2L+3} \frac{Z}{3} \Gamma_{\gamma W}. \quad (19)$$

For E1 transitions these estimates must be multiplied by an additional factor of N/A.

This sum rule is not rigorous and neglects the possible importance of certain types of nucleon correlation which cannot be evaluated without a specific model. They cannot however contribute a large factor and we could only expect this rule to be violated by a single transition in exceptional circumstances. It is also not a very strong rule for transitions of high energy. (We must always remember that compliance with a *particular* sum rule can never show that a transition is possible but only that it is not impossible.)

A stricter rule is available in self-conjugate nuclei for electric transitions of higher multipolarity than dipole that do not change the isotopic spin:

$$\Gamma_\gamma \leq \frac{3}{4} \frac{L+1}{[(2L+1)!!]^2} \frac{e^2}{\hbar c} \left(\frac{\omega R}{c}\right)^{2L-2} A \frac{E_\gamma^2}{Mc^2}$$
$$= \frac{L(L+3)^2}{24} \left(\frac{\omega R}{c}\right)^{-2} A \frac{E_\gamma}{Mc^2} \Gamma_{\gamma W}. \quad (20)$$

Here M is the nucleon mass. This rule is falsified only by possible velocity-dependent forces which are not linear in the momentum.

A useful sum rule that applies to E1 transitions only is:

$$\Gamma_\gamma \leq \frac{4}{3} \frac{e^2}{\hbar c} \frac{NZ}{A} \frac{E_\gamma^2}{Mc^2} \quad \text{or} \quad \frac{16}{3} \left(\frac{\omega R}{c}\right)^{-2} \frac{NZ}{A} \frac{E_\gamma}{Mc^2} \Gamma_{\gamma W}. \quad (21)$$

This sum rule is made uncertain by the need to evaluate an exchange term. This has been done liberally in the above statement of the rule (the limit for pure ordinary forces has been doubled).

It must be remarked that we cannot expect to find single E1 transitions that approach the strength suggested by the sum rule because in practice we see that sum rule being already virtually exhausted by the giant resonance of nuclear photodisintegration (*37*).

We may finally note a very crude rule for electric transitions imposed by the nuclear size and interesting only in exceptional cases:

$$\Gamma_\gamma \leq Z^2 \Gamma_{\gamma W}. \tag{22}$$

Sum rules for magnetic transitions are more uncertain in their evaluation and have not proved of much utility.[q]

Appendix

Fractional Parentage and the Parentage Coefficients

A number n of equivalent nucleons outside the closed shells are moving in what are generally thought of as independent orbits in a nucleus of mass A. Even in the total absence of any explicit forces coupling particle to particle, however, the orbits are not strictly independent because the net motion must be such that the over-all wave function is antisymmetric in the combined coordinates of space, spin, and isotopic spin. Since these n nucleons are equivalent it may seem an artificial device to single one out for special consideration. However, for many purposes a single nucleon is singled out by the reaction mechanism itself—a direct interaction for example at high energy may simply knock one or another of the n nucleons out of the nucleus, leaving the rest untouched. Another phenomenon which involves a single nucleon in the actual act is a radiative transition (for here only one nucleon is allowed to change its quantum numbers).

We are therefore led to ask how we can describe the collection of n nucleons so that one is separated from the others. Let us write the wave function of this one nucleon as $\phi(jm)$—we will discuss the problem in jj coupling for definiteness although any coupling scheme will do. We now ask how simultaneously to describe the other $n - 1$ equivalent nucleons. Since we are hypothesizing independent-particle motion in some potential well it is clear that the behavior of the $n - 1$ nucleons in the nucleus of mass A must be just the same as it would be in the nucleus of mass $A - 1$ that would result if we actually removed our last nucleon from nucleus A, instead of only mathematically setting it on one side by

[q] For general references to sum rules see Levinger and Bethe (*38*), also Gell-Mann and Telegdi (*8*).

writing a separate wave function for it. However, if the $n - 1$ nucleons were really making a state of mass $A - 1$ by themselves, this must be an eigenstate; whereas in mass A they can be simultaneously represented by any or all of these same eigenstates of $A - 1$ whose spins, parities, and isotopic spins can couple to the spin, parity, and isotopic spin of the last nucleon to give the correct spin, parity, and isotopic spin of the state we described in A. These various possible states of $A - 1$, that we find underlying the state of A when we single out one nucleon and then describe the remainder, are called the parent states of the particular state of A. They are not as a rule equally represented. In general then we expect to write:

$$\psi(JM_JTM_T\alpha) = \sum_p <\alpha_p |\} \alpha> \psi(J_pM_{J_p}T_pM_{T_p}\alpha_p)\phi(jm)$$

$$\times C^{J_pjJ}{}_{M_{J_p}mM_J} C^{T_p\frac{1}{2}T}{}_{M_{T_p}\pm\frac{1}{2}M_T}.$$

The $<\alpha_p |\} \alpha>$ are the respective amplitudes in which the various parent states $\psi(J_pT_p\alpha_p)$ are represented in this expansion. They are the fractional parentage coefficients. They are effectively the coefficients that are demanded in order that this expansion of $\psi(JT\alpha)$ should have over-all antisymmetry, namely that we should not pretend to know *which* of the n equivalent nucleons we are singling out for this special treatment. We normalize $\sum_p <\alpha_p |\} \alpha>^2 = 1$.

The fractional parentage coefficients then determine the course of any nuclear reaction of the single-particle sort (*39*). They are available in certain rather restricted regions of the periodic table for the pure jj and LS coupling schemes (*40,41*). With their aid the prediction of the independent-particle model in either extreme coupling scheme about a radiative transition may be quickly determined. Suggestions about possible identifications of real states with states of the models are found in various theoretical papers (*21,22,25*).

The use of the published tables is sometimes a little tricky and it may be worthwhile to give explicit directions.

The LS fractional parentage coefficients are found (*41*) as the product of an orbital fractional parentage coefficient (listed under $S, P, D \ldots$), a charge-spin fractional parentage coefficient (listed under $^{2T+1,2S+1}\Gamma$), and a weighting factor (listed under the spatial symmetry or partition— square bracketed $[XY \ldots]$). These three multiplied together give the total fractional parentage coefficient that is used in the computation.

Example: The parents of the ground state of C^{13}, N^{13}. In LS coupling this state is ^{22}P. Its parents include ^{11}S (the ground state of C^{12}), ^{11}D (the 2+ first excited state of C^{12} at 4.4 Mev), and several others both in C^{12} alone (the $T = 0$ parents) and also in the B^{12}-C^{12}-N^{12} system (the $T = 1$ parents). We find the parentage coefficients as shown in Table I.

TABLE I. PARENTAGE COEFFICIENTS

Parent State	Charge-Spin	Orbital	Weighting Factor	Total
^{11}S	1	$-\sqrt{\frac{2}{9}}$	$\frac{1}{\sqrt{6}}$	$-\frac{1}{3\sqrt{3}}$
^{11}D	1	$-\sqrt{\frac{7}{9}}$	$\frac{1}{\sqrt{6}}$	$-\frac{1}{3}\sqrt{\frac{7}{6}}$
^{13}P	$-\frac{1}{\sqrt{5}}$	$\sqrt{\frac{2}{3}}$	$\sqrt{\frac{5}{6}}$	$-\frac{1}{3}$
^{13}D	$-\frac{1}{\sqrt{5}}$	$-\frac{1}{\sqrt{3}}$	$\sqrt{\frac{5}{6}}$	$\frac{1}{3\sqrt{2}}$
^{31}P	$-\frac{1}{\sqrt{5}}$	$\sqrt{\frac{2}{3}}$	$\sqrt{\frac{5}{6}}$	$-\frac{1}{3}$
^{31}D	$-\frac{1}{\sqrt{5}}$	$-\frac{1}{\sqrt{3}}$	$\sqrt{\frac{5}{6}}$	$\frac{1}{3\sqrt{2}}$
^{33}P	$-\sqrt{\frac{3}{5}}$	$\sqrt{\frac{2}{3}}$	$\sqrt{\frac{5}{6}}$	$-\frac{1}{\sqrt{3}}$
^{33}D	$-\sqrt{\frac{3}{5}}$	$-\frac{1}{\sqrt{3}}$	$\sqrt{\frac{5}{6}}$	$\frac{1}{\sqrt{6}}$

Edmonds and Flowers (jj coupling) (*40*) list only the spin-orbital fractional parentage coefficients. To obtain the total fractional parentage coefficients we must multiply by the charge fractional parentage coefficient and the weighting factor taken from Jahn (*41*) (that weighting factor appearing between state and parent state of the same partition $[XY\ .\ .\ .]$). The jj charge fractional parentage coefficient we find as the fractional parentage coefficient for ordinary spin in the *atomic* fractional parentage coefficient tables of Jahn (*41*); that is, we interpret the entry as $^{2T+1}\Gamma$ (we here use the dual—the square bracketed symbol with the squiggle over it $[XY\ .\ .\ .]$ that reads the same as the partition $[XY\ .\ .\ .]$ of the jj tables and the LS weighting factor).

Example: The ground state of Li^7 (or Be^7) in jj coupling is the lowest $J = \frac{3}{2}$ state of $(1p_{\frac{3}{2}})^3$. Its parents are the 1+ ground state of Li^6 and the 3+ first excited state of Li^6 at 2.2 Mev ($T = 0$ parents), also the 0+ ground state of He^6 and the 2+ first excited state of He^6 at 1.7 Mev

($T = 1$ parents appearing also as the second and third excited states of Li^6 at 3.6 and 5.3 Mev, respectively). We take the spin-orbital fractional parentage coefficients from Edmonds and Flowers (40) and the charge fractional parentage coefficient and weighting factor from Jahn (41). See Table II.

TABLE II. PARENTAGE COEFFICIENTS

Parent State	Spin Orbital	Charge	Weighting Factor	Total
$J = 1$	$\sqrt{\frac{3}{10}}$	1	$\frac{1}{\sqrt{2}}$	$\frac{1}{2}\sqrt{\frac{3}{5}}$
$J = 3$	$\sqrt{\frac{7}{10}}$	1	$\frac{1}{\sqrt{2}}$	$\frac{1}{2}\sqrt{\frac{7}{5}}$
$J = 0$	$\sqrt{\frac{5}{6}}$	-1	$\frac{1}{\sqrt{2}}$	$-\frac{1}{2}\sqrt{\frac{5}{3}}$
$J = 2$	$\frac{1}{\sqrt{6}}$	-1	$\frac{1}{\sqrt{2}}$	$-\frac{1}{2}\frac{1}{\sqrt{3}}$

Edmonds and Flowers (40) list only fractional parentage coefficients for the lower half of each shell; those for the upper half may be derived using the relationship between particles and holes:

$$<j^{4j+2-n}(\alpha TJ) |\} j^{4j+1-n}(\alpha_p T_p J_p); j>$$
$$= (-)^{J_p - T_p - J - T + j + \frac{1}{2}} \left[\frac{(n+1)(2T_p + 1)(2J_p + 1)}{(4j + 2 - n)(2T + 1)(2J + 1)} \right]^{\frac{1}{2}}$$
$$\times <j^{n+1}(\alpha_p T_p J_p) |\} j^n(\alpha TJ); j>.$$

We have so far discussed only the evaluation of the fractional parentage coefficients within a single shell. For an E1 transition and for any other involving a change of parity we must also know in the simplest case the coefficients of a state of mixed configuration for a state of $n - 1$ equivalent particles. For example, for E1 transitions in the $1p$-shell our initial state is frequently $(2s, 1d)(1p)^{n-1}$ and the final state is $(1p)^n$. We know the $<\alpha_p |\} \alpha>$ for the final state but what of the initial state? In general they are very complicated but in practice it seems that such an odd nucleon is often very loosely bound (42) and that one state of the nucleus $A - 1$ in $(1p)^{n-1}$ is effectively a unique parent. In this circumstance we have $<\alpha_p |\} \alpha> = n^{-\frac{1}{2}}$ for this unique parent (which is revealed by the large reduced width for the emission to it of the odd nucleon).

REFERENCES

1. See, for example, N. F. Mott and I. N. Sneddon, *Wave Mechanics and Its Applications* (Oxford University Press, London and New York, 1948), p. 257.
2. See e.g. J. M. Blatt and V. F. Weisskopf, *Theoretical Nuclear Physics* (John Wiley and Sons, New York, 1952), p. 587.
3. See e. g. W. Heitler, *The Quantum Theory of Radiation* (Oxford University Press, London and New York, 1954), p. 401.
4. See e.g. E. H. S. Burhop, *The Auger Effect* (Cambridge University Press, London and New York, 1952), pp. 141, 156.
5. See e.g. National Research Council, U.S., Nuclear Data Cards.
6. See e.g. M. E. Rose, in *Beta- and Gamma-Ray Spectroscopy*, edited by K. Siegbahn (North Holland Publishing Co., Amsterdam, 1955), Chapter XIV; R. Wilson, *ibid.*, Chapter XX (II).
7. E. H. S. Burhop, *The Auger Effect* (Cambridge University Press, London and New York, 1952), Chapters V and VI.
8. L. A. Radicati, Phys. Rev. **87**, 521 (1952); M. Gell-Mann and V. L. Telegdi, Phys. Rev. **91**, 169 (1953).
9. L. A. Radicati, Proc. Phys. Soc. **A66**, 139 (1953); **A67**, 39 (1954); W. M. MacDonald, Phys. Rev. **100**, 51 (1955); **101**, 271 (1956).
10. W. E. Burcham, Progr. in Nuclear Phys. **4**, 171 (1955).
11. V. F. Weisskopf, Phys. Rev. **83**, 1073 (1951).
12. R. Hofstadter, Revs. Modern Phys. **28**, 214 (1956).
13. S. A. Moszkowski, in *Beta- and Gamma-Ray Spectroscopy*, edited by K. Siegbahn (North Holland Publishing Co., Amsterdam, 1955), Chapter XIII.
14. H. A. Bethe, Phys. Rev. **103**, 1353 (1956), and references to the work of K. A. Brueckner *et al.* contained therein.
15. See e.g. E. U. Condon and G. H. Shortley, *The Theory of Atomic Spectra* (Cambridge University Press, London and New York, 1951), Chapter III.
16. M. E. Rose, *The Elementary Theory of Angular Momentum* (John Wiley and Sons, New York, 1957), Chapter III.
17. See e.g. M. E. Rose, *The Elementary Theory of Angular Momentum* (John Wiley and Sons, New York, 1957), Chapter VI.
18. J. M. Kennedy and W. T. Sharp, Atomic Energy of Canada, Ltd., Report CRT-580.
19. A. M. Lane and L. A. Radicati, Proc. Phys. Soc. **A67**, 167 (1954).
20. J. P. Elliott and T. H. R. Skyrme, Proc. Roy. Soc. **A232**, 561 (1955).
21. D. Kurath, Phys. Rev. **101**, 216 (1956); **106**, 975 (1957).
22. A. M. Lane, Proc. Phys. Soc. **A66**, 977 (1953); **A68**, 189, 197 (1955); J. P. Elliott and B. H. Flowers, Proc. Roy. Soc. **A229**, 536 (1955); **A242**, 57 (1957).
23. D. H. Wilkinson, Phil. Mag. [8] **1**, 127 (1956).
24. D. H. Wilkinson, Proc. Rehovoth Conf. on Nuclear Structure, Amsterdam, 1957 (1958), p. 175.
25. See e.g. D. R. Inglis. Revs. Modern Phys. **25**, 390 (1953).
26. M. Goldhaber and A. W. Sunyar, in *Beta- and Gamma-Ray Spectroscopy*, edited by K. Siegbahn (North Holland Publishing Co., Amsterdam, 1955), Chapter XVI (II); M. Goldhaber and J. Weneser, Ann. Rev. Nuclear Sci. **5**, 13 (1955).
27. A. Bohr and B. R. Mottelson, Kgl. Danske Videnskab. Selskab, Mat.-fys. Medd. **27**, No. 16 (1953); in *Beta- and Gamma-Ray Spectroscopy*, edited by K. Siegbahn (North Holland Publishing Co., Amsterdam, 1955), Chapter XVII; A. Moszkowski, *Handbuch der Physik* (J. Springer Verlag, Berlin, 1957), Vol. 39.

28. G. Scharff-Goldhaber and J. Weneser, Phys. Rev. **98**, 212 (1955).
29. F. C. Barker, Phil. Mag. [8] **1**, 329 (1956).
30. R. D. Amado and R. J. Blin-Stoyle, Proc. Phys. Soc. **A70**, 532 (1957).
31. A. Bohr and B. R. Mottelson, Nuclear Phys. **4**, 529 (1957).
32. B. B. Kinsey in *Beta- and Gamma-Ray Spectroscopy*, edited by K. Siegbahn (North Holland Publishing Co., Amsterdam, 1955), Chapter XXV; *Handbuch der Physik* (J. Springer Verlag, Berlin, 1957), Vol. 40.
33. G. Alaga, K. Alder, A. Bohr, and B. R. Mottelson, Kgl. Danske Videnskab. Selskab, Mat.-fys. Medd. **29**, No. 9 (1955).
34. D. Strominger and J. O. Rasmussen, Nuclear Phys. **3**, 197 (1957).
35. A. E. Litherland, H. McManus, E. B. Paul, D. A. Bromley, and H. E. Gove, Atomic Energy of Canada, Ltd., Report PD-289 and references contained therein.
36. S. G. Nilsson, Kgl. Danske Videnskab. Selskab, Mat.-fys. Medd. **29**, No. 16 (1955).
37. See e.g. D. H. Wilkinson, Physica **22**, 1039 (1956).
38. J. S. Levinger and H. A. Bethe, Phys. Rev. **78**, 115 (1950).
39. A. M. Lane and D. H. Wilkinson, Phys. Rev. **97**, 1199 (1955).
40. A. R. Edmonds and B. H. Flowers, Proc. Roy. Soc. **A214**, 515 (1952).
41. H. A. Jahn and H. van Wieringen, Proc. Roy. Soc. **A209**, 502 (1951); H. A. Jahn, *ibid.* **A205**, 192 (1951).
42. A. M. Lane, Atomic Energy Research Establishment, Harwell Report T/R 1289 (1954).

V.G. The Analysis of Reduced Widths

by J. B. FRENCH

1. Single-Particle Widths.. 891
2. Selection Rules... 894
   a. General Discussion.. 894
   b. Configuration Selection Rules....................................... 895
   c. LS Selection Rules.. 900
   d. Miscellaneous Selection Rules....................................... 901
3. Quantitative Analysis via the Shell Model............................... 901
4. Miscellaneous Considerations.. 915
   a. Statistical Factors and Reduced-Width Variations.................... 915
   b. Spin Measurements, Single-Particle Levels, Etc...................... 916
   c. The Effective Interaction in Nuclei................................. 918
   d. Reduced Widths on the Collective Model.............................. 919
   e. Reduced Widths and Level Shifts..................................... 920
   f. Conclusion.. 921
   Appendix: Stripping Reaction Reduced Widths............................ 922
   References... 930

A reduced width connecting a state of one nucleus with a state of another tells us something about the structural relationship between these nuclei; how to translate this knowledge into a useful form is the basic question which enters in the analysis of the reduced width. From the outset it is clear that angular momentum considerations will be dominant in any reduced-width analysis for the angular momenta of the nuclear states involved and of the projectile particle (and in some cases too the channel spin) are all specified when we determine a reduced width. Quite apart from this we see, from the great success of the shell model, that the angular momenta of the individual nucleons in a nucleus are significant quantities. We will see in fact that it is the interplay between the angular momentum specifications due to the kinematics and those due to the internal structure of the nuclei which makes the reduced-width analysis a very useful one.

To make a quantitative analysis we must introduce a nuclear model. Our considerations will be largely based on the nuclear shell model since most of the data involve fairly light nuclei where the shell model has had its most detailed successes and since, in any case, it is ideally suited for describing the angular momentum coupling. We shall, however, discuss things in a sufficiently general manner so that the extension, for example, to the collective model would be fairly straightforward.

We will find too some cases in which a reduced-width measurement has a practical importance in determining the spin of a nuclear state or other of its important properties and give a little attention to the importance of reduced widths in helping to determine the effective interaction between the nucleons in a nucleus.

We discuss widths determined by deuteron stripping reactions on the same footing as widths determined in resonant reactions. We do this because of the great number and variety of such widths available and because of the all-important fact that a stripping width may connect bound states which cannot be connected by a resonant width. This is particularly important in the heavier nuclei where a compound state which can be reached in resonant reactions is liable to be of too high an excitation to be susceptible to detailed analysis. In an appendix we give formulae for extracting the stripping width from the experimental data.

1. Single-Particle Widths

A reduced width for a single-nucleon emission can be thought of as a product of two factors. One factor measures the probability that in a given compound state the nucleons will arrange themselves in a configuration corresponding to the final state; the other factor measures the intrinsic probability that when this happens the two components will actually separate. It is clear that a reasonable measure of the second factor should be obtainable by considering the simple problem of a single nucleon moving in a potential well produced by the other nucleons. Let us now briefly consider the reduced widths for this simple case, the so-called single-particle widths.

If f_l is the logarithmic derivative for the lth partial wave, defined by reduced width is simply $-(\partial f_l/\partial E)^{-1}$ evaluated at the resonance energy. For our present purpose it is convenient to write this in a different form. From the radial equation satisfied by R_l we can readily[a] derive the result $f_l R_l(R) = R \left(\dfrac{dR_l}{dr} \right)_R$ (where R_l is r times the radial wave function and R the nuclear radius) we have, as in Blatt and Weisskopf (1),[*] that the

---

[*] The reference list for Section V.G begins on page 930.

[a] Write the radial equations for $R_l^{(E)}$ and $R_l^{(E')}$ corresponding to two different energies. Multiply the first by $R_l^{(E')}$ and the second by $R_l^{(E)}$; on subtracting and applying Green's theorem over the nuclear volume the result follows in the limit $E' \to E$. For the later result, Eq. (3) note that the logarithmic derivative for arbitrary r but fixed energy, $f_l(r)$, satisfies the equation,

$$\frac{df_l(r)}{dr} = r\left[\frac{2M}{\hbar^2}(V(r) - E) + \frac{l(l+1)}{r^2}\right] + \frac{f_l(r) - [f_l(r)]^2}{r}.$$

But now for the neutron case outside the range of interaction the energy and radial dependence of $f_l(r)$ are expressed by the fact that f_l is a function of kr, where k is the wave number. The energy derivative is then simply given in terms of the radial derivative above.

that

$$\frac{\partial f_l}{\partial E} = -\frac{2MR}{\hbar^2 R_l^2(R)} \int_0^R R_l^2(r) dr \qquad (1)$$

where M is the reduced nucleon mass. The result is independent of the normalization of R_l; if we agree to normalize the radial function to unity over the interior of the nucleus (that is, $\int_0^R R_l^2(r) dr = 1$) we then have for θ_0^2, the single-particle width[b] in units of the Wigner-limit value $3\hbar^2/2MR^2$,

$$\theta_0^2 = \tfrac{1}{3} R R_l^2(R) \qquad (2)$$

evaluated of course at the resonance energy. This form can be applied equally well to a bound or unbound level.

We have a certain amount of freedom in defining the resonance energy. If we define it as the energy at which the lth-wave resonant scattering amplitude (without the potential scattering) has a maximum modulus, we find, as in Blatt and Weisskopf (1), $f_l^r = -\Delta_l$ where Δ_l is the real part of the logarithmic derivative of the corresponding outgoing wave and as such depends on the energy, radius, and charge. There are advantages in a definition which is independent of these external quantities and it is quite common instead to define the resonance by $f_l^r = -l$ which is the low-energy limit of Δ_l in the neutron case [note that Blatt and Weisskopf (1) instead use $f_l^r = 0$].

To see how this freedom can affect the θ_0^2-value it is convenient to take the simple case of a neutron scattered by a square well potential. An alternative form of Eq. (2) is possible in this case, namely

$$\tfrac{3}{2}\theta_0^2 = \left[1 - \frac{l(l+1) + f_l^r - (f_l^r)^2}{X^2} \right]^{-1} \qquad (3)$$

where $X j_l'(X) = (f_l^r - 1) j_l(X)$. In this equation X/R is the wave number inside the well and is connected to the logarithmic derivative by the equation involving the spherical Bessel functions. Using this we have from the $f_l^r = -l$ definition simply $\theta_0^2 = \tfrac{2}{3}$ for a neutron with a square well. Lane (2) has made a study of the allowable limits for θ_0^2 for both a square well and the harmonic oscillator potential using $f_l^r = -\Delta_l$ along with a physical assumption about the energy at which Δ_l should be evaluated. For our purposes it is sufficient that θ_0^2 is liable to be rather less than unity and may in actual cases be as small, say, as 0.15–0.2.

[b] In general we define θ^2 as the reduced width in units of $(C^{T_0 t T}{}_{M_{T_0}, m_t, M_T})^2 \cdot 3\hbar^2/2MR^2$ where the first factor is the isotopic spin coupling factor, and T_0 and T refer to the lighter and heavier nuclei, respectively. Many authors use as a basic unit $\hbar^2/MR^2$, many others do not divide out the isotopic spin factor (this step in fact is not appropriate for heavy nuclei).

This conclusion is unsatisfactory in its vagueness. It indicates to us first that, at least with our present understanding, widths measured in "single-particle" reactions (adding a nucleon to a closed-shell nucleus for example) will not in themselves tell us very much about nuclear structure.[c] Moreover for more complicated reactions, while we may write the reduced width as

$$\theta^2 = S\theta_0^2 \qquad (4)$$

we shall not be able to divide a measured θ^2 by a calculated θ_0^2 to obtain the "relative reduced width" S which contains in it the essential information about nuclear structure. For the most part one gets around this by comparing θ^2-values in cases where θ_0^2 might be reasonably regarded as being the same; of course if one of the comparison cases has a compound state with a single nucleon outside a closed shell (for which $S = 1$) we would have an empirical determination of θ_0^2. There is a whole variety of reactions too which will supply a lower limit to θ_0^2; these as we discuss later, are reactions where the transferred particle is not equivalent to any nucleon in the residual nucleus.

Most of the preceding discussion is, of course, specific for the resonant scattering widths. Besides these we have also the stripping widths (3). In the simple theory of a stripping reaction, say (d,p) [similarly for (d,n), also (d,t), (d,He^3) and perhaps even reactions like (He^3,α)], one considers that, on impact with the initial nucleus the deuteron breaks instantaneously, liberating a neutron with the same momentum which it has immediately before breakup; this distribution is easily calculated. Under the influence of the interaction the neutron may attach itself to the initial nucleus to form the final state. In this picture the proton has two functions; it serves as a source or sink of energy for the neutron and initial nucleus (and thus we can connect bound states) and it acts as an indicator to show (by its angular distribution) the l-value of the transferred neutron.

It is perhaps clear from this picture that the cross section is once again proportional to a reduced width which may be factored as in Eq. (4). For the same two states S should be the same factor in the resonant or stripping case; θ_0^2, on the other hand need not necessarily be the same since the radial boundary conditions are presumably very different in the two cases. Actually θ_0^2 is much more difficult either to define or to measure than in the resonant case. The simple theory outlined above

[c] The actual values of θ_0^2 measured in light nuclei by single-particle reactions [or in more complicated cases where S in Eq. (4) can be reasonably calculated] do lie in the predicted range. For obvious reasons little or nothing is known about θ_0^2 for heavier nuclei (say $A > 30$).

probably always overestimates the cross section;[d] it can be modified to take account of various corrections (4) but very little of this has been done; the calculations are very tedious and they introduce a large number of new parameters.

Fortunately while the significance of the absolute θ^2 (as deduced by using the simple theory) is at present quite obscure there is very good evidence that, just as with the resonant widths, comparative θ^2-values are quite meaningful when the reactions compared do show a good stripping angular distribution and when the atomic numbers and outgoing and incident energies are not too different. Then the same general procedure for analysis can be used as with resonant widths; the uncertainties are as yet somewhat larger but this is compensated for by the very wide range of application.

2. Selection Rules

a. General Discussion

The simplest way of analyzing reduced widths is by considering selection rules. If a measured reduced width is small (that is, if $\mathcal{S} \ll 1$) we are tempted to say that some selection rule is in operation and very often we can identify such a rule. This procedure divides widths into two classes.

(1) Intrinsically small because a valid selection rule imposed by the detailed nuclear structure is violated.

(2) Intrinsically large where no valid selection rule is violated.

We could of course add a third case where a reaction is forbidden by some absolute selection rule (total angular momentum or parity) but this would be pedantic (even though recent history does show the importance of questioning such "absolute" rules). The interesting selection rules are those then which are made to be broken.

If now a measured reduced width is large ($\mathcal{S} \simeq 1$) it must belong to class (2); but if it is small one cannot always be sure. For multiplicative numerical factors which enter into the theoretical $\mathcal{S}$ may give it a small value even though we can find no pertinent selection rule; or there may be an "accidental" cancellation between various amplitudes which must be summed to give the total reaction amplitude. Examples of the first type we will see later; a good case of the second is supplied by the vanish-

---

[d] It gives mostly quite small values for θ_0^2. Many p-shell measurements give $\theta_0^2(1p) \simeq 0.05$; at the low end of the (ds) shell $\theta_0^2(2s) \sim 0.2$, $\theta_0^2(1d) \sim 0.1$ but these values decrease quickly with increasing A; near $A = 40$ we have $\theta_0^2(f_{\frac{7}{2}}) \sim \theta_0^2(2p_{\frac{3}{2}}) \sim 0.01$.

ing of the stripping cross section in $N^{14}(d,p)$ to the 5.31-Mev level of N^{15}. At various times an absolute angular-momentum selection rule (high spin!) has been invoked to explain this or else a nonabsolute configuration selection rule (s^3p^{12}!). It seems probable now [see Halbert and French (5) for discussion and experimental references] that the spin is $\frac{1}{2}+$ (the N^{14} spin is $1+$) and that there is no configuration or other selection rule to discourage the $l = 0$ and $l = 2$ reactions. Instead they both seem to be very small because of cancellations. We should always bear in mind this uncertainty in interpreting a small reduced width. And of course a cancellation which is "accidental" from one point of view might be quite well motivated from another.

We now discuss examples of various kinds of selection rules.

b. Configuration Selection Rules

For a given single-nucleon width the l-value will be specified and in many cases too the j-value will be fixed by the total angular momenta. It may then happen that this l or j is inconsistent with the principal shell model configurations of the states involved; we may be attempting to force into a given nuclear state an unwelcome l or j or to take from it an l or j which is not readily available.

$$\begin{array}{lll} \text{Li}^6 + n \rightarrow \text{Li}^7 \ (7.46 \text{ Mev}) & & l_0 = 1 \\ 1+ & \frac{5}{2}- & l = 1 \ j = \frac{3}{2}; \ (\text{d.f.}) \ (l.\text{f.}) \\ & & l = 3 \quad \quad (\text{d.u.}) \ (l.\text{u.}) \end{array}$$
(I)

Here the configuration orbital angular momentum[e] is $l_0 = 1$ (p shell!) but we have no reason to specify a j_0-value since $p_{\frac{3}{2}}$ and $p_{\frac{1}{2}}$ particles should enter on the same footing [the states are well described by s^4p^n but not by $s^4(p_{\frac{3}{2}})^n$]. $l = 1$ is favored both by the dynamics (higher barrier penetration than for $l = 3$) and by the model. $l = 3$ is unfavored both ways and this width is in fact not measurable.[f] Experimentally[g] (9) $\theta_0^2(p) \simeq 0.2$.

$$\begin{array}{lll} \text{C}^{12} + p \rightarrow \text{N}^{13} \ (2.37 \text{ Mev}) & & (l_0, j_0) \equiv s_{\frac{1}{2}}, d_{\frac{3}{2}}, (d_{\frac{3}{2}}) \\ 0+ & \frac{1}{2}+ & (l, j) \equiv s_{\frac{1}{2}} \end{array}$$
(II)

---

[e] The notation here is that l, j are the angular momenta of the transferred nucleon; l_0, j_0 the angular momenta favored by the configuration; d.f., d.u., mean dynamically favored or unfavored; l.f. means favored by a single-particle orbital angular momentum rule; and so on.

[f] The reduced width of course is independent of barrier effects. But if the penetration is small the contribution to the reaction may simply not be detectable.

[g] The general experimental references are (6-8). We shall often, however, give an individual reference to an experiment.

In this case the final nucleus should belong to the configuration $s^4p^8d + s^4p^8(2s) + s^3p^{10}$. We might anticipate that $d_{\frac{3}{2}}$ is model-unfavored for such a low-lying state of N^{13} since the doublet d splitting is probably about 5 Mev [from the first and sixth levels of O^{17} (10)]. But the measured reduced width sheds no light on this since only $s_{\frac{1}{2}}$ is kinematically allowed. Experimentally (11) $\theta_0^2(s) \simeq 0.54$. This large width is consistent with the simple picture (12) that the state is effectively an s proton coupled to the ground state of C^{12}, though in fact it does not preclude the possibility of quite large amplitudes corresponding to a d particle and to excited states of C^{12}. However, this "weak-coupling" picture should be a very good one in this case because the C^{12} ground state is a stable structure not easily excited by an additional nucleon. This follows from the fact that it is reasonably well described as a jj-coupling closed shell (one piece of evidence for this is given in the next example) or, better still, it is obvious from the level structure of C^{12} which has a first excited state at 4.4 Mev and only three excited states below 10 Mev.

$$C^{12} + p \to N^{13} \ (3.51 \text{ Mev}) \qquad l_0 = 1 \ (j_0 = \tfrac{1}{2})$$
$$0+ \qquad \tfrac{3}{2}- \qquad\qquad l = 1 \quad j = \tfrac{3}{2}$$
$$\text{(III)}$$

We have here an l-favored transition ($l_0 = l = 1$) but, if jj coupling is at all a reasonable approximation, a j-unfavored one. Experimentally (11) we find the small value $\theta^2 \simeq 0.031$, suggestive that a jj approximation is reasonable (we would have $\mathcal{S} = \tfrac{1}{3}$ at the LS extreme).

$$Mg^{25} + p \to Al^{26} \ (7.20 \text{ Mev}) \qquad l_0 = 3$$
$$\tfrac{5}{2}+ \qquad 1- \qquad\qquad l = 1 \ (\text{d.f.}) \ (l.\text{u.})$$
$$\qquad\qquad\qquad\qquad l = 3 \ (\text{d.u.}) \ (l.\text{f.})$$
$$\text{(IV)}$$

In contrast with example (I) the higher l-value here is favored by the model. This comes about because the lowest odd-parity level in Al^{26} comes at or near 6.7 Mev. This level should mainly belong to configurations where a d or $2s$ particle is promoted to $f_{\frac{7}{2}}$; for a variety of reasons one expects then that the states of the next odd-parity configuration (a particle promoted to $p_{\frac{3}{2}}$) might begin at 8–8.5 Mev. Consequently we have the f-particle favoring.

The f–p reduced width ratio has been measured by Green et al. (13) (to whom also the l-assignment is due). They quote an f-wave reduced width about 16 times as large as the p-wave width.

$$P^{31}(d,p) \ P^{32}(\text{g.s.}) \qquad (l_0, j_0) \equiv d_{\frac{3}{2}}$$
$$\tfrac{1}{2}+ \qquad 1+ \qquad\qquad (l, j) \equiv s_{\frac{1}{2}} \ (\text{d.f.}) \ (l.\text{u.})$$
$$\qquad\qquad\qquad\qquad \equiv d_{\frac{3}{2}} \ (\text{d.u.}) \ (l.\text{f.})$$
$$\text{(V)}$$

We analyze this in terms of a jj picture which is known to be reasonable for $A \geq 30$. $l = 0$ is favored by the dynamics but $l = 2$ is favored by the model; in this way, just as in example (IV) we get an amplification of the effects due to a minor component in one of the wave functions, in this case an $(s_{\frac{1}{2}})^2 d_{\frac{3}{2}}$ component in a state whose major component would be $(s_{\frac{1}{2}})^3$. The importance of these cases in stripping reactions was first pointed out by Bethe and Butler (14). Example (IV) is a resonant-reaction analog.

Figure 1 shows the experimental results of Parkinson (15). Shown for comparison is the curve for the first excited state (2+ at 0.077 Mev) which allows only $l = 2$. The ground-state $l = 0$ cross section is quite prominent and the width measurable though the corresponding minor component has only about an 8% probability.

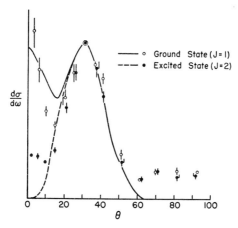

FIG. 1. The differential cross sections (in arbitrary units) for $P^{31}(d,p)P^{32}$, due to Parkinson. The ground-state Butler curve is $l = 2$ with an $l = 0$ admixture; the excited-state curve is pure $l = 2$. The ratio of peak cross sections is given as $d\sigma^*/d\sigma = 0.88 \times \frac{3}{5}$ where $\frac{3}{5}$ is the statistical ratio expected with pure jj coupling.

This technique for measuring a minor component is mainly of use with $l = 0 + 2$ mixtures, since a small $l = 0$ cross section is easily identifiable near 0°. With $l = 1 + 3$ it would often be difficult to obtain a reasonable measure of the widths.

$$Ca^{40}(d,p)Ca^{41}$$
$$Ca^{42}(d,p)Ca^{43}$$
$$(VI)$$

In Ca^{41} we expect low-lying states in which an $f_{\frac{7}{2}}$, $p_{\frac{3}{2}}$ . . . neutron is coupled to the Ca^{40} closed subshell core. In Ca^{43} we expect the lowest configuration to be $(f_{\frac{7}{2}})^3$ which, with isotopic spin $T = \frac{3}{2}$, has states $J = \frac{3}{2}, \frac{5}{2}, \frac{7}{2}, \frac{9}{2}, \frac{11}{2}, \frac{15}{2}$, while the ground state of Ca^{42} should be predominantly $(f_{\frac{7}{2}})^2$. The Ca^{41} results (16) are then useful in giving a measure of $\theta_0^2(f_{\frac{7}{2}})$, $\theta_0^2(p_{\frac{3}{2}})$, ... The Ca^{43} results (17,18) show several interesting features. We have effective here for the low-lying states a j-selection rule which must be violated in order to produce any cross section except for $J = \frac{7}{2}$. We thus would have

$J = \frac{3}{2}$: dynamically favored ($l = 1$) but j-unfavored
$J = \frac{5}{2}$: dynamically unfavored; j-unfavored
$J = \frac{7}{2}$: dynamically unfavored; j-favored
$J > \frac{7}{2}$: strongly unfavored by the dynamics ($l > 3$) and unfavored by the model

One sees in fact an $l = 3$ ground-state reaction comparable with that for Ca^{41}, a weak $l = 1$ reaction to the 0.6-Mev level ($\frac{3}{2}-$) and no stripping for the other levels below 2 Mev exactly as would be expected. On the other hand the $\frac{3}{2}-$ level at 2.05 Mev shows a strong $l = 1$ reaction with reduced width comparable to the single-particle $l = 1$ reaction in Ca^{41}.

We may profitably use these results as examples in defining "single-particle" and "multiparticle" levels for an odd-even nucleus since these may be distinguished most readily by reduced widths. A *multiparticle state* is one whose predominant configuration is $(j^{2n+1})_{J \neq j}$; examples are the low excited states of Ca^{43}. A *single-particle* level is profitably defined as one whose dominant configuration[h] is $(j^{2n+1})_{J=j}$ or $(j^{2n})_0 j'$. The 2.05-Mev level in Ca^{43} is an example of the second type and its reduced width to $(j^{2n})_0$ will have the single-particle value (as we shall see later). The ground state is an example of the first type. The reduced width of such a state to $(j^{2n})_0$ fluctuates with n, j but, as we shall see, commonly is of the same order of magnitude as the single-particle width.

$$Ne^{20}(d,p) \; Ne^{21}(g.s.)$$
$$0+ \qquad \tfrac{3}{2}+$$
$$(VII)$$

The transferred particle is $d_{\frac{3}{2}}$ which would be j-unfavored in jj coupling ($l_0, j_0 \equiv d_{\frac{3}{2}}$). Experimentally (19) no stripping is observed. It would be quite unsafe to use this one example as an argument for jj coupling (see

---

[h] According to convenience we may be referring here to the configuration of neutrons and protons together or only to the odd group of particles. For cases where j^{2n+1} contains $J = j$ more than once we mean here the so-called seniority-1 state.

the earlier discussion about "accidental" cancellations). In actual fact half a dozen cases of this sort have been found by the Liverpool group in this region of the periodic table and taken together they do support a rough jj picture.

$$N^{14}(p,d) \; N^{13} \; (2.37 \text{ Mev}) \quad (l,j) \equiv s_{\frac{1}{2}}, \, d_{\frac{3}{2}}$$
$$1^+ \qquad \tfrac{1}{2}+$$
(VIII)

In the j-forbidden reactions of (VI) we are attempting to insert a particle where there is plenty of room for it but where the particle is still unwelcome; for example, the $(f_{\frac{7}{2}})_0^2 f_{\frac{5}{2}}$ state is essentially unoccupied but it lies high. In these cases we implicitly measure an admixture in the state of the heavier nucleus. In the l-forbidden case of (V) we are trying to insert a particle into an already filled configuration; here the state of the lighter nucleus must make room for the added particle and the reaction measures an admixture in the lighter nucleus. The $N^{14}(p,d)$ case above is of the first variety and the small upper limit for the experimentally (20) observed $l = 0$ and $l = 2$ widths supplies an argument first given by Standing (20) that in the N^{14} state there is essentially no contribution from the $s^4p^8(2s)^2$ and $s^4p^8d(2s)$ configurations. The small widths could result from accidental cancellations but this is very improbable in view of the quite similar results found (21) for the $C^{14}(d,t)$ reaction to the mirror $\tfrac{1}{2}+$ state at 3.09 Mev in C^{13}. However, in the latter case a sizable width is found for the reaction leading to the "single-particle" $d_{\frac{5}{2}}$ level at 3.86 Mev, indicating a quite appreciable $s^4p^8d^2$ component in the C^{14} ground state.

$$B^{11}(d,n) - (1) \to C^{12} \; (7.65 \text{ Mev}) \leftarrow (2) - C^{13}(p,d)$$
$$\tfrac{3}{2}- \qquad\qquad\qquad 0+ \qquad\qquad\qquad \tfrac{1}{2}-$$
(IX)

This pair of reactions might shed some light on the structure of the 7.65-Mev level, believed to be 0+. This state is "anomalous" in that no excited 0+ state for s^4p^8 should lie below say 12 Mev (22). Possible contributing configurations are:

(i) The "normal" configuration s^4p^8. Then (1) and (2) are allowed except that (2) is forbidden in jj coupling.
(ii) A "breathing mode" configuration such as $s^4(1p)^7(2p)$ where one nucleon is radially excited (23). Then (1) is allowed but (2) is n-unfavored.

(iii) Two nucleon excitation; for example, $s^4 p^6 d^2$. Then (1) and (2) are both forbidden.

$$_{25}\text{Mn}^{55}(d,p)_{25}\text{Mn}^{56}$$
(X)

For heavier nuclei one often considers neutrons and protons separately; in this way one can (though often without good theoretical justification) simplify the configurations to be considered, for example, by ignoring the excitation of the even group in an odd-even nucleus. For odd-odd nuclei a very crude procedure is to consider only the two last nucleons; a model with a little more content is the "odd-group" model (*24*) which considers the interaction between the two groups but allows neither group to become excited in the process. In this model the five protons outside closed shells in any low-lying state of Mn^{56} would be in the state $(f_{\frac{7}{2}})_{\frac{5}{2}}^5$. In the same spirit however (ignoring the even group), the five protons in Mn^{55} are in $(f_{\frac{7}{2}})_{\frac{5}{2}}^5$ since this state is a "Mayer anomaly" in its spin. The odd-group model would then predict the cross sections to be zero.

c. *LS* Selection Rules

Just as we have been considering the selection rules for l, s, j of the single nucleon, we can also consider the rules appropriate to a defined L,S for the nuclear states involved. L and S are rather good quantum numbers in the lower part of the nuclear p shell (say $A \leq 9$) but not in general though, for heavier nuclei, there are a few special cases.

See example (I) above[i]
(XI)

The ground state of Li^6 should be almost completely ^3S. Thus p-wave resonances should be strongly discouraged except for intermediate ^2P or ^4P states. The large magnitude of the observed width is strong evidence against the assumption which has often been made that the 7.46-Mev level in Li^7 is $^{22}F_{\frac{7}{2}}$. Instead it would appear to be mainly $^{24}P_{\frac{3}{2}}$ (*25*).

$$d + \alpha \rightarrow \text{Li}^6 \ (5.4 \text{ Mev}; J = 1, T = 0)$$
(XII)

This resonant reaction should go mainly by d-waves since the 5.4-Mev level is mainly ^3D, while d, α are primarily ^1S. However s-wave is allowed by the kinematics. A measure of the s-wave width (which is dynamically favored but model-unfavored) could, in principle, be used to determine

[i] We number the examples consecutively throughout the chapter.

the sum of the D-state amplitude in the deuteron and the related S-state amplitude in the Li^6 excited state.

d. MISCELLANEOUS SELECTION RULES

We have dismissed the absolute selection rules (parity and angular momentum) as being without interest. Still there are a few cases where a combination will produce a result which at very first sight may be a trifle surprising. For example, in the scattering of α particles by a spin zero nucleus, we can reach only compound states with even π and J or with odd π and J; a result which is useful (?) in examining the levels of Mg^{24} which may be reached by Ne$^{20} + \alpha$ or Na$^{23} + p$. By the same argument we can find no resonance in $\alpha + d \rightarrow$ Li^6 (3.57 Mev; $J = 0$, $T = 1$), but note carefully that this should not be ascribed to an isotopic spin rule. And of course if a state decays into two identical nuclei the exclusion principle plays an important part; only the states of Be^8 with even π and even J can decay to two α's.

For displaying isotopic spin selection rules useful projectiles are the $T = 0$ deuteron and α-particle. For the most part the experiments performed that display such rules involve nonresonant compound states and are thus outside the province of this chapter.

Besides these there are space-symmetry selection rules especially appropriate to p-shell nuclei, and "seniority" selection rules that could be significant in heavier nuclei which are well described by jj coupling. There are few if any good examples of reduced widths inhibited by such rules and we shall not discuss them.

3. Quantitative Analysis via the Shell Model

In Section 2 we divided reduced widths into two classes favored and unfavored according as the corresponding reaction was or was not inhibited by some nonabsolute selection rule. We qualitatively discussed a number of cases particularly of the unfavored variety. In this section we treat the quantitative shell model analysis of reduced widths; we shall no longer be satisfied with a statement that a certain width is large or small but shall require an explicit expression for it. This expression will take a variety of forms depending, among other things, on the complexity of the configurations involved and whether or not the transferred particle is equivalent with others which are present. There are in fact more types of situations than we can treat but the general principles are pretty much the same and the reader should easily be able to make any necessary extensions.

But first we must learn something about the coupling of angular momenta and about the construction of antisymmetrized wave functions for many-nucleon systems. It is quite important too that we have a simple and flexible notation for dealing with these matters. The result of coupling together (by means of Clebsch-Gordan coefficients) two angular momenta J_1, J_2 to form a resultant J, M may be represented by Fig. 2a; it is convenient to ignore the M, which will always be specified, and all the arrows except the resultant, and thus we have Fig. 2b. If we couple together n angular momenta we must specify $2n$ quantum numbers, two for each

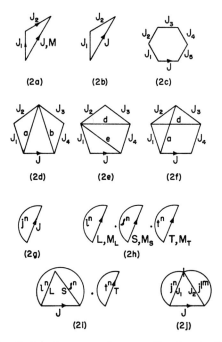

FIG. 2. Angular momentum coupling (see text).

angular momentum, whereas a diagram such as Fig. 2c displays only $(n+2)$ (an M is understood). Thus $(n-2)$ extra quantum numbers must be specified and these can be taken as internal-coupling angular momenta as for example in Fig. 2d or 2e for the case $n = 4$. One must be careful that the quantum numbers specified do belong to a set of *commuting* operators. The reader will prove easily that any set of *nonintersecting* coupling lines is allowed and that such a set does contain the desired $(n-2)$ angular momenta. On the other hand intersecting coupling lines are, in general, not compatible so that the scheme of Fig. 2f is not a proper one.

There are three basic recoupling rules[j] given by Equations 5–7.

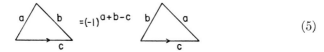

$$\text{(5)}$$

$$\text{(6)}$$

$$[(7)$$

The first of these follows from our phase convention for the Clebsch-Gordan coefficients which is the standard one of Condon and Shortley (*31*), the second may be regarded as the definition of the "normalized Racah coefficient" related to the usual Racah coefficient (*28*) by

$$U(abcd:ef) = [(2e+1)(2f+1)]^{\frac{1}{2}} \cdot W(abcd:ef),$$

and the third equation is the inverse of the second.

A wave function for many nucleons must be antisymmetrized either in all the nucleons if we use an isotopic spin formalism or else in neutrons and protons separately if we do not. It is convenient to represent antisymmetrization by a circular arc; for example, a state for n identical nucleons, each with angular momentum j coupled together to a resultant J, may be represented by Fig. 2g; a corresponding state for l^n with resultant $LSTM_LM_SM_T$ by Fig. 2h and with resultant $LSTJM_JM_T$ by Fig. 2i. It will be understood that all the antisymmetrization symbols in any diagram or product of diagrams must be understood together; 2h is not a product of three separately antisymmetrized functions. The exception

---

[j] The three rules which follow enable us to perform any angular-momentum recoupling. (For complicated cases the transformation coefficients may turn out to be one- or more-parameter sums of the product of U-functions which can be used to define the so-called X-functions or $9j$-symbols and others of greater complexity.) Equations 6 and 7 combine to give the unitarity sum rule

$$\sum_f U(abcd:ef)U(abcd:e'f) = \delta_{ee'}.$$

Diagrammatic wave functions seem to have been used first as above by Fano (*26*); several useful extensions are possible (Halbert, *27*). For a more detailed account of angular momentum coupling see Racah (*28*), Biedenharn *et al.* (*29*), or the recent books (*30*) of Edmonds or of Rose.

here is if we consider neutrons and protons without isotopic spin; in this case we may keep the two groups separate by an extra line as in Fig. 2j.

Next we must understand the notion of fractional parentage. The idea here was first used by Bacher and Goudsmit (*32*) in discussing the energy levels of an ion in terms of the energy levels of the systems with one less electron; it was developed in much more general form by Racah (*33*). Quite sophisticated techniques for calculating "coefficients of fractional parentage" (which are the essential fruit of the theory) have been developed by Racah (*33*) and by Jahn (*34*) for the case of equivalent particles and have been applied in important cases by Jahn (*34*) and by Edmonds and Flowers (*35*). Fortunately we can make good use of these coefficients without worrying about the procedures which produce them. No essential complications will arise when we consider the cases of interest to us involving inequivalent particles. In fact we begin with a definition and some results independent of any model.

The fractional parentage idea is simply that one should represent an antisymmetrized n-nucleon wave function as an expansion in terms of the states for the first $(n-1)$ particles, the nth particle being vector-coupled to these states. Explicitly we have

$$\left(\begin{array}{c}C\\J\end{array}\right)\cdot\left(\begin{array}{c}\\T\end{array}\right) = \sum_{\substack{C_0 J_0 \\ l\ j}} \langle CJT | C_0 J_0 T_0 \rangle_j \quad \cdots \qquad (8)$$

$$= \sum_{\substack{C_0 J_0 \\ l\ j\ z}} (-1)^{z+\mathscr{I}-j-J} U(J_0 \mathscr{I} J \mathit{l} : \bar{z} j) \langle CJT | C_0 J_0 T_0 \rangle_j \quad \cdots \qquad (9)$$

with obvious modifications if we have only identical particles (isotopic spin may be ignored) or if we choose to antisymmetrize neutrons and protons separately. On the left-hand side of Eq. (8) the symbol C stands for all the quantum numbers, apart from J, T (and their z-components), necessary to define the state. For example, C might define a shell model configuration along with L, S and the space symmetry if the n-particle state is well described as a pure LS-coupling shell model state; or it might define the so-called seniority or symplectic symmetry if the state is a pure jj state. Or if the state is not simply described in terms of some well-known scheme, C might stand for an entire set of expansion coefficients giving the state in terms of some complete set. C_0 on the right-hand side of Eq. (8) does the same thing for the $(n-1)$-particle states. On the right-hand side the particle numbered n has been separated out and its l, $s(=\frac{1}{2})$, and j are specified (and of course also the radial quantum number or other such necessary information is to be understood). Each term in the expansion is antisymmetric in particles #1 ... $(n-1)$ but not, in general, in all n particles, though of course the sum is. The quantity $<CJT|C_0J_0T_0>_{lj}$ is a generalized coefficient of fractional parentage[k] (c.f.p.) connecting the states (CJT) and $(C_0J_0T_0)$.

It is convenient to change from the representation of Eq. (8) to that given in Eq. (9); this is quite simply done by using the recoupling Eqs. (5–7). The advantage of the second representation is that it displays the channel spin[l] z. The relative reduced width for orbital angular momentum l and channel spin z corresponding to $(CJT) \to [(C_0J_0T_0) + l]_z$ now follows from Eq. (9) and is given by

$$S^{\frac{1}{2}}(l,z) = n^{\frac{1}{2}} \sum_j (-1)^{z+s-j-J} U(J_0 s J l: zj) <CJT|C_0J_0T_0>_{lj} \quad (10)$$

an entirely general result. Summing over channel spins by using the unitarity property of the U-functions[j] we see that the interference between different j-values vanishes and

$$S(l) = n \sum_j <CJT|C_0J_0T_0>_{lj}^2. \quad (11)$$

The factor n occurs because the c.f.p. describes the separation of particle #n whereas the reaction allows every particle to be emitted on the same footing but of course with no interference.

We now explicitly introduce the shell model. The jj coupling c.f.p. are defined by

$$\left(\widehat{j^n}\right)_J \cdot \left(\widehat{t^n}\right)_T = \sum_{J_0 T_0} \langle j^n: JT | j^{n-1}: J_0 T_0\rangle \; \left(\widehat{j^{n-1}}\widehat{j^{(n)}}\right)_J \left(\widehat{t^{n-1}}\widehat{t^{(n)}}\right)_T \quad (12)$$

and since this is a special case of Eq. (8) we have that for $(j^n)_{JT} \to (j^{n-1})_{J_0T_0} + j$

$$S = n <j^n: JT | j^{n-1}: J_0 T_0>^2. \quad (13)$$

If the states involved are not uniquely defined by J, T further quantum numbers will need to be specified and, as usual, if we have identical particles only, the isotopic spins may be ignored.

Things are a little more complicated in the LS representation. We

---
[k] The l-value is determined by the parities and the j-value.
[l] Essential because there is no interference between different channel spins.

write (omitting the l^n, l^{n-1} in writing the c.f.p. but including the space symmetry symbols α, α_0)

$$\left(\begin{smallmatrix}l^n & L \\ & \alpha\end{smallmatrix}\right) \cdot \left(\begin{smallmatrix}s^n & \\ & S\end{smallmatrix}\right) \cdot \left(\begin{smallmatrix}t^n & \\ & T\end{smallmatrix}\right)$$

$$= \sum_{\alpha_0 L_0 S_0 T_0} \langle \alpha LST | \alpha_0 L_0 S_0 T_0 \rangle \left(\begin{smallmatrix}l^{n-1} & L_0 \\ \alpha_0 & \end{smallmatrix}\, l(n)\right) \cdot \left(\begin{smallmatrix}s^{n-1} & S_0 \\ & \end{smallmatrix}\, s(n)\right) \cdot \left(\begin{smallmatrix}t^{n-1} & T_0 \\ & \end{smallmatrix}\, t(n)\right) \quad (14)$$

We would like to specify the initial J-value and for this we couple together L and S by Clebsch-Gordan coefficients (which depend only on angular momenta). The same coefficients enter on both sides of Eq. (14) and we immediately have

$$\left(\begin{smallmatrix}l^n & L \\ \alpha & S \\ & J \end{smallmatrix}\, s^n\right) \cdot \left(\begin{smallmatrix}t^n & \\ & T\end{smallmatrix}\right)$$

$$= \sum_{\alpha_0 L_0 S_0 T_0} \langle \alpha LST | \alpha_0 L_0 S_0 T_0 \rangle \cdot (\ldots) \cdot \left(\begin{smallmatrix}t^{n-1} & T_0 \\ & \end{smallmatrix}\, t(n)\right) \quad (15)$$

But, now unlike the jj case, the final-state total angular momentum is not specified in the diagram and neither is the channel spin. Thus before we can read off the reduced width we must use Eqs. 5–7 to exhibit these angular momenta. The right-hand side of Eq. (15) becomes then

$$\sum_{\substack{\alpha_0 L_0 S_0 T_0 \\ z}} \beta_z \, l(n) \left(\begin{smallmatrix}\alpha_0 & S_0 \\ L_0 & \\ & J_0 \\ & z\end{smallmatrix}\, s(n)\right) \cdot \left(\begin{smallmatrix}t^{n-1} & T_0 \\ & \end{smallmatrix}\, t(n)\right)$$

where

$$\beta_z = (-1)^{l+L_0+L} U(lL_0 JS : Lz) U(L_0 S_0 zs : J_0 S) \langle \alpha LST | \alpha_0 L_0 S_0 T_0 \rangle. \quad (16)$$

From this form [a special case of Eq. (9)] we have

$$\mathcal{S}(z) = n\beta_z^2 \qquad \mathcal{S} = n \sum_z \beta_z^2. \quad (17)$$

These formulae were given independently by Lane, Satchler, and Auerbach (36). As they stand they apply to the case of extreme LS coupling where each nuclear state has a specified $LS\alpha$. It is easy to relax this restriction; suppose that the nuclear states belong to l^n and have multiplet expansion coefficients $K(\alpha LST)$, $K(\alpha_0 L_0 S_0 T_0)$, respectively: for example

$$\Psi_{JT}(l^n) = \sum_{\alpha LST} K(\alpha LST)\Psi_{\alpha LSTJ}(l^n). \tag{18}$$

Then the formulae are unchanged if we redefine

$$\beta_z = \sum_{\substack{\alpha_0 L_0 S_0 T_0 \\ \alpha LST}} (-1)^{l+L_0+L} K(\alpha LST) K(\alpha_0 L_0 S_0 T_0) U(lL_0 JS:Lz) \cdot$$

$$U(L_0 S_0 zs:J_0 S) <\alpha LST|\alpha_0 L_0 S_0 T_0>. \tag{19}$$

So much for the case of equivalent particles. A different case of special interest has all the "active" particles[m] equivalent except one. The particle which is transferred in the reaction may then be the inequivalent one or one of the others and there are thus two rather distinct cases. We discuss this from the jj standpoint; except for the obvious modifications the LS case proceeds in the same way. For ease in writing let us at first assume that we have identical particles only and do not use the isotopic spin formalism.

The configuration $j^n j'$ has states which may be labeled $(j_{J_0}{}^{n-1})j'{}_J$ indicating that the $(n-1)$ equivalent particles are coupled to J_0, the inequivalent particle then being coupled on (of course further quantum numbers than J_0 may be needed in a particular case). Consider now

This is not antisymmetric but we can antisymmetrize it by operating with $n^{-\frac{1}{2}}\{1 - \Sigma_{i=1}^{n-1} P_{i,n}\}$ and the result is then normalized. We see immediately then (37) that the c.f.p. for separation of the inequivalent particle is simply $n^{-\frac{1}{2}}$ (remember that particle #n must always be separated out). To produce the c.f.p. for the separation of an equivalent particle we must carry out the antisymmetrization; as an exercise the reader may show that (if $j' \neq j$) Eq. (20) supplies us with the complete solution. The more general case with isotopic spin has of course the same structure but the phase factor is multiplied by $(-1)^{T_0-T_1+T_2-T}$, there is an extra U-factor, $U(\frac{1}{2}T_1 T\frac{1}{2}:T_0 T_2)$, and of course the $n \to (n-1)$ c.f.p. is replaced

[m] By active particles we mean those which do not remain in closed shells throughout the reaction.

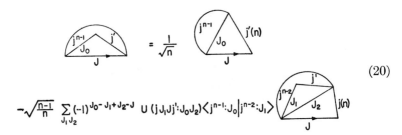

$$-\sqrt{\frac{n-1}{n}} \sum_{J_1 J_2} (-1)^{J_0 - J_1 + J_2 - J} \, U(jJ_1Jj';J_0J_2) \langle j^{n-1}:J_0|j^{n-2}:J_1\rangle \qquad (20)$$

by its generalization. The procedure above of using the tabulated c.f.p. to handle the equivalent particles of a group and direct antisymmetrization for an equivalent one may easily be extended to more complicated cases.

We see now that for separation of the inequivalent particle, the others being "undisturbed," namely $(j_{J_0T_0}{}^{n-1}j')_{JT} \rightarrow (j_{J_0T_0}{}^{n-1}) + j'$ we have simply $\mathcal{S} = 1$ and the width is the single-particle width. Now it will usually not happen that in the n- particle state the equivalent group is coupled to a unique J_0T_0, for the presence of the extra particle will usually produce excitations of this group[n] and instead we will have

$$(j^{n-1}j')_{JT} = \sum_{J_0 T_0} K(J_0T_0)(j_{J_0T_0}{}^{n-1}j')_{JT} \qquad (21)$$

and for the corresponding width

$$\mathcal{S} = |K(J_0T_0)|^2 \leq 1. \qquad (22)$$

We *will* have $\mathcal{S} = 1$ if it happens that there is a unique possible set (J_0T_0) and these cases are experimentally important; for example, $\mathcal{S} = 1$ for all cases where j^{n-1} corresponds to a single particle or a single hole or in a case like $(d_{\frac{3}{2}})^2 f_{\frac{7}{2}}$ ($J \neq \frac{7}{2}$) for identical particles. In situations like this the group is "inert" throughout the reaction; it cannot be ignored however for the coupling of the other particle to it will produce a whole family of states with single-particle reduced widths for separation of the inequivalent particle.

If several amplitudes occur in Eq. (21) but all except one of them are very small we again have $\mathcal{S} \simeq 1$. We could then say that the inequivalent particle is weakly coupled to the equivalent group. This could happen if the equivalent group is even and its excitation energies are high [see example (II) above]. Lane, who first discussed this weak coupling (*12*), has suggested that, for light nuclei at least, weak coupling might occur

---

[n] That is, excitation within the same configuration. We have already implied by using the single configuration that excitations outside the configuration may be ignored.

more generally but this is uncertain. Note that, because of the uncertainty in the single-particle width it is not easy to determine from a single measured width whether only one term enters appreciably in Eq. (21). Observe too that the odd-group model mentioned in example (X) assumes a weak coupling of two odd groups and because of their low excitation energies is probably a quite inadequate model.

The cases involving the transfer of the inequivalent particle supply us with either a measure of the single-particle width [if only one amplitude can occur in Eq. (21)] or at least with an upper limit for this width. The more complicated case described by $(j^{n-1}j')_{JT} \to (j^{n-2}j')_{J_0 T_0} + j$, where an equivalent particle is transferred, we discuss later by example (see XV following).

$$\text{Li}^7 + p \to \text{Be}^8 \text{ (17.63 Mev)}$$
$$\tfrac{3}{2}- \ (l=1) \quad 1+,\ T=1$$
$$\text{(XIII)}$$

The results of the elastic scattering and ($p\gamma$) reactions have been discussed by many authors, most thoroughly by Liberman and Christy (*38*); we will not review the arguments which lead to the T, J assignment for the compound level. There are two available channel spins, $z = 1, 2$. In terms of the channel-spin ratio $x = \beta_2^2/\beta_1^2$ it turns out that the ground-state M1 γ-ray angular distribution is $d\sigma/d\omega = 1 + \dfrac{5-x}{5+7x}\cos^2\theta$ and, since the angular distribution is almost isotropic, we have $x \simeq 5$; the most recent measurements at Hamburg (*39*) appear to establish $3.0 \leq x \leq 3.6$. This is one of the very few cases where a channel-spin ratio is known with some precision. The reduced width itself has a value $\theta^2 \simeq 0.13$ (*6*). We know also the stripping widths for the first two levels of Li^8; the values are 0.053 and 0.028, respectively (*40*). There is also a large amount of information concerning β and γ transitions but this we are obliged here to ignore.

If we are prepared to assume pure LS coupling (that is, specified L,S for Li^7 and Be^8) we can find, without further assumptions, the possible values of L,S which are compatible with the near isotropy of the γ-ray. This sort of thing was first done by Christy (*41*). It is thoroughly known (*22*) that Li^7 is essentially a $^{22}P^{[3]}$ state so we ask only for the L,S values for Be^8. From Eq. (16) we have°

$$[U(1\tfrac{1}{2}2\tfrac{1}{2}:\tfrac{3}{2}S)\,U(1S11:2L)]^2 = x[U(1\tfrac{1}{2}1\tfrac{1}{2}:\tfrac{3}{2}S)\,U(1S11:1L)]^2 \quad (23)$$

° But note carefully that Eq. (23) follows directly from Eq. (10) without any assumption concerning shell model configurations and similarly for the result quoted below for pure $p_{\frac{1}{2}}$ or $p_{\frac{3}{2}}$ capture.

and then for $Be^8 \equiv {}^3S$, ${}^1P$, ${}^3P$, ${}^3D$ (which are the only ones allowed by angular-momentum conservation) we have $x = 5, 0, 5, \frac{1}{5}$, respectively. The possible candidates then are ${}^3S$, ${}^3P$. At the other coupling extreme, for the proton being captured exclusively from a $p_{\frac{1}{2}}$ or $p_{\frac{3}{2}}$ state, we have from Eq. (10), $x = 5, \frac{1}{5}$, respectively; that $p_{\frac{1}{2}}$ capture should lead to γ-ray isotropy is of course obvious at sight.

TABLE I. THE β_z MATRICES FOR THE $Li^7 + p$ REACTION. IN EACH BLOCK THE UPPER NUMBER IS FOR $z = 1$, THE LOWER FOR $z = 2$.

	${}^{22}P[3]$	${}^{22}P[21]$	${}^{22}D[21]$	${}^{24}P[21]$	${}^{24}D[21]$
${}^{33}S[22]$	— —	$\sqrt{3}/9$ $-\sqrt{15}/9$	— —	$-\sqrt{30}/18$ $-\sqrt{6}/18$	— —
${}^{31}P[31]$	$-\sqrt{2}/3$ —	$-\sqrt{5}/6$ —	— $\sqrt{5}/10$	— —	— —
${}^{31}P[211]$	— —	$\sqrt{3}/6$ —	— $\sqrt{3}/6$	— —	— —
${}^{35}P[211]$	— —	— —	— —	$\sqrt{6}/24$ $-3\sqrt{30}/40$	$-\sqrt{6}/8$ $7\sqrt{30}/120$
${}^{33}D[31]$	$\sqrt{2}/12$ $\sqrt{10}/60$	— —	— —	$-5\sqrt{2}/24$ $\sqrt{10}/120$	$\sqrt{2}/8$ $-\sqrt{10}/40$
${}^{33}D[22]$	— —	$\sqrt{15}/36$ $\sqrt{3}/36$	$\sqrt{3}/4$ $\sqrt{15}/20$	$-5\sqrt{6}/72$ $\sqrt{30}/360$	$-\sqrt{6}/8$ $\sqrt{30}/40$
${}^{33}P[31]$	$-\frac{1}{6}$ $-\sqrt{5}/6$	— —	— —	$-\frac{5}{24}$ $\sqrt{5}/24$	$\frac{1}{8}$ $3\sqrt{5}/40$
${}^{33}P[211]$	— —	$-\sqrt{2}/12$ $-\sqrt{10}/12$	$-\sqrt{10}/12$ $\sqrt{2}/4$	$-\sqrt{5}/24$ $\frac{1}{24}$	$-\sqrt{5}/24$ $-\frac{1}{8}$

It would be quite unsafe to conclude from the foregoing that we do in fact have an extreme coupling scheme. We therefore make the further reasonable assumption that the states belong to s^4p^n and calculate the β_z matrices[p] using an LS representation. They are given in Table I. Note that the only ${}^3S$ state possible for Be^8 is ${}^{33}S[22]$ for which, as shown, the reduced width vanishes (because in fact of a symmetry selection rule that adding a nucleon to a state [3] produces [4] + [31] but not [22]); besides this, such a state should be about 10 Mev higher. There are now

[p] If we write the l^n and l^{n-1} states of Eq. (19) as column vectors ($\Psi^* \equiv$ a row vector) then the β_z matrix is $\beta_z^{(m)}$ defined by $\beta_z = \langle \Psi^*(l^n) \beta_z^{(m)} \Psi(l^{n-1}) \rangle$.

TABLE II. THE s^4p^4 $J = 1 = T$ ENERGY MATRIX FOR A ROSENFELD INTERACTION WITH $L/K = 6$ AND A SINGLE-PARTICLE SPIN ORBIT INTERACTION WITH $\zeta = a/K$. THE ENERGY UNIT IS K (~-1 MEV). THE MATRIX IS OF COURSE SYMMETRIC

	$^{31}P[31]$	$^{33}P[31]$	$^{33}D[31]$	$^{33}S[22]$	$^{33}D[22]$	$^{31}P[211]$	$^{33}P[211]$	$^{35}P[211]$
$^{31}P[31]$	$+15.3$	$-\frac{\sqrt{2}}{4}\zeta$	$-\frac{1}{2}\zeta$			-0.4	$-\frac{\sqrt{10}}{4}\zeta$	
$^{33}P[31]$	$-\frac{\sqrt{2}}{4}\zeta$	$+18.0 -\frac{3}{8}\zeta$	$+\frac{\sqrt{2}}{8}\zeta$	$-\frac{\sqrt{30}}{6}\zeta$	$+7\frac{\sqrt{6}}{24}\zeta$	$+\frac{\sqrt{30}}{12}\zeta$	$+0.4 +\frac{\sqrt{5}}{8}\zeta$	$-5\frac{\sqrt{6}}{24}\zeta$
$^{33}D[31]$	$-\frac{1}{2}\zeta$	$+\frac{\sqrt{2}}{8}\zeta$	$+16.4 -\frac{1}{4}\zeta$		$+\frac{\sqrt{3}}{4}\zeta$	$-\frac{\sqrt{15}}{6}\zeta$	$+\frac{\sqrt{10}}{8}\zeta$	$-\frac{\sqrt{3}}{12}\zeta$
$^{33}S[22]$		$-\frac{\sqrt{30}}{6}\zeta$		$+10.0$		$+\frac{2}{3}\zeta$	$+\frac{\sqrt{6}}{6}\zeta$	$-\frac{\sqrt{5}}{3}\zeta$
$^{33}D[22]$		$+7\frac{\sqrt{6}}{24}\zeta$	$+\frac{\sqrt{3}}{4}\zeta$		$+7.6 -\frac{3}{4}\zeta$	$-\frac{\sqrt{5}}{3}\zeta$	$-\frac{\sqrt{30}}{24}\zeta$	$+\frac{1}{12}\zeta$
$^{31}P[211]$	-0.4	$+\frac{\sqrt{30}}{12}\zeta$	$-\frac{\sqrt{15}}{6}\zeta$	$+\frac{2}{3}\zeta$	$-\frac{\sqrt{5}}{3}\zeta$	-2.9	$-\frac{\sqrt{6}}{4}\zeta$	
$^{33}P[211]$	$-\frac{\sqrt{10}}{4}\zeta$	$+0.4 +\frac{\sqrt{5}}{8}\zeta$	$+\frac{\sqrt{10}}{8}\zeta$	$+\frac{\sqrt{6}}{6}\zeta$	$-\frac{\sqrt{30}}{24}\zeta$	$-\frac{\sqrt{6}}{4}\zeta$	$+0.4 +\frac{1}{8}\zeta$	$+\frac{\sqrt{30}}{8}\zeta$
$^{35}P[211]$		$-5\frac{\sqrt{6}}{24}\zeta$	$-\frac{\sqrt{3}}{12}\zeta$	$-\frac{\sqrt{5}}{3}\zeta$	$+\frac{1}{12}\zeta$		$+\frac{\sqrt{30}}{8}\zeta$	$+7.0 -\frac{3}{4}\zeta$

so many undetermined amplitudes that we must make further assumptions in order to make any progress. One procedure, which has become conventional, is to adopt the single-particle spin orbit model of Inglis (discussed in Chapter VI.B by Kurath). Let us fix all but one of the central-interaction parameters by adopting a Rosenfeld interaction (42) with $L/K = 6$. The free Li^8 parameters are then K and $\zeta = a/K$ and of these we can hope to fix ζ by means of the reduced widths. The wave function, which depends on ζ, is now the lowest energy eigenfunction of the matrix given in Table II. The Li^7 wave function can be determined in the same way. For rough considerations, one could take Li^7 to be a pure

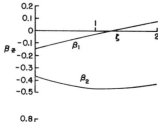

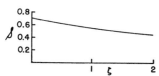

FIG. 3. For the 440-kev ($Li^7 + p$) reaction β_1, β_2, $\mathcal{S} = 4(\beta_1^2 + \beta_2^2)$ and $1/x = \beta_1^2/\beta_2^2$ are plotted against the Be^8 spin orbit parameter, for a Rosenfeld interaction with $L/K = 6$ and $\zeta(Li^7) = 1.1$.

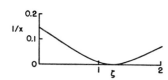

$^{22}P^{[3]}$ state (this approximation is allowable for Li^7 because the interfering multiplets must lie quite high having a lower space symmetry but would not be adequate for Li^8. In the following we do not make this approximation for Li^7 but use Rosenfeld wave functions with parameters fixed from other data).

We give in Fig. 3 a plot of β_1, β_2, $\mathcal{S} = 4(\beta_1^2 + \beta_2^2)$ and $1/x = \beta_1^2/\beta_2^2$ as determined from these amplitudes. Note the very strong dependence of x on ζ; coupling this with the experimental fact that $x < 5$ shows that we have in fact found a lower limit to the spin orbit parameter ζ. We see too that the close agreement with experiment of the LS value for the channel spin ratio is apparently without any significance at all. The total width value $\theta^2 \simeq 0.13$ is consistent with the calculated $\mathcal{S}$ but we cannot say more since θ_0^2 is unknown; in fact we might regard the measurement as yielding for θ_0^2 a reasonable value $\theta_0^2 \simeq 0.3$. The (d,p) widths to the first two states are known and the ratio of these supplies one more parameter. It turns out that this ratio and a large amount of other data too are consistent with $\zeta \simeq 2$.

The reader might well object that fixing the two-body interaction as completely as we have done is a high-handed procedure. Though it is beyond our present scope it is worth-while remarking that a general study of the effective interaction in p-shell nuclei results in fixing well enough all the multiplet positions which are important in the present problem with the single exception of the $^{31}P^{[31]}$. At the same time if he looks to find the source of the strong dependence of β_1 with ζ the reader will find from Tables I and II that this comes about because β_1 is very sensitive to the $^{31}P^{[31]} - {^{33}P^{[31]}}$ difference. Obviously then it might be useful to convert our two-parameter theory into a three-parameter one by taking the $^{31}P^{[31]}$ position as variable.

$$C^{13}(d,p)\ C^{14},\ C^{13}(d,p)\ C^{14*}$$
$$\tfrac{1}{2}-\qquad 0+\qquad\qquad 2+$$
$$(\text{XIV})$$

We are considering here the C^{14} ground state and lowest $J = 2+$ level which should be near 6 Mev. Shown, for these two cases, in Fig. 4

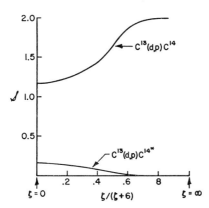

FIG. 4. For the transition between LS and jj coupling, S is shown for the reactions $C^{13}(d,p)$ to the ground state of C^{14} and the excited $J = 2$ state expected near 6 Mev.

is Auerbach's calculation (36) of the variation of S with ζ for the same model of example (XIII). C^{14*} demands a $p_{\frac{3}{2}}$ particle and consequently $\mathsf{S} \to 0$ as we move towards jj coupling. C^{14} demands a $p_{\frac{1}{2}}$ particle and in the jj extreme we have a transition $j^2 \to j + j$ for which, as the reader should easily see, $\mathsf{S} = 2$ independently of j.

$P^{31}(d,p)$ P^{32}, P^{32*} [see example (V), Fig. 1].
$\frac{1}{2}+$ $\quad$ $1+$ $2+$

(XV)

The selection rules have been discussed above. The principal configurations should be s^3 and $s^3d_{\frac{3}{2}}$ which have unique states $J,T = (\frac{1}{2},\frac{1}{2})$ and $J,T = (11), (21), (1,0), (2,0)$ the last two of which are of no interest to us. The $l = 2$ reactions proceed in lowest order as $s^3 + d_{\frac{3}{2}} \to s^3d_{\frac{3}{2}}$ and since s^3 has a unique state we have by Eq. (22), $\mathcal{S}(d) = 1$. The $l = 0$ reaction proceeds in lowest order according to $s^2d_{\frac{3}{2}} + s \to s^3d_{\frac{3}{2}}$ and, since only the state $(s^2)_{10}d_{\frac{3}{2}}$ of the $s^2d_{\frac{3}{2}}$ configuration can contribute to P^{31} the value is uniquely given in terms of a single amplitude. Formally we have, as in Eq. (20) but writing J,T together in the diagrams,

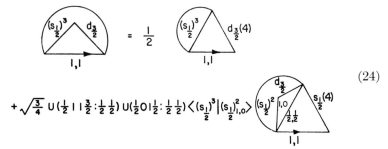

(24)

+ terms in which the antisymmetrized particles do not have $T,J = \frac{1}{2},\frac{1}{2}$.

The product of the numerical factors in the second term is $\sqrt{\frac{1}{6}}$ and since $n = 4$ we have to lowest order^q

$$\mathcal{S}(l = 2) = 1$$
$$\mathcal{S}(l = 0) = \tfrac{2}{3}[K(s^2d)]^2 \qquad (25)$$

The reduced-width ratio determined from experiment is $\theta^2(d)/\theta^2(s) \simeq 15$. Then taking $\theta_0^2(d)/\theta_0^2(s) \simeq 0.8$ as determined by the $\text{Si}^{28}(d,p)$ experiment (43) to the first two states of Si^{29} we deduce that $[K(s^2d)]^2 \simeq 8\%$.

The reader may find it instructive to calculate the relative reduced widths for this case without using the isotopic spin formalism, making the necessary small generalizations in Eq. (20). The $l = 0$ reaction is described, for example, as $\langle s|sd_{\frac{3}{2}}\rangle_{\frac{1}{2}} + \langle s|s_{\frac{1}{2}}\rangle \to \langle s|s^2d_{\frac{3}{2}}\rangle_1$ (where we use an obvious notation) and it will then be found that

$$\mathcal{S}(l = 0) = [K(s|sd)]^2 \cdot 2[\langle s_0^2|s_{\frac{1}{2}}\rangle U(\tfrac{1}{2}\tfrac{1}{2}\tfrac{3}{2}\tfrac{3}{2}:01) U(\tfrac{1}{2}11\tfrac{1}{2}:\tfrac{3}{2}\tfrac{1}{2})]^2. \qquad (26)$$

The c.f.p. has the value 1, as should be obvious, and we then find, just as before, $\mathcal{S}(l = 0) = \tfrac{2}{3}[K(s|sd)]^2$.

^{q} An analysis equivalent to this has been given by Parkinson (15).

4. Miscellaneous Considerations

a. STATISTICAL FACTORS AND REDUCED-WIDTH VARIATIONS

The c.f.p. for removing a particle from a closed jj shell is simply unity as is obvious from Eq. (12). Then for the relative reduced width we have $S = N$, the number of particles in the closed shell, and this may be a large number. Does this imply that reactions where we fill in a hole have much stronger cross sections than those involving a single particle (a "nature abhors a vacuum" type of thing)? The answer is, "No," and this is connected with the fact that the statistical factors [the $(2J + 1)$ factors and the isotopic spin Clebsch-Gordan factor] are not as devoid of significance as they are commonly regarded. [This same point has been well made in another connection by Wilkinson (*44*).]

Consider identical particles only; then, for a stripping reaction (and the same argument applies to the resonant case) we have, omitting immaterial constants and noting that the nuclear statistical factors do not enter for the (p,d) case,

$$\begin{aligned}\sigma^{dp}(j^{N-1} + j \to j^N) &\equiv (2j+1)^{-1} S(N \to N-1) = 1 \\ \sigma^{pd}(j^N \to j^{N-1} + j) &\equiv 1 \cdot S(N \to N-1) = (2j+1),\end{aligned} \quad (27)$$

while at the other end of the shell

$$\begin{aligned}\sigma^{dp}(j^0 + j \to j) &\equiv (2j+1) S(1 \to 0) = (2j+1) \\ \sigma^{pd}(j \to j^0 + j) &\equiv S(1 \to 0) = 1.\end{aligned} \quad (28)$$

We see that the statistical factors cooperate in such a way that there is a one-to-one correspondence between the cross sections themselves though not between the reduced widths. This is true in general as we can see by using the relationship between the relative reduced width connecting two levels and that connecting their complements (in the hole ⇌ particle sense). This is[r]

$$S(n, J, T \to n-1, J_0, T_0)$$
$$= \frac{(2J_0+1)(2T_0+1)}{(2J+1)(2T+1)} S(N-n+1, J_0, T_0 \to N-n, J, T) \quad (29)$$

where N is the number of particles in the closed shell and without the isotopic spin formalism we would ignore the $(2T + 1)$ factors. With Eq. (29) the reader will verify that the isotopic-spin Clebsch-Gordan factor is also nontrivial in the sense that the same S calculation made

---

[r] This formula follows for jj coupling by inserting the relationship between a c.f.p. and its complement (*28*) into Eq. (13). It is valid much more generally; namely, for any states involving equivalent particles (l^n or j^n) only.

with or without the isotopic spin formalism will in general give different results corresponding to this factor. This did not show up in our example (XV) above because there the isotopic-spin coupling factor was unity.

There are now two basic sum rules which we write for the case of jj coupling. The first of these simply reflects the normalization condition[s] in Eq. (12) and is

$$\sum_{J_0 T_0 \gamma_0} \mathcal{S}(n, JT\gamma \to n-1, J_0 T_0 \gamma_0) = n \tag{30}$$

where now we explicitly write the extra quantum numbers γ, introduced to allow for the fact that the states may not be unique.

Combining Eqs. (29) and (30) gives the second rule

$$\sum_{JT\gamma} (2J+1)(2T+1) \mathcal{S}(N-n+1, JT\gamma \to N-n, J_0 T_0 \gamma_0)$$
$$= n(2J_0+1)(2T_0+1). \tag{31}$$

We can use these to demonstrate fluctuations in $\mathcal{S}$ which occur as we go through a shell. Considering identical nucleons only (without the isotopic spin formalism) and assuming that all the states involved are unique[t] we have for the chain $(nJ) = (00) \leftarrow (1j) \leftarrow (20) \leftarrow (3j) \cdots$.

$$\mathcal{S}(n \to n-1) = n \qquad (n \text{ even})$$
$$= 1 - \frac{(n-1)}{(2j+1)} \quad (n \text{ odd}). \tag{32}$$

For $j = \frac{7}{2}$ the chain of $\mathcal{S}$-values is then $\mathcal{S} = 1, 2, \frac{3}{4}, 4, \frac{1}{2}, 6, \frac{1}{4}, 8$.

b. Spin Measurements, Single-Particle Levels, Etc.

There are many ways in which reduced-width measurements can be used to determine spins. There are of course the few trivial cases where a spin is determined uniquely by the kinematics but more interesting are the very many cases where combining a measurement with some general

---

[s] This may be regarded as a special case of the Teichman and Wigner approximate sum rule (45) which states that the sum of all the reduced widths for decay of a given level should be of order $3\hbar^2 A/MR^2$ where A = atomic number. For the reactions usually observable one expects instead a rule with A replaced by unity.

[t] The formula holds for the states of lowest seniority, and the numerical values given for $j = \frac{7}{2}$ apply to this case [not all the $(f_{\frac{7}{2}})^n$ states with $J = 0$ are unique]. It should be observed too, that the criterion of Section 2a to distinguish between unfavored and favored reactions ($\mathcal{S} \ll 1$, $\mathcal{S} \sim 1$, respectively) will not always be accurate. Experimentally one finds the large value $\mathcal{S} \sim 5$ in the $B^{11}(d,n)C^{12}$ ground-state reaction (46) while one would expect $\mathcal{S} = 8$ in jj coupling. It is puzzling, however, that a large value is apparently not found in the $Al^{27}(d,n)Si^{28}$ ground-state reaction (47).

knowledge about possible configurations will produce an answer. For example, there seems no doubt whatever (*17,18*) that the g.s. spin of Ca^{45} is $\frac{7}{2}^-$ since the $Ca^{44}(d,p)$ reaction shows a strong $l = 3$ reaction and similarly the 0.61-Mev level in Ca^{43} (with a weak $l = 1$ reaction) is surely $\frac{3}{2}^-$, not $\frac{1}{2}^-$ since j^3 for identical particles contains no state with $J = \frac{1}{2}$. The 0.97-Mev level of $_{24}Cr_{29}^{53}$ which shows a full-strength $l = 3$ cross section (*48*) in $Cr^{52}(d,p)$ is surely $\frac{5}{2}^-$ since the $f_{\frac{7}{2}}$ neutron subshell is closed in Cr^{52}.

Besides cases like these, we can use the reactions which begin with the same state and by adding an inequivalent particle lead to a set of final states with single-particle widths ($\mathcal{S} = 1$) and with a cross section therefore proportional (when the barrier effects are removed) to $(2J + 1)$. These are the cases, discussed earlier, where the initial particles necessarily remain inert throughout the reaction, and their utility in determining spins was first pointed out by Enge (*49*). The $l = 2$ reactions in $P^{31}(d,p)$ [example (V)] supply one example; another is given by the $K^{39}(d,p)$ reactions (*50*) leading to the first four states of K^{40} which should be describable as the addition of an inequivalent $f_{\frac{7}{2}}$ particle to the inert $(d_{\frac{3}{2}})^{-1}$ state of K^{39}. Because of the irreducible inaccuracy in the entire stripping-width procedure (perhaps $\pm 15\%$ in favorable cases) this procedure often will not give a unique spin determination, but it can sometimes be combined with another technique. For example, $J = 2, 3$ in K^{39} should admit an $l = 1$ admixture[u] while $J = 4, 5$ do not.

Cases where an inequivalent particle is added to a group of particles which do not remain inert throughout the reaction can also be of practical value since unity is an upper limit for the relative reduced width [Eq. (22)]. A good example here is found in the $N^{15}(d,p)$ experiment of Warburton and McGruer (*51*). They found that the 0.40-Mev state has, for $l = 0$, $(2J + 1)\theta^2 \simeq 0.54$; since $\theta_0^2(s) \sim 0.2$, as determined by the $O^{16}(d,p)$ reaction (*5*), and N^{15} has spin $\frac{1}{2}$, they conclude that the 0.40-Mev level has spin 1.

The inequivalent-particle reactions will very often have large cross sections since we may have $\mathcal{S} \sim 1$ and a large $(2J + 1)$-value even for small l. For $l = 0, 1$, peak cross-sections in a stripping reaction may be several hundred times the compound-nucleus background. The number of strong reactions will in many cases be limited; for example, four strong $l = 3$ reactions are expected in the $K^{39}(d,p)$ reaction. But we must not necessarily be surprised to find more strong reactions than simple counting would indicate, since the number can be increased by an interaction between two nuclear states. Consider, for example, the reaction $(j^n)_{J_0} + j' \to (j^n_{J_0'} j')_J$ where j^n necessarily remains inert. It could happen that in the final nucleus a state such as $(j^n_{J_0'} j'')_J$, with $J_0' \neq J_0$ or $l'' \neq l'$, falls

[u] In this particular case none seems to be observed.

in "zero order" quite close to the j' state, where by "zero order" we imply that the interaction matrix element connecting the two states is ignored. If the zero-order spacing is smaller, say, than the interaction matrix element $<H>$ then, when it is properly included, one state is raised, the other lowered, by an amount $\sim <H>$ and the wave functions both become approximately equal mixtures of the j' and j'' states.[v] Both then will give strong l' reactions though only one is allowed with pure coupling. To be sure we would have $\mathcal{S} \simeq \frac{1}{2}$ for each reaction instead of $\mathcal{S} = 1$ but this might well be disguised by the statistical factors if the spins are unknown or by the uncertainties in θ_0^2 if they are known.

This sort of phenomenon can, in light nuclei, usually occur only for fairly high excited states since for low excitations the spacings between interacting levels (same π, J, T) will usually be much larger than the admixing matrix element.[w] This, of course, is an essential requirement for a simple coupling scheme to be valid and explains why we have talked so often about *small* admixtures. There are several cases with states of higher excitation where this proliferation of strong levels is apparently occurring, in some of which authors have wrongly assumed that a simple coupling scheme is ruled out even for the target nucleus as well. A great deal more data will be useful in exploring the way in which the simple shell model for intermediate or heavier nuclei gradually becomes inapplicable as the excitation energy increases.

c. The Effective Interaction in Nuclei

We have seen that an unfavored reduced width is liable to measure some small admixture in a wave function whose major component has some simple coupling scheme. At the same time such a small admixture will often be susceptible to a simple calculation. If, for example, a state belongs predominantly to $j_{J_0}{}^n$, the interaction between the active particles will, among others, induce small amplitudes of the sort $(j_{J_0}{}^{n-1}j')_J$ belonging to the configuration $j^{n-1}j'$. Since the amplitudes are small they may be calculated by perturbation theory and

$$K[j_{J_0}{}^{n-1}j']_J \simeq \frac{<(j_{J_0}{}^{n-1}j')_J H(j_{J}{}^n)>}{E_J(j^n) - E_J(j_{J_0}{}^{n-1}j')}, \tag{33}$$

where the energies belong to the states whose dominant components are as indicated.

The unfavored reaction in $j_{J_0}{}^{n-1} + j'$ which ends in the state predominantly j^n will now measure the amplitude K to within a sign. The favored transition will locate for us the other state. In this way we meas-

---

[v] The reader might consider the 2×2 matrix problem.

[w] A characteristic shell model value for which is 0.1–0.3 Mev near $A = 40$.

ure an approximate absolute value for a single matrix element of the interaction H. A case for which this works very well is with the $\frac{3}{2}^-$ states in $Ca^{42}(d,p)$ [see French and Raz (18) and example (VI)]. l-unfavored amplitudes have been examined in a similar way by Okai and Sano (52) while several other authors have similarly examined other experimental quantities.

d. Reduced Widths on the Collective Model

There is essentially no data for heavier nuclei but the subject promises to become important because of the discovery [Litherland et al. (53)] that well-developed collective effects exist in quite light nuclei, in particular in the $A = 25$ region.

The first theoretical discussion is due to Yoshida (54), who considers reduced widths in both the weak- and strong-coupling approximations. The general principles of parentage and angular-momentum coupling which we discussed above also apply here, but since a whole new mathematical topic is required to discuss properly the strong coupling model (namely transformation of wave functions between fixed and rotating coordinate systems) we shall content ourselves with only a few comments about the rather uninteresting weak-coupling case. Here one imagines (55) that an interaction between the core and the outside particles will excite a spin-2 phonon of core oscillation at the same time perhaps changing the angular momentum of a single particle. For example, a state which without the collective motion is simply $(j^n)_J$ might now be written

$$\left(\begin{array}{c}j^n\\J\end{array}\right) + \sum_{J_1} A_{J_1} \left(\begin{array}{c}j^n\\J_1\\2\end{array}\right)_J + \sum_{J_1 J_2 j'} B_{J_1 J_2}(j') \left(\begin{array}{c}j^{n-1}\;j'\\J_2\\J_1\\2\end{array}\right)_J \qquad (34)$$

Here A_{J_1} and $B_{J_1 J_2}(j')$ are amplitudes for whose calculation we should resort to the interaction Hamiltonian [in exactly the same spirit as our example (XIII)]. The j' sum would be taken over as many of the unoccupied single-particle levels as would be appreciably excited (we could in fact invoke single-particle excitations of the core itself in which case things could become quite complicated).

The c.f.p. for separating out $j''(n)$ ($j'' = j$ or j') from a second or third term of Eq. (34) and ending with a one-phonon state of total angular momentum J_0 and particle angular momentum J_p is now seen to be simply the product of the corresponding particle c.f.p. and a recoupling factor $U(2J_p Jj'' : J_0 J_1)$, and thus we can calculate any reduced width connecting two states of the type in Eq. (34). It is obvious that, to first order in the particle-surface coupling, a reduced width connecting two states whose major configurations include no phonons is unchanged by this interaction since there are no cross terms. There is a second-order effect since the wave function of Eq. (34) is not normalized. If this effect is large, one should consider proceeding directly to the strong-coupling limit or else admit the excitation of more phonons than one. The calculation of the amplitudes might then become very complex but we could still calculate the reduced width by the above procedures. As a final remark it is probably fair to say that weak-coupling reduced-width considerations are of little interest for light and intermediate nuclei (say $A < 70$). They could be of interest for heavier nuclei if it should turn out that there really are some such nuclei where the phonon picture is a good one. For example a one-phonon state (in which the first term of (34) would not enter) would have a vanishing reduced width to a zero-phonon ground state. Concerning the strong-coupling extreme we remark that reduced-width calculations in this case are of considerable interest even for some quite light nuclei (56).

e. Reduced Widths and Level Shifts

Related to the freedom in defining the formal resonance energy (Section 1) there is a real effect pointed out first and discussed in detail by Thomas (57), which is directly observable if we compare level positions or reduced widths for corresponding levels in mirror nuclei. For the important case where only a single channel makes a significant contribution we have[x] for the "true" width and position

$$1/\theta^2_{\text{True}} \simeq 1/\theta^2_{\text{Obs}} - (3\hbar^2/2MR^2)(d\Delta_l/dE)$$
$$E_{\text{True}} \simeq E_{\text{Obs}} + (l + \Delta_l)(3\hbar^2/2MR^2)\theta^2_{\text{Obs}} \quad (35)$$

where, for either bound or unbound states, Δ_l is as defined in Section 1. WKB approximations for Δ_l and $d\Delta_l/dE$ have been given by Thomas (57) and by Lane (12).

The reduced-width correction may be quite large and, as stressed by Lane, should in these cases be made in order to eliminate a large error in the relative reduced width §. The level-shift correction has been fairly well verified in several cases, the most detailed example still being the original one for $A = 13$ [Thomas (57) and Ehrman (58)]. An unusual application of this has been made by Marion and Fowler[y] (59) who, in

---

[x] These follow very simply from Blatt and Weisskopf (1, p. 405), if one redefines the formal resonance energy as we have done in Section 1 and takes account of the first-order energy variation of Δ_l.

[y] The WKB level-shift formulae are given in convenient form in this paper. In their Eq. (8) a term $-2l\xi^2$ should be added in the bracket. The author is indebted to Dr. Marion for correspondence about level shifts.

examining neon reactions in the stars need to know the effect which the slightly bound 2.43-Mev level in Na^{21} may have on the $Ne^{20}(p,\gamma)$ reaction; this involves the proton width. This is estimated by assuming that the true proton reduced width is identical with the true neutron width for the mirror level, identified at 2.80 Mev in Ne^{21}. The neutron width is also unknown but ascribing the difference in the excitations to the Thomas shift (the ground-state shifts being negligible) they can use Eq. (35) to give a value.

Apart from quantitative applications of Eq. (35), which always suffer from uncertainties in radius, penetration factors, effects of other channels, etc., there are many cases where inspection of the level schemes for mirror nuclei show displacements in the direction given by Eq. (35) [both $(l + \Delta_l)$ and the energy derivative are positive functions]. Note finally that the form of the reduced-width correction is entirely unknown for the stripping widths because of the presumably very different boundary conditions, and Eq. (35) should not be applied to them.

f. CONCLUSION

For nuclear reactions in which the reaction mechanism is well understood (simple resonant or stripping reactions) the pertinent information about nuclear structure is contained in reduced widths. To show how one may extract this information we have, in the main, restricted ourselves to reactions involving the transfer of a single nucleon in cases where the nuclear shell model supplies an appropriate representation for the nuclear states involved. We have briefly mentioned analysis using the collective model.

But also one might ask about reactions where more than a single nucleon is transferred such as (He^3,n), (p,α), etc. Once again, if the reaction is recognizable as a simple resonant reaction, reduced widths can be extracted (though the uncertainties in penetration factors might be appreciably larger). The shell model analysis of such a width could proceed as before, one point of difference being that in many cases the excitation energies may be so high that we have as yet little knowledge about the nuclear states involved[z]; another difference is that more complex c.f.p. would be required. And, of, course the reaction itself might suggest the use of a different nuclear model; a transfer of an α-particle would be very simple to treat with an α-particle model.

If the reaction is identifiable as proceeding by a direct-interaction

---

[z] But, of course, reduced-width data would increase this knowledge. Note that if A is not too large (perhaps $A < 50$) there is usually an excitation region where the levels are still separated but where we are above the states belonging to the lowest configurations. These comments also apply, of course, to the simplest resonance reactions.

process the one difficulty with extracting a reduced width is that there is not available a reliable quantitative theoretical treatment. But if enough data become available it might well be possible to handle this in the same way as in the special case of the (d,p) reaction; namely, one might make comparisons between cross sections for quite similar reactions or more generally supplement the available crude formulae with determinations of the reduced widths for cases where the relative reduced widths $\mathcal{S}$ are calculable. The direct-interaction inelastic scattering (where angular momentum is transferred but no particles) might also be fitted into the same scheme if the projectile energies are sufficiently high. As yet there are not nearly enough data to make practicable this sort of analysis; the reader is referred, however, to recent papers by Bromley *et al.* (*60*), Butler (*61*), and Sherr (*62*) and to the contributions in the present book by Banerjee and Levinson (Section V.B).

We have mentioned earlier that reduced widths should eventually be of value in studying the strong admixing of configurations which should occur as the excitation energy increases. One should discuss here not so much specific reduced widths but rather the average of the reduced widths occurring in a certain band of excitation. In this way one would make contact with the remarkably successful optical model of Feshbach *et al.* (*63*); we are unable here to discuss this topic but refer the reader to papers of Lane *et al.* (*64*) and Vogt and Lascoux (*65*).

Appendix: Stripping Reaction Reduced Widths

In this appendix we give formulae for extracting, from a stripping or pickup reaction cross section, the reduced width for the nuclear transition. We are interested here not so much in deriving the results as in indicating in a schematic way from where the various factors come and in giving the results in a usable form. It seems worthwhile to do this because confusion has often arisen concerning the numerical constants appearing for even the simple (d,p) case and because in the future it will probably be worthwhile extracting widths from more complicated processes [(He^3,α) for example]. We do not treat the so-called heavy-particle stripping and ignore too the related exchange effects which would arise if we properly antisymmetrize between all nucleons. We use the simple Born approximation formulation of the theory.

In the following we are concerned then with properly handling the trivia (numerical constants and so on), but not at all with the more fundamental matters. Our approach is a purely empirical one. Extracting a Born approximation reduced width may be regarded simply as a method of reducing the data to useful form. We are not compelled to discuss the significance of the absolute stripping width and we shall not do so (beyond asserting that there is no physical reason why a stripping width should equal the resonant-reaction width connecting the same levels). We do assert, however, that it has been quite thoroughly demonstrated for the (d,p) and (d,n) cases (the situation is not settled for the more complicated ones) that the Born approximation stripping width is a very useful and significant quantity. The experimentalist who uses past theoretical discussion concerning the lack of meaning of such widths as an excuse to avoid measuring absolute cross sections is losing much of the value of his experiment.

We consider the process

$$A[J_0 T_0 M_0 M_{T_0}] + a[i_0 t_0 m_0 m_{t_0}] \rightleftarrows (A+1)[J_1 T_1 M_1 M_{T_1}]$$
$$+ (a-1)[i_1 t_1 m_1 m_{t_1}]$$

where a is the lighter of the two initial nuclei and the spins and isotopic spins are as indicated. We first define a number of quantities; we take $\hbar = 1$ and, where necessary, we measure energies in million electron volts and lengths in fermis (1 fermi = 10^{-13} cm).

$\dfrac{\overrightarrow{d\sigma}}{d\omega}, \dfrac{\overleftarrow{d\sigma}}{d\omega}$ = c.m. differential cross section for $A + a \rightarrow (A+1) + (a-1)$ and its inverse.

E_0 = kinetic energy of a in the reference frame with A at rest
E_1 = kinetic energy of $(a-1)$ in the reference frame with $(A+1)$ at rest
Q = c.m. Q-value for $A + a \rightarrow (A+1) + (a-1)$.
  $Q = \epsilon_{A+1} + \epsilon_{a-1} - \epsilon_A - \epsilon_a$
$\epsilon_A, \epsilon_{A+1}, \epsilon_a, \epsilon_{a-1} \equiv$ binding energies
  (For a deuteron, triton, He^3, α-particle, ϵ = 2.22, 8.48, 7.72, 28.3 Mev, respectively.)
M = nucleon mass
$\mathbf{k}_0$ = c.m. wave number for relative motion in the (A, a) system.
  $k_0^2 = 2aE_0 M A^2/(A+a)^2$
$\mathbf{k}_1$ = c.m. wave number for relative motion in the $(A+1, a-1)$ system.
  $k_1^2 = 2(a-1)E_1 M (A+1)^2/(A+a)^2$
r_0 = Butler radius
$\mathbf{q} = \mathbf{k}_0 - \dfrac{A}{A+1}\mathbf{k}_1$
$\boldsymbol{\kappa} = \dfrac{(a-1)}{a}\mathbf{k}_0 - \mathbf{k}_1$

iS = "wave number" (pure imaginary in all cases of present interest) corresponding to the binding energy in a of the transferred nucleon

$\dfrac{aS^2}{2(a-1)M} = \epsilon_a - \epsilon_{a-1} \cdot S^{-1}$ = 4.32 fermi [for (d,p) or (d,n)]; = 2.23 fermi [for (t,d)]; = 2.38 f [for (He^3,d)]; = 1.18 f [for α,t)]; = 1.16 f [for (α,He^3)]. For the (d,p) and (d,n) cases S is usually called α.

it = "wave number" corresponding to the binding energy in $(A+1)$ of the transferred nucleon.

$$\frac{(A+1)t^2}{2AM} = Q + \epsilon_a - \epsilon_{a-1} = Q + (2.22, \ 6.26, \ 5.50, \ 19.8, \ 20.6 \ \text{Mev})$$

respectively for the five cases listed above.

We have $\dfrac{a}{a-1}\kappa^2 - \dfrac{A+1}{A}q^2 = 2MQ$ and

$$\frac{a}{a-1}(\kappa^2 + S^2) = \frac{A+1}{A}(q^2 + t^2)$$

θ = c.m. scattering angle. $k_0 k_1 \cos\theta = \mathbf{k}_0 \cdot \mathbf{k}_1$

$x = qr_0.$ $\quad x^2 = \dfrac{2MA^2}{(A+a)^2}\{aE_0 + (a-1)E_1 - 2\sqrt{a(a-1)E_0 E_1}\cos\theta\}r_0^2$

$y = tr_0$

$(C^{T_0 \frac{1}{2} T_1}{}_{M_{T_0}, M_{T_1}-M_{T_0}})^2 \equiv (C)^2$ = isotopic spin coupling factor

$$= \frac{2T_1 \pm 4M_{T_0}(T-T_0) + 1}{2(2T_0+1)} \delta_{T,T_0 \pm \frac{1}{2}} \text{ if } M_{T_1} - M_{T_0} = \pm \frac{1}{2}$$

$W_l(x,y) = x\dfrac{d}{dx}j_l(x) - \dfrac{yj_l(x)}{h_l{}^{(1)}(iy)}\dfrac{d}{dy}h_l{}^{(1)}(iy)$

The cross section will involve the following factors:
(a) Density-of-states factor.
(b) Statistical factors (including isotopic-spin z-component factor)
(c) $a \rightarrow (a-1)$ factor.
(d) $A \rightarrow (A+1)$ factor.

The reaction is described as follows: a particle (say #i) whose coordinate is ϱ with respect to the center of mass of $(a-1)$ and λ with respect to the center of mass[a] of A encounters an interaction $J(\lambda)$ and is transferred from a to $(A+1)$. Since the final state reduced mass is $(A+1)(a-1)M/(A+a)$ we have for the unpolarized cross section

$$\frac{d\bar{\sigma}_i}{d\omega} = \frac{M^2}{4\pi^2}\frac{k_1}{k_0}\frac{a(a-1)A(A+1)}{(A+a)^2}\frac{1}{(2i_0+1)(2J_0+1)}\sum_{\substack{m_0, M_0 \\ m_1, M_1}}|<H>|^2 \quad (A1)$$

where the matrix element $<H>$, with the plane waves written in terms of ϱ, λ, is

$$\int \Psi^*(A+1)\Psi^*(a-1)e^{i(\mathbf{q}\cdot\boldsymbol{\lambda}-\boldsymbol{\kappa}\cdot\boldsymbol{\varrho})}J(\lambda)\Psi(A)\Psi(a)d\tau_A d\tau_{a-1}d\tau_i d\varrho d\lambda$$

where $d\tau$ represents an integration over internal coordinates.

We evaluate the integral by making a fractional-parentage expansion of $\Psi(a)$ and $\Psi(A+1)$. The precise procedure here depends on the form we give to the wave functions. If we are interested in $a \leq 4$ (as from now on, we assume we are) we can use the fact that the wave function for

---

[a] It is an approximation that the interaction should be centered at the center of mass of $(a-1)$.

these nuclei is, to a satisfactory approximation, an S-state, space symmetric in all nucleons, and is thus a simple product of a space and spin[b] function, the latter carrying all the angular momentum. Introducing now the spin c.f.p. $<a, i_0, t_0|a - 1, i_1, t_1>$ we have

$$\int \Psi^*(a - 1)\Psi(a)d\tau_{a-1} = <a, i_0, t_0|a - 1, i_1, t_1>$$
$$C^{i_1\frac{1}{2}i_0}{}_{m_1,m_0-m_1} C^{t_1\frac{1}{2}t_0}{}_{m_{t1},m_{t0}-m_{t1}} u(i)^{m_0-m_1} v(i)^{m_{t0}-m_{t1}} \phi(\rho) \quad (A2)$$

where u, v are spin and isotopic-spin single-particle functions and $\phi(\rho)$ (the unnormalized single-particle radial function) is defined in terms of the normalized space functions for $a, a - 1$ by

$$\phi(\rho) = \int \phi_{a-1}\phi_a d\tau_{a-1}.$$

If the nucleus A is also light we would use the same procedure but, say, if $A > 4$ our point of view would be slightly different and in terms of the more general c.f.p. we would write, assuming that only a single l-value is effective,[c]

$$\int \Psi^*(A + 1)\Psi(A)d\tau_A = <A + 1, J_1, T_1|A, J_0, T_0>{}_{lj}C^{J_0jJ_1}{}_{M_0,M_1-M_0}$$
$$\cdot C^{T_0\frac{1}{2}T_1}{}_{M_{T0},M_{T1}-M_{T0}} X_{l,j}{}^{M_1-M_0*}(i) v(i)^{M_{T1}-M_{T0}*} \quad (A3)$$

where $X_{l,j}$ is a jj-coupling single-particle wave function which may be decomposed into an orbital part and a spin part by a Clebsch-Gordan transformation.

$$X_{l,j}{}^{M_1-M_0} = C^{l\frac{1}{2}j}{}_{m,M_1-M_0-m} \phi_l{}^m u(i)^{M_1-M_0-m}. \quad (A4)$$

For the radial part of the space function $\phi_l{}^m$ we could substitute an appropriate function such as that of a square well or harmonic oscillator but it is better to leave it arbitrary. The space coordinate occurring in $\phi_l{}^m$ is of course λ.

We can now combine these two internal-coordinate integrals and integrate over the spin and isotopic spin of the transferred particle. The latter integration simply gives unity since T_z is unchanged during the reaction. For the angular momenta we can simplify things by taking the quantization axis along $\mathbf{q}$ in which case only $m = 0$ can contribute and $m_0 - m_1 = M_1 - M_0$. The reader will easily find on squaring the matrix element (which involves a sum over j) that the m-sums for the six angular-momentum Clebsch-Gordan coefficients (three with j, three with j') is simply $\delta_{jj'}(2J_1 + 1)(2i_0 + 1)/2(2l + 1)$. Besides this we find that in each case for $a = 2, 3, 4$ we have $<a|a - 1>^2(C^{t_1\frac{1}{2}t_0})^2 = \frac{1}{2}$. Finally we can eliminate $J(\lambda)$ from the integral by using the one-body Schrodinger

---

[b] We treat a in an isotopic spin formalism though this of course is unnecessary.

[c] Interference between l-values would disappear in the later Clebsch-Gordan summation.

equation describing the interaction of the transferred nucleon with the nucleus A. This tells us that

$$J(\lambda) \equiv \frac{1}{2\mu}(\nabla_\lambda^2 - t^2) \equiv -\frac{1}{2\mu}(q^2 + t^2) \tag{A5}$$

where the reduced mass $\mu = AM/(A + 1)$.

We have agreed to ignore the antisymmetrization between groups insofar as they produce a different sort of reaction (characterized really by different momentum transfers) but we must take account of the corresponding statistical factors. There are $N_0 = [(A + a)!/A!a!]$ ways of dividing $A + a$ nucleons into two groups A, a and thus the antisymmetrized wave function has N_0 orthogonal terms each with a multiplying factor $N_0^{-\frac{1}{2}}$. Similarly, there are N_1 terms in the final state each with a factor $N_1^{-\frac{1}{2}}$ and we have $N_1/N_0 = a/(A + 1)$. Each term in the initial function connects with a terms of the final function and thus (the phases cooperate properly) the antisymmetrized matrix element differs from that calculated above by $N_0^{-\frac{1}{2}} \cdot N_1^{-\frac{1}{2}} \cdot N_0 \cdot a = [(A + 1)a]^{\frac{1}{2}}$ and on squaring we have a factor $(A + 1)a$. We can now identify

$$(A + 1)\sum_j <A + 1, J_1, T_1|A, J_0, T_0>_{l,j}^2 = \mathcal{S}(l),$$

the relative reduced width for the transition [see Eq. 11)], and we have

$$\frac{d\vec{\sigma}}{d\omega} = \frac{1}{64\pi^2} \frac{a^2(a-1)(A+1)^3}{A(A+a)^2} \frac{(2J_1+1)}{(2l+1)(2J_0+1)} (C)^2 \frac{k_1}{k_0} \mathcal{S}(l)$$
$$\cdot (q^2 + t^2)^2 \left[\int \phi(\rho)e^{-i\kappa\cdot\varrho}d\varrho\right]^2 \cdot [X_l^0(\lambda)e^{i q \cdot \lambda}d\lambda]^2. \tag{A6}$$

Except for our restriction of a to the cases $a = 2, 3, 4$ (which the reader may easily remove) the result is symmetrical between the two transitions $(A + 1) \to A$ and $a \to (a - 1)$. We have several ways now of approximately evaluating the integrals appearing in the formula.

METHOD 1

Insert explicit wave functions and integrate over the entire region. We will write the result of this operation applied to the a-integral as follows:

$$\int \phi(\rho)e^{-i\kappa\cdot\varrho}d\varrho = P_a I_a(\kappa) \quad \text{where} \quad I_a(0) = 1. \tag{A7}$$

For example, with $a = 2$ the usual Hulthén function is

$$\sqrt{\frac{7S}{9\pi}}\frac{e^{-S\rho} - e^{-7S\rho}}{\rho}$$

where $S^{-1}(\equiv \alpha^{-1}) = 4.32$ fermi. Then we find

$$P_2 = [4096\pi/343S^3]^{\frac{1}{2}}; \quad I_2(\kappa) = [1 + (\kappa/S)^2]^{-1}[1 + (\kappa/7S)^2]^{-1}. \quad (A8)$$

For $a = 3$ we could use Irving's wave function (66)

$$\left[\frac{\sqrt{3}\,\gamma^5}{2\pi^3}\right]^{\frac{1}{2}} \frac{e^{-\gamma[w^2+u^2]^{\frac{1}{2}}}}{[w^2 + u^2]^{\frac{1}{4}}}$$

where $\mathbf{w} = (\mathbf{r}_1 - \mathbf{r}_2)3^{\frac{1}{2}}/2$, $\mathbf{u} = (\mathbf{r}_1 + \mathbf{r}_2)/2 - \mathbf{r}_3$, $d\tau_a = 4d\mathbf{w}d\mathbf{u}/3$,

$$\gamma^{-1} = 0.93 \text{ fermi},$$

and then using the Hulthén deuteron function we have

$$P_{3'} = 25.5\alpha^{\frac{1}{2}}/\gamma^2.$$

Values of I_3 for $\kappa = 0, 0.1, 0.2, \ldots 1.0$ fermi are 1.0, 0.98, 0.92, 0.84, 0.74, 0.63, 0.53, 0.44, 0.36, 0.29, 0.23. For $\kappa = 1.2, 1.4, 1.6, 1.8, 2.0$ fermi, $I_3 = 0.15, 0.10, 0.06, 0.04, 0.025$.

An approximation often used for the a-integral is to insert the asymptotic form of $\phi(\rho)$ and then to integrate over the entire space; that is,

$$\phi(\rho) = N_a \frac{e^{-S\rho}}{\rho} \quad (A9)$$

and then

$$P_a = 4\pi N_a/S^2, \quad I_a(\kappa) = [1 + (\kappa/S)^2]^{-1} \equiv I_a{}^0(\kappa). \quad (A10)$$

The function $I_a(\kappa)$ is now determined uniquely by the binding energies. The constant P_a is left to be determined by experiment. It would be hoped that values of P_3 and P_4 could be found so that the nuclear reduced widths found with $a = 3$ or 4 reactions would agree with those found with the usual (d,p) or (d,n) reactions. There is some evidence that this is indeed so but the question is not settled.[d]

It is probable that the asymptotic approximation $(I_a \to I_a{}^0)$ is more reasonable for $a = 3, 4$ than the explicit insertion of a wave function such as Irving's which is designed to be a reasonable approximation for small separation of the components and which does not have the proper asymptotic form (for $\rho \to \infty$) and consequently behaves wrongly for small κ. For $a = 2$ the only difference in the κ dependence between the asymptotic integral and the Hulthén integral is in the extra $[1 + (\kappa/7\alpha)^2]^{-1}$ factor occurring in the latter; this factor is, in practical cases, so close to unity that this difference is inconsequential.

[d] The evidence has been examined by A. I. Hamburger (private communication).

Method 2

Integrate over the outer region only ($\geq r_0$). Using this for the A-integral gives

$$(q^2 + t^2)\int \phi_l^0(\lambda)e^{i\mathbf{q}\cdot\boldsymbol{\lambda}}d\lambda = [4\pi(2l+1)]^{\frac{1}{2}}r_0 R_l(r_0) W_l(x,y)$$
$$= [12\pi(2l+1)]^{\frac{1}{2}}r_0^{-\frac{1}{2}}\theta_0 W_l(x,y). \qquad (A11)$$

Here $R_l(r_0)$ is the radial function occurring in $\phi_l^0(\lambda)$ and r_0 the Butler radius, and we have used Eq. (2) to introduce the dimensionless reduced width.

This is the standard procedure for dealing with $A > 4$. The integration over the exterior only has, in general, good physical justification though there are some cases where this may not be so.

Method 3

Evaluate by direct measurement. Here we have a possibility of measuring say the $a = 3$ integral by considering the experiment $d + d \to t + p$. We would assume that the $a = 2$ integral is known and, directly from experiment, would find the absolute value of the $a = 3$ integral as a function of κ. It would be of considerable interest to see whether the integral determined in this manner, when combined with the reduced width for $(A + 1) \to A$ as measured by a (d,p) experiment, would properly describe the $(A + 1) + d \to A + t$ experiment.[e] It is not really clear that it should, because the whole Born approximation procedure has very many uncertainties and crudities. For the same reason we should argue very strongly against the view that the a-integral tells us very much which can at present be interpreted concerning the properties of the a-nucleus.

We now collect together the results and write, evaluating the a-integral by the first method and the A-integral by the second,

$$\frac{\overrightarrow{d\sigma}}{d\omega} = \frac{3}{16\pi}\frac{a^2(a-1)(A+1)^3}{A(A+a)^2}\frac{(2J_1+1)}{(2J_0+1)}(C)^2 P_a^2 \frac{k_1}{k_0}$$
$$[I_a(\kappa)W_l(x,y)]^2 \frac{\theta^2}{r_0} \qquad (A12)$$

and

$$\frac{\overleftarrow{d\sigma}}{d\omega} = \frac{(2J_0+1)(2i_0+1)}{(2J_1+1)(2i_1+1)}\left(\frac{k_0}{k_1}\right)^2 \frac{\overrightarrow{d\sigma}}{d\omega}. \qquad (A13)$$

We can greatly simplify application of these formulae by making use of the tabulation given by Lubitz (67) who gives numerical values, for the

---

[e] This has been suggested by A. I. Hamburger (private communication), and is being examined by her.

cases where the transferred nucleon is either bound or unbound in the $A+1$ nucleus (it = pure imaginary or real), of a dimensionless quantity $\sigma'_{TAB}(x,y)$ which is related to the above quantities by

$$[W_l(x,y)]^2 = (x^2 + y^2)^2[1 + 0.008(x^2 + y^2)]^2 \sigma'_{TAB}(x,y). \quad (A14)$$

In terms of this[f] we have for the cross section in millibarns/steradian

$$\frac{\overrightarrow{d\sigma}}{d\omega} = 0.622 \frac{a^4}{(a-1)} \frac{(A+1)^2}{(A+a)^2} \left[\frac{(a-1)E_1}{aE_0}\right]^{\frac{1}{2}}$$
$$\frac{(2J_1+1)}{(2J_0+1)} (C)^2 P_a{}^2 r_0{}^3 S^4 \theta^2 D_a \sigma'_{TAB}(x,y) \quad (A15)$$

where, in the right-hand side, lengths are given in fermis. In this formula we have

$$D_a{}^{\frac{1}{2}} = \frac{48 I_a}{49 I_a{}^0}[1 + 0.008(x^2 + y^2)]. \quad (A16)$$

D_a is quite close to unity if we take the asymptotic form for I_a ($I_a \to I_a{}^0$) and is precisely unity for $a = 2$ if we use the Hulthén deuteron and $A \gg 1$, $r_0 = 5$. For almost all purposes in fact it will be satisfactory to take $D_a = 1$. In the most important (d,p) and (d,n) cases we have more explicitly

$$\frac{\overrightarrow{d\sigma}}{d\omega} = 61.2 \frac{(A+1)^2}{(A+a)^2} \left(\frac{E_1}{E_0}\right)^{\frac{1}{2}} \frac{(2J_1+1)}{(2J_0+1)} (C)^2 \theta^2 D\sigma'_{TAB}(x,y) r_0{}^3$$
$$\frac{\overleftarrow{d\sigma}}{d\omega} = 183.5 \frac{A^2}{(A+a)^2} \left(\frac{E_0}{E_1}\right)^{\frac{1}{2}} (C)^2 \theta^2 D\sigma'_{TAB}(x,y) r_0{}^3. \quad (A17)$$

The practical procedure for using these formulae is as follows: (Energies are in Mev.)

(a) Either E_0 or E_1 is given. Calculate the other by

$$(A+1)E_1 = (A+a)Q + AE_0.$$

(b) Choose values of l, r_0.
(c) Calculate $y = 0.22 r_0 [A/(A+1)]^{\frac{1}{2}} \{Q + \epsilon_a - \epsilon_{a-1}\}^{\frac{1}{2}}$.
(d) For each c.m. angle calculate x by

$$x = 0.22 \frac{Ar_0}{(A+a)} \{aE_0 + (a-1)E_1 - 2\sqrt{a(a-1)E_0 E_1} \cos\theta\}^{\frac{1}{2}}.$$

(e) Find $\sigma_l{}^{TAB}$ in Lubitz' tables.

As a general rule one should attempt by varying r_0 (or l) to fit best the cross section near the peak. The subtraction from the observed cross

[f] Note that $I_a{}^0(\kappa) = aAS^2 r_0{}^2 / \{(a-1)(A+1)(x^2+y^2)\}$.

section of an isotopic background representing the "compound nucleus" contribution does not seem to be a good idea. Finally in heavier nuclei one may analyze results without using the isotopic spin formalism; in such cases one simply places $(C)^2 = 1$ above but must then be careful in specifying that the reduced width has been determined in this manner.

REFERENCES

1. J. M. Blatt and V. F. Weisskopf, *Theoretical Nuclear Physics* (John Wiley and Sons, New York, 1952).
2. A. M. Lane, Atomic Energy Research Establishment Report AERE T/R 1289 (1954).
3. S. T. Butler, Proc. Roy. Soc. **A208**, 559 (1951); A. B. Bhatia, K. Huang, R. Huby, and H. C. Newns, Phil. Mag. [7] **43**, 485 (1952); P. B. Daitch and J. B. French, Phys. Rev. **87**, 900 (1952); F. L. Friedman and W. Tobocman, *ibid.* **92**, 93 (1953); Y. Fujimoto, K. Kikuchi, and S. Yoshida, Progr. Theoret. Phys. (Kyoto) **11**, 264 (1954).
4. J. Horowitz and A. M. L. Messiah, Phys. Rev. **92**, 1326 (1953); W. Tobocman and M. H. Kalos, *ibid.* **97**, 132 (1955).
5. E. C. Halbert and J. B. French, Phys. Rev. **105**, 1563 (1957).
6. F. Ajzenberg-Selove and T. Lauritsen, Nuclear Phys. **11**, 1 (1959).
7. P. M. Endt and C. M. Braams, Revs. Modern Phys. **29**, 683 (1957).
8. K. Way, R. W. King, C. L. McGinnis, and R. van Lieshout, *Nuclear Level Schemes*, $A = 40 - A = 92$ (U.S. Government Printing Office, Washington D.C., 1955).
9. H. B. Willard, J. K. Bair, J. D. Kington, and H. O. Cohn, Phys. Rev. **101**, 765 (1956).
10. C. K. Bockelman, D. W. Miller, R. K. Adair, and H. H. Barschall, Phys. Rev. **84**, 69 (1951).
11. H. L. Jackson and A. I. Galonsky, Phys. Rev. **89**, 370 (1953).
12. A. M. Lane, unpublished (1955).
13. L. L. Green, J. J. Singh, and J. C. Willmott, Proc. Phys. Soc. **A69**, 335 (1956).
14. H. A. Bethe and S. T. Butler, Phys. Rev. **85**, 1045 (1952).
15. W. C. Parkinson, Phys. Rev. **110**, 489 (1958).
16. J. R. Holt and T. N. Marsham, Proc. Phys. Soc. **A66**, 565 (1953).
17. C. K. Bockelman, C. M. Braams, C. P. Browne, W. W. Buechner, R. R. Sharp, and A. Sperduto, Phys. Rev. **107**, 176 (1957).
18. J. B. French and B. J. Raz, Phys. Rev. **104**, 1411 (1956).
19. H. B. Burrows, T. S. Green, S. Hinds, and R. Middleton, Proc. Phys. Soc. **A69**, 310 (1956).
20. K. G. Standing, Phys. Rev. **101**, 152 (1956).
21. W. E. Moore, J. N. McGruer, and A. I. Hamburger, Phys. Rev. Letters **1**, 29 (1958); E. Baranger and S. Meshkov, *ibid.* p. 30.
22. D. R. Inglis, Revs. Modern Phys. **25**, 390 (1953).
23. R. A. Ferrell and W. M. Visscher, Phys. Rev. **102**, 450 (1956).
24. C. Schwartz, Phys. Rev. **94**, 95 (1954).
25. S. Meshkov and C. W. Ufford, Phys. Rev. **101**, 734 (1956); J. B. Marion, Nuclear Phys. **4**, 282 (1957).
26. U. Fano, National Bureau of Standards Report #1214 (1951).
27. E. C. Halbert, Thesis, University of Rochester (1956).
28. G. Racah, Phys. Rev. **62**, 438 (1942).

29. L. C. Biedenharn, J. M. Blatt, and M. E. Rose, Revs. Modern Phys. **24**, 249 (1952).
30. A. R. Edmonds, *Angular Momentum in Quantum Mechanics* (Princeton University Press, Princeton, 1957); M. E. Rose, *Elementary Theory of Angular Momentum* (John Wiley and Sons, New York, 1957).
31. E. U. Condon and G. H. Shortley, *The Theory of Atomic Spectra* (Cambridge University Press, London and New York, 1935).
32. R. F. Bacher and S. Goudsmit, Phys. Rev. **46**, 948 (1934).
33. G. Racah, Phys. Rev. **63**, 367 (1943); **76**, 1352 (1949).
34. H. A. Jahn and H. van Wieringen, Proc. Roy. Soc. **A209**, 502 (1951); J. P. Elliott, J. Hope, and H. A. Jahn, Phil. Trans. Roy. Soc. **A246**, 241 (1953).
35. A. R. Edmonds and B. H. Flowers, Proc. Roy. Soc. **A215**, 120 (1952).
36. A. M. Lane, Proc. Phys. Soc. **A66**, 977 (1953); T. Auerbach, Thesis, University of Rochester (1954); G. R. Satchler, Proc. Phys. Soc. **A67**, 471 (1954).
37. A. M. Lane and D. H. Wilkinson, Phys. Rev. **97**, 1199 (1955).
38. D. A. Liberman and R. F. Christy, Phys. Rev. **92**, 1085A (1953); D. A. Liberman, Thesis, California Institute of Technology (1955).
39. H. Neuert, private communication (1957).
40. S. H. Levine, R. S. Bender, and J. N. McGruer, Phys. Rev. **97**, 1249 (1955).
41. R. F. Christy, Phys. Rev. **89**, 839 (1953).
42. L. Rosenfeld, *Nuclear Forces* (North-Holland Publishing Company, Amsterdam, 1948).
43. J. R. Holt and T. N. Marsham, Proc. Phys. Soc. **A66**, 467 (1953).
44. D. H. Wilkinson, Phil. Mag. [8] **1**, 127 (1956).
45. T. Teichman and E. P. Wigner, Phys. Rev. **87**, 123 (1952).
46. E. E. Maslin, J. M. Calvert, and A. A. Jaffe, Proc. Phys. Soc. **A69**, 754 (1956).
47. J. M. Calvert, A. A. Jaffe, A. E. Litherland, and E. E. Maslin, Proc. Phys. Soc. **A68**, 1008 (1955).
48. A. J. Elwyn and F. B. Shull, Phys. Rev. **111**, 925 (1958).
49. H. A. Enge, Univ. i Bergen Arbok, Naturvitenskap Rekke (1953).
50. H. A. Enge and D. H. Weaner, Bull. Am. Phys. Soc. [2] **2**, 179 (1957).
51. E. K. Warburton and J. N. McGruer, Phys. Rev. **105**, 639 (1957).
52. S. Okai and M. Sano, Progr. Theoret. Phys. (Kyoto) **15**, 203 (1956).
53. A. E. Litherland, E. B. Paul, G. A. Bartholomew, and H. E. Gove, Phys. Rev. **102**, 208 (1956).
54. S. Yoshida, Progr. Theoret. Phys. (Kyoto) **12**, 141 (1954).
55. A. Bohr, Kgl. Danske Videnskab. Selskab, Mat.-fys. Medd. **26**, No. 14 (1952).
56. G. R. Satchler, Ann. Phys. (N.Y.) **3**, 275 (1958); and J. Sawicki, Nuclear Phys. **6**, 575 (1958).
57. R. G. Thomas, Phys. Rev. **81**, 148 (1951); **88**, 1109 (1952).
58. J. B. Ehrman, Phys. Rev. **81**, 412 (1951).
59. J. B. Marion and W. A. Fowler, Astrophys. J. **125**, 221 (1957).
60. D. A. Bromley, E. Almqvist, H. E. Gove, A. E. Litherland, E. B. Paul, and A. J. Ferguson, Phys. Rev. **105**, 957 (1957).
61. S. T. Butler, Phys. Rev. **106**, 272 (1957).
62. R. Sherr, Proceedings of Pittsburgh Conference on Nuclear Structure (1957).
63. H. Feshbach, C. E. Porter, and V. F. Weisskopf, Phys. Rev. **96**, 448 (1954).
64. A. M. Lane, R. G. Thomas, and E. P. Wigner, Revs. Modern Phys. **98**, 693 (1955).
65. E. Vogt and J. Lascoux, Phys. Rev. **107**, 1028 (1957).
66. J. Irving, Phil. Mag. [7] **42**, 338 (1951).
67. C. R. Lubitz, University of Michigan Report (1957), unpublished.

V. H. Isotopic Spin Selection Rules

by WILLIAM M. MacDONALD

1. Isotopic Spin Vector Operator.. 933
2. Isotopic Spin of Nuclear States... 934
   a. Total Isotopic Spin... 934
   b. Charge Independence... 937
   c. Charge Parity.. 938
   d. Nuclear Isotopic Spin Assignment.. 939
3. Coulomb Perturbation of Isotopic Spin States.................................. 940
   a. Isotopic Spin Impurity... 941
   b. Dynamic Distortion.. 944
4. Nuclear Reactions.. 946
   a. Beta Decay.. 946
   b. Radiative Transitions.. 950
   c. Particle Reactions... 952
      (1) Selection Rules... 952
      (2) Relations between Cross Sections................................... 955
   d. Isotopic Spin from Particle Reactions.................................... 956
5. Validity of the Isotopic Spin Quantum Number................................. 957
   References... 959

The isotopic spin of a nucleon was first introduced by Heisenberg (*1*)* to write the nuclear Hamiltonian more symmetrically in proton and neutron coordinates. An isotopic spin formalism can be developed then which is the precise analog of Pauli spin theory. Cassen and Condon (*2*) showed that the formalism provides a simple representation for the four different types of nuclear exchange potentials. The nuclear potential assumes an especially simple form if the nuclear interactions are equal between two protons, two neutrons, and a neutron and a proton. This equality, or charge independence of nuclear forces, was suggested by Breit, Condon, and Present (*3*) on the basis of np and pp scattering data. The same conclusion was reached by Breit and Feenberg (*4*) from calculations of the relative stability and binding energies of nuclei. Wigner (*5*) observed that if the Coulomb interaction between protons can be neglected, this result would imply that a total isotopic spin quantum number can be used to characterize nuclear status.

Application of the total isotopic spin quantum number T to nuclear structure was made by Wigner (*5*) in his supermultiplet theory. Wigner (*6*) also formulated the theory of beta decay from the standpoint of this

* The reference list for Section V.H begins on page 959.

theory and derived selection rules on T for Fermi and Gamow-Teller nuclear matrix elements. The importance of T in other nuclear reactions was not generally realized until much later.

The existence of isotopic spin selection rules on particle reactions was first noted by Adair (*7*). These selection rules were merely the expression of simple restrictions on the coupling of angular momenta, combined with the order of levels of total isotopic spin derived much earlier by Wigner and discussed by Feenberg and Wigner (*8*). Nevertheless the rules provided a way to use experimental data to assign a value of T to many more excited states of nuclei. Conversely, the rules enabled one to understand the absence of or great reduction in cross sections for reactions not forbidden by angular momentum or parity conservation. Adair also noted that from charge independence follow relations between the cross sections for different reactions.

The derivation of a selection rule on electric dipole transitions also was given by Trainor (*9*), Christy (*10*), Radicati (*11*), and Morpurgo (*12*). The selection rule has made possible a quantitative determination of the validity of the isotopic spin quantum number or of charge parity for the states of light nuclei (*13*). The selection rule has been shown experimentally to be very well satisfied for transitions between the low lying states of nuclei (*14–23*). Violations of the selection rule, moreover, are within the theoretically calculated effects of the Coulomb interaction (*24,25*). Combining the information from the investigation of the electric dipole selection rule with the data on the location of the corresponding levels in isotopic spin triplets, Wilkinson (*26*) has set a limit on the deviation from charge independence of the interaction between nucleons in the nucleus.

An interesting application of the charge independence of the nuclear interaction and therefore of the isotopic spin characterization of nuclear states has been the use of lifetimes for $J = 0^+ \to J = 0^+$ positron decays to establish a condition on the beta decay coupling constants (*27–29*).

1. Isotopic Spin Vector Operator

The nuclear Hamiltonian and other nucleon operators are simply written by using an operator t_3 with the eigenvalue $\frac{1}{2}$ for the state which is a neutron, and $-\frac{1}{2}$ for the state which is a proton. The wave functions corresponding to these eigenvalues can be represented by the two-component vectors

$$\chi_+ = \begin{pmatrix} 1 \\ 0 \end{pmatrix} \quad \chi_- = \begin{pmatrix} 0 \\ 1 \end{pmatrix}. \tag{1}$$

In the representation provided by $\chi_\pm$ the operator t_3 is diagonal and has the form

$$t_3 = \frac{1}{2}\begin{pmatrix} 1 & 0 \\ 0 & -1 \end{pmatrix}. \tag{2}$$

Any other operator on $\chi_\pm$ can be expressed as a linear combination of the operator t_3, and two other operators t_1 and t_2 represented by

$$t_1 = \frac{1}{2}\begin{pmatrix} 0 & 1 \\ 1 & 0 \end{pmatrix}, \quad t_2 = \frac{1}{2}\begin{pmatrix} 0 & -i \\ i & 0 \end{pmatrix}. \tag{3}$$

The three operators t_1, t_2, t_3 can be taken as the three components of an isotopic spin vector $\mathbf{t}$ along axes ξ, η, and ζ in a three-dimensional isotopic spin space. The components of $\mathbf{t}$ satisfy the commutation relations

$$[t_i, t_j] = it_k \quad (i, j, k) = \text{cyclic permutation of 1, 2, 3}. \tag{4}$$

These relations differ only by a factor of $\hbar$ from the relations satisfied by the three components of the spin angular momentum. Consequently, the development of the isotopic spin theory follows the formalism of angular momentum.

An important point is that the use of the isotopic spin functions $\chi_\pm$ to distinguish a neutron and a proton does not require an assumption that these two particles have the same properties. In the total wave function for a nucleon the isotopic spin vector multiplies a function of the space, spin, and possible internal coordinates. The mass, magnetic moment, and charge differ for the neutron and proton, and therefore the internal wave functions certainly differ. Whether the space and spin dependent external wave functions are also different for neutrons and protons in the nucleus depends upon the nuclear and the Coulomb interaction. The Coulomb interaction is obviously not the same for all pairs of nucleons, and it will therefore produce differences in the wave function for neutrons and protons. Several kinds of evidence on the nuclear interaction will be considered.

2. Isotopic Spin of Nuclear States

a. Total Isotopic Spin

In a system consisting of a number of nucleons the total isotopic spin $\mathbf{T}$ can be introduced just as the total spin is defined.

$$\mathbf{T} = \sum_{i=1}^{A} \mathbf{t}^{(i)}. \tag{5}$$

The components of **T** satisfy the same commutation relations given by Eq. (4) for the components of **t**. The eigenvalues of $\mathbf{T}^2$ can be found from the commutation relations to be $T(T+1)$ where T can assume the value zero, or a positive half-integer or integer. The eigenvalues T_3 associated with an eigenstate of $\mathbf{T}^2$ characterize the components of the isotopic spin multiplet. For a given eigenstate of T^2 the eigenvalues of T_3 are T, $T-1, \cdots -T+1, -T$. In contrast to T^2 the eigenvalues of T_3 have a direct physical significance. For if N is the number of neutrons and Z is the number of protons in a system of nucleons the eigenvalue of T_3 is $\frac{1}{2}(N-Z)$.

A simple illustration of the construction of eigenstates of the total isotopic spin is provided by considering a system of two nucleons. The wave function of two nucleons can be factored into the product of a function of space and spin and a function of isotopic spin. The eigenfunctions of $\mathbf{T}^2$ and T_3 are four in number.

$$T = 1 \quad \begin{array}{ll} T_3 = 1 & \Xi(1, 1) = \chi_+(1)\chi_+(2) \\ T_3 = 0 & \Xi(1, 0) = 2^{-\frac{1}{2}}[\chi_+(1)\chi_-(2) + \chi_+(2)\chi_-(1)] \\ T_3 = 1 & \Xi(1, -1) = \chi_-(1)\chi_-(2) \end{array}$$
$$T = 0 \quad T_3 = 0 \quad \Xi(0, 0) = 2^{-\frac{1}{2}}[\chi_+(1)\chi_-(2) - \chi_+(2)\chi_-(1)]. \quad (6)$$

The raising and lowering operators $\mathfrak{I}_\pm = T_1 \pm iT_2$ can be used to generate all the components of the $T = 1$ multiplet from a single component. The relation which obtains for the isotopic spin eigenstates of a system of many particles is

$$\mathfrak{I}_\pm \Psi(T, T_3) = [(T \mp T_3)(T \pm T_3 + 1)]^{\frac{1}{2}} \Psi(T, T_3 \pm 1) \quad |T_3| \leq T. \quad (7)$$

These isotopic spin functions are all either symmetric or antisymmetric. Obvious symmetry properties also must be possessed by the total wave function in certain cases. The Pauli principle requires the antisymmetry of a wave function for identical particles. The isotopic spin functions $\Xi(1, 1)$ and $\Xi(1, -1)$ are states of two neutrons and two protons respectively. If $\Phi_a(\alpha, J)$ denotes a possible space and spin state of two neutrons or two protons characterized by the total angular momentum quantum number J and auxiliary quantum numbers α, then $\Phi_a(\alpha, J)$ must be antisymmetric with respect to exchange of the two nucleons. The total wave function can be written for these two states as

$$\Psi(\alpha J T T_3) = \Phi_a(\alpha, J)\Xi(TT_3) \quad (TT_3) = (1, 1), (1, -1) \quad (8)$$

and it will be antisymmetric with respect to exchange of the two nucleons.

On the other hand a neutron and a proton are nonidentical particles and the space and spin dependent wave function therefore can be a linear combination of the antisymmetric function $\Phi_a(\alpha J T T_3)$ and of the symmetric function $\Phi_s(\alpha J T T_3)$. Since both symmetric and antisymmetric isotopic spin functions, $\Xi(1, 0)$ and $\Xi(0, 0)$, are available, however, a totally antisymmetric function Ψ can be constructed using either Φ_s or Φ_a.

$$\begin{aligned}\Psi(\alpha J 00) &= \Phi_s(\alpha J)\Xi(0, 0)\\ \Psi(\alpha J 10) &= \Phi_a(\alpha J)\Xi(1, 0).\end{aligned} \qquad (9)$$

An immediate consequence of Eqs. (8) and (9) is that any state of nucleons which is compatible with the Pauli principle can be described by a linear combination of wave functions which are antisymmetric with respect to interchange of the space, spin, *and* isotopic spin coordinates of the nucleons.[a]

The reason for using the totally antisymmetrized wave functions provided by Eqs. (8) and (9) rather than a set of independent but unsymmetrized states, is that the application to the former of symmetric nucleon operators, such as the total isotopic spin raising or lowering operator, will never yield wave functions which violate the Pauli principle. Operator techniques of atomic spectroscopy then can be used to generate wave functions for different isotopic spin states and to evaluate matrix elements. The matrix elements for beta decay can be very simply expressed, for example.

This discussion of symmetry properties of the nuclear wave function has been for a system of two nucleons. However, general proofs have been given by Klein ([30]) and Rosenfeld ([31]) that only wave functions need be considered which are totally antisymmetric with respect to the interchange of coordinates of any two nucleons. The wave functions need not, of course, be eigenstates of $\mathbf{T}^2$, but only of T_3. This result is often called the *generalized Pauli principle*.

The use of totally antisymmetric wave functions simplifies the representation of nucleon potentials having each of the four different exchange characters ([2]). Let P_x^{12}, P_σ^{12}, P_τ^{12} denote the operators which exchange the space, spin, or isotopic spin of two nucleons. A simple expression for P_σ^{12} or P_τ^{12} has been given by Dirac ([32])

$$\begin{aligned}P_\sigma^{12} &= \tfrac{1}{2}(1 + \mathbf{\sigma}_1 \cdot \mathbf{\sigma}_2)\\ P_\tau^{12} &= \tfrac{1}{2}(1 + \mathbf{\tau}_1 \cdot \mathbf{\tau}_2).\end{aligned} \qquad (10)$$

From the generalized Pauli principle $P_x^{12}P_\sigma^{12}P_\tau^{12} = -1$. The expression for P_x^{12} follows

$$P_x^{12} = -\tfrac{1}{4}(1 + \mathbf{\sigma}_1 \cdot \mathbf{\sigma}_2)(1 + \mathbf{\tau}_1 \cdot \mathbf{\tau}_2). \qquad (11)$$

[a] The coordinates of a nucleon will mean the space, spin, and isotopic spin coordinates except where specifically stated otherwise.

b. CHARGE INDEPENDENCE

The usefulness of isotopic spin depends on the extent to which it is a constant of the motion for nuclear states. The condition for this is that $\mathbf{T}^2$ commute with the nuclear Hamiltonian. Of course, conservation of charge implies that T_3 also commutes with the nuclear Hamiltonian. These two conditions imply a nuclear interaction between any two nucleons which has a dependence on the isotopic spin variables of the form

$$V(r_{ij}) = A(r_i r_j \sigma_i \sigma_j) + B(r_i r_j \sigma_i \sigma_j)\mathbf{t}_i \cdot \mathbf{t}_j. \tag{12}$$

This potential provides equality of the interaction between two neutrons, two protons, or a neutron and a proton.

The first experimental evidence for such a potential was the cross section at low energy for the proton-proton and neutron-proton scattering in the ^1S state (*3*). At energies below 10 Mev the scattering cross section can be expressed in terms of the effective range and the scattering length (*33,34*). The comparison was not made between the experimental values of these quantities but between the values which these quantities would assume in the absence of the Coulomb interaction between the protons and of the differences in the interaction between the magnetic moments (*4*). After both these corrections are made the effective ranges for n-p and p-p scattering agree within experimental error (*35*). The scattering lengths, which are much more sensitive to the strength of nuclear interaction, can be brought into agreement by the use of a long-tailed potential. Unfortunately, for the potential with a hard core of radius greater than 3×10^{-14} cm the discrepancy reappears (*36*). Whether relativistic corrections affect this result is not certain.

The high-energy scattering data also can be analyzed in detail to test charge independence. A simple inequality can be derived, however, which is a necessary condition for the high-energy data to be compatible with experiment (*37*). The differential cross sections at 90° in the center-of-mass system for nucleon-nucleon scattering must satisfy the relation

$$\sigma_{pp}(90°) \leq 4\sigma_{np}(90°). \tag{13}$$

The data summarized by Hulthen and Sugawara (*38*) satisfy this inequality at energies up to 400 Mev.

A more stringent test is provided by the problem of fitting the data from nucleon scattering and polarization experiments with a charge independent interaction. The n-p and p-p data cross sections to 150 Mev and the n-p polarization data at 95 and 150 Mev have been fitted by a charge-independent potential which contains a spin-orbit interaction but which is otherwise energy independent (*39*). A charge independent and energy independent interaction with a spin orbit potential has been used successfully to fit the scattering and polarization data up to 310 Mev (*40*).

The evidence available from the two-nucleon data therefore supports charge independence although the interpretation of the experimental data is not completely established. A charge dependent interaction cannot exceed, however, a few percent of the charge-independent interaction.

Another source of information on the nuclear interaction is that provided by the experimental investigation of a selection rule on electric dipole transitions. These data can be combined with the information on the relative positions of corresponding levels in nuclei belonging to an isotopic spin triplet to deduce a limit on a charge dependent component of the nuclear interaction. We shall summarize these results in Section 5.

c. Charge Parity

The total isotopic spin quantum number can be assigned to nuclear states only if the interactions between all pairs of nucleons are equal. However, a useful quantum number for the states of $N = Z$ nuclei can be derived if only the neutron-neutron and proton-proton interactions are equal. This equality is often called the charge parity of nuclear forces. Charge symmetry is much less restrictive than the requirement of charge independence. Yet many, but not all, of the consequences of charge independence are also provided by charge symmetry.

The charge parity operator **P** is defined as an operator which changes neutrons into protons and protons into neutrons (*41*). A Hamiltonian which is charge symmetric will have an interaction containing the term $\mathbf{t}^{(1)} \cdot \mathbf{t}^{(2)}$ and $t_3^{(1)} t_3^{(2)}$. The charge parity operator will obviously commute with this Hamiltonian and the eigenvalues of P will be constants of the motion. The commutator of P with T_3, whose eigenvalues are $\frac{1}{2}(N - Z)$, however, can be seen to be

$$[P, T_3] = 2PT_3. \tag{14}$$

Therefore in a particular nucleus, which is of course an eigenstate of T_3, the eigenvalues of P cannot in general be assigned to the states of the nucleus. In a self-conjugate nucleus with $T_3 = 0$, however, the states can be assigned a definite charge parity.

The correspondence with the total isotopic spin quantum number can be established by noting that P commutes with $\mathbf{T}^2$, and that states which have a definite isotopic spin also have a definite charge parity. The charge parity of a state of isotopic spin T is $(-1)^T$. The coincidence of the consequences for nuclear reactions of charge independence and charge parity will be indicated when these occur.

d. NUCLEAR ISOTOPIC SPIN ASSIGNMENT

The nuclear interaction will be assumed to leave the total isotopic spin a good quantum number. Further the effect of the Coulomb interaction will be assumed to produce very little mixing of the ground and low lying states characterized by eigenvalues T of the total isotopic spin. The isotopic spin of a number of states then can be assigned using very simple arguments (8). These assignments can be used with selections rules which will be derived for various nuclear reactions to (a) test the validity of the isotopic spin quantum number and to (b) make further assignments to nuclear states.

The ground and low lying states can be identified by observing that a state of isotopic spin T will appear in all the nuclei with $|T_3| \leq T$. In order to compare states in different isobaric nuclei one must subtract all level shifts produced by the Coulomb energy and the neutron-proton mass difference. Without calculating this correction, we see that for "mirror" nuclei which have A odd and $N = Z \pm 1$, the level spectra should be identical. To each level with a given angular momentum J and parity there should correspond a level at the same excitation energy in the other nucleus. Levels of a number of such mirror nuclei are known and the correspondence has been well established (42). This correspondence is also a consequence of charge symmetry. If we apply the charge parity operator to the states of one nucleus, we obtain the states of the mirror nucleus.

A consequence of charge independence not derivable from charge parity is the existence of corresponding levels with $T = 1$ in three nuclei of the same even atomic number A with $N = Z - 2, Z, Z + 2$. Such nuclei constitute what we shall call a nuclear triad. The corresponding levels are the three $T_3 = 0, \pm 1$ components of an isotopic spin triplet state. The Coulomb energy differences can be calculated from mirror pairs, using charge-parity, and the position of the levels of a triad such as B^{12}, C^{12}, N^{12} can be compared: see Fig. 1. The ground states of the $N = Z \pm 2$ nuclei are found to have very nearly the same nuclear binding energy as one of the excited states of the $N = Z$ nucleus. All three states have the same angular momentum and parity. The ground state of a $T_3 = \pm 1$ nucleus can only be a state with $T = 1$.[b] The corresponding excited state of the $T_3 = 0$ nucleus is also $T = 1$, and all lower states in this nucleus must be $T = 0$. In the $A = 4n + 2$ triads the $T = 1$ state is the first or second excited state, but in the $A = 4n$ nuclei this state lies at quite a high excitation energy. In C^{12}, for example, the $T = 1$

[b] The possibility of $T \geq 2$ is ruled out by the fact that the ground states of the $N = Z \pm 4$ nuclei of the same A lie much higher after Coulomb corrections.

state lies at approximately 15 Mev, and there are a large number of $T = 0$ states below this energy.

The $T = 0$ states lie below the $T = 1$ states in the $T_3 = 0$ members of triads until $A = 34$ is reached. In Cl^{34} the ground state is unstable

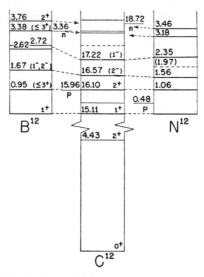

FIG. 1. The mass 12 isobaric triad. The levels whose energies are indicated (except in the case of the 4.4-Mev state) are believed to be $T = 1$ states. Corrections have been made for Coulomb energy differences and the n-p mass difference. From F. Ajzenberg-Selove, M. L. Bullock, and E. Almqvist, Phys. Rev. **108**, 1284 (1957).

against positron decay. From the Coulomb corrected position and the ft value for decay to the $J = 0^+$ ground state of S^{34}, the ground state of Cl^{34} is definitely established as a $T = 1$ level with total angular momentum $J = 0^+$.

3. Coulomb Perturbation of Isotopic Spin States

The Coulomb interaction between two nucleons vanishes unless both are protons. Such an asymmetrical potential violates the condition for the total isotopic spin T to be a good quantum number. The expression for the Coulomb interaction in the isotopic spin notation

$$C = e^2 \sum_{i<j} (\tfrac{1}{2} - t_{3i})(\tfrac{1}{2} - t_{3j}) r_{ij}^{-1} \tag{15}$$

can be used to verify that C does not commute with $\mathbf{T}^2$. The Coulomb interaction therefore perturbs and mixes states of different isotopic spin.

In order to determine the extent to which the isotopic spin is a good quantum number if the nuclear force is charge independent, the nature and extent of this perturbation of the nuclear levels must be evaluated. This is done more easily by decomposing C into terms having well-defined transformation properties for rotations of the isotopic spin space;

$$C = T^{(0)} + T_0^{(1)} + T_0^{(2)}$$

$$T^{(0)} = e^2 \sum_{i<j} (\tfrac{1}{4} + \tfrac{1}{3}\mathbf{t}_i \cdot \mathbf{t}_j) r_{ij}^{-1}$$

$$T_0^{(1)} = -\frac{e^2}{2} \sum_{i<j} (t_{3i} + t_{3j}) r_{ij}^{-1}$$

$$T_0^{(2)} = e^2 \sum_{i<j} (t_{3i} t_{3j} - \tfrac{1}{3}\mathbf{t}_i \cdot \mathbf{t}_j) r_{ij}^{-1}$$

$$\mathcal{C} = T_0^{(1)} + T_0^{(2)}. \tag{16}$$

The term $T^{(0)}$ is an isotopic spin scalar and commutes with $\mathbf{T}^2$; the $T^{(0)}$ can be included in the nuclear Hamiltonian. The part of C which produces a mixing of different isotopic spin states is $\mathcal{C}$, and we need consider only this operator. The Coulomb perturbation $\mathcal{C}$ has two effects on the nuclear states (a) the mixing of states of different isotopic spin and (b) the relative distortion of states of the same isotopic spin. We shall discuss these effects separately.

a. Isotopic Spin Impurity

The nuclear states will be eigenfunctions of the total isotopic spin $\mathbf{T}^2$ for a charge independent nuclear interaction if the Coulomb interaction is treated as a perturbation. One effect of the Coulomb interaction will be to mix into a state of isotopic spin T an amplitude $a_\nu(T')$ of a state which has the same angular momentum and parity but different isotopic spin T'. From perturbation theory we have

$$\Psi_0 = \sum_\nu a_\nu(T') \Psi_\nu(T') \qquad a_\nu(T') = (\Psi_\nu, \mathcal{C}\Psi_0)(E_0 - E_\nu)^{-1}.$$

The isotopic spin impurity arising from a particular state will be defined as $a_\nu^2(T')$. The impurity will be discussed from the point of view of the shell model.

The nuclear configuration will consist of filled shells and of nucleons in orbits which are not completely filled. The impurity of a state of this configuration can arise from mixing to states formed (1) by excitation of one of the nucleons in an open shell, (2) by excitation of a nucleon in a closed shell, or (3) by rearrangement of the angular momenta of nucleons in the same configuration. The impurity from states of the first kind was first calculated by Radicati (*43*) for two or four nucleons outside closed shells and the result was an impurity of the order of 10^{-4} to 10^{-6}. A simple calculation based on the uniform model of the nucleus revealed, however, that the contribution to the impurity from states of the second kind is actually larger than this (*24*). This crude calculation gave the result in fact that the total impurity $\Sigma_\nu a_\nu^2(T')$ of a nuclear ground state is proportional to the number $A(A-1)/2$ of interacting pairs. In a shell model calculation (*25*) the decreased overlap of different orbitals leads to a result which is no longer exactly proportional to the number of pairs of interacting nucleons. Nevertheless, in the shell model also the total impurity is contributed largely by states of excitation of closed shells.

Insight into the origin of the impurity and the composition of states making the greatest contribution to the impurity of a state is gained from the comparison of matrix elements between various two-particle configurations. Matrix elements can occur between configurations in which the individual orbital momentum l and total angular momentum j of one or both of the nucleons is different, providing the parity of the configurations is the same. However, the matrix elements are larger between configurations which do not differ in the l or j values of the individual nucleons. An example is the matrix element between the $(1p_{\frac{1}{2}})^2$ and $1p_{\frac{1}{2}}2p_{\frac{1}{2}}$ configurations. The reason for this is that the Coulomb interaction has a long range. The potential produced by a nucleon in a single particle state tends to simulate a central potential, and for a central potential l and j are good quantum numbers. Mathematically, the expansion of the Coulomb interaction into a series of Legendre polynomials gives rise to an expansion of the matrix element into Slater integrals $F^{(k)}$. For a long-range potential the largest term is $F^{(0)}$ (*44*) which arises from the spherically symmetric part of the Coulomb potential produced by a point charge.

The matrix element between two configurations of many nucleons can be expressed as a sum of two-particle matrix elements (*25*). This leads to the expectation, confirmed by calculation, that the Coulomb interaction between low-lying configurations is larger if the configurations differ only in the principal quantum number of one of the nucleons rather than if the configurations differ in the angular momentum of one or more nucleons (*25,43,45,46*). A change of the principal quantum number of one nucleon by one unit, a change of the orbital angular momentum of one nucleon by two units to conserve parity, or a change of the orbital angular momentum of two nucleons by one unit each all require the same energy within less than an order of magnitude. Consequently the actual mixing coefficient should be larger also between two configurations not differing in the angular momenta of the individual nucleons. From this result the origin of the isotopic spin impurity is seen to be the differential expansion of neutron and proton radial wave functions caused by the interaction of an individual nucleon with the Coulomb potential produced by all the other nucleons.

The contribution to impurity of the mixing of states of the lowest configurations for the nucleons in an open shell has been calculated and found to be smaller than the impurity arising from the excitation of nucleons from closed shells. For the states of an excited configuration where the energy separations are expected to be smaller, the impurity arising from such mixing may be relatively more important.

The total isotopic spin impurity of the ground states of a number of light nuclei have been calculated on the jj coupling shell model with harmonic oscillator wave functions and are listed in Table I. From the

TABLE I. ISOTOPIC SPIN IMPURITIES OF GROUND STATES FROM SHELL MODEL

Nucleus	He^4	Li^6	Be^8	B^{10}	C^{12}	N^{14}	O^{16}	Cl^{34}
$E_0 - E$	35	33	33	33	33	30	30	24
Impurity	1×10^{-5}	1×10^{-4}	3×10^{-4}	7×10^{-4}	1×10^{-3}	2×10^{-3}	4×10^{-3}	2×10^{-2}

way in which the impurity arises we do not expect the results to depend sensitively on the coupling. The absolute values, however, are subject to some uncertainty stemming from an uncertainty in the separation. The states which were considered in calculating the impurity are formed by single nucleon excitation from an (nlj) orbit to an $(n + 1lj)$ orbit, n being the principal quantum number. On the harmonic oscillator model, neglecting spin-orbit effects, all such energy separations are equal to $2\hbar^2\nu/M$. The quantity ν appears in the Gaussian factor $\exp(-\nu r^2/2)$ which multiplies all harmonic oscillator radial functions. The value of ν used to calculate matrix elements and energy separations was determined from the Coulomb energy differences of mirror nuclei (47).

The energy separation of states of different isotopic spin which differ only in the principal quantum number of a single nucleon should decrease with excitation energy less rapidly than the average separation of states of different isotopic spin with the same angular momentum and parity. Consequently the total isotopic spin impurity of low-lying levels should be approximately the same as those of the ground states.

The significance of these results is that under the Coulomb perturbation the isotopic spin can be used to characterize the ground and low-lying states of nuclei. However, the impurities will certainly have observable effects on isotopic spin selection rules.

b. Dynamic Distortion

The introduction of isotopic spin impurity into a nuclear state by the Coulomb interaction is the effect which determines the extent of violation of isotopic spin selection rules. In beta decay, however, the evaluation of a certain matrix element depends on the validity of Eq. (7). This relation is changed not only by the introduction of impurity, but also by a "dynamic distortion" of one charge component of a nuclear state which is different from the "dynamic distortion" of the other charge component. This distortion is produced by the mixing of states of the same total angular momentum, parity, and isotopic spin, but with the amplitudes and phases different in the different charge components.

An illustration of this effect is obtained by comparing the wave function of a neutron and of a proton outside a closed shell which produces an effective central nuclear potential $V(r)$ and a Coulomb potential $g(r)$. The Schrödinger equation for the extra nucleon is

$$\left\{ -\frac{\hbar^2}{2M} \Delta + V(r) + g(r) \left(\frac{1}{2} - t_3 \right) \right\} \Psi = E\Psi. \tag{17}$$

A suitable charge-independent unperturbed Hamiltonian is

$$H_0 = -\frac{\hbar^2}{2M} \Delta + V(r) + \frac{1}{2} g(r) \tag{18}$$

with the perturbation being $H_1 = -g(r)t_3$. Eigenfunctions of H_0 will be $f(r)\begin{pmatrix}1\\0\end{pmatrix}$ for a neutron and $f(r)\begin{pmatrix}0\\1\end{pmatrix}$ for a proton. Excitation of the closed shell is not being considered, and therefore the perturbation cannot change the total isotopic spin of $t = \frac{1}{2}$. Nevertheless, the perturbation will distort the radial function for a proton differently from that of a neutron and it will no longer be true that

$$(t_1 + it_2)\Psi_{\text{proton}} = \Psi_{\text{neutron}}. \tag{19}$$

The effect of the Coulomb perturbation upon Eq. (7) can be determined by evaluating the matrix element

$$M_F = \int \Psi(T, T_3')(T_1 \pm iT_2)\Psi(T, T_3). \qquad T_3' = T_3 \pm 1. \tag{20}$$

The effect of the Coulomb perturbation is to mix the state $\Psi^0(T, T_3)$ with other states $\Psi^\nu(T', T_3)$ which have the same angular momentum and parity. For example, consider the case with $T = 1$, $T_3 = 0$, $T_3' = -1$ and expand the nuclear states into isotopic spin eigenstates:

$$\Psi(1, 0) = a_0\Psi^0(1, 0) + \sum_\nu [a_\nu^{(1)}\Psi^\nu(1, 0) + a_\nu^{(2)}\Psi^\nu(2, 0) + a_\nu^{(3)}\Psi^\nu(0, 0)]$$

$$\Psi(1, -1) = b_0\Psi^0(1, -1) + \sum_\nu [b_\nu^{(1)}\Psi^\nu(1, -1) + b_\nu^{(2)}\Psi^\nu(2, -1)]. \quad (21)$$

The $a_\nu^{(2)}$ and $a_\nu^{(3)}$ are isotopic spin impurity coefficients, and $a_\nu^{(1)}$ will lead to dynamic distortion. The matrix element M_F is given by

$$M_F = \left(a_0 b_0 + \sum_\nu a_\nu^{(1)} b_\nu^{(1)}\right) \sqrt{2} + \sum_\nu a_\nu^{(2)} b_\nu^{(2)} \sqrt{6}. \quad (22)$$

Normalization of the states gives the conditions

$$a_0^2 + \sum_\nu [(a_\nu^{(1)})^2 + (a_\nu^{(2)})^2 + (a_\nu^{(3)})^2] = 1$$

$$b_0^2 + \sum_\nu [(b_\nu^{(1)})^2 + (b_\nu^{(2)})^2] = 1. \quad (23)$$

The coefficients $a_\nu^{(i)}$ and $b_\nu^{(i)}$ are small compared to unity and an expansion in powers of these coefficients gives the result to lowest order

$$M_F^2 = 2(1 - \delta)$$
$$\delta = \sum_\nu [(a_\nu^{(1)} - b_\nu^{(1)})^2 + (a_\nu^{(3)})^2 + (a_\nu^{(2)})^2 + (b_\nu^{(2)})^2$$
$$- 2\sqrt{3}\, a_\nu^{(2)} b_\nu^{(2)}]. \quad (24)$$

The δ measures the total effect upon Eq. (7) of the Coulomb interaction. The $T = 0$ impurity causes a reduction of $|M_F|^2$, while the $T = 2$ impurity causes an increase in $|M_F|^2$. The dynamic distortion is represented by the terms in $a_\nu^{(1)}$ and $b_\nu^{(1)}$ and always produces a reduction of $|M_F|^2$ unless $a_\nu^{(1)} = b_\nu^{(1)}$. This condition is satisfied only if the amplitudes and phases of perturbation admixtures are the same in $\Psi(1, 0)$ and $\Psi(1, -1)$. This will be true of the admixtures produced by a charge-independent interaction, but not of the admixtures produced by the Coulomb potential.

The unperturbed states $\Psi^0(1, 0)$ and $\Psi^0(1, -1)$ differ only in the replacement of one of the outside neutron states by a proton state. The dynamic distortion of one state relative to the other must arise from the difference in the distortions of all the other nucleon states produced by the outside neutron or proton. As with impurity the contribution to distortion made by states formed by promoting a nucleon from a state (nlj) to a state $(n + 1lj)$ is greater than that of states formed by exciting a nucleon to a state of different orbital angular momentum. The interaction of different nuclear states with the same total angular momentum, parity, and isotopic spin *within* a configuration must be considered also.

This is to be contrasted to the mixing of states of different isotopic spin within a configuration. This could be neglected in the impurity calculations. From the fact that dynamic distortion is produced by just the outside nucleons, the dependence on the number of particles is roughly as A. Since the impurity is roughly proportional to A^2, as A increases the effect of isotopic spin impurity on M_F is expected to become more important than the reduction of M_F by dynamic distortion. With a nucleus having a number of particles in an open shell and several states of the same configuration which can interact through small energy separations, however, the dynamic distortion could be relatively more important.

4. Nuclear Reactions

The isotopic spin formalism provides selection rules, matrix elements, and relations between certain cross sections. By use of these rules, isotopic spin quantum numbers can be assigned to nuclear states, or conversely, the amplitudes for certain processes can be predicted. Important relations can be derived for each of three reactions of interest in nuclear physics; beta decay, radiative transitions, and particle reactions.

a. Beta Decay

The nuclear matrix element for beta decay can be written for both allowed and forbidden transitions through any of the ten possible couplings of the electron and the neutrino fields in the form

$$M = \int \Psi_f^* \sum_{k=1}^{A} (t_1^{(k)} \pm i t_2^{(k)}) O_k \Psi_i. \tag{25}$$

The Ψ_i and Ψ_f are wave functions for the initial and final nuclear states. The O_k is a space- and spin-dependent operator on the wave function of the kth nucleon, and the isotopic spin operator is simply that which effects the transformation of the kth nucleon between a neutron and a proton state. The entire operator is a linear combination of the components of a vector $\sum_{k=1}^{A} \mathbf{t}^{(k)} O_k$ in isotopic spin space, and the nuclear states are eigenstates of $\mathbf{T}^2$ and T_3. The selection rule follows

$$\Delta T = 0, \pm 1 \qquad T_3 = \pm 1. \tag{26}$$

In the nonrelativistic limit for the nucleon velocities the matrix elements for beta decay reduce to just two forms, the Fermi and the Gamow-Teller matrix elements.

Fermi:
$$M_F = \int \Psi_f{}^* \mathfrak{I}_\pm \Psi_i = (T \mp T_3)(T \pm T_3 + 1)^{\frac{1}{2}} \qquad \Delta T = 0$$

Gamow-Teller:
$$M_G = (\mathbf{M}_G{}^* \cdot \mathbf{M}_G)^{\frac{1}{2}} \qquad \mathbf{M}_G = \int \Psi_f{}^* \sum_{k=1}^{A} (t_1{}^{(k)} \pm i t_2{}^{(k)}) \mathbf{\sigma}^{(k)} \Psi_i. \quad (27)$$

In general the Gamow-Teller matrix element can be evaluated only by recourse to a model which provides the nuclear wave function. The Fermi matrix element, however, can be evaluated independently of any model by using Eq. (7). This fact leads to a way of obtaining a relation between the scalar and vector coupling constants for beta decay (27,28).

The Gamow-Teller matrix element for an allowed beta decay vanishes when the transition is between two states with $J = 0$. Such transitions are pure Fermi transitions. The isotopic spin of the initial and final states is $T = 1$ for all known $J = 0^+ \to J = 0^+$ transitions, giving $M_F = \sqrt{2}$. Of the ten possible beta interactions only four give rise to Fermi matrix elements in the nonrelativistic limit. These are the scalar and axial vector interactions in both parity conserving and nonconserving couplings (48). The *ft* value for an allowed positron decay is

$$ft = 2\pi^3 \ln 2 [\xi + \xi b <W^{-1}>]^{-1} \qquad \hbar = c = m = 1$$
$$\xi = [(|C_s|^2 + |C_s{}^1|^2)\kappa^{-2} + (|C_v|^2 + |C_v{}^1|^2)]M_F{}^2$$
$$\xi b = -2(1 - \alpha^2 Z^2)^{\frac{1}{2}} |M_F|^2 \operatorname{Re} [\kappa^{-1}(C_s C_v{}^* + C_s' C_v'^*)]; \qquad \alpha = e^2/\hbar c \quad (28)$$

where W is the electron energy in units of mc^2 and

$$f(Z, W_0) = \int_1^{W_0} F(Z, W) p W (W_0 - W)^2 \, dW$$
$$<W^{-1}> = \int_1^{W_0} F(Z, W) p (W_0 - W)^2 \, dW. \quad (29)$$

The W_0 is the maximum energy of the emitted positrons and p is the positron momentum. $F(Z, W)$ is the function which represents the effect of the Coulomb interaction of the final nucleus of charge Ze and the positrons upon the positron spectrum. The function $F(Z, W)$ approaches unity as Z approaches zero. For the transitions being considered the nuclear matrix element of the axial vector interaction is precisely M_F, while the scalar matrix element has relativistic correction terms in the nucleon velocities. We have introduced therefore the quantity κ defined by

$$M_F = \kappa \int \Psi_f{}^* \sum_{k=1}^{A} (t^{(k)} + i t_2{}^{(k)}) \beta_k \Psi_i. \quad (30)$$

The nuclear states must be constructed from Dirac spinors and β_k is the Dirac matrix for the kth nucleon.

The term ξb is called the Fierz interference term between the scalar and the axial vector interactions. A limit on the corresponding term for the vector and tensor beta interactions has been established by analyzing the positron energy spectrum for pure allowed Gamow-Teller transitions (49). The Fierz term has a different energy dependence from the other terms and its presence alters the shape of the spectrum. No spectrum for the pure Fermi transitions has been measured and a similar analysis is not possible at this time. However, the Fierz term also affects the ft value. In the absence of the Fierz term the ft values of all $J = 0^+ \rightarrow J = 0^+$ transitions should be equal since M_F is a constant independent of the nucleus undergoing decay. Since the Fierz term depends on W_0 through $<W^{-1}>$, a limit can be placed on the quantity b by using the ft values for several transitions with very different maximum positron energies.

The ft values for a number of these pure Fermi transitions are known. But only for the decays of $O^{14} \rightarrow N^{14*}$, $Al^{26} \rightarrow Mg^{26}$, $Cl^{34} \rightarrow S^{34}$, $V^{46} \rightarrow Ti^{46}$, $Mn^{50} \rightarrow Cr^{50}$, and $Co^{54} \rightarrow Fe^{54}$ are the positron maximum energies and the lifetimes known sufficiently accurately to determine a limit on b. The experimental information on these decays is given in Table II. The limit on b given from these decays is $b = 0.01 \pm 0.06$ (49a).

The assumptions used in this analysis are that (a) the nucleon interaction is charge independent, (b) relativistic correction terms are negligible and $k = 1$. A priori, neither of these approximations are justified to the accuracy of a few percent with which M_F must be calculated. Nevertheless, the important question is not the absolute value of corrections to M_F or k, but the variation in M_F and k produced by the corrections. The effect of the Coulomb interaction can be found from Eq. (24) using the coefficients of admixture of $T = 0$, $T = 1$, and $T = 2$ states to the unperturbed $T = 1$ states which are the principal component of states between which the decay occurs (29). The dynamic distortion arising from the excitation of nucleons from $1p_{\frac{3}{2}}$ to $1p_{\frac{1}{2}}$ orbits is not important in the decay of O^{14}. This contribution was calculated by using the wave functions found by Ferrell and Visscher (49b) in the analysis of the C^{14} decay. Average energy separations again were used for the calculation of mixing coefficients to states of $n \rightarrow (n + 1)$ nucleon excitation. The effect upon the ft value is given by δ, which is quite small but depends sensitively on the average excitation $E_0 - E_2$ of contributing $T = 2$ states *relative* to the average excitation $E_0 - E_1 \approx E_0 - E_3 = E_0 - \bar{E}$ of all the contributing $T = 0$ and $T = 1$ states. If the average energy of excitation of all three types of states were the same, the greater value $M_F = \sqrt{6}$ for the matrix element between $T = 2$ states produces a *decrease* in the ratio of the *ft* value for $Cl^{34} \to S^{34}$ to that for $O^{14} \to N^{14*}$ by 0.5 percent. If the contributing $T = 2$ states lie above the contributing $T = 0$ and $T = 1$ states by about 20 percent, or 5 Mev in S^{34}, the *ft*

TABLE II. EXPERIMENTAL INFORMATION ON THE $J = 0^+ \to J = 0^+$ POSITRON DECAYS

Decay	Half-life (sec)	End-point kinetic energy (Mev)		*ft* (sec)		$<W^{-1}>$
$C^{10} \to B^{10**}$	1160 ± 150	a,b 1.08 ± 0.10		b 5900 ± 2700		0.573
		0.827 ± 0.050	i	2020 ± 570		0.640
$O^{14} \to N^{14*}$	7.25 ± 0.5	c,d 1.8097 ± 0.0078	j	3101 ± 62		0.438
$Al^{26*} \to Mg^{26}$	6.60 ± 0.06 e	3.202 ± 0.010	k	3122 ± 52		0.301
$Cl^{34} \to S^{34}$	1.53 ± 0.02 f	4.50 ± 0.03	l	3155 ± 103		0.225
$K^{38} \to A^{38}$	0.935 ± 0.025 f	5.06 ± 0.11	m	3140 ± 400		0.215
$Sc^{42} \to Ca^{42}$	0.62 ± 0.05 g					
$V^{46} \to Ti^{46}$	0.424 ± 0.002 e	6.052 ± 0.028	h	3098 ± 70		0.184
$Mn^{50} \to Cr^{50}$	0.304 ± 0.007 e	6.579 ± 0.027	h	3202 ± 100		0.172
$Co^{54} \to Fe^{54}$	1.194 ± 0.003 e	7.337 ± 0.041	h	3290 ± 100		0.156

a $t = 19.1 \pm 0.8$ sec (*50*).
b Branching 1.65 ± 0.20 (*51*).
c $t = 72.1 \pm 0.4$ sec (*52*).
d Branching $= 99.40 \pm 0.10$ (*53*).
e Sutton (*54*).
f Kline and Zaffarano (*55*).
g See Morinaga (*56*).
h See Miller (*49a*).
i See Ajzenberg-Selove and Lauritsen (*57*), and C. F. Cook, J. B. Marion, and T. W. Bonner (unpublished, cited in *57*).
j $C^{12}(He^3,n)O^{14}$ threshold (*58*).
k See Kington *et al.* (*59*).
l See Green and Richardson (*60*).
m See King (*61*), and S. E. Hunt, R. M. Kline, and D. J. Zaffarano (unpublished, cited in *61*).

values for the two decays will be equal. In Table III the ratio of the *ft* values for the two decays is given as a function of the average excitation of the contributing $T = 2$ states in terms of the ratio $\dfrac{(E_0 - E_2)}{(E_0 - \bar{E})}$. For this illustrative calculation the ratio of the energies is taken to have the same value in both the $A = 14$ and the $A = 34$ nuclei. The ratio of the *ft* values differ from unity by less than 1 percent for reasonable values of the ratio of the average excitation energies for $T = 2$ and $T = 0, 1$ states.

The scalar matrix element is also subject to a correction arising from relativistic terms in the nucleon velocities (62). The matrix element in Eq. (30) can be written in terms of nonrelativistic wave functions as

$$k \left\{ \int \Psi_f^* \mathfrak{J}_+ \Psi_i - \int \Psi_f^* \sum_{k=1}^{A} (t_1^{(k)} + it_2^{(k)})(p_k^2/2M^2C^2) \Psi_i \right\} = M_F. \quad (31)$$

The second term is the correction term which reduces to the kinetic energy of the decaying nucleon divided by nucleon rest mass energy. A calculation on the harmonic oscillator model gives the correction to the

TABLE III. COULOMB CORRECTION TO THE RATIO OF ft VALUES FOR O^{14} AND Cl^{34} ($J = 0 \to J = 0$)

$R = \dfrac{E_0 - E_2}{E_0 - E_1}$	$\alpha = \dfrac{(ft)_{Cl^{34}}}{(ft)_{N^{14}}} - 1$
1.0	-5×10^{-3}
1.25	2
1.5	6
1.75	8
2.0	10

matrix element to be about 2 percent. The nuclear binding energies of the last nucleon in N^{14} and Cl^{34} are so nearly equal, however, that the relativistic correction to the ratio of the ft values for decay by the scalar interaction is about 0.5 percent (62).

Summarizing the results of these calculations, we have the result that the ratio of the ft values for the $O^{14} \to N^{14*}$ and $Cl^{34} \to S^{34}$ decays should differ from unity by less than 1 percent as a result of Coulomb and relativistic corrections. The quoted experimental uncertainties in the ft values are larger than this, however, and the limit on b remains unchanged.

b. RADIATIVE TRANSITIONS[c]

The interaction of a system of nucleons with an electromagnetic field is given by the nonrelativistic Schrödinger equation as

$$H_I = -\sum_{i=1}^{A} \left\{ \frac{e}{Mc} \mathbf{p}_i \cdot A(\mathbf{r}_i) \left(\frac{1}{2} - t_3^{(i)} \right) + \mu_p \left(\frac{1}{2} - t_3^{(i)} \right) \right.$$
$$\left. + \mu_N \left(\frac{1}{2} + t_3 \right) \boldsymbol{\sigma}_i \times A(\mathbf{r}_i) \right\}. \quad (32)$$

[c] See Section V.F.

The H_I is the sum of a scalar and the third component of a vector in isotopic spin space. The selection rule on all nuclear transitions by gamma ray absorption or emission follows (Wigner)

$$\Delta T = 0, \pm 1 \qquad T = 0 \nleftrightarrow T = 0. \tag{33}$$

This selection rule can be observed only in radiation from states of isotopic spin $T \geq 2$. A much more useful selection rule can be derived for electric dipole transitions.

The electric dipole selection rule was first derived by Trainor (9) on the basis of the symmetry of the nuclear wave functions of the Wigner supermultiplet theory. The assumption of spin independent nuclear forces used in this theory is really not necessary, however, and a simple derivation can be given on the basis of charge independence. Indeed, Morpurgo (12) has shown that the selection rule is also a simple consequence of the charge parity of nuclear states.

The matrix element for electric dipole transitions is easily derived in the long wavelength approximation

$$M_m(\text{E1}) = \int \Psi_f \left[\sum_{s=1}^{A} (\tfrac{1}{2} - t_3^{(s)}) r_s Y_{1m}(\theta_s, \varphi_s) \right] \Psi_i. \qquad kR \ll 1 \tag{34}$$

The Ψ_i and Ψ_f are initial and final nuclear states, R is the nuclear radius, and k is the photon wave number. The part of the operator not containing $t_3^{(s)}$ is simply a (spherical vector) component of the center-of-mass vector. Since the motion of the center-of-mass does not change significantly as the result of gamma emission, the matrix element of this part of the dipole operator is zero. The second part of the operator is the third component of a vector in isotopic spin space. The selection rule on such an operator, and therefore on the matrix element, is

$$\begin{aligned} \Delta T &= 0, \pm 1 & T_3 &\neq 0 \\ \Delta T &= \pm 1 & T_3 &= 0. \end{aligned} \tag{35}$$

The selection rule in $T_3 = 0$ ($N = Z$) nuclei is very important because of the large number of $T = 0$ levels which can be established in the even-even nuclei.

Violations of the electric dipole selection rule can be caused by the following effects:

(a) isotopic spin impurity in the initial or final nuclear states,
(b) retardation terms of order $(kR)^2$ in the electric dipole matrix element,

(c) the neutron-proton mass difference which displaces the center of mass from the centroid,
(d) the electric dipole matrix element of the spin dependent part of the interaction,
(e) recoil effects caused by emission and absorption of virtual mesons.

The corrections (b), (c), and (d) have been shown to be much smaller than the effect of isotopic spin impurity (*13*). The calculation of (e) has been carried out by Morpurgo (*12*) who obtained an effect as large as (a). However, the result is unsatisfactory because of the necessarily approximate nature of the calculation, so no further consideration can be given it. Therefore the violations of the electric dipole selection rule will be attributed entirely to isotopic spin impurity.

The investigation of the selection rule on electric dipole transitions has been carried out principally by D. H. Wilkinson (see Section V.F) and collaborators (*14–23*).

The isotopic spin impurities found for the transitions between the low-lying states of the nuclei are in general agreement with the impurities calculated to result from the Coulomb interaction. As has been noted this fact supports the charge symmetry of nuclear forces but does not imply charge independence. The charge symmetry established by the electric dipole selection rule, however, can be used in the analysis of corresponding levels in the isotopic triplets to establish charge independence.

c. Particle Reactions

(1) *Selection Rules*

The assumption of charge independence of the nuclear interaction provides both selection rules and relations between cross sections for reactions involving particles (*7*). The charge symmetry of nuclear interactions also provides some of the relations.

The isotopic spin selection on particle reactions is merely an expression of the triangular inequality for adding vectors. Two systems in states of isotopic spin T_1 and T_2 can be coupled to form a system in a state of isotopic spin T only if the following inequality is satisfied.

$$|T_1 - T_2| \leq T \leq T_1 + T_2. \tag{36}$$

The simplest application of this rule is to the reaction $O^{16}(d,\alpha)N^{14}$ with N^{14} being left in the ground state or the first excited state at 2.31 Mev. The deuteron has an ^3S wave function in space and spin, and the isotopic spin must be $T = 0$ to provide a totally antisymmetric wave function. The alpha particle and the ground state of O^{16} are also both $T = 0$

states. The ground state of N^{14} has $T = 0$, but the first excited state has the correct binding energy, angular momentum, and parity to be the $T_3 = 0$ component of the $T = 1$ ground state of C^{14} and O^{14}. The isotopic spin of the initial system consisting of a deuteron and O^{16} is $T = 0$. The final state must also have $T = 0$. The reaction to the first excited state of N^{14} can proceed only through the isotopic spin impurity of the initial and final states, or through the impurity of states of the compound nucleus if the reaction proceeds in that manner.

In this reaction the charge parity will give the same selection rule as charge independence (41). The charge parity of the initial state is even, therefore the charge parity of the final state must be even. The charge parity of a $T = 1$ state is odd, however, so the final state of the reaction cannot be the first excited state of N^{14}. The charge parity and the isotopic spin selection rules do not coincide for all reactions. In fact, in the above reaction the charge parity selection rule would permit the final stage of N^{14} to be a $T = 2$ state, while the isotopic spin selection rule would forbid such a reaction.

The relative yields of alpha particles which leave N^{14} in various excited states in the reaction $O^{16}(d,\alpha)N^{14}$ have been measured by Browne (63) for incident alpha particle energies of 5.5 to 7.5 Mev. The forbidden reaction to the first excited state has been observed, but the relative intensity is less than a few percent of the allowed ground and excited state alpha-particle groups.

The $Be^9(p,d)Be^8$ and $Be^9(p,\alpha)Li^6$ reactions have been analyzed by Weber et al. (64) for evidence of isotopic spin impurity. The (p,d) reaction shows no contribution of the $J = 2^+$, $T = 1$ state at 8.89 Mev in B^{10}, although a weak anomaly is seen in the (p,α) differential cross section at 90°. A limit is derived for the isotopic spin impurity of either the Li^6 ground state or the 8.89 Mev excited state of B^{10} of $2.2 \times 10^{-5} \leq \alpha^2 \leq 1.1 \times 10^{-3}$.

These estimates of isotopic spin impurity are in order of magnitude agreement with the theoretical calculations of the effect of the Coulomb interaction in mixing states of different isotopic spin (24).

The particle selection rules have been combined with electric dipole selection rules by Gell-Mann and Telegdi (65) in a discussion of the (γ,α) reactions where the initial nucleus is even-even with $N = Z$. If the first $T = 1$ state in the initial nucleus occurs at an energy E_1, then the following statements can be made if T is a good quantum number.

(a) For $E_\gamma < E_1$ all (γ,α) reactions must proceed by M1 absorption through a $J = 1^+$, $T = 0$ state or by E2 through a $J = 2+$, $T = 0$ state.

(b) If $E_\gamma > E$ absorption with a transition to a $T = 1$ state is possible by E1, M1, or E2, but deuteron or alpha-particle emission from this state can occur only if the energy is sufficient to leave the final nucleus in a $T = 1$ state.

(c) For any E_γ, a (γ,α) or (γ,d) reaction to a $T = 0$ final state can proceed only as in (a).

(d) A (γ,α) reaction to a $J = 0$ ground state can occur only through a $J = 2^+$, $T = 0$ state.

These selection rules also follow from charge parity selection rules.

An application of these rules can be made to the absorption of γ rays by C^{12}.[d] The first $T = 1$ state is at $E = 15.11$ Mev and therefore the only contribution to the cross section for $E_\gamma < 15$ Mev must be M1 or E2. The threshold for E1 absorption comes at $E_\gamma = 15.11$ Mev, but the energy E_γ required to reach the first $T = 1$ state in Be^8 at about 16.8 Mev is about 26 Mev. In the region 15.09 Mev $< E_\gamma < 26$ Mev the electric dipole absorption will occur in the reactions (γ,n) and (γ,p) and lead to the giant resonance observed in C^{12} around 22 Mev. If the absorption cross section does not fall the (γ,α) cross section should be larger in the highest energy region. A peak does occur at about 29 Mev in C^{12}, and 90 percent of the (γ,α) reactions lead to two $T = 1$ states at about 17 Mev in Be^8.

A different type of selection rule which does not follow from charge parity and tests charge independence has been given by Peaslee and Telegdi (66) for (γ,t) reactions. The rule applies to (γ,t) reactions on $A = 4n + 3$ nuclei and can be tested by comparison with (γ,n) reactions on the same nuclei. The selection rule on gamma absorption given by Eq. (35) shows that $T = \frac{1}{2}$, $\frac{3}{2}$ states are reached in the initial nucleus. The $T = \frac{1}{2}$ states can decay by neutron or triton emission to $T = 0$ states. The $T = \frac{3}{2}$ states can decay only to $T = 1$ states. The final nucleus is $A = 4n + 2$ for the (γ,n) reaction, and $A = 4n$ for the (γ,t) reaction. The $T = 1$ states in $A = 4n + 2$ nuclei begin at low energies, while in $A = 4n$ light nuclei these states first occur at high excitation energies. Consequently the (γ,t) reactions through $T = \frac{3}{2}$ states have a higher threshold, and the lowest such states can be observed in the (γ,n) cross section but not in the (γ,t) cross section. The available experimental evidence of the reactions $Li^7(\gamma,n)Li^6$ and $Li^7(\gamma,t)He^4$ supports the existence of this selection rule (57).

Although experimental evidence of a more qualitative nature exists

---

[d] The experimental information on this and other reactions discussed is taken from the article by Ajzenberg-Selove and Lauritsen (57).

for the operation of these selection rules, considerably more remains to be done in the way of quantitative investigations.

(2) *Relations between Cross Sections*

We shall consider first mirror reactions which can be obtained from each other by the process of turning all protons into neutrons, and vice versa. Both charge independence and charge parity predict that the total cross sections for two mirror reactions should be equal except for barrier penetration and phase space factors. However, not only are the experimental measurements of the total cross sections difficult, but in the case of low-energy particles the Coulomb barrier penetration factor is extremely sensitive to the choice of the nuclear radius.[e] Barker and Mann (*68*) have pointed out, however, that for (γ,p) and (γ,n) reactions proceeding through a $T = 1$ state of the compound nucleus the ratio of the cross sections is very sensitive to the $T = 0$ impurity of the state. In fact, if α is the amplitude of the $T = 0$ component in the $T = 1$ state and R is the ratio of cross sections after reduction by penetration and phase space factors, then

$$R = [(1 + \alpha)/(1 - \alpha)]^2. \qquad (37)$$

For example, the (γ,p) and (γ,n) cross sections in C^{12} exhibit giant resonances at nearly the same energy, 22 Mev, with nearly the same widths, but the ratio R is approximately 2. This value for R requires only an impurity α^2 of about 4 percent. A careful measurement of the $Li^6(d,n)Be^7$ and $Li^6(d,p)Li^7$ cross sections has been made by Wilkinson (*67*) for deuteron energies between 200 kev and 1.8 Mev. The cross sections agree within 15 percent. The impurity α^2 calculated from Eq. (37) is 1.6×10^{-3}.

Although the equality of the "adjusted" or intrinsic cross sections for mirror reactions can also be derived by charge parity, additional relations are provided by charge independence. For example, the ratio of the intrinsic cross sections for the reactions $Be^9(d,p)Be^{10}$ to the ground state and $Be^9(\alpha,n)B^{10*}$ to the second excited $T = 1$ state is predicted to be 2. This result is true whether the reaction proceeds by compound nucleus formation or direct interaction. A simple argument can be given.

On a compound nucleus picture (*6*) a $T = \frac{1}{2}$, $T_3 = \frac{1}{2}$ is formed which breaks up into two systems, each having quantum numbers T and T_3. Let $\Phi(1, T_3)$ denote the ground state of $Be^{10}(T_3 = 1)$ and the 1.74 Mev level of $B^{10}(T_3 = 0)$, and let $\phi(t_3)$ denote a neutral ($t_3 = \frac{1}{2}$) or a proton

---

[e] A discussion of the difficulties in testing the equality of (γ,p) and (γ,n) cross sections is given by Wilkinson (*67*).

($t_3 = -\frac{1}{2}$) state. The isotopic spin state of the compound system is

$$\Psi(\tfrac{1}{2}, \tfrac{1}{2}) = \sqrt{\tfrac{2}{3}}\,\Phi(1, 1)\varphi(\tfrac{1}{2}, -\tfrac{1}{2}) - \sqrt{\tfrac{1}{3}}\,\Phi(1, 0)\varphi(\tfrac{1}{2}, \tfrac{1}{2}). \tag{38}$$

The intrinsic cross sections for proton and neutron emission are in the ratio of the square of the amplitudes for the components ($Be^{10} + p$) and ($B^{10} + n$), if we assume separability of the space and spin states of the compound system. This ratio is 2.

In the direct interaction picture either the neutron is captured by Be^9 to form Be^{10}, or the proton is captured to form B^{10*} in the 1.74 Mev state. The cross sections in this case are in the ratio of the amplitudes with which the ($Be^9 + n$) or the ($Be^9 + p$) appear in Be^{10} and B^{10*}. If $\Psi(\tfrac{1}{2}, \tfrac{1}{2})$ now denotes the isotopic spin function of the ground state of Be^9 we have

$$\Phi(1, 1) = \Psi(\tfrac{1}{2}, \tfrac{1}{2})\varphi(\tfrac{1}{2}, \tfrac{1}{2})$$

$$\Phi(1, 0) = \frac{1}{\sqrt{2}}\left[\Psi\left(\frac{1}{2}, \frac{1}{2}\right)\varphi\left(\frac{1}{2}, -\frac{1}{2}\right) + \Psi\left(\frac{1}{2}, -\frac{1}{2}\right)\varphi\left(\frac{1}{2}, \frac{1}{2}\right)\right]. \tag{39}$$

The ratio of the intrinsic cross sections for neutron capture to proton capture is again 2.

The relations of this kind between cross sections are extremely difficult to establish with the required accuracy for the same reasons as are mentioned in the discussion of mirror reactions. At this time there is no experimental confirmation of these consequences of charge independence.

d. Isotopic Spin from Particle Reactions

Although deviations from the predictions of charge independence for nuclear reactions have been observed experimentally, the isotopic spin impurity required to explain these is usually in the range of 10^{-5} to a few percent. If the particle reactions are understood to proceed through a level of a compound nucleus, then the excitation of this level is usually 20 Mev or higher. In the Li^6 mirror reactions measured by Wilkinson (67), the excitation of the compound Be^8 system is greater than 22.6 Mev.

Although considerably more mixing of different isotopic spin states might be expected at such high excitation energies of the compound nucleus, it has been suggested that only an effective isotopic spin impurity is determined by the reaction (69,70,71). This effective isotopic spin impurity arises whenever the reaction proceeds through a region of excitation of the compound nucleus in which the density of states is very high. A number of different overlapping states with different isotopic spins can contribute to the cross section then, and the state of the compound *system* will be a linear combination of such states. That state will be formed which has the greatest overlap with the initial state, and the isotopic spin of the compound system can be rather well defined initially. After formation of the compound state the Coulomb perturbation C will act to introduce isotopic spin impurity. From the time dependence of a Schrödinger state, $\exp -i(H_0 + C)t/\hbar$, we see that the characteristic time for Coulomb effects is $\hbar/<C>$. The time available for such mixing is the lifetime $\hbar/\Gamma$ of the states, where Γ is the width of the states. If $\Gamma \gg <C>$ the time will be insufficient for the introduction of appreciable isotopic spin impurity. Consequently the selection rule for particle reactions will be very well obeyed for particle energies where the density of states of the compound nucleus is sufficiently high.

Wilkinson (20) has suggested that the isotopic spin impurity determined from nuclear reactions will exhibit a maximum at energies where $<C> \geq \Gamma$, but at which the energy separation of states of different isotopic spin is small. Whether a maximum actually appears in the isotopic spin impurity depends on the energy separation of states between which the Coulomb interaction has large matrix elements. Complete theoretical calculations on the mixing of excited states have not yet been made. However, the ratios of (γ,p) to (γ,n) cross sections have often been taken to indicate large impurities in the emitting $T = 1$ levels. The Eq. (37) shows, however, that this ratio is very sensitive to the amount of impurity; ratios of the (γ,p) and (γ,n) cross sections less than 2 imply an impurity of less than 5 percent. Conversely, we have learned that we must beware of using the impurity α^2 rather than the amplitude α as a measure of the effect of isotopic spin impurity upon certain reactions.

5. Validity of the Isotopic Spin Quantum Number

The existence of the isotopic spin quantum number for nuclear states has been shown to be implied by the invariance of the interaction under rotations in isotopic spin space. More commonly we call this property of the nuclear interaction "the charge independence" of nuclear forces, by which we mean the equality of the interaction between any two nucleons. We have given selection rules and relations between cross sections which follow from charge independence, and we have discussed the effect of the Coulomb perturbation on these relations. Although the experimental evidence for most of these relations is very incomplete, all the deviations from the predictions of charge independence can be interpreted to result from the Coulomb interaction.

Unfortunately many of the relations, such as the electric dipole selection rule, follow not only from charge independence but also from charge symmetry. Charge symmetry, or the equality of the interaction between two neutrons or two protons, does not predict the lack of $\Delta T = 2$ electromagnetic transitions or certain of the relations between particle reactions. However, these consequences of charge independence are the most difficult to test experimentally.

Wilkinson (26) has observed that the experimental evidence for the electric dipole selection rule at least establishes charge symmetry. Using charge symmetry a more careful calculation can be made of the relative position of the different T_3 components of a $T = 1$ state in an even A isobaric triad. The $(Z + 1, A - 1)$ and $(Z, A - 1)$ nuclei are mirror pairs, and the energy difference between the ground states, after the subtraction of the neutron-proton mass difference, must be the Coulomb energy difference. This follows from the charge symmetry of nuclear forces. Wilkinson corrects for the decrease in Coulomb energy resulting from the expansion of the radius from an $A - 1$ to an A nucleus by multiplying by a factor of $(1 - A^{-1})^{\frac{1}{3}}$. This Coulomb energy is used to calculate the position of the first $T = 1$ state in the $(Z = N, A)$ nucleus.

The $T = 1$ level in 14 $(N = Z)$ even A nuclei can be identified and the position compared with the theoretically calculated position. The average difference between the calculated and measured positions is 35 Kev. If the difference were taken seriously, this would indicate an n-p interaction stronger than the n-n interaction by only about 1 percent (20). Thus even if the information from the two-body scattering data is ignored, the experimental evidence for the validity of the electric dipole selection rule and the position of corresponding levels of mirror nuclei suffices to establish the charge independence of nuclear forces to within 1 percent or less.

The conclusion to be drawn from the present theoretical and experimental results, therefore, is that the specific nuclear forces are charge independent to a very good approximation. In addition the Coulomb perturbation of ground and low excited states in the nuclei of $A < 50$ is small, and these states are nearly eigenstates of the total isotopic spin. The impurities range from 10^{-5} to 10^{-2} in the squared amplitude of states of different isotopic spin.

Selection rules and relations between cross sections given by the isotopic spin formalism have been examined experimentally and these results indicate that many of the excited states of nuclei also have very little admixture of states of different isotopic spin. Isolated states appear to exhibit large isotopic spin impurities, but the theoretical analyses of such cases have not been completed. Arguments from collision theory indicate that for reactions proceeding through very high excited states, the isotopic spin selection rules will be well satisfied in any case.

The total isotopic spin appears to be a useful quantum number for characterizing nuclear states and for understanding nuclear reactions. The experimental and theoretical investigation of the subject is far from complete, however, and we can expect that our understanding of the subject will continue to grow.[f]

REFERENCES

1. W. Heisenberg, Z. Physik **77**, 1 (1932).
2. B. Cassen and E. U. Condon, Phys. Rev. **58**, 953 (1940).
3. G. Breit, E. Condon, and R. D. Present, Phys. Rev. **50**, 825 (1936).
4. G. Breit and E. Feenberg, Phys. Rev. **50**, 850 (1936).
5. E. P. Wigner, Phys. Rev. **51**, 106, 947 (1937).
6. E. P. Wigner, Phys. Rev. **56**, 519 (1939).
7. R. K. Adair, Phys. Rev. **92**, 1491 (1952).
8. E. Feenberg and E. P. Wigner, Repts. Progr. in Phys. **8**, 274 (1941).
9. L. E. H. Trainor, Phys. Rev. **85**, 962 (1952).
10. R. F. Christy, Conf. on Medium Energy Nuclear Phys., Pittsburgh (1952).
11. L. A. Radicati, Phys. Rev. **87**, 521 (1952).
12. G. Morpurgo, Nuovo cimento **12**, 60 (1954).
13. W. M. MacDonald, Phys. Rev. **98**, 60 (1955).
14. D. H. Wilkinson and G. A. Jones, Phil. Mag. [7] **44**, 542 (1953).
15. D. H. Wilkinson, Phil. Mag. [7] **44**, 1019 (1953).
16. A. B. Clegg and D. H. Wilkinson, Phil. Mag. [7] **44**, 1269 (1953).
17. D. H. Wilkinson and A. B. Clegg, Phil. Mag. [7] **44**, 1322 (1953).
18. G. A. Jones and D. H. Wilkinson, Phil. Mag. [7] **45**, 703 (1954).
19. D. H. Wilkinson and A. B. Clegg, Phil. Mag. [8] **1**, 291 (1956).
20. D. H. Wilkinson, Phil. Mag. [8] **1**, 379 (1956).
21. S. D. Bloom, B. J. Toppel, and D. H. Wilkinson, Phil. Mag. [8] **2**, 57 (1957).
22. B. J. Toppel, S. D. Bloom, and D. H. Wilkinson, Phil. Mag. [8] **2**, 61 (1957).
23. D. H. Wilkinson and S. D. Bloom, Phil. Mag. [8] **2**, 63 (1957).
24. W. M. MacDonald, Phys. Rev. **100**, 51 (1955).
25. W. M. MacDonald, Phys. Rev. **101**, 271 (1956).
26. D. H. Wilkinson, Phil. Mag. [8] **1**, 1031 (1956).
27. J. B. Gerhart and R. Sherr, Bull. Am. Phys. Soc. [2] **1**, 195 (1956).
28. J. B. Gerhart, Phys. Rev. **109**, 897 (1958).
29. W. M. MacDonald, Phys. Rev. **110**, 1420 (1958).
30. P. Klein, J. Phys. radium **9**, 1 (1938).
31. L. Rosenfeld, *Nuclear Forces* (Interscience Publishers, New York, 1948), Chapter IV.
32. P. A. M. Dirac, *Quantum Mechanics* (Clarendon Press, Oxford, 1947), 3rd ed., Chapter 9.
33. L. Landau and J. Smorodinskii, J. Phys. Acad. Sci. U.S.S.R. **8**, 154 (1954).
34. J. Schwinger, unpublished notes, Harvard (1947).
35. J. Schwinger, Phys. Rev. **78**, 135 (1950).
36. E. Saltpeter, Phys. Rev. **91**, 944 (1953).
37. B. Jacobsohn, Phys. Rev. **89**, 881 (1953); D. Feldman, *ibid.* **89**, 1159 (1953).

[f] For an excellent review of the isotopic spin in nuclear reactions see the article by Burcham (*71*).

38. L. Hulthen and M. Sugawara, in *Handbuch der Physik* (J. Springer Verlag, Berlin, 1957), Vol. 39, p. 1.
39. R. S. Signell and R. E. Marshak, Phys. Rev. **106**, 832 (1957).
40. J. L. Gammel and R. M. Thaler, Phys. Rev. **107**, 291, 1337 (1957).
41. M. Kroll and L. Foldy, Phys. Rev. **88**, 1177 (1952).
42. W. E. Burcham, in *Handbuch der Physik* (J. Springer Verlag, Berlin, 1957), Vol. 40, p. 1.
43. L. A. Radicati, Proc. Phys. Soc. (London) **A66**, 139 (1953); **A67**, 39 (1954).
44. E. U. Condon and G. H. Shortley, *The Theory of Atomic Spectra* (Cambridge University Press, London and New York, 1953), p. 177.
45. F. C. Barker, Phil. Mag. [8] **2**, 386 L (1957).
46. V. G. Neudachin, J. Expl. Theoret. Phys. (U.S.S.R.) **31**, 892 (1956).
47. I. Talmi and B. C. Carlson, Phys. Rev. **96**, 436 (1954); I. Talmi and R. Thieberger, *ibid.* **103**, 719 (1956).
48. T. D. Lee and C. N. Yang, Phys. Rev. **104**, 254 (1956); D. C. Peaslee, *ibid.* **91**, 1232 (1953).
49. H. M. Mahmoud and E. J. Konopinski, Phys. Rev. **88**, 1266 (1952).
49a. J. Miller and D. C. Sutton, Bull. Am. Phys. Soc. [2] **3**, 206 (1958), and private communication from J. Miller.
49b. W. M. Visscher and R. A. Ferrell, Phys. Rev. **107**, 781 (1957).
50. R. Sherr, H. R. Muether, and M. G. White, Phys. Rev. **75**, 282 (1949).
51. R. Sherr and J. B. Gerhart, Phys. Rev. **91**, 909 (1953).
52. J. B. Gerhart, Phys. Rev. **95**, 288 (1954).
53. R. Sherr, J. B. Gerhart, H. Horie, and W. F. Hornyak, Phys. Rev. **100**, 945 (1955).
54. D. C. Sutton, private communication to J. Miller.
55. R. M. Kline and D. J. Zaffarano, Phys. Rev. **96**, 1620 (1954).
56. H. Morinaga, Phys. Rev. **100**, 431 (1955).
57. F. Ajzenberg-Selove and T. Lauritsen, Nuclear Phys. **11**, 1 (1959).
58. D. A. Bromley, E. Almqvist, H. E. Gove, A. E. Litherland, E. B. Paul, and A. J. Ferguson, Phys. Rev. **105**, 957 (1957).
59. J. D. Kington, J. K. Bair, H. O. Cohn, and H. B. Willard, Phys. Rev. **99**, 1393 (1955).
60. D. Green and J. R. Richardson, Phys. Rev. **101**, 776 (1956).
61. R. W. King, Revs. Modern Phys. **26**, 327 (1954).
62. A. Altman and W. M. MacDonald, Phys. Rev. Letters **1**, 456 (1958).
63. C. P. Browne, Phys. Rev. **104**, 1598 (1956).
64. G. Weber, L. Davis, and J. B. Marion, Phys. Rev. **104**, 1307 (1956).
65. M. Gell-Mann and V. L. Telegdi, Phys. Rev. **91**, 172 (1953).
66. D. C. Peaslee and V. L. Telegdi, Phys. Rev. **92**, 126 (1953).
67. D. H. Wilkinson, Phil. Mag. [8] **2**, 83 (1957).
68. F. C. Barker and A. K. Mann, Phil. Mag. [8] **2**, 5 (1957).
69. E. P. Wigner, private communication (1953).
70. H. Morinaga, Phys. Rev. **97**, 444 (1955).
71. W. E. Burcham, Progr. in Nuclear Phys. **4**, 171 (1955).

VI.

Nuclear Models

VI. A. The Nuclear Shell Model*

by R. D. LAWSON

1. Evidence for Magic Numbers................................... 964
2. The Shell Model... 968
   a. Ground State Spins....................................... 971
   b. Parity of Nuclear Levels.................................. 975
   c. Beta Decay.. 976
   d. Nuclear Isomerism....................................... 978
   e. Nuclear Moments.. 978
   f. Concluding Remarks...................................... 980
   References.. 981

It is well known that the electronic energy levels in an atom show a distinct shell structure. The closure of an electron shell is marked by the occurrence of an inert noble-gas atom; for example, helium and neon. Immediately after shell closure, the ionization potential of the atom shows a considerable decrease; that is, atoms like lithium and sodium readily form positive electrolytic ions. On the other hand, the atoms preceding the noble gases tend to pick up electrons and form negative ions. Such evidence clearly indicates a grouping of the electronic energy levels and is readily explained by assuming that the electrons in an atom move in a central, spherically symmetric potential.

A similar situation occurs in nuclei. There is now overwhelming evidence which shows that when a nucleus contains 2, 8, 20, 50, 82, or 126 neutrons or protons a shell closure occurs. Expressed differently, one can say that the binding energy of the 3rd, 9th, 21st, 51st, 83rd, and 127th neutron or proton is less than would be expected from a consideration of nuclei in the immediate neighborhood. In addition, there is some indication that the 29th nucleon exhibits a decrease in binding energy. However, we shall see that a close inspection of the experimental data shows that the number 28 is not as well marked as the previously mentioned ones. The experimental evidence supporting shell structure in nuclei will be presented in Section 1 of this paper.

The numbers 2, 8, 20, 50, 82, and 126 are commonly referred to as the "magic numbers" and the nuclear shell model has its origin in an attempt to understand them. As we shall see in Section 2 a model of the nucleus in which particles are assumed to move independently in a smooth potential gives rise to the desired shell structure, provided a strong spin-

* Work supported by the U.S. Atomic Energy Commission.

orbit force is assumed. Further the ground state spins and parities of odd-even nuclei (odd number of neutrons and even number of protons or vice versa) are correctly predicted when a residual short-range two-body force is included. This force, of course, perturbs the shell model wave functions and, when this effect is considered, one obtains an adequate explanation of the transition probabilities in β-decay, the magnetic moments, and in many cases the quadrupole moments of nuclei. Finally it should be mentioned that in the few places where detailed calculations have been made, it has been found that the shell model also provides an excellent description of the spins and parities of low-lying excited nuclear states.

1. Evidence for Magic Numbers[a]

A great deal of evidence for the fact that certain nucleon numbers are energetically favored comes from a close inspection of the nuclide chart.[b] One observes that, in general, the heaviest stable isotopes of neighboring even Z nuclei differ by at least two neutrons. For example, the heaviest stable isotope of zirconium is $_{40}Zr_{56}^{96}$, whereas the stability limit of its neighbor molybdenum is $_{42}Mo_{58}^{100}$, showing an increase in N by two. In another example, the pair $_{58}Ce_{84}^{142}$ and $_{60}Nd_{90}^{150}$ exhibit a neutron change of six when Z changes by two. Thus we have the rule: if the heaviest stable isotope of an even Z nucleus has N neutrons, then the stability limit of the neighboring nucleus with $Z + 2$ protons will have N increased by at least two. There are three exceptions to this rule; namely, ($_{20}Ca_{28}^{48}$, $_{22}Ti_{28}^{50}$); ($_{36}Kr_{50}^{86}$, $_{38}Sr_{50}^{88}$); and ($_{54}Xe_{82}^{136}$, $_{56}Ba_{82}^{138}$), indicating that the binding of the 29th, 51st and 83rd neutrons is smaller than usual. In a similar way the lightest isotopes of neighboring even Z nuclei generally differ by $N = 2$. There are two exceptions to this rule: ($_{40}Zr_{50}^{90}$, $_{42}Mo_{50}^{92}$) and ($_{60}Nd_{82}^{142}$, $_{62}Sm_{82}^{144}$).

For nuclei with odd Z and odd A numbers, there are only three cases where the pairs (Z,N) and $(Z + 2,N)$ are both stable (or at least very long lived). These are ($_{17}Cl_{20}^{37}$, $_{19}K_{20}^{39}$); ($_{37}Rb_{50}^{87}$, $_{39}Y_{50}^{89}$) and ($_{57}La_{82}^{139}$, $_{59}Pr_{82}^{141}$). This is exceptional since one expects that the Coulomb repulsion of the protons would tend to favor making the nuclei $(Z + 1, N + 1)$ occur and in all three cases this is not so—indicating the anomalous position of the 21st, 51st, and 83rd neutrons.

The average number of naturally occurring elements with given N generally lies between three and four. However for $N = 20$ and 28 there

[a] For more details see Mayer and Jensen (1).*
* The reference list for Section VI.A begins on page 981.
[b] Chart of the Nuclides, Knolls Atomic Power Laboratory, 5th ed. (1956).

are five stable nuclei, $N = 50$ has six naturally occurring ones, and $N = 82$ has seven stable elements. Further, the isotopic abundance of nuclei with the latter neutron numbers is abnormally high.

The evidence showing that $Z = 50$ is energetically favored is quite marked. Tin is the element with the greatest number of naturally occurring isotopes; namely ten, ranging from $_{50}Sn_{62}^{112}$ to $_{50}Sn_{74}^{124}$. One would expect that if $N = 82$ has "magic number" properties, then $Z = 82$ also should. This is borne out by a study of α-decay energies in that region

TABLE I

The binding energies of the last odd neutron in the regions $N = 20, 28, 50, 82$, and 126 are shown in Tables Ia–d for various neutron excesses. In Table Ie the odd proton binding energies near $Z = 28$ are given. The tabulated neutron binding energies are defined by B.E.$(N + 1,Z)$ − B.E.(N,Z), where B.E.(N,Z) is the total binding energy of the nucleus with Z protons and N neutrons. A similar definition holds for the proton binding energies.

TABLE Ia

Neutron Excess $(N - Z)$	Binding Energy of Odd Neutron (Mev)									
	$N = 15$	17	19	21	23	25	27	29	31	33
1	8.47	8.65	8.84	8.37						
2	7.73	7.93	8.57	7.79						
3	6.44	6.60	7.02	6.64	7.94	8.79	9.11	9.29	9.00	8.91
4			6.54	6.11	7.37	8.91	9.13	9.10	8.40	8.64
5				5.01	5.96	7.43	8.02	7.93	7.64	8.51
6						6.91	8.21	7.30	7.26	7.50

(2,3). The fact that $_{82}Pb_{126}^{208}$ is the heaviest stable even-even nucleus also leads one to expect that perhaps $N = 126$ is abnormal. This we shall see is indeed the case.

From this very brief discussion we find that particular attention should be paid to the nucleon numbers 20, 28, 50, 82 and 126. In addition, the tightly bound structure exhibited by $_2He_2^4$ and $_8O_8^{16}$ leads one to expect that the numbers 2 and 8 fall into the same category.

The binding energy information available for neutron numbers 20, 28, 50, 82, and 126 is shown in Tables Ia–d. This is taken from the recent review articles by Wapstra (4) and Huizenga (5), and embodies the data available from α-decay, β-decay, and various nuclear reactions. If the odd neutron binding energy (B.E.) is defined as

$$\text{B.E.}(N + 1,Z) - \text{B.E.}(N,Z),$$

one sees that there is a progressive increase in the separation energy as N

TABLE Ib

Neutron Excess ($N - Z$)	Binding Energy of Odd Neutron (Mev)							
	$N = 41$	43	45	47	49	51	53	55
10	6.94	7.32	7.81		8.95	8.39	7.64	
11	6.12	6.48	6.98	7.49	8.41	7.16	7.09	7.37
12			7.07	7.81	8.70	7.32	7.19	
13			5.86	6.88	6.93	6.60	6.96	6.91
14					6.79	6.11	6.71	6.91

TABLE Ic

Neutron Excess ($N - Z$)	Binding Energy of Odd Neutron (Mev)					
	$N = 75$	77	79	81	83	85
23	6.45	6.51		7.81	6.00	6.40
24		6.45	6.73		5.82	
25		6.99	6.82		5.74	
26			6.36		5.14	
27			6.30		5.24	
28				6.84	5.57	
29				5.96	3.57	

TABLE Id

Neutron Excess ($N - Z$)	Binding Energy of Odd Neutron (Mev)						
	$N = 121$	123	125	127	129	131	133
42	5.25	6.76	6.85	5.13			
43		7.31	6.73	4.55	5.09	5.55	5.97
44			6.54	4.67	5.10	5.56	6.21
45			5.54	3.87	4.32	4.59	5.33

TABLE Ie

Neutron Excess ($N - Z$)	Binding Energy of Odd Proton (Mev)				
	$N = 23$	25	27	29	31
0	5.14	4.60	3.91	1.50	
2	6.80	6.44	5.76	4.34	3.79
3	6.83	6.55	6.34	5.50	3.97
4	7.84	7.73	7.10	5.63	5.12
5	8.09	8.16	7.15	5.84	5.13

increases, then a sudden drop as the "magic number" is crossed. In the case of $N = 20$ the decrease is distinct for any neutron excess, but does vary as a function of $(N - Z)$. At $N = 50$ and 126 the data are undeniable and although at $N = 82$ the experimental information is sparse, what there is of it indicates a sharp drop in binding energy. The proton results at $Z = 20, 50,$ and 82 show similar structure indicating a decrease in binding energy of the 21st, 51st, and 83rd protons.

Thus one may conclude that immediately after the magic numbers N or $Z = 2, 8, 20, 50, 82$ and $N = 126$ there is a sharp decrease in binding energy. This is analogous to the situation encountered in atomic structure; and although the energy discontinuities in nuclear physics are not as marked, they still exist and require an explanation.

Turning to the data for nuclei with 28 neutrons or protons, we see from Table Ia that in the neutron case there is no appreciable discontinuity in the binding (or separation) energy except when the neutron excess is six. Thus binding energies lead one to doubt the validity of a shell structure here. Indeed, of the two pieces of evidence presented in the preceding paragraphs for the $N = 28$ shell, one of them, namely, the fact that there are five stable nuclei with $N = 28$, depends on the existence of $_{20}Ca_{28}^{48}$. This may equally well be interpreted as indicating that the 21st proton is very loosely bound.

The proton data near $Z = 28$ are presented in Table Ie and show that the separation energy decreases with increasing Z due to the Coulomb repulsion of the particles. There is some tendency towards a slight decrease in binding energy for the 29th proton although the effect is certainly not as marked as it is for the other proton magic numbers. One could, however, interpret the small drop in the separation energy as evidence for a subshell at $Z = 28$. We shall see later that the difference between N and Z at 28 may be qualitatively explained by Coulomb effects.

Finally it is instructive to see that the energies of the first excited states of nuclei also indicate increased stability at the magic numbers. The experimental information regarding this has been presented by Scharff-Goldhaber (6). In the case of $_8O_8^{16}$, the first excited state lies at 6.05 Mev and in $_{20}Ca_{20}^{40}$ is at 3.35 Mev—both of these are considerably higher than exhibited by other nuclei in the neighborhood. At 50 neutrons the energy of the first excited state rises to approximately 2 Mev, whereas other nuclei in the region have levels at approximately 1 Mev excitation. A factor of two shows up again at $N = 82$. For $_{82}Pb_{126}^{208}$ the first level is at 2.6 Mev—an increase by a factor of four to five over other nuclei in this neighborhood.

The increased stability also shows up as a decrease in the nuclear level densities (7) at 8 Mev excitation. This low level density manifests itself in a reduction of the capture cross section of low-energy neutrons on nuclei with magic numbers (8).

2. The Shell Model

The simplest model one can make of the nucleus is that of particles moving independently in a spherically symmetric potential well. For orientation purposes it is convenient to consider cases where the energy levels and eigenfunctions are well known. Both the harmonic oscillator and the infinite square well fall into this category (9). In the former case the levels group themselves as follows: $1s$, $1p$, $(2s,1d)$, $(2p,1f)$, $(3s,2d,1g)$, $(3p,2f,1h)$, $(4s,3d,2g,1i)$, etc., where degenerate levels are bracketed and s, p ... refer to orbital angular momentum 0, 1 .... In the latter potential the degeneracy is split, the higher angular momentum states being lowered, but a grouping of levels still persists. From this one sees that indeed a shell structure (or grouping of levels in energy) is obtained. For example, one can put two identical particles in the $1s$ state, one with spin up and the other with spin down. After this an energy gap occurs; then one can add six particles to the $1p$ level giving a total of eight particles in the well. The next set of levels can hold 12 particles making a total of 20. Therefore, from this extremely simple picture one obtains a shell structure at 2, 8, and 20 nucleons. However, proceeding in this way we run into trouble since the next shell break should occur after the $2p$ and $1f$ levels are filled with twenty particles so that an energy gap would occur at Z or $N = 40$—in contradiction with experiment.

To overcome this difficulty, Elsasser (2) and Feenberg and Hammack (10) have used a square well potential with a central elevation which increases with increasing mass number A. For small A the shell structure remains the same, but as the number of nucleons increases the central rise becomes more important and the $2s$ and $2p$ levels are pushed upward and out of binding. In this way the fifty shell is obtained from the configuration $(1s)^2(1p)^6(1d)^{10}(1f)^{14}(1g)^{18}$ and the structure at $N = 82$ from the addition of $(1h)^{22}$ and $(2d)^{10}$. Nordheim (11) has obtained the desired magic numbers by still another grouping of the levels.

The preceding considerations, in which a nucleon is assumed to move independently in a spherical potential generated by all the other particles, neglects one extremely important fact; namely, a nucleon has an intrinsic spin[c] of $\frac{1}{2}$ which may interact with its orbital motion and give rise to a

[c] Throughout this article the unit of angular momentum $\hbar$, will be suppressed when speaking of spins. Thus spin $\frac{1}{2}$ means an angular momentum of $\frac{1}{2}\hbar$.

spin-orbit coupling. This additional interaction energy may be written as

$$H' = f(r)\mathbf{\sigma}\cdot\mathbf{l} \tag{1}$$

where $\mathbf{\sigma}$ is the Pauli spin operator of the nucleon, $\mathbf{l}$ the orbital angular momentum, and $f(r)$ is an arbitrary function of r which has the dimensions of energy.

A term of the above form is encountered in atomic spectroscopy, but there it is quite small except for the heaviest atoms. However, in the case of nuclei, since the forces are not primarily electrostatic, there is no reason to consider this term to be only a small correction.

It was pointed out independently by Mayer (*12*) and Haxel, Jensen, and Suess (*13*) that if the spin-orbit energy is large, then one indeed obtains a nuclear shell structure at precisely the magic numbers. To see this, let us consider the effect of Eq. (1) on the motion of a particle. Since $\mathbf{l}$ is a good quantum number and the eigenvalues of the total angular momentum $\mathbf{j} = \mathbf{l} + \tfrac{1}{2}\mathbf{\sigma}$ are well defined, it follows that

$$j^2 = j(j+1) = l(l+1) + \mathbf{\sigma}\cdot\mathbf{l} + \tfrac{3}{4}.$$

Thus for $j = l + \tfrac{1}{2}$, $\mathbf{\sigma}\cdot\mathbf{l} = l$; and for $j = l - \tfrac{1}{2}$, $\mathbf{\sigma}\cdot\mathbf{l} = -(l+1)$. Therefore, the potential experienced by a nucleon depends on the relative orientation of its spin and orbital angular momenta and is

$$V(r) + lf(r) \qquad \text{if } j = l + \tfrac{1}{2}$$

and

$$V(r) - (l+1)f(r) \qquad \text{if } j = l - \tfrac{1}{2}$$

where $V(r)$ is the central potential (similar to the harmonic oscillator or square well mentioned previously).

Since the added potential is proportional to l, its effect is greatest for high angular momentum states. Further, $V(r)$ is negative, and hence if $f(r)$ is also negative, a particle whose spin and orbital angular momenta are parallel will experience a stronger binding than will a nucleon whose spin is antiparallel to its orbital angular momentum. As a guide, let us consider the harmonic oscillator level sequence given at the beginning of this section. For N or $Z \leq 20$, low angular momentum states are involved and hence the shell structure previously obtained should persist. Immediately after 20 nucleons the $2p$ and $1f$ levels, which can accommodate twenty particles, start to fill. The inclusion of a strong spin-orbit force depresses the $1g$ state with $j = l + \tfrac{1}{2}$ (the $1g_{9/2}$ level) until it lies near the $2p$ and $1f$ levels. Since ten particles can be added to the $1g_{9/2}$ state, we see that now shell closure occurs at N or $Z = 50$. The same arguments hold for the 82 and 126 shells which owe their origin to a depression of the $1h_{\frac{11}{2}}$ and $1i_{\frac{13}{2}}$ states, respectively.

A closer inspection of these arguments based on the harmonic oscillator level sequence shows that although, for example, the $1g_{\frac{9}{2}}$ state is depressed due to the spin-orbit interaction, the $1f_{\frac{5}{2}}$ level is raised a comparable amount (that is, the lowering of the $1g_{\frac{9}{2}}$ level is proportional to l, where $l = 4$, whereas the $1f_{\frac{5}{2}}$ state is raised an amount proportional to $l + 1$, where $l = 3$). Thus one can conclude that the harmonic oscillator fails to give the correct shell structure and its failure can be traced to the fact that the high angular momentum states are not bound tightly enough in this well. On the other hand, a potential which binds the high angular momentum states more tightly should work. As stated earlier, the square well does precisely this. However, here difficulty in the opposite extreme is encountered; the high angular momentum states lie too deep in this well! These two negative results do, however, lead to the positive conclusion that, if a potential intermediate between a square well and an harmonic oscillator were considered, the correct level sequence and shell structure would be obtained. This has recently been verified (14,15).

A potential intermediate between the square well and the harmonic oscillator—one which is flat at the center and falls smoothly to zero at the edge of the nucleus—can be represented by the function

$$V(r) = -V_0/[1 + \exp \alpha(r-a)] \qquad (2)$$

where V_0 is the well depth, α is a parameter which is related to the fall-off distance of the potential (that is, for $\alpha \to \infty$, this potential becomes the square well, $V(r) = -V_0$ for $r < a$, $V(r) = 0$ for $r > a$), and "a" may be loosely referred to as the nuclear radius. This potential alone does not, of course, lead to the correct shell structure. However, if one includes a spin-orbit potential which is a multiple, λ, of the Thomas term (16); that is

$$f(r) = \frac{-\lambda \hbar^2}{4m^2c^2} \frac{1}{r} \frac{dV}{dr} \qquad (3)$$

then the desired magic numbers are obtained.

In Fig. 1 we have plotted the neutron level sequence obtained assuming the above forms of $V(r)$ and $f(r)$ (15). In these calculations, α was taken to be 1.45×10^{13} cm^{-1} (corresponding to a fall-off distance of the potential of approximately 3×10^{-13} cm), $\lambda = 39.5$, $V_0 = 42.8$ Mev, and $a = 1.3A^{\frac{1}{3}} \times 10^{-13}$ cm. For comparison, the level sequence for a square well with the same parameters, except that $\alpha \to \infty$, is also shown. From this diagram it is seen that the shell structure at 126 is impossible to obtain with the square well because the $1i_{\frac{13}{2}}$ state lies too low.

Finally, it should be noted that there is now a great deal of evidence which indicates that the spin-orbit force in nuclei is very strong and has the sign assumed by the shell model (that is, lowers the state with $j = l + \frac{1}{2}$) (17). However, there is no strong evidence that the Thomas form, $\frac{1}{r}\frac{dV}{dr}$, where V is the central potential, is necessarily correct.

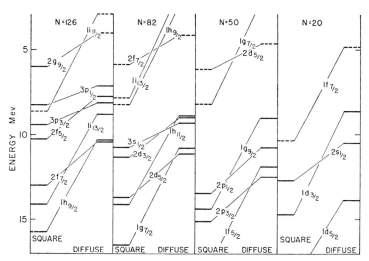

FIG. 1. The level sequence near the neutron shells 20, 50, 82, and 126 are shown. The diffuse-well calculations assume a potential, $V(r) = -V_0/\exp \alpha(r - a)$ with $\alpha = 1.45 \times 10^{13}$ cm^{-1}, $a = 1.3A^{\frac{1}{3}} \times 10^{-13}$ cm, and $V_0 = 42.8$ Mev; and a spin-orbit coupling which is 39.5 times the Thomas term. For illustrative purposes, the level sequence for a square well with the same parameters (except $\alpha \to \infty$) is plotted on the same diagram. Levels which are unfilled when the shell closes are shown by dotted lines. The occurrence of the $1i_{\frac{13}{2}}$ state below the $3p_{\frac{1}{2}}$ level at $N = 126$ clearly shows that the correct shell structure in this region cannot be obtained with a square-well potential.

a. Ground State Spins

The model as it stands at this stage of development is capable not only of predicting the magic numbers but also the spins of nuclei which are one particle away from a doubly closed shell or subshell (that is, closed-shell or subshell ± 1 nuclei). For example, the nuclei $_6C_7^{13}$ and $_7N_6^{13}$ both have a single particle outside the closed $p_{\frac{3}{2}}$ subshell and the model implies they should both have ground state spins of $\frac{1}{2}$—a prediction borne out by experiment. The nucleus $_8O_9^{17}$ should have spin $\frac{5}{2}$; $_{14}Si_{15}^{29}$ and $_{15}P_{14}^{29}$, $J = \frac{1}{2}$; $_{16}S_{17}^{33}$, $\frac{3}{2}$; and $_{20}Ca_{21}^{41}$, $\frac{7}{2}$. All of these agree with experiment.

The reason why these predictions are limited to a single nucleon outside a closed shell or subshell is easily seen as follows: Consider the simplest shell; namely, the $(1s_{\frac{1}{2}})$ level. This state can accommodate two identical particles. By the Pauli exclusion principle, one must go into a state with m_j, the z component of angular momentum, equal to $+\frac{1}{2}$, and the other into a state with $m_j = -\frac{1}{2}$. Thus the only possible value for the total z-component of angular momentum is zero, and hence the total spin must be zero. The same argument, of course, holds true for shells or subshells which can accommodate more than two particles. Therefore, the total spin of a closed shell + one nucleus must be $0 + j = j$, where j is the angular momentum of the last nucleon—and there is no ambiguity. For the case of one hole in a shell the spin is again uniquely determined to be j, where j is the angular momentum of the level which is being filled.

On the other hand, if an unfilled level has more than one particle or hole, the total angular momentum is not uniquely determined. For example, two identical particles in the $f_{\frac{7}{2}}$ level, which we shall denote by $(f_{\frac{7}{2}})^2$, can couple their spins to 0, 2, 4, and 6 (the values 1, 3, 5, and 7 are forbidden by the exclusion principle); and the more complicated configuration, $(f_{\frac{7}{2}})^3$ leads to spins of $\frac{3}{2}, \frac{5}{2}, \frac{7}{2}, \frac{9}{2}, \frac{11}{2}$, and $\frac{15}{2}$. In the model thus far proposed, all of these states are degenerate in energy, and hence one cannot make a prediction about the ground state spins of these nuclei.

To proceed further the model must be refined to include the possibility of internucleon forces acting between particles. Let us, therefore, assume that, although in zeroth order the nucleons move independently in a central potential with strong spin-orbit coupling, there is also a residual two-body force acting between the particles in a given level which it is legitimate to treat by perturbation methods. It is important to notice that the situation encountered here is just the opposite of the case in atomic physics. In nuclei it is assumed that the spin-orbit force is more important than the residual two-body interaction, and hence j-j coupled wave functions form the basis vectors of our unperturbed states; that is the orbital and spin angular momenta of each nucleon are coupled to give a good quantum number j, and then the individual j's are added to given the total angular momentum J of the aggregate of particles. On the other hand, in atomic spectroscopy except for the heaviest atoms the two-body force is more important than the spin-orbit interaction, and then the L-S coupling scheme is more nearly correct.[d] Because of the fact that the spin-orbit force predominates in nuclei, one often calls the nuclear shell model the j-j coupling shell model.

Due to the residual two-body force, the first order correction to the

[d] In light nuclei the coupling rules are intermediate between the j-j and L-S limits. These cases are discussed in detail by D. Kurath in Section VI.B.

energy of a state, ψ_J, is given by (18)

$$\Delta E = \int \psi_J^* U \psi_J \, d\tau \tag{4}$$

where U is the two-body force, ψ_J is the wave function of the state in question and is compounded from the individual particle wave functions taking into account the Pauli principle, and the integration is over all space. The ground state will then be that state for which the pairing energy, given by Eq. (4), has its greatest negative value. Of course, the magnitude of this pairing energy depends not only on the potential U, but also on the shape of the nuclear wave functions; that is, although for a spherically symmetric potential the angular and spin dependent part of ψ_J is unambiguously determined, the radial part depends on the assumed form of $V(r)$ and $f(r)$. However, the main features of this pairing energy are easily found if one assumes certain limiting forms for U.

Mayer (19) and Racah (20) have shown that if a short-range spin-independent force acts between identical nucleons in the same level [that is, $U(r) = -V_0 \delta(r_1 - r_2)$], then the following rules result:

(1) An even number of particles have lowest energy when they are in a $J = 0$ state. Thus the ground state spins of all even-even nuclei are zero.

(2) An odd number of particles filling a level j have maximum negative pairing energy when their spins couple to $J = j$. Thus the ground state spin of an odd-even nucleus should be j, where j is the angular momentum of the level which the odd particles are filling. This rule leads to the extreme single particle model in which all the angular momentum of the nucleus is assumed carried by the last odd nucleon.

(3) The interaction energy of two particles coupling their individual angular momenta j to a total spin J, depends not only on J but also on j. For $J = 0$, the pairing energy increases with increasing j.

On the other hand, attractive space exchange forces with a long range predict that the lowest energy state of a configuration is the one in which J has its minimum value (1,21).

Actually in all pairing energy calculations the parameter of importance is not the range of the forces, but is the ratio of the force range to the radius of the nucleus. Since in all but the lightest nuclei the two-body force range is much less than the size of the nucleus, one expects that the delta function approximation to the forces should generally give results in accord with experiment. This is, indeed, found to be true—the ground state spins of almost all odd-even nuclei are found to be j, where j is the angular momentum of the level which the odd particles are filling. For example, the nucleus $_{20}Ca_{23}{}^{43}$ which has a configuration $(1f_{\frac{7}{2}})^3$ has measured spin $\tfrac{7}{2}$; $_{42}\text{Mo}_{53}{}^{95}$ is $(2d_{\tfrac{5}{2}})^3$ with spin $\tfrac{5}{2}$; and $_{60}\text{Nb}_{85}{}^{145}$ is represented by $(2f_{\tfrac{7}{2}})^3$ with angular momentum $\tfrac{7}{2}$.

There are many instances where rule (3) is operative. For example, a glance at the level scheme shown in Fig. 1 shows that the $1h_{\tfrac{11}{2}}$ and $3s_{\tfrac{1}{2}}$ levels lie close together. Thus, according to rule (3), it is energetically favorable for the $1h_{\tfrac{11}{2}}$ state to fill in pairs, and therefore the region of nuclei where these two levels are filling should exhibit a spin of $\tfrac{1}{2}$. This is precisely the situation encountered from $N = 65$ to 75. It is interesting to note that even when the spins differ by only a small amount rule (3) may be operative. For example, in the region $Z = 29$–38, the two levels $(2p_{\tfrac{3}{2}})$ and $(1f_{\tfrac{5}{2}})$ appear to lie quite close together with a ground state spin of $\tfrac{3}{2}$ being measured in all cases except $_{37}\text{Rb}_{48}{}^{85}$ which is $\tfrac{5}{2}$. In the region $N = 29 - 38$, there are many fewer data. The spin of $_{24}\text{Cr}_{29}{}^{53}$ is $\tfrac{3}{2}$ and that of $_{30}\text{Zn}_{37}{}^{67}$ is $\tfrac{5}{2}$. For $_{26}\text{Fe}_{31}{}^{57}$ the experiments indicate either $\tfrac{3}{2}$ or $\tfrac{1}{2}$ (22), and the other odd neutron nuclei in the region have ground state spins which are not known with certainty.

For lighter nuclei there are some instances where the odd neutron or odd proton nuclei have spin $j - 1$ instead of j. For example, $_{10}\text{Ne}_{11}{}^{21}$ and $_{11}\text{Na}_{12}{}^{23}$ both exhibit spins of $\tfrac{3}{2}$, whereas the shell model says the configuration should be $(d_{\tfrac{5}{2}})^3$. This result is not surprising since certainly the delta function approximation for the interaction is less valid for these light nuclei. The spins are easily explained when forces of finite range are considered.

It is interesting to note that the level order for neutrons and protons is not always identical. For example, the last neutron to be added before the closure of the 82 shell is a $(2d_{\tfrac{3}{2}})$ particle, as shown by the ground state spin of $_{56}\text{Ba}_{81}{}^{137}$ which is $J = \tfrac{3}{2}$. On the other hand, the proton shell closes with the addition of a $3s_{\tfrac{1}{2}}$ particle (the ground state spin of $_{81}\text{Tl}_{124}{}^{205}$ is $\tfrac{1}{2}$). Similarly the 83rd neutron goes into a $(2f_{\tfrac{7}{2}})$ state ($_{60}\text{Nd}_{83}{}^{143}$ has spin $\tfrac{7}{2}$) whereas $_{83}\text{Bi}_{126}{}^{209}$ has $J = \tfrac{9}{2}$, indicating the $1h_{\tfrac{9}{2}}$ level. Detailed calculations (15) in which the Coulomb repulsion between the protons is incorporated shows that these level crossovers are easily explained. This alteration in level sequence may be understood by noting that for a uniform charge distribution the Coulomb potential is a maximum when $r = 0$. On the other hand, the high angular momentum states are kept away from the origin by the centrifugal barrier (that is, the radial eigenfunction for a state of angular momentum l goes like r^l for $r \to 0$) and hence these states are less affected by the repulsive force.

This immediately provides a qualitative explanation of the difference in proton and neutron shell structure at N or $Z = 28$. Here the two competing levels are $1f_{\tfrac{7}{2}}$ and $2p_{\tfrac{3}{2}}$. While these may be reasonably close together for neutrons, they tend to be pushed apart in the case of protons because of Coulomb repulsion.

b. Parity of Nuclear Levels

In addition to predicting the ground state spins of nuclei, the shell model also gives the orbital angular momentum of the topmost nucleon. Therefore, the transformation properties of the wave function under space reflection are determined; that is, when $x \to -x, y \to -y, z \to -z$, the eigenfunction changes sign or not depending on whether l is odd or even

$$\psi(xyz) = (-1)^l \psi(-x,-y,-z).$$

The parity of a state is defined by its transformation under space reflection and is said to be even $(+)$ if the wave function is unchanged and odd $(-)$ if the eigenfunction changes sign. Thus a nuclear state is designated not only by its spin but also by its parity. For example, the notation $\frac{7}{2}^+$ would imply a state of angular momentum $\frac{7}{2}$ and even parity, the $g_{\frac{7}{2}}$ state, whereas $\frac{7}{2}^-$ would mean a level with spin $\frac{7}{2}$ and odd parity, that is an $f_{\frac{7}{2}}$ level.

When two nucleons are in the same state, for example the $(f_{\frac{7}{2}})^2$ configuration discussed previously, although each particle has odd parity, the two of them together have even properties under space inversion. Thus the complete designation of the ground state of this configuration is 0^+, and the possible excited states would be 2^+, 4^+, and 6^+. On the other hand, a $d_{\frac{3}{2}}$ particle and an $f_{\frac{7}{2}}$ nucleon can have spins 5, 4, 3, and 2 (that is, $|j_1 - j_2| \leq J \leq j_1 + j_2$) but in this case the resultant parity would be negative so that the complete description of the $J = 5$ state arising here would be 5^-.

Since the shell model makes this additional prediction about the nuclear levels, it is interesting to see if it checks with experiment. Perhaps the most beautiful experimental confirmation of the shell model comes from deuteron stripping (23) experiments.[e] For example, the d-p stripping on $_{20}Ca_{22}^{42}$ leading to the ground state of $_{20}Ca_{23}^{43}$ shows that the angular momentum of the captured particles is $l = 3$ (24). Thus the ground state arises from adding an f-state neutron to $_{20}Ca_{22}^{42}$ in agreement with the shell model prediction.

It is interesting to note that the magnitude of the cross section obtained in the stripping reaction depends on the l_n value of the accepted particle and roughly obeys the rule $\sigma_{l_{n+2}}/\sigma_{l_n} \sim \frac{1}{10}$. This provides a stringent test for the shell model, as pointed out by Bethe and Butler (25). Suppose we consider the d-p stripping using as target nucleus $_{17}Cl_{18}^{35}$ which has spin $\frac{3}{2}^+$. The resulting nucleus, $_{17}Cl_{19}^{36}$ has a ground state of 2^+ (26). The allowed (23) values of l_n are 0, 2, and 4 ($l_n = 1$ and 3

[e] For a discussion of direct interactions, see M. K. Banerjee and C. A Levinson, Section V.B.

are excluded since parity is conserved in the interaction). On the other hand, the shell model predicts that in this region a $d_{\frac{3}{2}}$ particle is being added so that l_n should be 2. Since $l_n = 0$ stripping is approximately ten times more favorable, when it is allowed, one sees that this reaction provides a critical test of the model. The experiments of King and Parkinson (27) show that the $l_n = 2$ curves agree excellently with the experimental data and that there is at most something of the order of 4% mixing of the $2s_{\frac{1}{2}}$ level in the ground state wave function. There are other d-p reactions which provide a similar test for the model, and the experiments are all in excellent agreement with the shell model predictions.

Thus an analysis of the stripping reactions leads to the conclusion that not only does the shell model predict the correct ground state spins for nuclei, but also that it gives an excellent account of the orbital angular momentum of the states.

c. Beta Decay

The preceding sections have shown that the shell model when applied to stable nuclei gives results in excellent accord with experiment. The results of beta decay provide a test of the predictions of the model when unstable nuclei are considered. The theory of beta decay, discussed in Section V.D, shows that there are certain selection rules governing the lifetime of the process which depend on the relative parity and angular momenta of the nuclear states involved. The data on lifetimes and spectrum shapes have been tabulated in many places (1,28). A study of these results shows that the beta-decay selection rules indicate spins and parities in agreement with the shell model predictions.

However, in the matter of ft values, the extreme single particle model—which satisfactorily explains the spins of the majority of odd-even nuclei and assumes that all the angular momentum is carried by the last odd nucleon—has much less success. This is not surprising since the matrix elements involved depend on a detailed knowledge of the nuclear wave functions and certainly one does not expect the extreme single particle description to be more than qualitatively correct.

As an example of the degree of variance between theory and experiment, consider the two decays (1) $_{21}Sc_{22}^{43} \rightarrow {}_{20}Ca_{23}^{43}$ and (2) $_{22}Ti_{21}^{43} \rightarrow {}_{21}Sc_{22}^{43}$, which have log ft values of 4.8 and 3.4 respectively. The latter decay falls into the category of a Wigner superallowed transition (29) since it involves mirror nuclei, and hence one expects a high degree of overlap of the two wave functions used in calculating the nuclear matrix element. The extreme single particle shell model predicts that both log ft values should be 3.37. On the other hand, we have already encountered cases in explaining the spins of nuclei, where it is necessary to depart from this picture and include all the nucleons outside a closed shell in determining the spin. This effect should also be operative in calculating the beta-decay matrix elements. Indeed, Grayson and Nordheim (30) have shown that when wave functions involving both the neutrons and protons in the $f_{7/2}$ shell are used, the log ft values become 3.97 and 3.58 respectively. Thus some improvement is obtained with this type of consideration.

To see if better results can be obtained, let us look at the matrix elements involved in allowed beta decay. These are two in number, and in the nonrelativistic limit become

$$\int \psi_f^* \sum_k t^k \psi_i \, d\tau$$

and

$$\int \psi_f^* \sum_k t^k \sigma^k \psi_i \, d\tau$$

where ψ_i and ψ_f are the initial and final nuclear wave functions, t^k is the operator which changes a neutron to a proton, σ^k is the Pauli spin operator for particle k, the sum on k is over all nuclear particles, and the integration is over all space.

Now, as we have stated previously, in order to explain the observed ground state spins of nuclei, one has to introduce a residual two-body force between nucleons. Due to the influence of this perturbation, the ground state wave function will no longer have its unperturbed form, but will be modified by the inclusion of other configurations. Thus if ψ_0 is the unperturbed wave function and χ_n denotes another function which has the same spin and parity as ψ_0, the wave function as a result of the perturbation will become (18)

$$\psi = \psi_0 + \sum_n \alpha_n \chi_n$$

where
$$\alpha_n = \frac{\int \chi_n^* U \psi_0 \, d\tau}{E_0 - E_n} \tag{5}$$

with U the residual two-body force and $E_n - E_0$ the energy difference between the two configurations.

In particular, let us consider those χ_n's which can give a contribution to the beta decay matrix element which is linear in α_n. Obviously these must correspond to functions in which one and only one of the particles has been elevated from the state with $j = l + \frac{1}{2}$ to the state with $j = l - \frac{1}{2}$. In the example considered in this section, the shell model configuration is $(f_{\frac{7}{2}})^3$ coupled to $J = \frac{7}{2}$. The type of mixing we want to consider would arise from a configuration $[(f_{\frac{7}{2}})_{J'}{}^2 f_{\frac{5}{2}}]_{J=\frac{7}{2}}$, by which we mean two $f_{\frac{7}{2}}$ particles coupled to spin J' and then J' and $f_{\frac{5}{2}}$ coupled to $J = \frac{7}{2}$. Although the amount of this type of admixture is generally small, say of the order of a few percent, the additional contribution we calculate is linear in α_n and hence the effects may be quite large (that is, even a 4% probability corresponds to $\alpha_n = 0.2$).

Blin-Stoyle and Caine (*31*) have shown that it is possible to choose a two-body interaction which gives log ft values for the above two decays of 4.8 and 3.38 respectively, in excellent agreement with experiment.

d. Nuclear Isomerism

A feature of the strong spin-orbit coupling shell model is the fact that shell structure is obtained by the lowering of a high angular momentum state with $j = l + \frac{1}{2}$ from one harmonic oscillator shell to another. Thus we have states of high angular momentum and given parity occurring very close in energy to levels which have the opposite parity and perhaps a low angular momentum. One example of this was mentioned when the ground state spins of nuclei were discussed; namely, the close occurrence of the $3s_{\frac{1}{2}}$ and $1h_{\frac{11}{2}}$ levels in the neighborhood of N or $Z = 65$. In addition the states $1g_{\frac{9}{2}}$ and $2p_{\frac{1}{2}}$ in the region N or $Z = 38$–50 occur close together, and near $N = 100$ the $1i_{\frac{13}{2}}$ state is in competition with lower angular momentum states.

In much the same way as discussed in the theory of beta decay, gamma-ray transitions between these neighboring states is markedly inhibited because of the large angular momentum change involved. Thus when one of the levels occurs as ground state and the other as the first excited level, nuclear reactions which populate this latter state will give rise to excited nuclei which have a long half-life. These levels, whose lifetimes range from a fraction of a second to years, are known as isomeric states. The shell model predicts the existence of these isomeric states in odd A nuclei in the above mentioned regions and these are precisely where they are found to exist (*32*).

In principle the model should also predict the lifetimes of these states, but again a detailed knowledge of the nuclear wave function is required, and the calculations made to date have yielded no more than qualitative agreement with experiment (*33*).

e. Nuclear Moments

In addition to the spins and parities already discussed, another property of the nuclear states is the distribution of the charge and current making up the system. These are characterized by electric and magnetic multipole moments.

The single-particle magnetic dipole moment, in units of the nuclear magneton, $e\hbar/2mc$, is defined as

$$\mu = <gl_z + \mu_i\sigma_z>m_{j=j}$$

where l_z and σ_z are the z-components of the orbital and spin angular momenta of the particle; μ_i is the magnetic moment of the free particle and is equal to 2.79 for a proton and -1.91 for a neutron; and $<...>m_{j=j}$ means the expectation value of the operator taken in the state in which the z-component of angular momentum m_j has its maximum value j. The number g is unity for protons and zero for neutrons corresponding to the fact that a charged particle can give rise to a current.

The above matrix element is easily evaluated (1) and gives the so-called Schmidt (34) or single particle value

$$\mu = (j - \tfrac{1}{2})g + \mu_i \qquad \text{for } j = l + \tfrac{1}{2}$$
$$\mu = jg + \frac{j}{j+1}\left(\frac{1}{2}g - \mu_i\right) \qquad \text{for } j = l - \frac{1}{2}. \qquad (6)$$

Comparison of the calculated and observed magnetic moments (35) shows that in general the experimental values differ greatly from those predicted by Eq. (6), again vividly illustrating the shortcomings of the single-particle model wave functions. However, with only three exceptions, $_2\text{He}_1{}^3$, $_1\text{H}_2{}^3$, and $_6\text{C}_7{}^{13}$, the measured moments do fall between the two possible Schmidt values for a particle of given total angular momentum, and invariably lie closer to the value derived from shell model considerations. For example, $_{54}\text{Xe}_{77}{}^{131}$ has a measured spin of $\tfrac{3}{2}$ and a magnetic moment of $+0.7$. The angular momentum $\tfrac{3}{2}$ could arise from a $d_{\frac{3}{2}}$ state, as predicted by the shell model, or from a $p_{\frac{3}{2}}$ level. The theoretical moments for these two configurations are $+1.15$ and -1.91 respectively, showing that indeed the shell model configuration gives results in better accord with experiment.

Recently it has been shown (36,37) that configuration mixing of the sort discussed in beta decay brings the experimental and theoretical values of the magnetic moments into good agreement. In addition Blin-Stoyle and Perks (37) have pointed out that correction terms linear in α_n, where α_n is given by Eq. (5), should be very small for nuclei with doubly closed shells in both L-S and j-j coupling plus or minus one odd nucleon. There are five nuclei satisfying this criterion; namely, $_8\text{O}_9{}^{17}$, $_9\text{F}_8{}^{17}$, $_{20}\text{Ca}_{21}{}^{41}$, $_{20}\text{Ca}_{19}{}^{39}$, and $_{19}\text{K}_{20}{}^{39}$. Experimentally the moments of the first and last of these are known and both have small deviations from the Schmidt limit—0.02 and 0.27 nuclear magnetons respectively.

There have been a few measurements of higher magnetic multipole moments (the next nonvanishing magnetic moment is the magnetic octopole moment), and they deviate from the single-particle equations (38) in much the same manner as did the magnetic dipole moments from the Schmidt lines.

Turning to the electric moments, the lowest order one that occurs is the electric quadrupole moment which is defined as

$$Q = e<3z^2 - r^2>m_{j=j}$$

where e is the electric charge and z and r are the coordinates of the particle. In general the measured quadrupole moments again agree very poorly with the shell model predictions. The inadequacy of the single-particle assumption is clearly borne out by the fact that odd neutron nuclei have nonvanishing moments. Thus it is clear that although the general characteristics of the nuclear ground state seem to be determined by the single-particle model, it is necessary to consider mixed configurations to obtain an adequate description. Near closed shells the theoretical values of the quadrupole moments are in reasonable accord with experiment. Indeed, when a small amount of configuration mixing is considered (39), the shell model can account for many nuclear quadrupole moments. However, in the rare earth region the moments become as high as thirty times the single-particle value. It is precisely this difficulty which leads one to the conclusion that this is a region in which collective effects are important and it is not a valid approximation to consider nucleons as moving independently in a spherically symmetric potential.[f]

f. Concluding Remarks

From the preceding discussion we see that the shell model leads not only to the desired shell structure, but also is an excellent tool for predicting the ground state spins and parities of both stable and unstable nuclei. However, the actual nuclear wave function is bound to differ from the shell model one because of the influence of the residual two-body forces. Indeed, we have seen that this is necessary if we are to explain the observed moments and transition probabilities. When more experimental information becomes available on magnetic octopole moments and the moments of excited states, it will be interesting to see if these too are explainable in terms of the model.

[f] For a detailed discussion of collective effects in nuclei, see A. Bohr and B. Mottelson, Section VI.C.

For the most part the early shell model calculations were concerned with explaining the anomalous ground state spins of nuclei (*40*). Recently however, even more ambitious projects have been undertaken; namely, the understanding of excitation energies and spins of low-lying excited nuclear states. Of necessity these calculations are limited to regions near closed shells since here the number of competing configurations is not overwhelming. In general the results of these calculations (*41*) are in excellent accord with experiment. However, in many cases the theoretical spin assignments cannot be compared with experiment since only the excitation energies are known. Measurements of the spins of these low-lying states would do much to clarify the situation.

Finally no mention has been made of the validity of the assumption that the nucleons in a nucleus can be assumed to move independently in a smooth central potential. Certainly the well known fact that the nucleon-nucleon force is strong and of short range would indicate that this was not a good approximation. However, such a conclusion does not take into account the Pauli principle and hence, despite the existence of a strongly fluctuating potential, a nucleon in nuclear matter may often have a smooth path since there are no empty states into which it can scatter (*42*). Recently, attempts to understand the basic assumptions of the shell model have been made (*43*). However, these calculations are still in progress and the tentative results obtained will not be discussed here.

REFERENCES

1. M. G. Mayer and J. H. D. Jensen, *Elementary Theory of Nuclear Shell Structure* (John Wiley and Sons, New York, 1955).
2. W. J. Elsasser, J. phys. radium **4**, 549 (1933); **5**, 389, 635 (1934).
3. I. Perlman, A. Ghiorso, and G. T. Seaborg, Phys. Rev. **77**, 26 (1950).
4. A. H. Wapstra, Physica **21**, 367, 385 (1955).
5. J. R. Huizenga, Physica **21**, 410 (1955).
6. G. Scharff-Goldhaber, Phys. Rev. **90**, 587 (1953).
7. A. A. Ross, Phys. Rev. **108**, 720 (1957).
8. D. J. Hughes and D. Sherman, Phys. Rev. **78**, 632 (1950).
9. H. A. Bethe and R. F. Bacher, Revs. Modern Phys. **8**, 82 (1936).
10. E. Feenberg and K. C. Hammack, Phys. Rev. **75**, 1877 (1949).
11. L. W. Nordheim, Phys. Rev. **75**, 1894 (1949).
12. M. G. Mayer, Phys. Rev. **75**, 1969 (1949); **78**, 16 (1950).
13. O. Haxel, J. H. D. Jensen, and H. E. Suess, Phys. Rev. **75**, 1766 (1949); Z. Physik **128**, 295 (1950).
14. A. E. S. Green and K. Lee, Phys. Rev. **99**, 772 (1955).
15. A. A. Ross, H. Mark, and R. D. Lawson, Phys. Rev. **102**, 1613 (1956).
16. L. H. Thomas, Nature **117**, 514 (1926); D. R. Inglis, Phys. Rev. **50**, 783 (1936).
17. For example, M. Heusinkveld, and G. Freier, Phys. Rev. **85**, 80 (1952); R. M. Sternheimer, *ibid.* **97**, 1314 (1955).
18. L. I. Schiff, *Quantum Mechanics* (McGraw-Hill Book Co., New York, 1949).
19. M. G. Mayer, Phys. Rev. **78**, 22 (1950).

20. G. Racah, Research Council Israel, L. Farkas Memorial Vol. (1952).
21. G. Racah, Phys. Rev. **78**, 622 (1950).
22. G. Trumpy, Nuclear Phys. **2**, 664 (1956); Nature **176**, 507 (1955).
23. S. T. Butler, Proc. Phys. Soc. (London) **A208**, 559 (1951).
24. C. M. Braams, Phys. Rev. **95**, 650 (1954).
25. H. A. Bethe and S. T. Butler, Phys. Rev. **85**, 1045 (1952).
26. P. B. Sogo and C. D. Jeffries, Phys. Rev. **98**, 1316 (1955).
27. J. S. King and W. C. Parkinson, Phys. Rev. **88**, 141 (1952).
28. E. Feenberg, *Shell Theory of the Nucleus* (Princeton University Press, 1955).
29. E. P. Wigner, Phys. Rev. **56**, 519 (1939).
30. W. C. Grayson, Jr., and L. W. Nordheim, Phys. Rev. **102**, 1084, 1093 (1956).
31. R. J. Blin-Stoyle and C. A. Caine, Phys. Rev. **105**, 1810 (1957).
32. D. E. Alburger, *Handbuch der Physik* (J. Springer Verlag, Berlin, 1957), Vol. 42, p. 1.
33. V. F. Weisskopf, Phys. Rev. **83**, 1073 (1951); B. Stech, Z. Naturforsch. **7a**, 401 (1952); S. A. Moszkowski, Phys. Rev. **89**, 474 (1953).
34. T. Schmidt, Z. Physik **106**, 358 (1937).
35. R. J. Blin-Stoyle, *Theories of Nuclear Moments* (Oxford University Press, London and New York, 1957).
36. A. Arima and H. Horie, Progr. Theoret. Phys. **11**, 509 (1954).
37. R. J. Blin-Stoyle, Proc. Phys. Soc. (London) **A66**, 1158 (1953); R. J. Blin-Stoyle and M. A. Perks, *ibid.* **A67**, 885 (1954).
38. C. Schwartz, Phys. Rev. **97**, 380 (1955).
39. H. Horie and A. Arima, Phys. Rev. **99**, 778 (1955).
40. D. Kurath, Phys. Rev. **80**, 98 (1950), **88**, 804 (1952), **91**, 1430 (1953); B. H. Flowers, Proc. Roy. Soc. (London) **A212**, 248 (1952); **A215**, 398 (1952); A. R. Edmonds and B. H. Flowers, *ibid.* **A214**, 515 (1952); **A215**, 120 (1952); I. Talmi, Helv. Phys. Acta **25**, 185 (1952).
41. K. W. Ford and C. A. Levinson, Phys. Rev. **100**, 1, 13 (1955); D. Kurath, Phys. Rev. **101**, 216 (1956); S. Goldstein and I. Talmi, *ibid.* **102**, 589 (1956); S. P. Pandya, *ibid.* **103**, 956 (1956); J. B. French and B. J. Raz, *ibid.* **104**, 1411 (1956); I. Talmi, *ibid.* **107**, 326 (1957); R. D. Lawson and J. L. Uretsky, *ibid.* **106**, 1369 (1957); **108**, 1300 (1957); R. K. Sheline, Proc. Conf. on Nuclear Structure, Pittsburgh (1957); W. W. True and K. W. Ford, Phys. Rev. **109**, 1675 (1958); M. J. Kearsley, Phys. Rev. **106**, 389 (1957); Nuclear Phys. **4**, 157 (1957).
42. E. Fermi, *Nuclear Physics* (University of Chicago Press, Chicago, 1950); V. F. Weisskopf, Helv. Phys. Acta **23**, 187 (1950).
43. K. A. Brueckner, C. A. Levinson, and H. M. Mahmoud, Phys. Rev. **95**, 217 (1954); K. A. Brueckner and J. L. Gammel, *ibid.* **105**, 1679 (1957) and the references quoted therein; L. C. Gomes, J. D. Walecka and V. F. Weisskopf, Ann. Phys. (N.Y.) **3**, 241 (1958).

VI. B. Nuclear Coupling Schemes

by D. KURATH

1. Fundamental Coupling Procedure.................................... 984
   a. Definition of a Configuration.................................... 984
   b. Coupling of Two Angular Momenta............................ 985
   c. Two Nucleons in the Same Level................................ 986
   d. Energy Separation of States of Different J................. 988
   e. Application of Results to Zr^{90}.................................... 990
   f. Application of Results to Pb^{206}.................................... 993
   g. Summary.. 997
2. Many-Particle Spectra.. 997
   a. Coupling of Three Nucleons.................................... 997
      (1) Pertinent Configurations.................................... 997
      (2) Spectrum of $(j_1)^2(j_2)$....................................... 998
      (3) Spectrum of $(j_1)^3$.. 998
      (4) Experimental Evidence...................................... 999
   b. Extrapolation to More Complex Nuclei......................... 1000
      (1) Seniority Concept and Application........................ 1001
      (2) Experimental Evidence..................................... 1002
3. Intermediate Coupling in Light Nuclei............................. 1002
   a. Two Identical $1p$ Nucleons.................................... 1003
      (1) The jj Description.. 1003
      (2) The LS Description... 1005
   b. General Features and Experimental Comparison................ 1006
   References... 1008

The nuclear shell model treated in Section VI.A is able to describe many features of nuclei, particularly concerning the ground states, with a few simple assumptions. In this chapter we shall consider to what extent these assumptions must be modified if we want to understand the spectra of excited states of nuclei in greater detail. The basic assumptions which lead to what we shall call the single-particle shell model are that the nucleus can be described in terms of particles moving in a spherically symmetric potential well, that there is a strong force coupling together the spin angular momentum and orbital angular momentum of each particle, and that there is a large pairing energy. The first assumption will be retained in this chapter and is discussed in Chapter VI.C. The assumption of strong spin-orbit coupling will also be retained in the first part of this chapter since it seems to be valid for most of the periodic table. The deviations will be discussed in the latter part of this chapter where the light nuclei are treated.

The assumption of a large pairing energy means that when there are two neutrons or two protons in the same shell model level they will find it energetically very favorable to couple their angular momenta to a resultant of zero. The consequences of this are that even-even nuclei have zero angular momentum in the ground state, and that the properties of odd-even nuclei can be described in terms of the possible states available to the single odd nucleon. The latter consequence is the reason for calling it a single-particle model. Modification of this picture is what we shall consider first.

There is much experimental information available about the states of low excitation in the spectra of nuclei. One often knows the energy of excitation, angular momentum, parity, and something about the probability for transition between states by emission of gamma radiation. If we attempt to interpret these data, it is soon clear that the single-particle shell model is inadequate. The spectra of odd-even nuclei generally contain many more states than are available from the single-particle picture. This is evidence that we must refine the assumption that an even number of particles in the same shell model level always couple their angular momenta to zero. The hypothesis that it takes much more energy to break this coupling than to excite a single particle from one shell model level to the next, is strictly true only for the case that the even number fills a complete shell to one of the "magic" numbers. The spectra of nuclei having a single particle beyond a closed shell or a single hole below a closed shell are in agreement with such a simple picture. However, when we do not have such special cases, the experimental evidence indicates that we must consider the possibility of forming excited states by recoupling the angular momenta of the nucleons beyond closed shell, to obtain resultants other than those given by the assumptions of the single-particle model. The energy required to form such states is apparently no more than is needed to go from one shell model level to the adjacent level. The basic material for this chapter on nuclear coupling schemes is the description of these new states, and the modifications of shell model interpretations of spectra and other nuclear properties brought about by the inclusion of such states.

1. Fundamental Coupling Procedure

a. Definition of a Configuration

In order to investigate the relative importance of the two possible ways of obtaining excited states, we shall study those nuclei where either the neutrons or the protons are in an especially stable configuration such as occurs at one of the magic numbers. Then the states of low excitation energy must be attributed to the remaining group of N identical nucleons in the unfilled shell. The term shell is used in the sense of referring to all the levels occurring between two adjacent magic numbers. The states available to each individual nucleon of this group are still those arising from a spherically symmetric potential well with strong spin-orbit coupling. The levels are labelled by the shell model identification (nlj), and we distribute our N nucleons among these levels as:

$$(n_1 l_1 j_1)^{N_1} (n_2 l_2 j_2)^{N_2} \cdots (n_k l_k j_k)^{N_k}$$

where $\Sigma_{i=1}^{k} N_i = N$. Such a distribution is called a configuration, and the number of configurations depends on the number of single-particle levels (nlj), that we consider, and also on the number of nucleons N. It is clear that the number of possible configurations is very large unless either N is small or else the number of levels that must be considered is small. Both of these limited cases have been studied, and we shall first consider the limitations of a small value of N in order to understand the basic procedure.

b. COUPLING OF TWO ANGULAR MOMENTA

The combining of two angular momenta is the basic process in the theory of angular momentum coupling. If we consider the two-nucleon case there will in general be two single-particle levels involved, $(n_1 l_1 j_1)$ and $(n_2 l_2 j_2)$ although the possibility that both nucleons are in the same level is not excluded. Within a level the nucleon can be in any of the $2j + 1$ possible states with different m, the z-component of angular momentum, which has values $j, j - 1, j - 2, \cdots -j$. There are then $(2j_1 + 1)(2j_2 + 1)$ possible products we can form from the single-particle functions:

$$\varphi_{m_1}{}^{j_1}(a) \varphi_{m_2}{}^{j_2}(b).$$

Here a and b refer to the two nucleons and we have omitted the (nl) labels which are not relevant to the process of coupling angular momenta. We can just as well consider linear combinations of these products, and we want to do this in such a way that the resulting combinations are states with total angular momentum J and z-component M. This can be done, giving functions which are written in the form:

$$\psi_M{}^J(a,b) = \sum_{m_1=-j_1}^{j_1} \sum_{m_2=-j_2}^{j_2} (j_1 j_2 m_1 m_2 | j_1 j_2 J M) \varphi_{m_1}{}^{j_1}(a) \varphi_{m_2}{}^{j_2}(b). \quad (1)$$

In Eq. (1) the symbol $(j_1 j_2 m_1 m_2 | j_1 j_2 J M)$ merely indicates that this is a coefficient relating a product state identified by $j_1 m_1$ and $j_2 m_2$ to a state of total angular momentum J and z-component M, formed by coupling j_1 and j_2. We also want the resulting functions to be normalized, and these requirements determine the coefficients in the linear combinations to within a phase factor. To avoid confusion a conventional choice of phase is used by physicists and the coefficients are called Clebsch-Gordan coefficients, or more appropriately, vector-addition coefficients. These coefficients are the building blocks in the theory of angular momentum coupling and they have been tabulated (1)* for the most frequently occurring arguments. The name vector-addition coefficients comes from the fact that the coefficients are zero unless the angular momenta satisfy the relationship:

$$\mathbf{j}_1 + \mathbf{j}_2 = \mathbf{J}$$
$$m_1 + m_2 = M. \tag{2}$$

From Eqs. (2) we see that the allowed values of J which can arise in such linear combinations (1), are given by the range:

$$J = j_1 + j_2, j_1 + j_2 - 1, j_1 + j_2 - 2, \cdots |j_1 - j_2|. \tag{2'}$$

The vector-addition coefficients enable us to form $(2j_1 + 1)(2j_2 + 1)$ independent linear combinations of the two-nucleon product functions which have different total angular momentum quantum numbers J, and z-components M with $-J \leq M \leq J$.

The wave function described by Eq. (1) does not satisfy the Pauli Exclusion Principle for identical particles. In order to do this, it should be antisymmetric to exchange of nucleons a and b. However this property is simply obtained by subtracting $\psi_M{}^J(b,a)$ from $\psi_M{}^J(a,b)$ to obtain:

$$\Phi_M{}^J(a,b) = \frac{1}{\sqrt{2}} \sum_{m_1,m_2} (j_1 j_2 m_1 m_2 | j_1 j_2 J M) \{\varphi_{m_1}{}^{j_1}(a)\varphi_{m_2}{}^{j_2}(b) - \varphi_{m_1}{}^{j_1}(b)\varphi_{m_2}{}^{j_2}(a)\}. \tag{3}$$

The factor $1/\sqrt{2}$ has been added to give a normalized function. When the two nucleons are in different levels their wave function is given in final form by Eq. (3). The possible resultant angular momenta are given by Eqs. (2) with the limiting values contained in the vector-addition coefficients since they vanish outside these limits.

c. Two Nucleons in the Same Level

If the nucleons are in the same (nlj) level, the Exclusion Principle has some further effects on the wave functions and the determination of

---

* The reference list for Section VI.B is on page 1008.

which angular momenta are allowed. The wave function from Eq. (3) for $j_1 = j_2 = j$ is:

$$\Phi_M^J(a,b) = \frac{1}{\sqrt{2}} \sum_{m_1, m_2} (jjm_1m_2|jjJM)\{\varphi_{m_1}^j(a)\varphi_{m_2}^j(b) - \varphi_{m_1}^j(b)\varphi_{m_2}^j(a)\}. \quad (4)$$

Now m_1 and m_2 both run through the same range of values, so there are some terms in the summation which vanish and some which are repeated. First of all, when $m_1 = m_2$ the wave function inside the summation sign is zero.[a] This just demonstrates the requirement of the Exclusion Principle that no two particles can occupy the same state.

The remainder of the double summation is now split into two parts having the same number of terms; namely, those for which m_1 is greater than m_2 and those for which m_1 is less than m_2:

$$\sum_{m_1 > m_2} + \sum_{m_1 < m_2} = \sum_{m_1 = m_2+1}^{j} \sum_{m_2 = -j}^{j-1} + \sum_{m_1 = -j}^{m_2-1} \sum_{m_2 = -j+1}^{j}$$

Each term in the first sum has a partner in the second sum and we can combine them as follows. The term for $m_1 = \mu$ and $m_2 = \gamma$ in the first sum contributes:

$$(jj\mu\gamma|jjJM)\{\varphi_\mu^j(a)\varphi_\gamma^j(b) - \varphi_\mu^j(b)\varphi_\gamma^j(a)\}.$$

There is a term in the second sum coming from $m_1 = \gamma$ and $m_2 = \mu$ which contributes:

$$(jj\gamma\mu|jjJM)\{\varphi_\gamma^j(a)\varphi_\mu^j(b) - \varphi_\gamma^j(b)\varphi_\mu^j(a)\}.$$

These two terms can be combined[b] to give:

$$\{(jj\mu\gamma|jjJM) - (jj\gamma\mu|jjJM)\} \times \{\varphi_\mu^j(a)\varphi_\gamma^j(b) - \varphi_\mu^j(b)\varphi_\gamma^j(a)\}.$$

From the properties of the vector-addition coefficients under exchange of symbols[c] we find:

$$(jj\gamma\mu|jjJM) = (-1)^{2j+J}(jj\mu\gamma|jjJM)$$

[a] Notice that it is essential that the particles be in the same level for this to happen. For even if $j_1 = j_2$ but $n_1 \neq n_2$ or $l_1 \neq l_2$ the single-particle wave function would be different, cancellation would not occur, and the wave function is still given completely by Eq. (3). The labels (nl) must be kept in mind even though they are not explicitly shown.

[b] The possibility of combining terms in this way again depends on having particles in the same level with not only $j_1 = j_2$ but $n_1 = n_2$, $l_1 = l_2$ as well.

[c] See, for example, M. E. Rose (2).

so that the μ, γ contribution is finally:

$$(1 - (-1)^{2j+J})(jj\mu\gamma|jjJM)\{\varphi_\mu{}^i(a)\varphi_\gamma{}^i(b) - \varphi_\mu{}^i(b)\varphi_\gamma{}^i(a)\}.$$

In this way the partners from the two summations are combined, and we see that they either cancel each other if $(2j + J)$ is an even number or add equal contributions if $(2j + J)$ is an odd number. Using the fact that $2j$ is an odd number we can write Eq. (4) as:

$$\Phi_M{}^J(a,b) = \frac{(1 + (-1)^J)}{\sqrt{2}}$$
$$\times \sum_{m_1 > m_2} (jjm_1m_2|jjJM)\{\varphi_{m_1}{}^i(a)\varphi_{m_2}{}^i(b) - \varphi_{m_1}{}^i(b)\varphi_{m_2}{}^i(a)\}. \quad (4')$$

Therefore the wave function vanishes when J is odd, so that for two identical particles in the same level, the possible resultant angular momenta are restricted to be:

$$J = 0, 2, 4, \cdots 2j - 1. \quad (5)$$

In order to obtain a normalized wave function we must include another $1/\sqrt{2}$ factor, so we have finally:

$$\Phi_M{}^J(a,b) = \sum_{m_1 > m_2} (jjm_1m_2|jjJM)\{\varphi_{m_1}{}^i(a)\varphi_{m_2}{}^i(b) - \varphi_{m_1}{}^i(b)\varphi_{m_2}{}^i(a)\} \quad (6)$$

with $J = 0, 2, 4, \cdots 2j - 1$. Alternatively since for even J we have seen that $\Sigma_{m_1 > m_2} = \Sigma_{m_2 > m_1} = \frac{1}{2}\Sigma_{m_1,m_2}$, where m_1 and m_2 assume all possible values, it is sometimes more convenient to replace $\Sigma_{m_1 > m_2}$ by $\frac{1}{2}\Sigma_{m_1,m_2}$ in Eq. (6).

d. Energy Separation of States of Different J

Insofar as the forces acting on a nucleon can be represented by the spherically symmetric potential well of the shell model, the states of different angular momentum J which are obtained by coupling two nucleons should be degenerate in energy. However, replacing the interaction of all other nucleons with a given nucleon by an effective potential well is not an adequate representation of all the interactions. It seems to be equivalent when one is talking about interactions with nucleons in states of quite different energy, and since the majority of the interactions are of such nature, it is reasonable to speak of the nucleon as being in a potential well as a first approximation. However, the interactions of a nucleon with nucleons in the same energy level or in a level of nearly the same energy cannot be replaced completely by an effective potential, and it is these residual interactions which split up the levels of different J coming from a configuration. In the single-particle model of Chapter VI.A the effect of such interactions was included by making the assumption about pairing energy. This says that for two identical nucleons in the same (nlj) level the state of angular momentum zero is greatly favored. As far as determining the ground state is concerned, this is sufficient since putting the two nucleons in different levels would give higher energy.

Since we now wish to consider excited states as well as ground states we require more information about the nature of the residual interactions which split the levels of a configuration. This question touches on the problem of the theoretical foundations of the shell model, and though the foundations are much more nearly apparent than they were when the model was first successfully applied, explicit information on many points is still lacking. The nature of the residual interactions is one of these points, so treatments of the problem are generally based on using an interaction of relatively simple form with parameters chosen so that calculations fit certain simple cases.

The interaction between identical nucleons is taken to be a two-body potential of the form:

$$V_{12} = [A + B(\mathbf{\sigma}_1 \cdot \mathbf{\sigma}_2)]J(r_{12}) \qquad (7)$$

where $\mathbf{\sigma}_1$ and $\mathbf{\sigma}_2$ are the Pauli spin operators for the coordinates of the respective nucleons, $J(r_{12})$ is some function of the distance between nucleons, and A and B are constants. While there is considerable uncertainty about the nuclear interaction parameters, some results have been obtained which are relatively insensitive to the parameters one chooses. These can be stated in the form of coupling rules, the first of which concerns nucleons in the same level.

(a) For two nucleons in the same shell model level, the states listed in order of increasingly large excitation are:

$$J = 0, 2, 4, 6, \cdots (2j - 1)$$

with $J = 0$ lowest by a considerable amount.

This more detailed treatment therefore contains as a first approximation the assumption of the single-particle model that there is a large pairing energy favoring the state $J = 0$.

When the two single-particle levels are not the same no such general rule can be stated. The states of the configuration lie much closer together than they do when the single-particle levels are alike. However, a general tendency can be stated which is useful for a first orientation of the level ordering, though it is much less firmly established than the rule cited above for nucleons in the same level. Based on results (3) obtained when the spatial dependence of the residual interaction is of very short range, one can say:

(b) The possible J's tend to split into two groups, the one of higher energy comprising those states for which $[l_1 + l_2 + J]$ is odd, while the lower energy group has $[l_1 + l_2 + J]$ even. Furthermore the states in the lower group will be ordered with the largest J lowest in energy if $j_1 + j_2 = l_1 + l_2$. On the other hand, if $j_1 + j_2 = l_1 + l_2 \pm 1$, the state of smallest J will be lowest in energy.

Since $[l_1 + l_2]$ determines the parity of the two-nucleon state, the first part of the rule can be rephrased to say that the low-lying states of the configuration obey the rule:

Odd parity → odd J
Even parity → even J.

The second part of the rule says that if one nucleon is of the type $j = l + \frac{1}{2}$ and the other has $j = l - \frac{1}{2}$, the ground state will be the one having the largest J available in the lower energy group. Finally if both nucleons are of the same type, both $j = l + \frac{1}{2}$ or both $j = l - \frac{1}{2}$, the ground state will be the smallest J available in the low-energy group.

Now that we have refined the coupling rules for two nucleons, let us look at some two-nucleon spectra to see whether such states are observed.

e. APPLICATION OF RESULTS TO Zr^{90}

In looking for nuclei which should fit our simple coupling picture one would expect to find them near doubly magic nuclei so that either the neutrons or the protons can be ignored. For reasons of simplicity we should also like to find a region where the number of levels available to the nucleons is not large so that we need consider only a few configurations.

The simplest example that has been found is $_{40}Zr^{90}$, where the states populated by the positron decay of $_{41}Nb^{90}$ can be interpreted as arising from the configurations of two protons in two single-particle levels. The experimental information on single-particle levels is given in Fig. 1, where the numbers on the left indicate the number of protons present when this level and all the ones below it are full. Since the 50 neutrons form a closed major shell, the excited states of $_{40}Zr^{90}$ should arise from the proton configurations. The ground configuration is $(2p_{\frac{1}{2}})^2$ and the states of low excitation should come from the $(2p_{\frac{1}{2}})^1(1g_{\frac{9}{2}})^1$ and $(1g_{\frac{9}{2}})^2$ configurations which have a basic excitation of 0.9 Mev and 1.8 Mev, respectively, due to the separation of the $(2p_{\frac{1}{2}})$ and $(1g_{\frac{9}{2}})$ levels. There are two configurations giving rise to states of even parity, and one giving states of odd parity.

The effects of including the interaction energy between the particles in the same configuration can be estimated from the coupling rules of the previous subdivision to obtain a qualitative picture of the spectrum. The ground configuration has only one state, $J = 0$, since it is a full level, and the interaction will lower the energy of this state. The $(2p_{\frac{1}{2}})(1g_{\frac{9}{2}})$ configuration can form only two states, $J = 4$ and 5, and the parity is odd since $l_1 = 1$ and $l_2 = 4$. The rule which told us that odd parity goes with odd angular momentum for the lowest state tells us that $J = 5^-$ is lower in energy. The lowering of the energy by the interaction is also much less in this case than in the identical level configuration. Finally, the $(1g_{\frac{9}{2}})^2$ configuration will have states $J = 0, 2, 4, 6$, and 8 of even parity, lowered in energy and spread out by a considerable amount as the result of the interaction. In order to make this picture quantitative

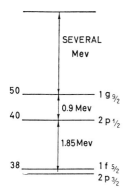

Fig. 1. Experimental evidence about proton single-particle levels near $A = 90$. Number of protons present when level is full is given on the left.

one would need to make a calculation, but we can see what sort of magnitudes must be given to the effects we have mentioned to fit the experimental results (4). This is shown in Fig. 2, and the experimental evidence is summarized in Fig. 3.

The experimental evidence (5) shows that the beta decay of $_{41}Nb^{90}$ exhibits all of the excited states we expect from our configuration except for $J = 4^-$. There is good reason to believe that this level is skipped in the de-excitation of $_{40}Zr^{90}$ because of gamma-ray selection rules, so there is no evidence that it cannot be present. It is very fortunate that $_{41}Nb^{90}$ has apparently such a high spin that it prefers to decay to the $J = 8^+$ state. It should be mentioned that there are two other states below 3.6 Mev that have been seen by inelastic neutron scattering but are not seen in this beta decay. They occur at 3.2 Mev and 2.76 Mev, and their spin assignments are uncertain.

However the experiment is good evidence that low excited states come from within configurations. The amount of energy between adjacent states from the $(1g_{\frac{9}{2}})^2$ configuration is of the same order of magnitude as

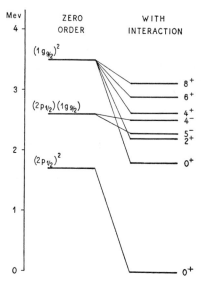

FIG. 2. Interpretation of the Zr^{90} spectrum in terms of two-proton configurations. [After K. W. Ford, Phys. Rev. **98**, 1516 (1955).]

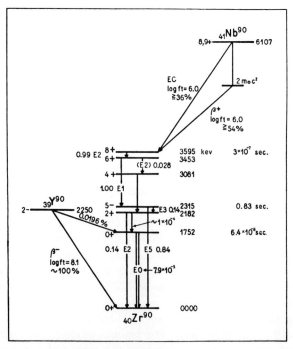

FIG. 3. Summary of experimental evidence in Zr^{90}. [From S. Bjørnholm, O. B. Nielsen, and R. K. Sheline, Phys. Rev. **115**, 1613 (1959).]

the separation in energy of the single-particle levels $(2p_{\frac{1}{2}})$ and $(1g_{\frac{9}{2}})$. The different degree to which the interaction spreads apart the states of various configurations can also lead to overlapping of configurations as $(1g_{\frac{9}{2}})^2$ and $(2p_{\frac{1}{2}})(1g_{\frac{9}{2}})$. There is one further point which arises in this example, namely that both the $(2p_{\frac{1}{2}})^2$ and the $(1g_{\frac{9}{2}})^2$ configurations contain a state $J = 0^+$. States of the same angular momentum arising from different configurations do interact by means of the two-nucleon potential V_{12}, of Eq. (7), and this interaction is important when the two configurations lie close together as they do here. In order to obtain a quantitative picture, this effect, which is called configuration interaction, must be included. The result in Zr^{90} is that the two $J = 0^+$ states effectively push each other apart, and the resultant wave functions are mixtures of the $(2p_{\frac{1}{2}})^2{}_{J=0}$ and the $(1g_{\frac{9}{2}})^2{}_{J=0}$ functions. A quantitative treatment of $_{40}Zr^{90}$ has been carried out (6), and expectation values of the various observed quantities have been calculated using the coupled wave functions of the form of Eq. (3) and Eq. (6) with quite encouraging results.

f. Application of Results to Pb^{206}

The other example of a two-nucleon spectrum which has been studied in detail is $_{82}Pb^{206}$. Here there are two neutrons missing from the doubly magic $_{82}Pb^{208}$, and the states one can form from two such "holes" in a

Table I. Single Particle Levels from $_{82}Pb^{207}$

Energy of Excitation (Mev)	Assigned Single-Particle Level	Parity
0	$(3p_{\frac{1}{2}})^{-1}$	−
0.57	$(2f_{\frac{5}{2}})^{-1}$	−
0.90	$(3p_{\frac{3}{2}})^{-1}$	−
1.63	$(1i_{\frac{13}{2}})^{-1}$	+
2.35	$(2f_{\frac{7}{2}})^{-1}$	−

shell are the same as those one obtains from having just two particles in the shell. Furthermore, for a two-body interaction such as Eq. (7), the spectrum of excited states one obtains is the same as that for two particles, so we shall talk as though there were two particles to be coupled instead of two holes.

The single-particle levels that are available can be obtained from the states of $_{82}Pb^{207}$, where there is a single hole in the neutron shell. They are given in Table I, for states below 2.5 Mev excitation. The energy values show how much is needed to raise a neutron from a given filled level into the highest level $(3p_{\frac{1}{2}})$. When treating the particle picture which is equivalent to this "hole" situation we then talk as though the level order were $(3p_{\frac{1}{2}})$, $(2f_{\frac{5}{2}})$, $(3p_{\frac{3}{2}})$, $(1i_{\frac{13}{2}})$, and $(2f_{\frac{7}{2}})$ with the excitation energies above $(3p_{\frac{1}{2}})$ given in column 1 of Table I.

There are more close-lying single-particle levels present than occurred in Zr^{90}, so that we have many more configurations for two neutrons in $_{82}Pb^{206}$. However, the fact that there is a level of parity different from that of the other levels, provides a gap above the $(3p_{\frac{3}{2}})$ level. The states with one neutron in $(1i_{\frac{13}{2}})$ and the other neutron in a different level will all be negative parity states, and we shall consider these. The states of positive parity which involve the $(2f_{\frac{7}{2}})$ level or the $(1i_{\frac{13}{2}})$ level are omitted since they will be considerably higher than those from the lower three levels. The configurations and their resultant spins are given in Table II, where the separation of the configurations due to the energy differences of the single-particle levels is listed in the second column.

TABLE II. STATES EXPECTED IN Pb^{206} FROM THE LOW CONFIGURATIONS

Configuration	Rel. Energy from Single-Particle Levels	Allowed J's	Parity	Lowest State
$(3p_{\frac{1}{2}})^2$	0	0	+	0
$(3p_{\frac{1}{2}})(2f_{\frac{5}{2}})$	0.57	2, 3	+	2
$(3p_{\frac{1}{2}})(3p_{\frac{3}{2}})$	0.90	1, 2	+	2
$(2f_{\frac{5}{2}})^2$	1.14	0, 2, 4	+	0
$(2f_{\frac{5}{2}})(3p_{\frac{3}{2}})$	1.47	1, 2, 3, 4	+	4
$(3p_{\frac{3}{2}})^2$	1.80	0, 2	+	0
$(3p_{\frac{1}{2}})(1i_{\frac{13}{2}})$	1.63	6, 7	−	7
$(2f_{\frac{5}{2}})(1i_{\frac{13}{2}})$	2.20	4, 5, 6, 7, 8, 9	−	9
$(3p_{\frac{3}{2}})(1i_{\frac{13}{2}})$	2.53	5, 6, 7, 8	−	5

The rules for estimating the effect of the nuclear interaction in splitting the states of a configuration give lowest states for each configuration as listed in the last column of Table II. A quantitative picture of the way the configurations are split by the interaction is shown in Fig. 4 from the work of True and Ford (7). The effect of the interaction is seen to be greater for configurations of two nucleons in the same level, which are those on the left, except for the $(3p_{\frac{1}{2}})(3p_{\frac{3}{2}})$ configuration. Here the effect is just as large, and this is because the two levels have the same n and l, so that the radial wave functions are the same. The reduction of the interaction effect should be expected only in states of different nl.

The states from the various configurations overlap considerably and one must expect that configuration interaction will be important. Indeed the two extensive treatments of $_{82}Pb^{206}$ which have been carried out (7), have demonstrated that configuration interaction must be included in order to obtain a quantitative fit to the experimental spectrum. This means that the states of the same parity and angular momentum coming from the various configurations must be considered together. The effect of the two-body interaction V_{12}, must be evaluated within each such

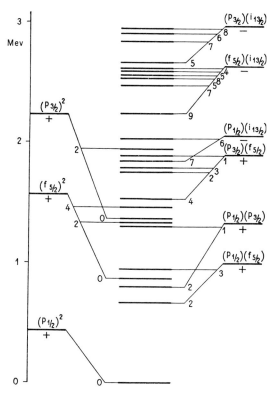

FIG. 4. Effect of the interaction within the configurations of Pb^{206}. First and last columns give the configuration, parity, and relative positions of the configurations resulting from the single-particle level spacings. The interaction spreads the states, identified by J's in the intervening sections, to give the spectrum in the center. Note confirmations of the coupling rules. [After W. W. True and K. W. Ford, Phys. Rev. **109**, 1675 (1958).

group of states. If the matrix elements of V_{12} between the states are small with respect to the separation of the states which is shown in Fig. 4, then the effect of this configuration interaction can be evaluated by perturbation theory. Usually this is not the case and one must diagonalize energy matrices as in the treatment of the effect of a perturbation on degenerate states.

In either case, the wave functions which result from including configuration interaction are then linear combinations of the pure configuration states given by Eqs. (3) and (6). For example, the three $J = 0^+$ states, in order of increasing energy, contain the following fractions

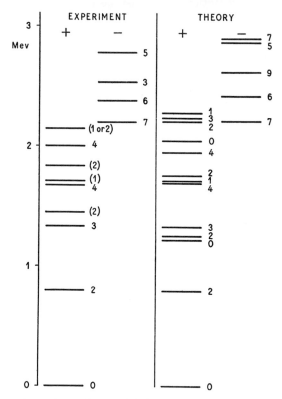

FIG. 5. Comparison of experiment with theory for Pb^{206}. Positive parity states are in the (+) columns, negative in the (−) columns. Note the change in the theoretical spectrum from that of Fig. 4, caused by including configuration interaction.

(squares of coefficients) of $J = 0^+$ states from pure configurations:

$$\psi_1(J = 0^+) \rightarrow 0.74(p_{\frac{1}{2}})^2 + 0.11(f_{\frac{5}{2}})^2 + 0.15(p_{\frac{3}{2}})^2$$
$$\psi_2(J = 0^+) \rightarrow 0.18(p_{\frac{1}{2}})^2 + 0.79(f_{\frac{5}{2}})^2 + 0.03(p_{\frac{3}{2}})^2$$
$$\psi_3(J = 0^+) \rightarrow 0.08(p_{\frac{1}{2}})^2 + 0.11(f_{\frac{5}{2}})^2 + 0.81(p_{\frac{3}{2}})^2.$$

Each still comes chiefly from one configuration, but its wave function contains admixtures which can have large effects on matrix elements for things like gamma-transition probabilities. Other states like those with

$J = 2^+$ are even more mixed, while those with $J = 6^-$, for example, stay pure configurations to a few tenths of a percent. The amount of mixing depends on the size of the interaction matrix elements between configurations, relative to the energy separation when no configuration interaction is included as in Fig. 4.

The agreement between the calculated and experimental (8) spectra is shown in Fig. 5. The comparison for the positive parity states is cut off above about 2 Mev since states resulting from configurations that have been neglected will be important. The over-all agreement is excellent, with some seven excited states well fitted and some others tentatively identified. There are some states not seen, like the excited $J = 0^+$ levels, but these would be difficult to populate experimentally, especially in beta decay from which most of this experimental information comes. There are also a few states seen experimentally which are not accounted for in the calculation, but these can be attributed to configurations not considered. The states of low excitation in $_{82}Pb^{206}$ can therefore be considered well understood.[d]

g. SUMMARY

From the study of these two examples we can conclude that the shell model is able to account for states of low excitation in these quite simple even-even nuclei. However, in order to do this one must consider all the likely configurations, so that if there are many single-particle levels close in energy, the spectrum is complicated. An idea of which states will be low can be obtained from the coupling rules of Section 1.d. Finally a quantitative fit must include the effect of interactions between configurations if the simple ordering results in having states of the same J and parity close together. So while a qualitative picture can be rather easily obtained, a quantitative picture requires calculation.

2. Many-Particle Spectra

a. COUPLING OF THREE NUCLEONS

(1) *Pertinent Configurations*

The discussion of the two-nucleon coupling has exhibited all of the basic ingredients which enter into the analysis of nuclear spectra. Since odd-A nuclei are also observed to have levels which are not explained by the single-particle model, we wish to extend the ideas about angular momentum coupling to their spectra. The coupling of three identical

[d] See also Chapter VI.C.

nucleons shows a possible modification to help in the interpretation of odd-A spectra. We put these nucleons into single-particle levels (nlj), indicating the lowest available level by j_1, the next by j_2 and so on with the understanding that the (nl) differ also. Then since we learned in the two-nucleon case that the effect of the residual interactions is much stronger if the nucleons are in the same level, the lowest configuration will[e] certainly be $(j_1)^3$, followed by $(j_1)^2(j_2)$. The next configurations would be $(j_1)^2(j_3)$ and $(j_2)^2(j_1)$, so that there are two types of configurations producing the low odd-A states.

(2) Spectrum of $(j_1)^2(j_2)$

The ordering of states from the $(j_1)^2(j_2)$ type should be determined principally by the strong interaction between the two nucleons in the same level. We found that these two couple to states $J' = 0, 2, 4, \cdots (2j_1 - 1)$, with $J' = 0$ considerably lower in energy than the others. The states of $(j_1)^2(j_2)$ are formed by vector-addition of j_2 to the various J', and since the effect of the residual interaction is much weaker for nucleons in different levels, the lowest state comes from coupling j_2 to $J' = 0$, giving the resultant $J = j_2$. The many other states which arise from coupling j_2 to the remaining values of J' will lie considerably higher, by an amount of the same order of magnitude as the two-nucleon separation between $J' = 0$ and $J' = 2$. In the odd-A spectra such configurations therefore provide states of low excitation of single-particle nature with J equal to that of the odd nucleon. Only in higher excitation would their other states occur.

(3) Spectrum of $(j_1)^3$

The $(j_1)^3$ configuration differs in two ways from $(j_1)^2(j_2)$. First of all we cannot simply vector-couple j_1 to the possible states of $(j_1)^2$ to obtain the allowed J for $(j_1)^3$, because the Pauli Principle must be included. This requires that no two nucleons be in the same single-particle state $\varphi_{m_i}{}^{j_1}$, and severely restricts the allowed values of J. Secondly, the effect of the residual interactions in splitting the states of allowed J cannot be estimated from the two-nucleon results as in $(j_1)^2(j_2)$. There the interaction between the nucleons in j_1 is much stronger than between j_1 and j_2, so we effectively ignored the latter, but when all three nucleons are in j_1 such an approximation is not valid.

The question of what resultant angular momenta are possible for $(j_1)^3$ can be answered by a straight-forward procedure (9) of classifying

[e] This is not true for $j_1 = \frac{1}{2}$ since the Pauli Principle allows only two nucleons in such a level.

the possible three-nucleon states one can form consistent with the Pauli Principle. The results, in configurations for which there is some experimental evidence, are given in Table III.

The problem of how the states of $(j_1)^3$ are ordered in energy requires evaluation of the effect of the residual interaction, Eq. (7). In the limit of very short range, the ground state is $J = j_1$, well below the other states. However as the range of the spatial dependence of the interaction is taken to be more nearly that expected in nuclei, the isolation of the ground state becomes less pronounced, and the states with $J < j_1$ are

TABLE III. POSSIBLE ANGULAR MOMENTA FOR $(j)^3$ CONFIGURATIONS

j	Allowed J
$\frac{3}{2}$	$\frac{3}{2}$
$\frac{5}{2}$	$\frac{3}{2}, \frac{5}{2}, \frac{9}{2}$
$\frac{7}{2}$	$\frac{3}{2}, \frac{5}{2}, \frac{7}{2}, \frac{9}{2}, \frac{11}{2}, \frac{15}{2}$
$\frac{9}{2}$	$\frac{3}{2}, \frac{5}{2}, \frac{7}{2}, \frac{9}{2}$ (twice), $\frac{11}{2}, \frac{13}{2}, \frac{15}{2}, \frac{17}{2}, \frac{21}{2}$

much closer to it. This contrasts with the range effect in the $(j_1)^2$ configuration where $J = 0$ is always well below the other states. Therefore one can conclude that while for the $(j_1)^2(j_2)$ type of configuration the ground state has J equal that of the odd nucleon and is well isolated, the $(j_1)^3$ configuration has J equal that of the odd nucleon, but states with $J < j_1$ are not far above it.

(4) Experimental Evidence

In order to check these conclusions experimentally one would like a region where there is a fairly well isolated single-particle level of sufficiently large j to allow several states from $(j)^3$. Also the remaining nucleons should be in a closed shell configuration to simplify the spectrum. Such is the case for the calcium isotopes where the neutrons are filling the $(1f_{\frac{7}{2}})$ level which lies some 2 Mev below the $(2p_{\frac{3}{2}})$ level. A quantitative study of the $_{20}\text{Ca}^{43}$ spectrum has been carried out (10), but there is some difficulty in making such calculations since other single-particle levels in this region are not well located, so including the effect of configuration interaction is difficult.

However, there is ample evidence for the presence of several low-lying states from $(1f_{\frac{7}{2}})^3$ as well as the higher $J = \frac{3}{2}^-$ state from $(1f_{\frac{7}{2}})^2(2p_{\frac{3}{2}})$. In $_{20}\text{Ca}^{43}$ there are three states below 0.6 Mev and these are $J = \frac{7}{2}^-, \frac{5}{2}^-$, and $\frac{3}{2}^-$, in that order, from $(1f_{\frac{7}{2}})^3$. The same three states are the lowest found in $_{23}\text{V}^{51}$ where there is a closed neutron shell and the $(1f_{\frac{7}{2}})^3$ proton configuration. Such states are also seen in $_{20}\text{Ca}^{45}$ and $_{25}\text{Mn}^{53}$ where the $(1f_{\frac{7}{2}})^{-3}$ configuration,—namely, three holes—is expected. The fact that such states are not seen in the single-particle nucleus $_{20}\text{Ca}^{41}$ is good evidence that they are the expected many-particle states from $(1f_{\frac{7}{2}})^{3}$, but there is another very convincing way to identify them.

The basis is of such an identification is provided by the relative intensities with which such states are seen in the deuteron stripping reactions. If we apply that discussion to Ca^{43}, we know that the stripping reaction intensity from $\text{Ca}^{42}(d,p)$ to a state of given J in Ca^{43} will depend primarily on how well that state is described as two $(1f_{\frac{7}{2}})$ neutrons coupled to angular momentum zero together with a third neutron whose single-particle angular momentum is $j = J$. If this is a good description, as it is for $(1f_{\frac{7}{2}})^3$ with $J = \frac{7}{2}^-$, or $(1f_{\frac{7}{2}})^2(2p_{\frac{3}{2}})^1$ with $J = \frac{3}{2}^-$, then the cross section will be large. These have strong similarity to single-particle states, and are seen in Ca^{43} with large intensity and about the same energy separation as in $\text{Ca}^{40}(d,p)\text{Ca}^{41}$, where $2p_{\frac{3}{2}}$ at about 2 Mev is the only well identified single-particle excited state.

The other states arising from $(1f_{\frac{7}{2}})^3$ with $J \neq \frac{7}{2}$ do not have such a single-particle nature, so they would not be populated in a stripping reaction. However, if there is some configuration interaction so that for example the lowest $J = \frac{3}{2}^-$ state in Ca^{43} is predominantly $(1f_{\frac{7}{2}})^3$ with a small admixture of $(1f_{\frac{7}{2}})^2(2p_{\frac{3}{2}})^1$, then it would be populated weakly. Therefore these deuteron-stripping reaction intensities can be used first of all to decide from what configuration the state comes and second to obtain a measure of the amount of admixture from other configurations and thus find the importance of configuration interaction. Such an analysis has been made in the Ca isotopes (*11*) and it confirms the $(1f_{\frac{7}{2}})^{\pm 3}$ nature of the low-lying states of Ca^{43} and Ca^{45}.

b. Extrapolation to More Complex Nuclei

The interpretation of excited states by means of nuclear coupling schemes has been tested for cases involving only two or three identical nucleons or holes outside of closed shells. Here one obtains a very good account of the experimental results, even when the presence of many easily available single-particle levels complicates the calculation as in Pb^{206}. Adding more nucleons outside of closed shells presents a prohibitive amount of labor if one were to attempt a complete calculation of the effects of residual interactions, but we can extrapolate the previous results to get a rough idea of what to expect. Since the low states should come from the configuration with as many nucleons as possible in the lowest single-particle level, we must first look at the question of what to expect from the configuration $(j)^n$ with $n > 3$.

(1) *Seniority Concept and Application*

In going from the configuration $(j)^1$ to $(j)^3$ we found that the state $J = j$ was still lowest in energy. This is essentially because we can form such a state from $(j)^1$ by adding a pair of nucleons coupled in the energetically favored way of angular momentum zero. Similarly, if we add another pair we expect that the low states of $(j)^5$ will be those which were low for $(j)^3$ and that $J' = j$ is lowest. Such a generalization of the pairing energy assumption from the single-particle model is the basis of the "seniority" classification, introduced by Racah (*12*) as a method of distinguishing those states of $(j)^{n+2}$ which are formed from states of $(j)^n$ by adding a pair of nucleons coupled to $J' = 0$. This method enables one to classify the states (*13*) according to the number of such paired nucleons they contain; the ones with the larger numbers are said to have the lower seniority and should have lower energy. Actually, we saw that for $(j)^3$ there was not a large separation between the state of lowest seniority, $J = j$, and the states of higher seniority which have $J < j$. Nevertheless the low states of $(j)^{2k+3}$ should contain those which are low in $(j)^3$ and similarly $(j)^{2k+2}$ should resemble $(j)^2$.

If the seniority classification were strictly obeyed, then there should be a quantitative similarity between the lower states of spectra of nuclei which can have states of the same seniority, such as $(1f_{\frac{7}{2}})^3$ and $(1f_{\frac{7}{2}})^5$. However, in actual cases there are various factors which tend to destroy the quantitative similarity, even though the level order may remain similar. The chief disturbance is configuration interaction which has a strong effect on the quantitative spacing of the states, and in a region such as we have around Pb^{206} there would be so many possible nearby configurations that without calculation one can do little more than list the likely low states.

A second factor is the interaction between the neutron and proton configurations when both are away from closed shells. For the heavier nuclei this should not be so strong since different levels are being filled, but below $A \sim 60$, the same levels are being filled by neutrons and protons, so there will be stronger interaction. So in an odd-A nucleus the quantitative spacing of the states from the odd group of nucleons will depend on how many nucleons there are in the even group.

The essence of what we can hope to extrapolate from the simple cases to the more complicated nuclei in the region is then the following. In odd-A nuclei the states of low excitation come from the odd group of nucleons, and according to the idea of seniority we expect them to be those present in the three-nucleon case since they come from the related configurations $(j)^{2k+3}$ and $(j)^{2k+2}(j')$. For even-even nuclei if one group is much harder to excite than the other, the spectrum should be like that for $(j)^2$ of the more easily excited group. If both neutrons and protons can be easily excited, the spectrum may still be $J = 0, 2, 4, \ldots$, but the states will be mixed by the interaction, reflecting the fact that we cannot tell whether the neutrons or protons are being excited.

(2) *Experimental Evidence*

Experimental evidence for the validity of such extrapolation is rather limited. It is most easily found in regions where configuration interaction is not of major importance, such as the filling of the $(1f_{7/2})$ level between 20 and 28 particles and the $(1g_{9/2})$ level between 40 and 50 particles. For odd-A nuclei with the odd group between 41 and 49, a low-lying $J = \frac{7}{2}^+$ state is found as well as $J = \frac{9}{2}^+$ only for 43, 45, and 47 nucleons. Similarly, low-lying $J = \frac{5}{2}^-$ and $\frac{3}{2}^-$ levels are found in addition to $\frac{7}{2}^-$ only for 23 and 25 particles in odd-A nuclei of the $1f_{7/2}$ region. The even-even nuclei also show some spectra similar to what one expects from an interpretation via seniority. In both $_{22}\text{Ti}_{26}{}^{48}$ and $_{24}\text{Cr}_{28}{}^{52}$ the $J = 0^+, 2^+, 4^+, 6^+$ series of states characteristic of $(1f_{7/2})^2$ has been identified.

While these spectra seem to show the many-particle shell model features there are further factors which come in. Besides configuration interaction there is also the possibility of exciting the closed shell core, which is configuration interaction of a more complex form. This is a sort of collective excitation which is most strongly shown by the presence of electric quadrupole transition probabilities which are much greater than those computed with just the extra-core nucleons. Another fact which requires explanation is that the state $J = j - 1$ often receives a small additional amount of binding energy so that it comes close to and even below the level $J = j$ when configurations of the j^3 type are involved.

The regions of the $1f_{7/2}$ and $1g_{9/2}$ shells are ones in which the many-particle shell model should be a good approximation. There are also regions in which interaction between the closed shell core and the extra-core nucleons is being developed, and as more nucleons are added there is a blending of the many-particle and collective descriptions of the nucleus. For this reason these regions deserve special study, both experimentally and theoretically, to determine how this transition takes place.

3. Intermediate Coupling in Light Nuclei

The assumption of strong spin-orbit coupling which was introduced to account for the magic numbers above 20, appears to be a good approximation for most of the periodic table. The single-particle levels with the same (nl) are split so that $j_+ = l + \frac{1}{2}$ lies considerably below $j_- = l - \frac{1}{2}$. Then states from the configuration $(j_+)^n$ lie considerably below those from $(j_+)^{n-1}(j_-)^1$ which in turn are below those from $(j_+)^{n-2}(j_-)^2$ and so on.

Counteracting this tendency is the effect of the residual interactions which spreads apart the states from two such configurations so that they are not in well-separated clusters. This we have seen in Fig. 4 for the $(3p_{\frac{3}{2}})^2$ and $(3p_{\frac{3}{2}})(3p_{\frac{1}{2}})$ configurations of Pb^{206}. Furthermore, the interactions between such configurations where the same (nl) is involved are strong, so that the residual interactions tend to mix the states of same J and destroy the validity of labeling a state as coming from a single jj-coupling configuration.

The question then is which of these competing factors is stronger, and the empirical evidence indicates that above $_{20}Ca_{20}^{40}$ spin-orbit splitting is stronger by an amount such that jj-coupling is a good approximation.

Between $A = 16$ and $A = 40$ the $1d$ and $2s$ levels are being filled, and while it appears that jj-coupling may be good near the end of the shell this is not true for the beginning. Calculations for $A = 18$ and 19 have shown (14) that the residual interaction mixes the configurations thoroughly. Although this region is of great interest,[f] the presence of the $(2s_{\frac{1}{2}})$ level complicates the study of deviations from jj-coupling. The simplest region for this purpose lies between $A = 4$ and $A = 16$ where the $(1p_{\frac{3}{2}})$ and $(1p_{\frac{1}{2}})$ levels are all that need be considered.

a. Two Identical $1p$ Nucleons

(1) *The jj Description*

The study of the relative importance of the residual interactions compared to the spin-orbit splitting can be exhibited in its simplest form by the case of two identical nucleons in the $1p$ shell. We can use jj-coupled configurations to discuss the problem, but since we wish to allow the possibility that the interactions are much more important than the spin-orbit splitting we must consider all the possible ways of putting two nucleons in the two single-particle levels. From our previous investigation we know that the possible states and their configurations are $J = 0, 2$ from $(1p_{\frac{3}{2}})^2$, then $J = 2, 1$ from $(1p_{\frac{3}{2}})(1p_{\frac{1}{2}})$ and finally $J = 0$ from $(1p_{\frac{1}{2}})^2$.

The operator which gives the spin-orbit energy is

$$a\Sigma \mathbf{l}_i \cdot \mathbf{s}_i \tag{8}$$

where the sum is over the nucleons in the $1p$ shell.

[f] Both the individual particle model and the collective model can be used to interpret F^{19}, and the study of the reason for this overlap is of great importance.

Since we are using individual particles with definite j, and

$$\mathbf{j}_i = \mathbf{l}_i + \mathbf{s}_i,$$

with $l_i = 1$, $s_i = \frac{1}{2}$, we get:

$$2\mathbf{l}_i \cdot \mathbf{s}_i = j_i(j_i + 1) - l_i(l_i + 1) - s_i(s_i + 1)$$
$$= +1 \text{ for } j_i = \tfrac{3}{2}, \text{ or } -2 \text{ for } j_i = \tfrac{1}{2}.$$

Therefore the spin-orbit energy is $+a$, $-\tfrac{1}{2}a$ and $-2a$ for states of the $(1p_{\frac{3}{2}})^2$, $(1p_{\frac{3}{2}})(1p_{\frac{1}{2}})$, and $(1p_{\frac{1}{2}})^2$ configurations respectively.

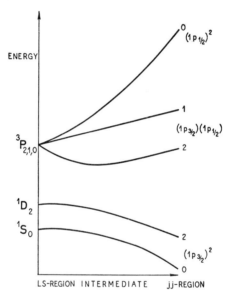

Fig. 6. Spectrum for two identical $(1p)$ nucleons as a function of the strength of spin-orbit coupling relative to the interaction. Angular momentum J, and jj-configuration given on the right, weak-coupling label on the left.

If the interactions are relatively weak, we know from our previous results that the order of the states is that in which they are listed above, and we get the simple separation of the spectrum into states from three distinct configurations as on the right of Fig. 6. If the interaction energy becomes comparable to or greater than the spin-orbit energy, states of the same J will be strongly mixed, and pure jj-configurations are no longer good approximations for the wave functions. Then diagonalization of the energy matrices from the two contributing sources must be carried out to find the spectrum and the configuration mixing in the wave functions.

By varying the spin-orbit strength, a, and carrying out such diagonalizations we find the variation of the spectrum as a function of the strength of spin-orbit coupling relative to the interaction, as shown in Fig. 6. As the spin-orbit term becomes less important the wave functions become quite complicated in terms of jj-configurations. In the limit that $a = 0$ the ground state, $J = 0$, consists of $[\frac{2}{3}(1p_{\frac{3}{2}})^2 + \frac{1}{3}(1p_{\frac{1}{2}})^2]$, followed by $J = 2$ which is $[\frac{1}{3}(1p_{\frac{3}{2}})^2 + \frac{2}{3}(1p_{\frac{3}{2}})(1p_{\frac{1}{2}})]$ and then a degenerate triplet of states $J = 0, 1, 2$. Although we can still describe the wave functions in terms of mixed jj-configurations—which is called using a jj-representation—we can no longer understand the reason for the ordering of the states in terms of such a description.

(2) *The LS Description*

However, if the wave functions for the case of no spin-orbit term are written out completely, they are seen to be separable into product of the spatial dependence of the two nucleons times their spin dependence. The spatial function has orbital angular momentum L, the resultant of coupling two individual $1p$ functions, having $l = 1$, to $L = 0, 1,$ or 2. The spin function has spin angular momentum $S = 0$ or 1, the possible resultants of coupling the two $s = \frac{1}{2}$ spins. Although L and S are vector-coupled to a total angular momentum J, the energy of the state depends only on L, and we find the ground state is $(L = 0, S = 0, J = 0)$, followed by $(L = 2, S = 0, J = 2)$, and finally the three degenerate states, $(L = 1, S = 1$ with $J = 0, 1, 2)$. These labels are customarily given in the notation of atomic spectroscopy as $(^{2S+1}L_J)$ where states with $L = 0, 1,$ and 2 are called S, P, and D states. This identification of states is given on the left of Fig. 6.

The possibility of labeling states by L and S, and having their order in energy be determined by L are consequences of the properties of the interaction potential. When the spin-orbit term is dominant it is appropriate to have states constructed with individual-nucleon j's into a jj-representation. Similarly, when the interaction is dominant it is preferable to start with states constructed in accordance with its requirements of separate space and spin dependence, instead of finding these wave functions only after a difficult calculation with an inappropriate representation, as we did above. The derivation of such a system of classification of states into an LS representation was carried out and applied (15) to the $1p$ shell long before jj-coupling was postulated, so our discussion of the two-nucleon problem in terms of the jj-representation did violence to the historical order of development.

The properties of the interaction potential between two nucleons which are important for understanding the spectrum in the LS approximation, are that it is an attractive short-range interaction which depends primarily on the spatial coordinates of the nucleons. The nature of this dependence makes it appropriate to separate the wave functions into a space factor times a spin factor, with the energy determined by the space factor, characterized by L. Furthermore the short-range attractive nature means that the lower states are those for which the nucleons have the greater probability of being close together. For two nucleons the S state is best followed by D, and the least likely is the P state since this wave function is antisymmetric to the exchange of space coordinates of the two nucleons and so must go toward zero when the nucleons come close and the coordinates become alike. The S and D functions are symmetric to exchange, so they allow the nucleons to be closer together, and we then understand the ordering of the spectrum when the LS approximation is good.

If we express our states in the LS representation and then see what happens when a small spin-orbit term is present we will again find mixing of states of the same J. The ground state $^1S_0$ will be mixed with a little $^1P_0$, and the $^1D_2$ will mix with $^1P_2$. Notice that states of different L are being mixed and also that states symmetric under permutation of space coordinates are mixing with antisymmetric states. As the spin-orbit term becomes larger the admixture will grow and the arguments we used for ordering the states in energy in the LS approximation are no longer valid.

At the limits of weak or dominant spin-orbit coupling we can understand the ordering of the spectrum, and it helps to use wave functions of the appropriate representation. However in the intermediate region while wave functions from either representation can be used, calculation is required and the description of the situation in terms of either representation will be complicated and difficult to interpret.

b. General Features and Experimental Comparison

The treatment of the problem of competition between the residual interaction and spin-orbit coupling which we have outlined for the case of two identical nucleons can be applied for any number of nucleons in the $1p$ shell. At the extremes when one of the two competing factors is much stronger than the other, the wave functions have properties which afford rather simple interpretations of the spectra. For strong spin-orbit coupling this just means ordering according to jj-configurations. In the LS approximation it means favoring the states of greatest possible symmetry[g] under exchange of the space coordinates of any pair of nucleons,

[g] The Pauli Principle limits the possibilities. For discussion see Blatt and Weisskopf (16).

and favoring most the state of smallest angular momentum with such symmetry. These are generalizations of the traits we saw in the two-nucleon case, and characterize the LS-model of Wigner and Feenberg (15).

The wide region when one is not close to either extreme is called the region of intermediate coupling. Since the two extremes emphasize different features of the wave functions, this transition region represents a quite complicated situation. When there are more than three nucleons in the shell the magnitude of the calculation drives one to use high-speed electronic computing machines to diagonalize the energy matrices which give the spectrum, extract wave functions, and compute expectation values of physical quantities to compare with experiment.

The question of whether the LS model, the jj-model or intermediate coupling represents the actual situation in the $1p$ shell can be decided by comparison with experiment. The most obvious feature is the level spectrum since there is often a different ordering of levels between the two extremes. However, this is not always a sensitive feature since there are usually only a few states identified experimentally in any given nuclear spectrum and they may not be the critical ones. Much additional evidence can be found by comparing computed observable quantities like magnetic dipole moments, gamma-transition probabilities between states, beta-decay matrix elements, and reduced widths in nuclear reactions. Such quantities are often quite sensitive to the nuclear wave function and change quite radically with different relative strength of spin-orbit coupling even though the spectrum may be rather insensitive. They are of great importance in deciding where the best approximation lies.

A comparison with experiment soon shows that things are not as simple as they might be. While the LS approximation seems to be good at the beginning of the shell, neither extreme model gives an adequate picture of the experimental facts elsewhere. This has led to intermediate coupling studies (17) which have provided a great improvement toward agreement with experiment. The result is that by choosing the interaction so that one can fit the simple two-nucleon spectrum of Li^6, a nucleus which is close to the LS limit, one can then fit a large majority of the features of the low excitation spectra for the rest of the shell (18) by simply increasing the spin-orbit strength a, of expression (8), as more nucleons are added.

While intermediate coupling seems to provide a good general picture in the light nuclei, there are many unsettled points and signs that further modifications are needed (19). Since the nuclei are relatively simple it is a region where experimental results are most likely to be fruitful in improving our understanding of nuclear structure.

REFERENCES

1. E. U. Condon and G. H. Shortley, *The Theory of Atomic Spectra* (Cambridge University Press, London and New York, 1953), pp. 76–77; R. Saito and M. Morita, Progr. in Theoret. Phys. **13**, 540 (1955); A. Simon, Oak Ridge National Laboratory Report, ORNL 1718 (1954).
2. M. E. Rose, *Elementary Theory of Angular Momentum* (John Wiley and Sons, New York, 1957), Chapter 3.
3. I. Talmi, Phys. Rev. **90**, 1001 (1953).
4. K. W. Ford, Phys. Rev. **98**, 1516 (1955).
5. S. B. Bjørnholm, O. B. Nielsen, and R. K. Sheline, Phys. Rev. **115**, 1613 (1959).
6. B. Bayman, A. S. Reiner, and R. K. Sheline, Phys. Rev. **115**, 1627 (1959).
7. M. J. Kearsley, Nuclear Phys. **4**, 157 (1957); W. W. True and K. W. Ford, Phys. Rev. **109**, 1675 (1958).
8. D. E. Alburger and M. H. L. Pryce, Phys. Rev. **95**, 1482 (1954); R. Day, A. Johnsrud, and D. Lind, Bull. Am. Phys. Soc. [2] **1**, 56 (1956); J. A. Harvey, Can. J. Phys. **31**, 278 (1953).
9. M. G. Mayer and J. H. D. Jensen, *Elementary Theory of Nuclear Shell Structure* (John Wiley and Sons, New York, 1955), p. 64.
10. C. A. Levinson and K. W. Ford, Phys. Rev. **100**, 13 (1955).
11. J. B. French and B. J. Raz, Phys. Rev. **104**, 1411 (1956).
12. G. Racah, Phys. Rev. **63**, 367 (1943); G. Racah and I. Talmi, Physica **17**, 1097 (1952).
13. B. H. Flowers, Proc. Roy. Soc. (London) **A212**, 248 (1952); A. R. Edmonds and B. H. Flowers, *ibid.* **A214**, 515 (1952); **A215**, 120 (1952).
14. J. P. Elliott and B. H. Flowers, Proc. Roy. Soc. (London) **A229**, 536 (1955); E. B. Paul, Phil. Mag. [8] **2**, 311 (1957).
15. E. Feenberg and E. Wigner, Phys. Rev. **51**, 95 (1937); E. Feenberg and M. Phillips, *ibid.* **51**, 597 (1937).
16. J. M. Blatt and V. F. Weisskopf, *Theoretical Nuclear Physics* (John Wiley and Sons, New York, 1952), Chapter 6.
17. D. R. Inglis, Revs. Modern Phys. **25**, 390 (1953); A. M. Lane, Proc. Phys. Soc. (London) **A66**, 977 (1953); **A68**, 189, 197 (1955); A. M. Lane and L. A. Radicati, *ibid.* **A67**, 167 (1954).
18. D. Kurath, Phys. Rev. **101**, 216 (1956); **106**, 975 (1957).
19. D. Kurath, Proc. Rehovoth Conf. on Nuclear Structure, Amsterdam (1958), p. 46.

VI. C. Collective Motion and Nuclear Spectra[a]

by A. BOHR and B. R. MOTTELSON

1. Vibrational Spectra.. 1012
2. Deformed Nuclei... 1014
   a. Rotational Spectra.. 1014
   b. Intrinsic Spectra.. 1019
3. A Survey of the Low-Energy Nuclear Spectra......................... 1023
   References.. 1032

The starting point for the nuclear shell model, as described in the two previous sections, is the introduction of a spherically symmetric binding field which represents the average effect of the interactions between the nucleons. To a first approximation one then considers the independent motion of the nucleons in this binding potential. Such a description of a nuclear state is, for many purposes, a rather good approximation in the case of closed-shell configurations, because of their high stability.

For nuclear configurations with several particles outside of closed shells, however, it is important to consider also the part of the nucleonic interactions which is not contained in the average field. In fact, because of the degeneracy of the solutions in the independent-particle approximation, even relatively small residual interactions introduce important correlations in the motion of the particles outside closed shells. For configurations with only a few particles outside of closed shells, one may attempt to treat the correlations by considering in detail the coupling in the motion of these few nucleons, as in the case of atomic spectra (see Section VI.B). However, a detailed treatment along these lines becomes very complicated if the number of these particles is not very small. Thus, in addition to the coupling between the angular momenta of the particles, it is often necessary to include configuration mixings, and the interaction with the closed-shell configurations may also play an important role.

In spite of the complexity of the problem, as viewed in this manner,

---

[a] The theoretical picture given in this chapter is described in somewhat more detail in the review article by Alder *et al.* (*1*).* This article also contains references to the original literature. A survey of the theory of rotational spectra has been given recently by Kerman (*2*). A discussion of the spectra of odd-*A* nuclei in the regions of deformed nuclei may be found in Mottelson and Nilsson (*3*).

* The reference list for Section VI.C is on page 1032.

one finds empirically that the energy spectra of nuclei with many-particle configurations show many strikingly simple features which vary in a systematic manner from nucleus to nucleus. These regularities are associated with the fact that a major part of the correlations between the particles may be described in terms of ordered collective motion of the nucleons, corresponding to variations in the shape of the nucleus. One is thus led to a generalization of the shell model in which the binding field is no longer considered as a static isotropic potential, but as a variable field which may take shapes differing from spherical symmetry.

The first problem is to determine the nuclear equilibrium shape. If the nucleus had an amorphous structure like a liquid drop, the spherical shape would have the lowest energy, but shell structure implies a systematic tendency for distortions of the nuclear shape. In fact, the orbits of the individual particles are strongly anisotropic (for $j > \frac{1}{2}$), and the self-consistent field must therefore in general be expected to deviate significantly from spherical symmetry. A simple description of the deformation effect is the one first given by Rainwater. Consider a nucleon moving in a nuclear potential with deformable surface; in the orbital plane, the nucleon exerts a centrifugal pressure which tends to produce an oblate deformation of the nuclear surface. In a closed-shell configuration, the nucleonic orbits are oriented equally in all directions and the nucleus remains spherical, but particles in unfilled shells tend to deform the nuclear shape, so as to adjust it to their own density distribution.

The introduction of a variable anisotropic binding potential includes an increased part of the nucleonic interactions in the average field, but significant residual interactions still remain. These are, for instance, responsible for the characteristic differences between the binding energies of even-even, odd-A, and odd-odd nuclei (the pairing effects) and are sometimes referred to as the pairing forces. These forces tend to couple two equivalent nucleons to a state of zero total angular momentum, that is, a spherically symmetric state, and they thus to some extent counteract the tendency of the individual nucleons to deform the nuclear shape.

The nuclear equilibrium shape and the character of the collective modes of motion of the nucleons may be understood as the result of the competition between the deforming power of the individual nucleons and the effect of the pairing forces. In the region of the closed shells, the latter dominate and the nuclear equilibrium shape remains spherical. With the addition of nucleons in unfilled shells, the tendency to deformation, which is a coherent effect of all these nucleons, increases in importance. Initially, the equilibrium shape remains spherical, but the nucleus becomes gradually softer against deformation. This softening manifests itself in a decrease of the frequency for collective vibrations about the spherical equilibrium shape. Finally, for sufficiently many particles outside closed shells, the spherical shape becomes unstable, and the nucleus acquires an ellipsoidal equilibrium shape. For such nuclei, the collective motion separates into rotational and vibrational modes. The former corresponds to a rotation of the nuclear orientation with preservation of shape, and possesses especially low excitation energies, while the latter corresponds to oscillations about the anisotropic equilibrium shape.

The equilibrium shape of nuclei of the deformed type can be obtained by calculating the energy of the nucleons as a function of the shape of the nuclear field, and seeking the shape that gives minimum energy. It is found that the equilibrium shape tends to have axial symmetry and to be of approximately spheroidal type. The deformation may be described by the parameter

$$\epsilon = \Delta R/R_0 \tag{1}$$

where ΔR is the difference between the nuclear radii in the direction of the symmetry axis and perpendicular to this direction, and R_0 is the mean nuclear radius. The order of magnitude of ϵ is n/A where n is the number of nucleons outside closed shells and A is the total number of particles in the nucleus. Thus the largest values of ϵ occur for light nuclei; on the other hand, the deformations are most sharply defined in heavy nuclei, for which the relative fluctuations in ϵ are smaller.

Fundamentally, ϵ is a small quantity, corresponding to the fact that the spherical field represents a first approximation and already reproduces many of the gross features of the nuclear structure. The collective features associated with the deformations, however, have an important influence on the low-energy spectra of nuclei.

For the unified description of nuclear structure one thus considers the dynamics of the nucleus in terms of collective and intrinsic modes of excitation. The former are associated with deformations, and the lowest modes of this type correspond, in the case of nuclei with sufficiently many particles outside closed shells, to rotations of the spheroidal shape. For nuclei in the regions nearer closed shells, the collective motion corresponds to vibrations about the spherical equilibrium shape. The intrinsic modes represent the motion of the nucleons in a fixed field, corresponding to the nuclear equilibrium shape. This motion is subjected to the influence of the pairing forces; in the case of odd-A nuclei the low lying intrinsic states may often be described in terms of the orbits of the last, unpaired particle.

In the following sections, we first consider some of the simple properties of the vibrational and rotational spectra (Sections 1 and 2); later, in Section 3, we attempt to give a brief survey of the low-energy nuclear spectra.

1. Vibrational Spectra

If a dynamical system, such as the nucleus, can perform small oscillations about the equilibrium shape, one expects quite generally to be able to describe these in terms of a set of normal modes. For small amplitudes, these modes of oscillation are independent, and each is equivalent to a harmonic oscillator. A simple example from classical physics is provided by the surface vibrations of a liquid drop, the theory of which was given by Rayleigh. The general method can be directly carried over into quantum mechanics.

When the equilibrium shape of the oscillating system is spherical, the normal modes can be labelled by their multipole order λ. Thus, a deformation of order λ has the same angular dependence as a spherical harmonic of this order; there are $(2\lambda + 1)$ independent modes of this type, corresponding to the different harmonics $Y_{\lambda\mu}$ of order λ, having

$$\mu = \lambda, \lambda - 1, \ldots, -\lambda.$$

The lowest mode of deformation is of quadrupole type ($\lambda = 2$), since a deformation of order $\lambda = 1$ is equivalent to a translation of the whole system. The quadrupole oscillations are thus expected to be of special importance for the low-energy spectra of nuclei; a deformation of this type gives rise to an ellipsoidal nuclear shape.

There are five independent modes of quadrupole vibrations, corresponding to the different values of μ. Each of these modes is equivalent to a harmonic oscillator. Thus, if x denotes the amplitude of one of these modes of deformation, the corresponding energy of vibration can be written

$$E = \tfrac{1}{2}B\dot{x}^2 + \tfrac{1}{2}Cx^2. \tag{2}$$

The parameters B and C, which are independent of μ, characterize, respectively, the inertia of the vibrational motion and the restoring force of the nucleus towards a quadrupole deformation. These parameters depend on the intrinsic structure of the oscillating system, and in the nuclear case are strongly influenced by the shell structure, in the manner qualitatively described above.

The frequency of vibration is given by

$$\omega = \sqrt{C/B} \tag{3}$$

and the quantum mechanical excitation spectrum consists of a series of equidistant levels, separated by the quantum energy $\hbar\omega$. Each excitation quantum of quadrupole type possesses a total angular momentum of two units ($\lambda = 2$), whose component along the z-axis equals $\mu\hbar$.

The nuclear energy levels corresponding to quadrupole excitation are shown on Fig. 1. The first excited state, of energy $\hbar\omega$ above the ground state, has an angular momentum of two units and positive parity, since the nuclear parity is not affected by a quadrupole deformation. The second excited states with energy $2\hbar\omega$ contain two quanta of excitation. The total angular momentum values may be obtained by coupling the two quanta, each having an angular momentum of two units; however,

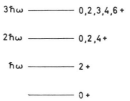

FIG. 1. Quadrupole vibrational spectrum for even-even nuclei with spherical equilibrium shape. The quadrupole vibrational quanta each have an energy $\hbar\omega$ and carry two units of angular momentum. The resulting spectrum is indicated in the figure in which the total angular momentum values are shown on the right. The equality of the energy spacings and the degeneracies are a consequence of the harmonic oscillator approximation and will be removed by higher order terms in the nuclear energy.

only the symmetric combinations ($J = 0, 2, 4$) occur, since the excitation quanta are to be considered as particles obeying Bose statistics.

Equation (2) for the energy of vibration corresponds to the assumption of very small amplitudes. The finite amplitudes encountered in even the lowest vibrational states of the nucleus will in general imply corrections to this harmonic spectrum.

A characteristic feature of the quadrupole deformations are the associated large electric quadrupole (E2) moments. Because of the collective character of the deformation, these moments may greatly exceed the moments produced by individual particles. The large E2 moments manifest themselves in the short lifetimes for radiative decay of the excited vibrational levels and in the large cross sections for Coulomb excitation of these levels. The matrix element for an E2 transition between vibrational levels is proportional to the corresponding matrix element of the coordinate x; for a harmonic oscillator this matrix element is proportional to $(BC)^{-\frac{1}{4}}$. From a measurement of the lifetime or excitation cross section for vibrational levels, together with a knowledge of the excitation energy, it is thus possible to determine both the parameters B and C which characterize the vibrational motion.

The well-known selection rule for harmonic oscillators, that the matrix elements of x vanish except for transitions between neighboring levels, implies that the vibrational levels tend to decay by cascade transitions. Thus, the second vibrational level should decay predominantly to the first excited level rather than by a crossover transition to the ground state, even when the last transition is not unfavored by angular momentum selection rules. Such a decay pattern is also observed for the 2+ second vibrational levels.

The accuracy of the vibrational description depends essentially on the condition that the vibrational motion be slow compared to the motion of the individual nucleons, in order that these may follow adiabatically the varying shape of the nucleus. Thus the excitation energies of the vibrational states should be much less than those of the intrinsic states. While it is difficult to estimate theoretically the ratio of these two energies, it can be seen from Figs. 8 and 9 that the vibrational energies are usually of the order of two to five times smaller than the lowest intrinsic excitations. Thus the vibrational description appears to be a very useful though often qualitative approximation.

In an odd-A nucleus the low-energy excitations may be described in terms of the shell model states of the odd-particle and the vibrational excitations of the even-even core. In order to obtain the low lying spectrum one must treat the coupling between these two degrees of freedom, and the resulting spectra are often found to be quite complex.

2. Deformed Nuclei

a. Rotational Spectra

For nuclei whose shape deviates essentially from spherical symmetry it is possible to distinguish between the degrees of freedom which describe the orientation of the system in space and the intrinsic degrees of freedom which describe the motion of the nucleons with respect to a body-fixed coordinate system.[b] The excitation spectrum associated with the reorientation of the system in space can be obtained from very simple and general arguments. Just as for a classical rigid body we can express

---

[b] The situation is quite similar to that encountered in the treatment of molecules, where one separates between rotational motion, on the one hand, and intrinsic motion associated with vibration of the nuclei and the electronic motion, on the other hand. For an elementary discussion of the molecular rotational spectra, much of which can be carried over to the present problem, see Herzberg (4).

the rotational energy in terms of the components R_i of the rotational angular momentum with respect to a coordinate system fixed in the nucleus. For an appropriate choice of the intrinsic coordinate system we have,

$$E_{\text{rot}} = \sum_{i=1}^{3} \frac{R_i^2}{2\mathfrak{J}_i}, \qquad (4)$$

where the constants $\mathfrak{J}_i$ are the effective moments of inertia which measure the additional kinetic energy which the nucleons must have in order to follow the rotation of the system. The form of Eq. (4) depends only on the possibility of separating between the intrinsic and rotational motions. However, the magnitude of the parameters $\mathfrak{J}_i$ depends on more detailed features of the nuclear shape and intrinsic structure.

All the nuclei which deviate strongly from spherical shape appear to remain axially symmetric (compare below). Of course it is then impossible to distinguish orientations of the nucleus which differ only by rotations about the symmetry axis. If we choose the intrinsic 3-axis to coincide with the nuclear symmetry axis (compare Fig. 2), we then have $R_3 = 0$. The effective moments of inertia $\mathfrak{J}_1$ and $\mathfrak{J}_2$ will be equal because of the axial symmetry. Choosing $\mathfrak{J}_1 = \mathfrak{J}_2 = \mathfrak{J}$ we may thus rewrite Eq. (4) in the form

$$E_{\text{rot}} = \frac{1}{2\mathfrak{J}} (\mathbf{R})^2. \qquad (5)$$

For a state in which the intrinsic angular momentum vanishes, the total angular momentum J is equal to the rotational angular momentum R and we have

$$E_{\text{rot}} = \frac{\hbar^2}{2\mathfrak{J}} J(J+1). \qquad (6)$$

The extra factor $\hbar^2$ appears in Eq. (6) because we now measure the total angular momentum in units of $\hbar$.

In the general case, the intrinsic angular momentum will not vanish, but will contribute a component, K, of angular momentum along the symmetry axis (see Fig. 2). Because of the axial symmetry, K will be a constant determined entirely by the intrinsic motion and will be independent of the state of rotational motion of the system. Since we now have $(\mathbf{R})^2 = J(J+1) - K^2$, $J \geq K$, we again obtain the rotational spectrum Eq. (6) aside from a constant.

From the above considerations we expect that the spectra of nuclei possessing a nonspherical shape will be characterized by the occurrence of a rotational band structure. In each nucleus there will be different intrinsic states characterized by K and perhaps other quantum numbers; associated with each such intrinsic state there will be a rotational band within which the energy follows approximately the expression (6).

In order to decide which values of J appear in a rotational band it is necessary to consider another symmetry property which characterizes the observed nuclear shapes: if we had to do with a single nucleon moving

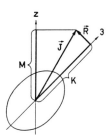

FIG. 2. Coupling scheme for deformed nuclei. For strongly nonspherical nuclei, possessing axial symmetry, the angular momentum properties may be characterized by the three constants of the motion J, M, and K. While J and M represent the total angular momentum and its component along the fixed z-axis, the component of J along the nuclear symmetry axis, 3, is denoted by K. The collective rotational angular momentum, **R**, is perpendicular to the 3-axis; thus K represents an intrinsic angular momentum.

in an orbit characterized by a definite value of l and m, the angular part of its wave function would be

$$\psi \sim Y_{lm}(\theta,\varphi)$$

and the average value of its density distribution would be

$$\rho \sim |Y_{lm}(\theta,\varphi)|^2.$$

This density distribution is not only axially symmetric but is also symmetric under reflection in a plane perpendicular to the symmetry axis and passing through the origin. A nucleon moving in this orbit would thus induce a deformation which also possessed these symmetry properties. Although the actual orbits in the nucleus are somewhat more complicated than the one considered here, these symmetry properties persist. We can also express the reflection symmetry by saying that the nuclear shell structure is much more effective in inducing deformations of even multipole order (mainly quadrupole) than of odd order.

In a rotational band with $K = 0$ it is impossible, because of the reflection symmetry, to distinguish two orientations of the nucleus which differ only by a rotation through 180° about an axis perpendicular to the symmetry axis. This invariance effectively eliminates the odd values of J from the rotational spectrum and so we get

$$J = 0, 2, 4, \ldots \quad \text{for } K = 0. \tag{7}$$

For $K \neq 0$ we must have $J \geq K$ since one of the components of J is equal to K, and so we have

$$J = K, K+1, K+2, \ldots \quad (K \neq 0). \tag{8}$$

Rotational bands with $K = \frac{1}{2}$ require a somewhat special discussion because of the possibility that the intrinsic spin of the last odd nucleon may not be able to follow the rotation of the nucleus. We can see the necessity for an additional term in the rotational energy by considering

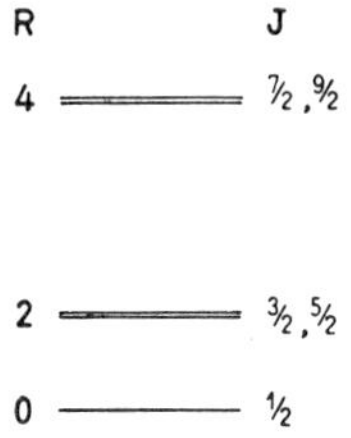

FIG. 3. Theoretical rotational spectrum for an $s_{\frac{1}{2}}$ particle coupled to an even-even core. The total angular momentum J is shown on the right, and the angular momentum of the core R is shown on the left.

	R	J
	4 ————	7/2, 9/2
	2 ————	3/2, 5/2
	0 ————	1/2

the case of a single nucleon in an $s_{\frac{1}{2}}$ state coupled to an even-even nucleus with $K = 0$. Since the $s_{\frac{1}{2}}$ state has a spherically symmetric density distribution, the energy of the system must be independent of the manner in which the angular momentum of the added nucleon is coupled to the rotational angular momentum of the even-even core. Thus, the spectrum is as shown in Fig. 3, which differs considerably from Eq. (6). This example is a special case of the spin decoupling which can occur for $K = \frac{1}{2}$ bands and which leads to the more general energy expression

$$E_{\text{rot}} = \frac{\hbar^2}{2\mathfrak{J}} \left\{ J(J+1) + a(-1)^{J+\frac{1}{2}} \left(J + \frac{1}{2} \right) \right\} \quad \text{for } K = \frac{1}{2}. \tag{9}$$

The additional parameter a appearing in Eq. (9) is called the decoupling parameter and depends entirely on the intrinsic structure [compare Eq. (15) below]. The example considered above corresponds to $a = 1$.

Since the different states in a rotational band all have the same intrinsic structure it is possible to obtain "intensity rules" which relate a nuclear property of one state to that of another state in the band. The nuclear properties concerned may be either transition rates leading to different states in the rotational band or the static moments of the levels. Perhaps the simplest of these intensity rules to visualize is that which relates the static quadrupole moments of the different states in a rotational band.

We consider first the intrinsic quadrupole moment Q_0 which measures the eccentricity of the charge distribution as seen in the body-fixed coordinate system. Since the intrinsic coordinate system is rotating with respect to the laboratory system, the charge distribution seen in the laboratory system will be somewhat smeared out and thus the static quadrupole moment Q measured in the laboratory will be smaller than Q_0. Using the proper quantum mechanical wave functions to describe the rotational motion we can calculate the amount of the smearing out and obtain

$$Q = \frac{3K^2 - J(J+1)}{(2J+3)(J+1)} Q_0. \qquad (10)$$

Equation (10) is typical of all the intensity rules in that it expresses some nuclear matrix element Q in terms of a matrix element between intrinsic states Q_0 which is thus the same for all the states in the rotational band, multiplied by a factor that depends on J and K. In general this last factor can be expressed in terms of vector addition coefficients. Such intensity rules have been given not only for nuclear moments, but also for alpha, beta, and gamma ray transition intensities and for reduced widths describing (d,p) and nucleon scattering reactions.

The different features of the nuclear band structures which we have discussed so far follow directly from the existence of a nonspherical equilibrium shape and from the symmetry properties of the equilibrium distortion. It is characteristic of such results that they give only relations between the different states of a rotational band. There occur in each of the expressions of this section parameters such as the moment of inertia and the intrinsic quadrupole moment, which are not determined by these arguments.

To gain an orientation in order of magnitude which one might expect for the moment of inertia $\Im$ we may consider two extreme situations. If the nucleus were to rotate as a rigid body, the collective flow would be as in Fig. 4a and the moment of inertia would be

$$\Im_{\text{rig}} = \tfrac{2}{5} A M R_0^2, \qquad (11)$$

where A is the number of nucleons, M the nucleon mass, and R_0 the mean nuclear radius. An opposite extreme from the rigid rotation is provided by the flow of an irrotational fluid contained within a rotating ellipsoidal boundary (Fig. 4b). For such irrotational flow the moment of inertia is much smaller than for rigid flow since each particle has only to move a small distance in order to make the distortion move around.

The moment of inertia is given by

$$\mathfrak{J}_{\text{irrot}} = \tfrac{2}{5} A M R_0^2 \epsilon^2, \qquad (12)$$

where ϵ is the eccentricity parameter defined by Eq. (1). It is another important characteristic of the irrotational flow that the moment, Eq. (12), depends strongly on the deformation, vanishing as ϵ goes to zero.

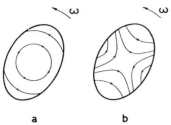

FIG. 4. Collective flow patterns; (a) rigid rotation and (b) irrotational flow.

Detailed calculations of the manner in which nucleons follow the rotation of the average field in which they move have shown that if the nucleons moved completely independently they would have a moment of inertia quite close to that given in Eq. (11). However, correlations in the motion of different nucleons tend to reduce the moment below this value and to make the moment decrease with decreasing ϵ. For correlations so strong that the shell structure is completely destroyed, the moment of inertia would approach the value in Eq. (12). The actual moments of inertia in nuclei are thus intermediate between the values given by Eqs. (11) and (12).

b. INTRINSIC SPECTRA

In the last section we considered the general features of the rotational band structure associated with deformed nuclei. Each rotational band is characterized by the quantum number K which represents the contribution of the intrinsic motion to the total angular momentum. In order to know what K-values may be expected in the spectrum of a given nucleus or to calculate any other intrinsic property we must attempt a more detailed description of the intrinsic motion.

The simplest possibility for the intrinsic motion would be to assume the nucleons to move independently in an average nuclear field. Since the nuclei we are considering have intrinsic shapes which deviate essentially from spherical symmetry, though retaining axial symmetry, the nuclear potential should have a similar shape. We are thus led to consider a single-particle Hamiltonian which contains, besides the usual kinetic energy and spin orbit terms, a potential energy which can be written in first approximation as

$$V = V_0(r) + V_2(r)Y_{20}(\vartheta). \quad (13)$$

The first term in Eq. (13) represents the central field which we know from the usual shell model should be somewhere between a square well and a harmonic oscillator. The second term corresponds to a quadrupole distortion of the nuclear field. The magnitude of the distortion is determined by the magnitude of V_2. We expect that V_2 will be approximately proportional to the deformation parameter ϵ.

In order to gain a qualitative understanding of the intrinsic spectrum implied by Eq. (13) we may consider deformations which are sufficiently small that we are permitted to treat the distortion as a perturbation. Evaluating the expectation value of the second term in Eq. (13) for a state characterized by definite values for the particle angular momentum, j, and its component K along the symmetry axis, we obtain

$$E_{j,K} = -\sqrt{\frac{5}{4\pi}} <V_2>_j \frac{3K^2 - j(j+1)}{4j(j+1)}, \quad (14)$$

where $<V_2>_j$ represents the value of V_2 averaged over the radial wave function corresponding to the state j. In a spherical potential there is a degeneracy of all $2j + 1$ states characterized by different values of K, but the same value of j. According to Eq. (14) the distortion of the nuclear field removes this degeneracy. However, since the energy depends only on K^2, states with $+K$ and $-K$ will still be degenerate. This is a general result for arbitrary deformations and simply reflects the fact that there can be no energy difference between two orbits which differ only in the sense (right-handed or left-handed) in which the particle rotates around the symmetry axis.

It is also easy to understand the sign of the splittings implied by Eq. (14). If V_2 is positive then Eq. (13) implies an oblate distortion of the nuclear field. In that case Eq. (14) tells us that states with large values of K are lowest in energy; this just corresponds to the fact that for large values of K the particle is moving approximately in the equatorial plane and is thus best able to exploit the oblate distortion.

As the distortion increases it is essential to recognize that the potential [Eq. (13)] contains matrix elements which mix single-particle states with different values of j. When this effect becomes important, a perturbation calculation may no longer be possible. However, solutions of single-particle Hamiltonians with potentials of the type given by Eq. (13) have been calculated on electronic computing machines and tabulated for a considerable range of distortions (5) (for example, see Figs. 5 and

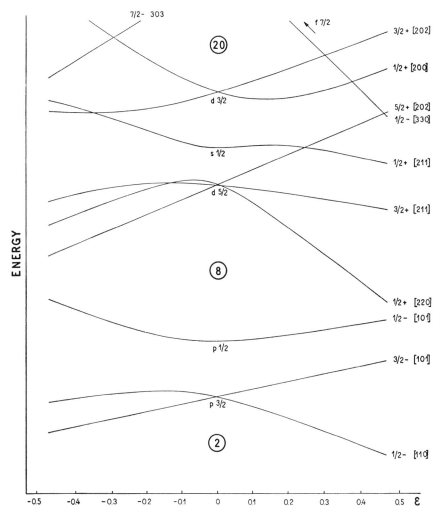

FIG. 5. Energy spectrum for single-particle motion in an ellipsoidal potential. The spectra are drawn as a function of the eccentricity ϵ. For $\epsilon = 0$, the potential is spherical and we have the usual sequence of orbitals corresponding to a spherical shell model, that is $1s_{\frac{1}{2}}$ (which comes lower than the lowest orbit drawn in this figure), $1p_{\frac{3}{2}}$, $1p_{\frac{1}{2}}$, $1d_{\frac{5}{2}}$, $2s_{\frac{1}{2}}$, $1d_{\frac{3}{2}}$, and then a little above the region covered in the figure, $1f_{\frac{7}{2}}$. For $\epsilon \neq 0$ the states are labelled by the quantum numbers K and π which are indicated on the right. A complete characterization of the states requires additional quantum numbers besides K and π and these are given in the square brackets (see Nilsson, 5, for a discussion of these additional quantum numbers). As mentioned in the text each of these orbits is twofold degenerate.

12). It is found that these single-particle spectra represent rather well the observed spectrum of K-values and other intrinsic properties associated with the last odd nucleon in an odd-A nucleus.

As one of the simplest intrinsic properties which can be expressed in terms of the orbit of the last odd nucleon, we give the expression for the decoupling parameter appearing in Eq. (9)

$$a = -\sum_j (-1)^{j+\frac{1}{2}}(j + \tfrac{1}{2})|c_j|^2 \tag{15}$$

where c_j is the amplitude of the state j in the intrinsic wave function. If the potential is almost spherical as in the example above, then only a single term will contribute to Eq. (15) (the example considered in Section 2a had $j = \tfrac{1}{2}$ and thus $a = +1$). However, for large deformations it is necessary to employ the coefficients c_j obtained from the more exact calculations of the intrinsic wave functions.

It should be emphasized that although this one-particle model provides a rather good description of the low-energy intrinsic excitations in the nonspherical odd-A nuclei, it fails quite seriously to account for the even-even nuclei.

Because of the twofold degeneracy of the orbits in an axially symmetric field we would expect the lowest configuration of an even-even nucleus to have $K = 0$, corresponding to a pairwise filling of all the lowest orbits. Indeed the even-even nuclei always have $K = 0$ for the ground-state rotational band.

If the nucleons were moving truly independently there would also be quite a number of low lying excited states in even-even nuclei corresponding to the breaking of pairs or the pairwise excitation of nucleons. The excitation energy of these states would then be expected to be of the same order of magnitude as that of the low lying intrinsic states of neighboring odd-A nuclei. Actually one does not find any intrinsic excitations in the even-even nuclei until one comes to excitation energies five to ten times greater than this amount. (The first intrinsic excitations are usually at about 1 Mev in the heavy elements.) This rather striking deviation from the independent-particle model clearly reveals an important correlation in the motion of a pair of nucleons with K and $-K$. We are dealing here with an effect of the residual forces, which also manifest themselves in the systematic difference between the total binding energies of even-even and odd-A isotopes. As mentioned at the beginning of this Section, these pairing forces are also responsible for the transition from rotational to vibrational types of spectra with the approach to closed-shell regions.

3. A Survey of the Low-Energy Nuclear Spectra

In the present section we shall briefly discuss a number of nuclear level schemes with a view to illustrating the extent to which it is possible to interpret the available data in terms of the theoretical ideas described in the previous sections. The spectra are selected to provide typical examples from the different regions of the periodic table.[c]

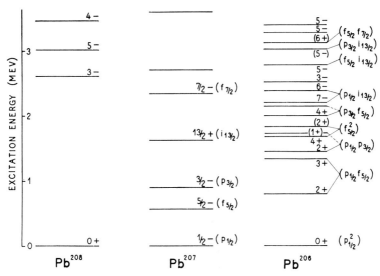

FIG. 6. Spectra of lead isotopes. The figure shows the observed levels labelled by the experimentally determined spins and parities. The configuration assignments are shown in parentheses (see ref. 8).

We first consider a few examples of spectra for nuclei with almost closed-shell configurations. The spectra in the neighborhood of Pb^{208} are illustrated in Fig. 6. The ground state of Pb^{208} is characterized by closed shells of both neutrons ($N = 126$) and of protons ($Z = 82$). It is consistent with this interpretation that no low lying excited states are found in this nucleus.

---

[c] The experimental data contained in the figures of this chapter represent the results of a very large number of laboratories employing a wide variety of experimental techniques. The compilation "Nuclear Data Cards" contains the original references to this work. There have recently appeared a number of surveys, discussing the various experiments which have established these level schemes. See especially Perlman and Rasmussen, Burcham, and Kinsey (6). For a discussion of Coulomb excitation see Heydenburg and Temmer (7) as well as the review article by Alder et al. (1).

It is expected that the low lying excited states of Pb^{207} may be described in terms of the one-particle states accessible to the single neutron hole in the closed shell. Indeed, the observed spectrum of Pb^{207} exhibits just the one-particle states, $p_{1/2}$, $f_{5/2}$, $p_{3/2}$, $i_{13/2}$, and $f_{7/2}$, which are expected in the region below $N = 126$. Above 2.5 Mev additional states are observed in Pb^{207} which presumably represent more complicated configurations, such as are expected to occur at energies comparable with the lowest excitation energies in Pb^{208}.

The simplest states of Pb^{206} may be constructed by coupling together two of the available hole states observed in the Pb^{207} spectrum. The spectrum of Pb^{206} has been quite extensively studied[d] both in the decay of Bi^{206} and by particle reaction studies [Pb^{207} (p,d) and Pb^{206} (n,n')], and it is found that such a two-particle description can account very well for the observed levels up to about 2.5 Mev of excitation. Above this energy, several states are observed which appear to represent more complicated configurations and in the figure no configuration assignment has been indicated for these levels. When one attempts to account in more quantitative detail for the energies and transition probabilities of the states of Pb^{206}, it is found that, although the configuration assignments indicated in the figure provide a useful first approximation, the actual states involve considerable configuration mixing. Thus the ground state wave function contains not only the indicated configuration $(p_{1/2}^2)_{J=0}$ but also components with $(f_{5/2}^2)_{J=0}$, $(p_{3/2}^2)_{J=0}$, and so forth.

One may conclude from the above that it is indeed possible to go quite far in describing the low-energy states of Pb^{206} and Pb^{207} in terms of the configurations of the neutron holes in the closed shell at $N = 126$. However, in the observed E2 transition rates one sees clearly that the protons are not inert as they would be if they stayed in their spherically symmetric closed-shell configuration, but rather that they also to some extent participate in the observed excitation. Thus, while pure neutron transitions would be expected to have very long E2 lifetimes, the observed E2 rates for the decay of the $\frac{5}{2}-$ state in Pb^{207} and the first $2+$ state in Pb^{206} are found to be about as short as would be expected for a single proton transition. This effect can be described in terms of the collective polarization of the proton closed shell by the motion of the neutron holes. The magnitude of the polarization is such that in E2 transitions there is about one unit of charge following the motion of each neutron hole. Since each proton in the closed-shell core is excited only a small fraction of the time, the polarization effect is in some sense only a small perturbation on the wave function and does not essentially affect the energy spectrum.[e]

[d] See also Fig. 5 in Section VI.B.

[e] For a detailed analysis of the Pb^{206} spectrum, see reference (8).

Similar analyses as in the Pb^{208} region have been made of the spectra of the very light elements ($4 < A < 16$), as well as of the spectra in the regions of the closed-shell nuclei O^{16}, Ca^{40}, and Zr^{90} (9).

Figure 7 illustrates the systematics of the even-even nuclei ranging from the closed-shell region of Pb down to the region of rotational spectra which begins around W and Os.

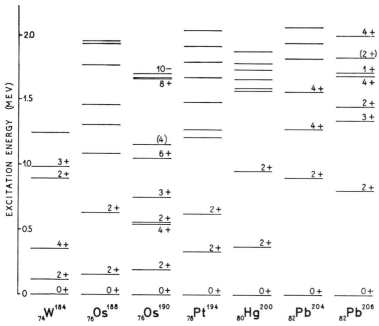

FIG. 7. Systematics of the spectra of even-even nuclei in the region from $A = 208$ to $A = 184$. The figure shows the observed levels of a number of relatively well-studied nuclei in the region between the closed-shell configurations at Pb^{208} and the region beginning around Os and W, where the rotational level structure appears.*

As we move away from the closed-shell configuration of Pb^{208} the description in terms of the holes in the closed shells involves an increasing number of degrees of freedom which might at first be expected to lead to very complicated spectra. However, the first few excited states of these nuclei are seen in Fig. 7 to exhibit a simple systematics which can be described in terms of the collective vibrations of quadrupole type (Section 1).

The Coulomb excitation reaction has provided a powerful tool for studying the first vibrational excitation. It is found that the E2 matrix elements for the excitation of these states are systematically an order of magnitude greater than would correspond to a single-particle transi-

* *Added in proof:* The study of the β-decay of Ir^{188} has established the existence of a 4+ level in Os^{188} at 0.478 Mev. It is seen that this level contributes a valuable element to the systematics in Fig. 7.

tion. While the first excited state is always of 2+ type, the second is at about twice the energy of the first and has a spin of either 0+, 2+, or 4+. Also the decay of the second 2+ state is found to be governed by strong selection rules which can be understood simply in terms of the vibrational quantum number.

It has not been possible so far to follow the vibrational spectra in this region of elements above about 1 Mev of excitation because above

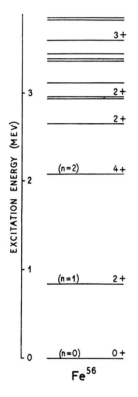

FIG. 8. Energy levels for Fe^{56}. The levels have been located primarily by β-decay studies and by inelastic proton scattering. The quantum number n gives the number of vibrational quanta.

this excitation energy there occur a large number of intrinsic excitations whose detailed interpretation is not as yet understood.

As we go away from the closed-shell configuration of Pb^{208} the frequency of the vibrational excitations decreases corresponding to the approach to the point at which the spherical shape becomes unstable. Finally in the region of the Os isotopes the nuclei acquire a nonspherical equilibrium shape and thus begin to exhibit rotational spectra, described by the simple properties discussed in Section 2. Vibrational spectra, of roughly the type illustrated in the intermediate region of Fig. 7, are found in all even-even nuclei excluding the immediate neighbors of closed shells and excluding the regions $A \sim 25$, $150 < A < 190$, and $A > 220$ where the rotational spectra are found.

Figure 8 illustrates a nuclear vibrational spectrum taken from a quite different region of the periodic table from that of Fig. 7. Again we find a simple vibrational spectrum in the neighborhood of the ground state

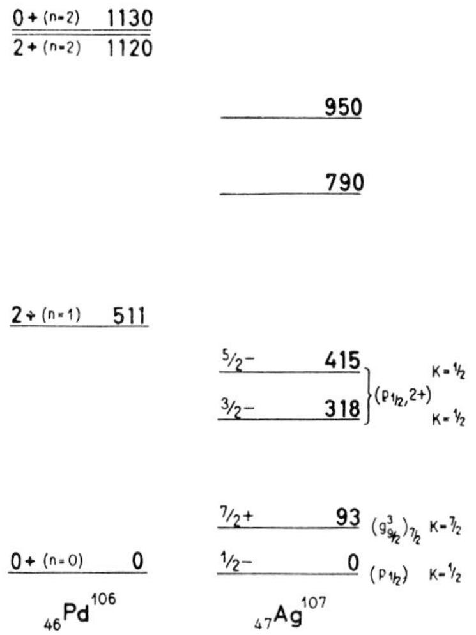

```
0+ (n=2)   1130
2+ (n=2)   1120

                              950

                              790

2+ (n=1)   511
                      5/2−    415  ⎫          K=½
                      3/2−    318  ⎬ (p½,2+)  K=½

                      7/2+    93   (g_{9/2}^3)_{7/2}  K=7/2
0+ (n=0)    0         1/2−    0    (p½)   K=½
   ₄₆Pd¹⁰⁶              ₄₇Ag¹⁰⁷
```

FIG. 9. Energy levels of Ag^{107} and Pd^{106}. The observed levels are labelled by the excitation energy (in kev) as well as by the experimentally determined spins and parities. The Pd^{106} spectrum is interpreted in terms of vibrational excitations with the vibrational quantum number indicated in parentheses. For Ag^{107} the level assignments correspond to the two simple coupling schemes which may be employed in describing the coupling of the last odd nucleon to the even-even core. The shell model configurations together with the core excitations are appropriate to a weak coupling while the rotational quantum number K applies to the case of strong coupling.

and then a complex spectrum of intrinsic excitations beginning at about 2.5 Mev.

One expects that the low-energy spectra of the odd-A nuclei in the intermediate region may be described in terms of the degrees of freedom of the odd nucleon coupled to the vibrations of the even-even core. A qualitative interpretation of the spectra on this basis is often possible, as illustrated by the spectrum of Ag^{107} in Fig. 9. One may attempt to analyze such a spectrum in terms of either of two simple extreme coupling schemes. If we assume that the degrees of freedom of the odd particle are very weakly coupled to those of the even-even core, then we may describe the ground state of Ag^{107} as simply a $p_{\frac{1}{2}}$ proton coupled to the 0+ ground state of Pd^{106}. Similarly the $\frac{7}{2}+$ isomeric state represents a $(g^3_{9/2})_{7/2}$ state coupled to an even-even core in its ground state. In this interpretation the $\frac{3}{2}-$ and $\frac{5}{2}-$ states at about 300 and 400 kev are described as a $p_{1/2}$ proton coupled to the first vibrational 2+ state of Pd^{106}. As an opposite extreme we could assume that the added odd particle in Ag^{107} is so strongly coupled to the even-even core that it gives rise to a nonspherical equilibrium shape. Then the intrinsic configuration is described by the quantum number K. The excited states at about 300

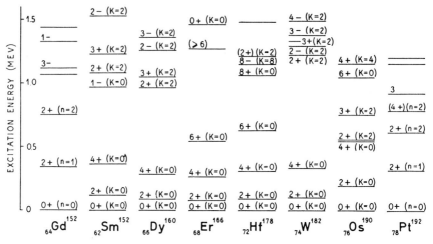

FIG. 10. Systematics of the spectra of even-even nuclei in the region from Gd to Pt. At the two ends of this region the spectra are interpreted as vibrational and the levels are labelled by the vibrational quantum number n. From Sm^{152} to Os^{190}, the spectra are best described as rotational and the quantum number K is used to characterize the intrinsic state.

and 400 kev are in this picture interpreted as rotational excitation based on the lowest ($K = \frac{1}{2}-$) intrinsic state; the value of the decoupling parameter a needed to fit the energy levels is about 0.7 and corresponds approximately to that calculated from Eq. (15).

The difference between the two coupling schemes for Ag^{107} will manifest itself mainly in the pattern of the higher excited states, but it is probable that in the present case the weak coupling interpretation is somewhat more appropriate than the description in terms of a strong coupling.

The systematics of the nuclear rotational spectra in even-even nuclei are illustrated in Fig. 10. The figure begins in the region where the low lying states are best described in terms of vibration. As we move further away from the closed-shell configurations the spectra change to rotational type. Within the region of the rotational spectra it is found from

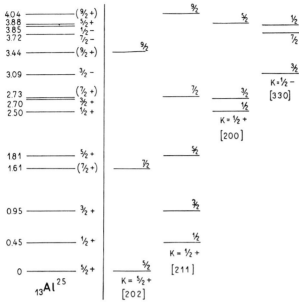

FIG. 11. Level spectrum of Al^{25}. On the left are drawn all the levels of Al^{25} that have been observed up to 4 Mev together with the measured spins and parities. The properties of these levels have been established mainly from the study of the Mg$^{24}(p,\gamma)$ and Mg$^{24}(d,n)$ reactions (see Litherland et al., 10). On the right these levels have been classified in terms of the rotational bands associated with the different intrinsic states. The intrinsic states correspond well with those expected, assuming the last proton ($Z = 13$) to move independently in an ellipsoidal potential. Thus, the eccentricity may be estimated from the measured quadrupole moment of Al^{27}, $Q = 0.15 \times 10^{-24}$ cm^2. Using Eq. (10) and the relation $Q_0 = \frac{4}{5}ZR_0^2\epsilon$ we find $\epsilon = 0.23$. Then employing Fig. 5 we see that for this value of ϵ we might expect either $K = \frac{5}{2}+$ or $K = \frac{1}{2}+$ for the ground state of Al^{25}. Indeed the ground state rotational band is well described by the first of these states while the rotational band beginning at 0.45 Mev corresponds with the second. The detailed calculation of the wave function of this state yields $\psi_{K=\frac{1}{2}} = 0.53\varphi(s_{\frac{1}{2}}) - 0.68\varphi(d_{\frac{3}{2}}) - 0.51\varphi(d_{\frac{5}{2}})$. It is seen that this wave function contains appreciable contributions from all three of the available shell model configurations, $s_{\frac{1}{2}}$, $d_{\frac{3}{2}}$, and $d_{\frac{5}{2}}$. Employing Eq. (15) we calculate $a = 0.1$ which agrees reasonably well with the value $a = -0.02$ which is obtained from the observed energies by means of Eq. (9).

In addition Fig. 5 indicates that the next intrinsic states should have either $K = \frac{1}{2}+$ or $K = \frac{1}{2}-$ as is observed for the rotational bands beginning at 2.50 Mev and 3.09 Mev, respectively. The decoupling constants observed in these bands are also in good agreement with those calculated from the one-particle wave functions by means of Eq. (15). The decoupling constant is so large in the $K = \frac{1}{2}-$ band ($a = -3.2$) that the $J = \frac{3}{2}$ and $J = \frac{7}{2}$ levels are observed to have lower energies than the $J = \frac{1}{2}$ member of this band.

measurements of the E2 transition rates that the distortion of the nucleus continues to increase as one moves further away from a closed shell. The moments of inertia describing the rotational bands of Fig. 10 reflect this variation in the nuclear eccentricity; the energy of the rotational excitations continues to decrease until the middle of the region is passed

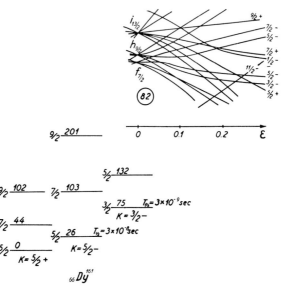

FIG. 12. Level spectrum of Dy^{161}. The ground state rotational band has been studied especially by Coulomb excitation while the other known levels have been found in the study of the β-decay of Tb^{161} and the K-capture of Ho^{161}. It is seen that the spectrum is well described in terms of three rotational bands having $K = \frac{5}{2}+$, $K = \frac{5}{2}-$, and $K = \frac{3}{2}-$ in increasing order of excitation. The insert in the upper right shows the levels expected for the last odd particle moving in an ellipsoidal potential. The notation is similar to that in Fig. 5. Employing the measured eccentricity $\epsilon = 0.3$ observed for this nucleus, we see that the three observed intrinsic states correspond well with those expected for $N = 95$.

The 26-kev and 75-kev levels, which can only decay by transitions involving a change of intrinsic state, have relatively long lifetimes which have been measured directly by delayed coincidence experiments. All the other levels, which may decay by rotational transitions, are expected to have appreciably shorter lifetimes.

and then begins to increase with the approach to the next closed shell. The rotational description of the excitation is most accurate for the nuclei possessing the largest deformations. For these nuclei, the ratio of the energy of the 4+ state to that of the 2+ state agrees with the value $\frac{10}{3}$ predicted by Eq. (6) to within a fraction of a percent. In several cases rotational levels with as much as 8 units of angular momentum have been located. Some of the first systematic information about the nuclear rotational level structure was obtained from a study of the fine structure of the α-decay.

The absence of any low lying intrinsic excitations is a conspicuous feature of Fig. 10. When the intrinsic excitations are observed, usually beginning at about 1 Mev, it is found that there is a rotational band associated with each intrinsic state.

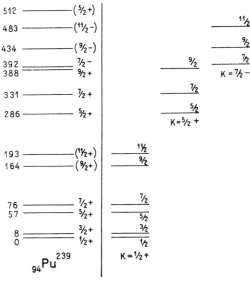

FIG. 13. Level spectrum of Pu^{239}. On the left are shown all the levels of Pu^{239} which have been found so far. These levels have been established mainly as the result of studies of the Np^{239} beta decay and of the Cm^{243} alpha decay. Coulomb excitation has added information on the two lowest excited states. On the right these levels are arranged in rotational bands. It is seen that the spectrum is well described in terms of rotational bands built on intrinsic states having $K = \frac{1}{2}+$, $\frac{5}{2}+$, and $\frac{7}{2}-$ in order of increasing excitation energy. A new band, which may also have $K = \frac{5}{2}+$, appears to start with the observed level at 512 kev. As in the spectra of Al^{25} and Dy^{161} discussed above, it is possible to identify these intrinsic states with the orbitals of the last odd nucleon assumed to move independently in an ellipsoidal potential. On this basis it is also possible to account approximately for the decoupling constant ($a = -0.58$) observed for the ground state band.

In Figs. 11, 12, and 13 we give some examples of odd-A spectra in the regions where the nuclei are known to possess a nonspherical equilibrium shape. It is seen that these spectra can be classified into rotational bands. Within each band the energies follow the relation of Eq. (6) with the spin sequence of Eq. (8) except for the $K = \frac{1}{2}$ bands where the rotational energies are given by Eq. (9). The intrinsic state associated with each band is found to be rather well described in terms of the orbits of the last odd particle assumed to move in a nonspherical nuclear field.

The ground state bands have been explored especially by Coulomb excitation while the higher bands are populated in α- and β-decay processes. For the study of the spectra of the lighter elements, other charged-particle reactions have played an important role.

May 1958.

REFERENCES

1. K. Alder, A. Bohr, T. Huus, B. R. Mottelson, and A. Winther, Revs. Modern Phys. **28**, 432 (1956).
2. A. K. Kerman, in *Nuclear Reactions*, edited by Endt and Demeur (North-Holland Publishing Co., Amsterdam, 1959), Chapter X.
3. B. R. Mottelson and S. G. Nilsson, Kgl. Danske Videnskab. Selskab, Mat.-fys. Skr. **1**, No. 8 (1959).
4. G. Herzberg, Spectra of Diatomic Molecules (D. Van Nostrand Co., New York, 1950), pp. 66 ff., 115 ff., 128 ff., and 218 ff.
5. See, e.g., S. G. Nilsson, Kgl. Danske Videnskab. Selskab, Mat.-fys. Medd. **29**, No. 16 (1955); K. Gottfried, Phys. Rev. **103**, 1017 (1956).
6. I. Perlman and J. O. Rasmussen, Alpha Radioactivity, in *Encyclopedia of Physics* (Springer Verlag, Berlin, 1957), Vol. 42; W. E. Burcham, The Spectra of Light Nuclei, *ibid.*, Vol. 40; B. B. Kinsey, The Spectra of Heavy Nuclei, *ibid.*, Vol. 40.
7. N. P. Heydenburg and G. M. Temmer, Ann. Rev. Nuclear Sci. **6**, 77 (1956).
8. D. E. Alburger and M. H. L. Pryce, Phys. Rev. **95**, 1482 (1954); M. J. Kearsley, Phys. Rev. **106**, 389 (1957); Nuclear Phys. **4**, 157 (1957); W. W. True and K. Ford, Phys. Rev. **109**, 1676 (1958).
9. See J. P. Elliot and A. M. Lane, in *Encyclopedia of Physics* (Springer Verlag, Berlin, 1957), Vol. 39.
10. A. E. Litherland, H. McManus, E. B. Paul, D. A. Bromley, and H. E. Gove, Can. J. Phys. **36**, 378 (1958).

VI. D. The Complex Potential Model

by HERMAN FESHBACH

1. Average Cross Sections and Average Wave Amplitudes.................. 1034
   a. Scattering of Spinless Particles by Spinless Nuclei..................... 1035
   b. The General Case... 1037
2. Empirical Determination of the Complex Potential..................... 1045
   a. Form of the Potential.. 1045
   b. Scattering by Spin Zero Nuclei..................................... 1048
   c. Fluctuation Cross Section and Elastic Scattering..................... 1050
   d. Empirical Values for the Complex Potential Parameters............... 1054
   References... 1061

The success of the shell model indicates that it is a good approximation to represent the interaction between a nucleon and a nucleus by a potential well. However, the naive use of such a potential well model in scattering and reaction problems is not correct since such a model cannot lead to the narrow compound nuclear resonances. The point at issue here is that the shell model potential gives, speaking very roughly, the average of the nucleon-nucleus interaction. It is the fluctuations away from the average which are responsible for the compound nuclear resonances. To obtain a potential model for scattering and absorption it is clearly necessary to average over these fluctuations in some fashion. The procedure which has been adopted and which is dictated to a large extent by the requirement of linearity involves first averaging the elastic scattering amplitude over a narrow energy band. The complex potential of the complex potential model is then that potential which generates an elastic scattering amplitude equal to the aforementioned energy average of the exact scattering amplitude. This potential is not necessarily equal to the shell model potential except in the limit of very weak fluctuations but the two are, of course, closely related.

The theoretical possibility of generating the average scattering amplitude by means of a complex potential was first shown by Bethe (1)* who considered only the case of overlapping resonances. The recent development of the complex potential model stems in great part from the experiments of Barschall and his collaborators (2) which were interpreted in terms of this model by Feshbach, Porter, and Weisskopf (3).

* The reference list for Section VI.D begins on page 1061.

In the Barschall experiments the energy average of neutron cross sections was measured (see Section III.E.2). The most relevant feature revealed by these experiments was the broad maxima, of the order of 1 Mev wide, in the average total cross section when the latter was plotted against the average neutron energy E. These peaks, usually referred to as *giant resonances* vary in a smooth fashion upon changing the mass number A of the target nucleus. It was found that as A increases the energy at which they occur decreases. This behavior suggests a potential well model since, in qualitative agreement with the experimental results, the model predicts resonances near particular values of the parameter KR where K is the average wave number inside the well and R is the radius of the well (or more accurately KR is the phase change of the wave over the potential well). Moreover, in order that the constructive interference required for the resonances exist it is necessary that the imaginary part of the potential well (or equivalently the mean free path of a neutron inside the well) be small, contrary to the assumption made in early theoretical work (*1,4*). Quantitative calculation (*3*) showed that this description of the Barschall giant resonances is correct and thus established the validity of the complex potential model on an empirical basis.

In this chapter we shall be primarily interested in those aspects of the complex potential model which are relevant for nuclear spectroscopy. For discussions of the meaning of the model we refer the reader to various review articles (*5,6,7*). We should also mention some recent articles (*8,9*) published since the appearance of these reviews. In Section 1 we shall be concerned with the relation of the parameters of the complex potential model and those describing the compound nuclear resonances. In Section 2 the types of complex potentials employed, the corresponding expressions for the cross sections, and the determination of the complex potential from experiment will be reviewed.

1. Average Cross Sections and Average Wave Amplitudes

We define the energy average $\bar{\eta}$ of a quantity η as follows:

$$\bar{\eta}(E) = \int \rho(E',E)\eta(E') \, dE' \qquad (1)$$
$$\int \rho(E',E) \, dE' = 1.$$

The width of the weight function $\rho(E',E)$ must be large compared to D, the distance between the compound nuclear levels, so that many levels are included in the average. Then $\bar{\eta}$ is insensitive to the precise value of the width or the shape of ρ. It is customary to take ρ as square in which event Eq. (1) becomes

$$\bar{\eta}(E) = \frac{1}{\Delta} \int_{E-\Delta/2}^{E+\Delta/2} \eta(E')\, dE' \equiv \int_\Delta \eta(E')\, dE'. \tag{2}$$

Here Δ is the width of the weight function.

We now apply this definition to such quantities as the total cross section, the scattering amplitude, etc. In order to make the physics of the averaging process clearer we shall first discuss the problem of the scattering of spinless particles by spinless nuclei (for example, alpha particles on even-even nuclei).

a. Scattering of Spinless Particles by Spinless Nuclei

In this case the scattering can be described by a single function, the scattering amplitude, $f(\vartheta)$, where ϑ is the angle of scattering. The total cross section σ_t is just

$$\sigma_t = \frac{4\pi}{k}\, \text{Im}\, f(0), \tag{3}$$

while the total elastic scattering is

$$\sigma_{el} = \int |f(\vartheta)|^2\, d\Omega. \tag{4}$$

The potential of the complex potential model is defined in terms of the average amplitude $\bar{f}(\vartheta)$. It is that potential for which $\bar{f}(\vartheta)$ is the scattering amplitude. The corresponding total cross section $<\sigma_t>$ is (we shall use the symbol $<\ >$ to indicate complex potential cross sections) according to Eq. (3),

$$<\sigma_t> = \frac{4\pi}{k}\, \text{Im}\, \bar{f}(0)$$

or

$$<\sigma_t> = \bar{\sigma}_t. \tag{5}$$

This is the fundamental equation of the complex potential model. It states that the total cross section predicted by the model and the energy average of the cross section are identical. It is a consequence of the linear relation between the total cross section and the scattering amplitude. No such simple relation exists between $\bar{\sigma}_{el}$ and the elastic cross section $<\sigma_{el}>$ calculated from the model:

$$<\sigma_{el}> = \int |\bar{f}(\vartheta)|^2\, d\Omega. \tag{6}$$

The difference is the fluctuation cross section σ_{fl}:

$$\sigma_{fl} = \bar{\sigma}_{el} - <\sigma_{el}> = \int [\overline{|f(\vartheta)|^2} - |\bar{f}(\vartheta)|^2]\, d\Omega. \tag{7}$$

The identical effect is present in the reaction cross section σ_r,

$$\sigma_r = \sigma_t - \sigma_{el}. \tag{8}$$

We find

$$\langle\sigma_r\rangle = \langle\sigma_t\rangle - \langle\sigma_{el}\rangle$$
$$= \bar{\sigma}_t - \bar{\sigma}_{el} + \sigma_{fl}.$$

Or
$$\langle\sigma_r\rangle = \bar{\sigma}_r + \sigma_{fl} \tag{9}$$
$$\langle\sigma_{el}\rangle = \bar{\sigma}_{el} - \sigma_{fl}. \tag{10}$$

We note that even when there are no reactions ($\bar{\sigma}_r = 0$) that $\langle\sigma_r\rangle$ is not zero indicating an absorption even when only elastic scattering can occur. It is for this reason that the potential is complex. A qualitative understanding of this effect will be given below.

Clearly σ_{fl} is needed before we can compare the scattering angular distribution and the reaction cross section as computed from the complex potential model with the energy average of these cross-sections. σ_{fl} is known in two limiting situations (3). At sufficiently high energies of the incident nucleon the widths of the compound nuclear levels become comparable with the distance between levels. Then the cross sections no longer fluctuate wildly but are smooth. The fluctuation cross section then tends to zero:

$$\sigma_{fl} \to 0. \quad E \text{ large} \tag{11}$$

The value of E for which the limiting case holds will vary from nucleus to nucleus. Present experimental evidence indicates that σ_{fl} is still important for energies as high as 6 Mev. In this limit [Eq. (11)], the complex potential model gives $\bar{\sigma}_r$ and $\bar{\sigma}_{el}$ directly.

The fluctuation cross section is also known at low energies where the compound nuclear resonances are extremely narrow and well separated. Then by direct evaluation of the averages $\bar{\sigma}_{el}$ and $\bar{f}$ one can show [see ref. 3 as well as Section 2 of this chapter] that

$$\langle\sigma_r\rangle \simeq \sigma_c \tag{12}$$

where σ_c is the cross section for formation of the compound nucleus while

$$\sigma_{fl} \simeq \sigma_c - \bar{\sigma}_r. \quad \text{low } E \tag{13}$$

That is, in this limit σ_{fl} is that part of the cross section for the formation of the compound nucleus which does not lead to reactions. It is therefore just the average compound elastic cross section $\bar{\sigma}_{ce}$

$$\sigma_{fl} \simeq \bar{\sigma}_{ce} \quad \text{low } E \tag{14}$$

$\bar{\sigma}_{ce}$ describes that part of the elastic scattering in which the incident particle and the target form a compound nucleus which then decays, the incident particle being emitted with its original energy. We see that in this low-energy limit the imaginary part of the complex potential describes the rate at which particles in the incident beam combine with the target nucleus to form a compound nucleus.

A qualitative understanding of this phenomenon can be obtained from an argument of Friedman and Weisskopf (7) and of Adair (10). This argument will also indicate how σ_{fl} varies with energy and the relation σ_{fl} and $\bar{\sigma}_{ce}$. The energy average of the wave function introduces an uncertainty in the energy so that it becomes possible to give a temporal description of the scattering process. A wave packet striking the target nucleus will be separated by the interaction with the nucleus into two parts. One part is delayed because of the formation of compound nuclear states, the delay being of the order of the lifetime of these states. The other part is undelayed as would also be the case for scattering by a static potential well. A potential model can only describe the undelayed fraction with the consequence that in the model the delayed portion will appear as an absorption which from the point of view of observation is fictitious. The absorption is then just given by the cross section for compound elastic scattering, $\bar{\sigma}_{ce}$. The separation into two portions is not exact. It is valid only when the lifetime of the compound nuclear states is very long; that is when the widths of the resonances are extremely narrow. These conditions are usually met only at low energies of the incident particle. At higher energies the widths increase and the compound nuclear states have a sufficiently short lifetime so that the probability that a particle is undelayed in spite of the formation of a compound nucleus becomes appreciable. Thus as the energy increases the effective absorption which arises from formation of a compound nucleus decreases. This is just the statement that σ_{fl} tends to zero as the energy increases. It is also clear that σ_{ce} need not decrease as rapidly. Instead it will contribute more and more to $<\sigma_{el}>$ and less and less to $<\sigma_r>$. This description is completely in accord with the limits described by Eq. (11) and Eq. (12).

b. The General Case

We now consider the more general situation in which the target nucleus has a spin I, the incident particle an intrinsic angular momentum i as well as an orbital angular momentum l. As discussed in Chapter V.A, we combine I and i to form the channel spin s and finally combine s with l to form J the angular momentum of the entire system:

$$\mathbf{s} = \mathbf{I} + \mathbf{i} \tag{15}$$
$$\mathbf{J} = \mathbf{s} + \mathbf{l} = \mathbf{I} + \mathbf{i} + \mathbf{l}. \tag{16}$$

The transition matrix $\mathfrak{T}$ which describes reactions as well as scattering in which J the total angular momentum and the parity Π stays constant but in which the channel spin, the orbital angular momentum, and the character of the particle (include here isotopic spin, energy, etc.) change from s to s', l to l', α to α', respectively, is

$$\Im = \Im(\alpha's'l'|\alpha sl|J\Pi). \tag{17}$$

The energy average of $\Im$ can be readily obtained from Eq. (132, Section V.A). We state it for the diagonal case $\alpha = \alpha'$, $s = s'$, and $l = l'$:

$$\bar{\Im}(\alpha sl|\alpha sl|J\Pi) = 1 - \exp[2i\delta_{sl}{}^{(J)}] + \pi \exp[2i\delta_{sl}{}^{(J)}]\left\langle\frac{\Gamma(\alpha sl|J\Pi)}{D_{J\Pi}}\right\rangle \tag{18}$$

where $\delta_{sl}{}^{(J)}$ is the potential scattering phase shift[a] and $\Gamma(\alpha sl|J\Pi)$ is the particle width for producing a compound state of total angular momentum J and parity Π when the orbital angular momentum is l and s is the channel spin. $D_{J\Pi}$ is the distance between levels of the compound nucleus of angular momentum J and parity Π.

We now compare this description of the scattering process with that given by the complex potential model. The transition matrix

$$<\Im(\alpha sl|\alpha sl|J\Pi)>$$

predicted by that model must equal the energy average given in Eq. (18). Let

$$<\Im(\alpha sl|\alpha sl|J\Pi)> = 1 - \exp[2i(\xi_{sl}{}^{(J)} + i\eta_{sl}{}^{(J)})], \tag{19}$$

where $\xi + i\eta$ is the complex phase shift. Then

$$1 - \exp[2i(\xi_{sl}{}^{(J)} + i\eta_{sl}{}^{(J)})]$$
$$= 1 - \exp[2i\delta_{sl}{}^{(J)}] + \pi \exp[2i\delta_{sl}{}^{(J)}]\left\langle\frac{\Gamma(\alpha sl|J\Pi)}{D_{J\Pi}}\right\rangle. \tag{20}$$

Before obtaining the relation between ξ and η and δ and $<\Gamma/D>$ we need to know the reality properties of δ. The phase shift δ will be complex if the determining potential produces changes in the channel spin or angular momentum l or if nonresonant inelastic scattering ("direct interaction") is present. The first two of these can be present only when I, the spin of the target nucleus differs from zero; the third is important for sufficiently high incident neutron energies. To allow for these possibilities we permit δ to be complex:

$$\delta_{sl}{}^{(J)} = \alpha_{sl}{}^{(J)} + i\beta_{sl}{}^{(J)}, \tag{21}$$

[a] We have abbreviated our notation somewhat here. The phase shift δ should depend on α and Π as well as J, l, and s. In Section V.A we therefore used the more elaborate notation $\delta(\alpha ls|J\Pi)$.

Eq. (20) becomes

$$\exp[2i\xi_{sl}^{(J)} - 2\eta_{sl}^{(J)}] = \exp[2i\alpha_{sl}^{(J)} - 2\beta_{sl}^{(J)}]\left(1 - \pi\left\langle\frac{\Gamma(\alpha sl|J\Pi)}{D_{J\Pi}}\right\rangle\right). \quad (22)$$

It follows that when $\langle\pi\Gamma/D\rangle$ is less than one that

$$\xi_{sl}^{(J)} = \alpha_{sl}^{(J)} \quad (23)$$

$$\exp[-2\eta_{sl}^{(J)}] = \exp[-2\beta_{sl}^{(J)}]\left[1 - \pi\left\langle\frac{\Gamma(\alpha sl|J\Pi)}{D_{J\Pi}}\right\rangle\right]. \quad (24)$$

However, when $(\pi\Gamma/D)$ is greater than one

$$\xi_{sl}^{(J)} = \alpha_{sl}^{(J)} + \pi/2 \quad (25)$$

$$\exp[-2\eta_{sl}^{(J)}] = \exp[-2\beta_{sl}^{(J)}]\left[\pi\left\langle\frac{\Gamma(\alpha sl|J\Pi)}{D_{J\Pi}}\right\rangle - 1\right]. \quad (26)$$

Equation (24) demonstrates that the consequence of the averaging process is the introduction of additional effective absorption which is directly related to the possibility of forming the compound nucleus as represented by the $\pi\Gamma/D$ term. Of course, part of this absorption is an experimental absorption if compound inelastic scattering or reaction is possible. But even when the energy is so low that only elastic scattering can occur, an effective absorption which is not experimental is present because of compound elastic scattering. Again for pure elastic scattering $[\beta_{sl}^{(J)} = 0]$ Eq. (26) shows that as $(\pi\Gamma/D)$ increases, $\eta_{sl}^{(J)}$ and therefore the imaginary part of the complex potential decreases becoming zero when $(\pi\Gamma/D)$ equals two. Moreover, since $\eta_{sl}^{(J)}$ must be positive, we obtain an upper limit (6) on (Γ/D):

$$\left\langle\frac{\Gamma(\alpha sl|J\Pi)}{D_{J\Pi}}\right\rangle \leq \frac{2}{\pi}. \quad \text{pure elastic scattering} \quad (27)$$

This inequality does not hold if nonresonant inelastic scattering occurs $[\beta_{sl}^{(J)} \neq 0]$.

Equations (23) to (26) relate the parameters of the complex potential model and those of the theory of nuclear reactions. When the resonances are well separated so that resonance parameters can be determined these relations can be used to determine the complex well parameters. On the other hand when the resonances overlap, the predictions of the complex well can be employed to obtain some information on the average properties of the resonance parameters.

As we stated earlier, all of the predictions of the complex potential are completely observable in the limits of low and high energy. Rigorously, however, the only quantity which allows direct comparison with experiment is the average total cross section:

$$\langle \sigma_t \rangle = \bar{\sigma}_t = \sum_{sJl} \frac{2J+1}{(2I+1)(2i+1)} [4\pi\lambda^2 \sin^2 \xi_{sl}^{(J)} + 2\pi\lambda^2 \cos 2\xi_{sl}^{(J)}(1 - \exp[-2\eta_{sl}^{(J)}])]. \quad (28)$$

This quantity will of course show "size" resonances as discussed in the introduction whenever $\xi_{sl}^{(J)}$ comes close to an odd multiple of $\pi/2$. The corresponding value of the term in square brackets is

$$4\pi\lambda^2[1 - \tfrac{1}{2}(1 - \exp[-2\eta_{sl}^{(J)}])]$$

showing the reduction caused by absorption from the maximum possible value $4\pi\lambda^2$. The corresponding wave function will have its greatest possible amplitude inside the nucleus relative of course to a unit incident amplitude. Consequently we can expect that the effective absorption cross section, $\langle \sigma_r \rangle$, will simultaneously show a resonance. The justification for this last statement is given by the following two equations. The first expresses the cross section $\langle \sigma_r \rangle$ in terms of phase shifts:

$$\langle \sigma_r \rangle = \sum_s \sum_{J,l} \frac{(2J+1)\pi\lambda^2}{(2I+1)(2i+1)} [1 - \exp(-4\eta_{sl}^{(J)})]. \quad (29)$$

The second equation connects the bracket in Eq. (29) with the imaginary part of the complex potential W and the wave function:

$$1 - \exp[-4\eta_{sl}^{(J)}] = 4k \int (u_{sl}^{(J)})^* \left(\frac{2m}{\hbar^2} W\right) u_{sl}^{(J)} dr \quad (30)$$

where $u_{sl}^{(J)}$ is just r times the radial wave function describing the partial wave with angular momentum properties given by J, s, and l, normalized as follows:

$$u_{sl}^{(J)} \xrightarrow[r \to \infty]{} \frac{\exp[i\delta_{sl}^{(J)}] \sin(kr - \tfrac{1}{2}l\pi + \delta_{sl}^{(J)})}{k}. \quad (31)$$

A description of these resonances which is qualitatively (but not quantitatively!) correct is obtained by considering the square well case in which the effective potential between a neutron and nucleus is taken to be:

$$V = -\frac{\hbar^2}{2m} K_0^2(1 + i\zeta) \quad r < R$$
$$= 0. \quad r > R \quad (32)$$

It is easy to show that for a large nucleus, small absorption, and sufficiently low energies, resonances occur whenever

$$KR \simeq \pi[n + \tfrac{1}{2}(l+1)], \qquad K^2 = K_0^2 + k^2, \qquad n \text{ integer.} \quad (33)$$

The width of these resonances, Γ_{sp}, is just the "single particle" width:

$$\Gamma_{sp}/2kR \simeq \hbar^2/mR^2. \quad (34)$$

From Eq. (33) we learn that the energy at which a given type of resonance (specified n and l) occurs decreases as the nuclear radius R increases. We also see that energies at which the odd l resonances occur will interleave the resonance energies for even l. These results can be given approximate validity for other potentials when the WKBJ method is employed to obtain the appropriate generalization of Eq. (33).

Thus by examining the position of the peaks in the total cross section and the way they vary in going from element to element one can obtain a rough value of R and K_0. However, to specify the complex potential with greater precision further areas in which the predictions of the complex potential model can be fitted to the data are required. Two exist. One is at sufficiently high energy where σ_{fl} is zero and then $<\sigma_r> = \bar{\sigma}_r$. The other is at low energy where it becomes correct to identify $<\sigma_r>$ with σ_c, the cross section for the formation of the compound nucleus.

At sufficiently high energy, then, the average inelastic cross section is given by Eq. (29). One can also now calculate the angular distribution $<d\sigma_{el}/d\Omega>$ which in this limit equals $\overline{d\sigma_{el}/d\Omega}$. This angular distribution is given by replacing $\mathfrak{I}(\alpha's'l'|\alpha sl|J\Pi)$ in Eq. (5) of Section V.A by the corresponding quantity computed from the complex potential model:

$$\left\langle \frac{d\sigma_{el}}{d\Omega} \right\rangle = \frac{\lambda^2}{(2i+1)(2I+1)} \sum_L P_L(\cos\vartheta) \sum \frac{(-)^{s'-s}}{4} \bar{Z}(l_1 J_1 l_2 J_2; sL)$$
$$\cdot \bar{Z}(l_1' J_1' l_2' J_2'; s'L) \text{ Re } <\mathfrak{I}(\alpha s' l_1'|\alpha s l_1|J_1\Pi_1)> <\mathfrak{I}^*(\alpha s' l_2'|\alpha s l_2|J_2\Pi_2)>. \quad (35)$$

We shall give more explicit formulae for $<\mathfrak{I}>$ in the next section.

At low energies $<\sigma_r>$ equals σ_c. The latter may be expressed in terms of "transmission" coefficients $T_{sl}^{(J)}$ as follows:

$$\sigma_c = \sum_s \sum_{J,l} \frac{(2J+1)\pi\lambda^2}{(2i+1)(2I+1)} T_{sl}^{(J)}. \quad (36)$$

We therefore obtain from Eq. (29)

$$T_{sl}^{(J)} = 1 - \exp[-4\eta_{sl}^{(J)}] = 1 - \exp[-4\beta_{sl}^{(J)}]\left(1 - \pi\left\langle\frac{\Gamma(\alpha sl|J\Pi)}{D_{J\Pi}}\right\rangle\right)^2. \quad (37)$$

When Γ/D is small compared to one, Eq. (37) is

$$T_{sl}^{(J)} \simeq 1 - \exp[-4\beta_{sl}^{(J)}] + 2\pi \left\langle \frac{\Gamma(\alpha sl|J\Pi)}{D_{J\Pi}} \right\rangle \exp[-4\beta_{sl}^{(J)}]. \quad (37')$$

The first two terms are just the nonresonant ("direct interaction") inelastic scattering terms which suggests that we rewrite σ_c as follows:

$$\sigma_c = (\sigma_c)_p + (\sigma_c)_{CN} \quad (38)$$

where the subscript p stands for potential [since the nonresonant inelastic scattering is just a generalization of the potential (nonresonant) elastic scattering] and the subscript CN for compound nucleus. The quantity $(\sigma_c)_p$ is just the cross section for inelastic scattering via the direct interaction

$$(\sigma_c)_p = \sum_s \sum_{J,l} \frac{(2J+1)\pi\lambda^2}{(2I+1)(2i+1)} (1 - \exp[-4\beta_{sl}^{(J)}]) \quad (39)$$

while

$$(\sigma_c)_{CN} = \sum_s \sum_{J,l} \frac{(2J+1)\pi\lambda^2}{(2I+1)(2i+1)} (T_{sl}^{(J)})_{CN}. \quad (39')$$

Here $(T_{sl}^{(J)})_{CN}$ is the transmission coefficient for the formation of the compound nucleus:

$$(T_{sl}^{(J)})_{CN} = 2\pi \left\langle \frac{\Gamma(\alpha ls|J\pi)}{D_{J\Pi}} \right\rangle \exp[-4\beta_{sl}^{(J)}]. \quad (40)$$

This is the familiar relation (11) between T and $\langle\Gamma/D\rangle$ except for the factor which is required by the existence of nonresonant inelastic scattering as was first emphasized by Yoshida (12). Relations (38)–(40) are significant because they permit (with some additional assumptions) the calculation of inelastic total and differential cross sections as we have discussed in Section V.A.

Finally we turn to very low energies where only the $l = 0$ partial wave plays an important role. We assume inelastic scattering is not possible. Then the complex potential model cross sections are

$$\langle\sigma_t\rangle = \bar{\sigma}_t = g_+[4\pi(a_+^2 - b_+^2) + 4\pi\lambda b_+] + g_-[4\pi(a_-^2 - b_-^2) + 4\pi\lambda b_-] \quad (41)$$

where

$$g_\pm = \frac{1}{2}\left[1 \pm \frac{1}{2I+1}\right] \quad (42)$$

$$a_\pm = -\lim_{k\to 0} \lambda \xi_{so}^{(J)}, \quad J = s = I \pm \tfrac{1}{2} \quad (43)$$

VI.D. THE COMPLEX POTENTIAL MODEL

and
$$b_\pm = \lim_{k\to 0} \lambda \eta_{so}{}^{(J)}. \qquad J = s = I \pm \tfrac{1}{2} \tag{44}$$

The quantities, b, may be expressed in terms of the strength function since from Eq. (24) ($\beta = 0$) we have

$$b_\pm = \frac{\pi\lambda}{2}\left\langle\frac{\Gamma(\alpha 0s|J\Pi)}{D_{J\Pi}}\right\rangle. \qquad J = s = I \pm \frac{1}{2} \tag{45}$$

Equation (41) can then be written as follows:

$$\bar\sigma_t = g_+ \left\{ 4\pi\left[a_+{}^2 - \frac{\pi^2\lambda^2}{4}\left\langle\frac{\Gamma_+}{D_+}\right\rangle^2\right] + 2\pi^2\lambda^2\left\langle\frac{\Gamma_+}{D_+}\right\rangle\right\}$$
$$+ g_- \left\{4\pi\left[a_-{}^2 - \frac{\pi^2\lambda^2}{4}\left\langle\frac{\Gamma_-}{D_-}\right\rangle^2\right] + 2\pi^2\lambda^2\left\langle\frac{\Gamma_-}{D_-}\right\rangle\right\} \tag{46}$$

where
$$\left\langle\frac{\Gamma_\pm}{D_\pm}\right\rangle = \left\langle\frac{\Gamma(\alpha 0s|J\Pi)}{D_{J\Pi}}\right\rangle. \qquad J = s = I \pm \frac{1}{2}$$

The significant point here is that the quantities $\lambda\langle\Gamma/D\rangle_\pm$ and $|a_\pm|$ can be separately measured. The parameters describing the complex potential model must then be so chosen as to predict these quantities correctly.

Actually it is also possible to measure the sign of the scattering lengths $a_\pm$ by the appropriate interference experiments. The complex potential model predicts that for most nuclei $a_\pm$ will be positive corresponding to repulsive sphere scattering. There will, however, be narrow regions in A in the neighborhood of the giant resonances for which the scattering lengths will be negative. The reason for this can be understood on the basis of a qualitative argument given by Weisskopf (13) in another connection. Let us again employ the square well, Eq. (32) and let us note (this will be discussed in the next section) that $(\hbar^2 K_0{}^2/2m)$ is of the order of 42 Mev. This has the consequence that in most nuclei, the wave function for $r < R$ will perform several oscillations inside R. This wave function is to join on smoothly to the outside wave function $(r > R)$ which at low energy is of very long wave length. This join can occur only if the amplitude of the inside wave function is very small, a condition which is of course approximately equivalent to repulsive sphere scattering. In other words by making the amplitude of the inside wave function small its slope is reduced sufficiently so that it can join on to the outside wave function. There is one exception to this argument which occurs whenever the inside wave function happens to have a small slope at R; that is when $K_0 R$ is an odd number times $(\pi/2)$. This is just the condition for a size resonance so that over the width in A of such a resonance it is possible for the scattering length to be negative but everywhere else it is positive.

As we have just indicated, there are size resonances in the scattering of neutrons by nuclei which in the square well model will occur whenever the nuclear radius is such that K_0R is roughly an odd multiple of $\pi/2$. The quantity $\lambda <\Gamma_\pm/D_\pm>$ will be a maximum for these nuclei while the scattering length will decrease from its positive value possibly becoming negative and then increase again to a positive value once the resonance region has been traversed. The width, Γ_A, of these resonances in A, the mass number, is

$$\Gamma_A \simeq 2kR \frac{3A}{K_0^2 R^2} \qquad (47)$$

which shows that the resonance is broader the larger the value of A.

In this section, we have discussed the relation between experiment and the predictions of the complex potential model. At the high energies at which the fluctuation cross section σ_{fl} goes to zero, each of these predictions, $<\sigma_t>$, $<\sigma_{el}>$, and $<\sigma_r>$ can be directly compared with experiment. At lower energies, it becomes necessary to obtain σ_{fl}. This is possible at very low energies where only $l = 0$ neutrons are involved and where we can pick out $<\sigma_r>$ and therefore $<\sigma_{fl}>$ because of its characteristic energy dependence. It is also possible in this energy range to make direct measurements of the widths of individual levels and thus find their energy average and to determine the scattering lengths a by measuring the cross sections between resonances. Such measures become less effective with increasing energy. Several different angular momenta for the neutrons need to be included in the analysis. Picking out $<\sigma_r>$ either by its energy variation or by direct measurement of the widths becomes more involved and ambiguous. In addition if inelastic scattering or reactions are possible, the effect of compound elastic scattering will be very sensitive to the properties of the inelastic levels. As a consequence of these difficulties direct evaluation of the complex potential model phase shifts have been made only with the aid of some fairly strong assumptions on the effect of compound elastic scattering. These ambiguities do not prevent an over-all broad understanding of neutron scattering data since they are not large once the neutron energy is of the order of Mev and these results taken together with the low-energy analysis are sufficient. However, for a detailed analysis of the experimental data for a particular nucleus for neutrons in the intermediate energy region more information than that provided by the measured total cross section and angular distribution are needed. As we shall show in the next section, with the aid of double and triple scattering experiments it is possible to determine the complex model phase shifts, subject of course to the usual ambiguities of phase shift analysis.

2. Empirical Determination of the Complex Potential (6)

In this section the various parametric forms employed for the complex potential and the determination of the parameters from experiment are discussed. We shall conclude by giving the "best" values for these parameters available at the present time.

a. Form of the Potential

The most general form for the potential which has been employed to represent the interaction between a nucleon and a *spherical* nucleus in the sense discussed in detail in Section 1 is a combination of central and spin orbit potentials. Omitting for the moment electromagnetic terms the potential has the following form:

$$V = -[V_c \rho_c(r) + iW_c \rho_c'(r)] + \left(\frac{\hbar}{\mu c}\right)^2 \left[V_{so}\frac{1}{r}\frac{d\rho_{so}}{dr} + iW_{so}\frac{1}{r}\frac{d\rho_{so}'}{dr}\right] \mathbf{\sigma} \cdot \mathbf{L}, \tag{48}$$

where V_c, W_c, V_{so}, W_{so} are constants; ρ_c, etc. are functions of the radial distance r only and are normalized to be unity at the origin. The operators $\mathbf{\sigma}$ and $\mathbf{L}$ are the Pauli spin operator and the orbital angular momentum operator divided by $\hbar$ for the incident neutron. The constant $(\hbar/\mu c)$ is the π-meson Compton wave length so that $(\hbar/\mu c)^2$ is 2×10^{-26} cm^2. If the target nucleus is not spherical the principal change in Eq. (48) is in the replacement of the spherical ρ's by spheroidal ones, a procedure which is correct for low neutron energies.

It is also assumed that the constants V_c, W_c, etc. are independent of the nature of the target nucleus. The functions ρ_c depend on the radius of the target nucleus. In going from a nucleus of small radius to a nucleus of large radius the depths of the potential remain the same, the radius of the potential increasing in direct proportion to the increase in radius. The shape factors ρ contain in addition a surface region whose thickness is independent of the target nucleus. These assumptions are of course central to the whole idea of the complex potential model. Their success is suggestive of and consonant with the concept of nuclear matter.

Form (48) and the accompanying assumptions are rather sweeping and some possible exceptions to their general validity will now be briefly discussed. It should be pointed out in advance that these effects, in virtue of the successes of form (48) are small. The real part of the potential is for the most part just the shell model potential extrapolated to a higher energy and, like it, is determined from the fundamental nucleon-nucleon potential by a self-consistent method. This has the consequence that the potential depends on the configuration of the nucleons in the target nucleus as well as that of the incident nucleon. There should therefore be some dependence of the potential on the character of the target nucleus. For example, several authors (14,15,16) have suggested that the potential should depend on $(N - Z)$, the number of neutrons minus the number of protons. The evidence is at present not conclusive. On the positive side there are the binding energy calculations of Green and Sood (14), and the d,p measurements of Schiffer (17). On the other hand Fulmer (18) finds no evidence for such effects in the scattering of 22-Mev protons from Ni^{64}, Zn^{64}, and natural copper.

Form (48) also assumes that V is a local potential. A nonlocal central potential operating on the wave function, ψ, has the following form:

$$\int v(\mathbf{r} + \mathbf{r}')K(\mathbf{r} - \mathbf{r}')\psi(\mathbf{r}')\, d\mathbf{r}'.$$

That V is in general a nonlocal potential follows from (a) that nuclear matter forms a dispersive medium and that (b) the Pauli principle plays an important role in determining the neutron-nucleus interaction. In particular, the Bethe-Goldstone equation (19) involves a nonlocal potential. How important are the effects of nonlocality? They can be completely simulated by permitting the parameters of the potential form (48) to be energy dependent and by making the ρ's sufficiently flexible. However, if this last condition in particular is not satisfied, characteristic difficulties dependent on the range of the nonlocality in fitting the data will appear. Nonlocal effects should be important whenever the change in momentum $\Delta \mathbf{p}$ of the particle by scattering is so large that the associated wave length $(\hbar/|\Delta \mathbf{p}|)$ is of the order of the range of the nonlocality. We may therefore expect effects for large angles of scattering.

Form (48) also assumes that the potential is independent of the spin of the target nucleus, I. However, there are several additional invariants involving I which can be formed and these, therefore, could be included in the empirical complex potential. For example, the form

$$\mathbf{I} \cdot \mathbf{\sigma} f(r) \tag{49}$$

could be employed. It would give rise to different scattering for the various possible channel spins, an effect which is certainly observed with low-energy neutrons. A potential which would lead to changes in the channel spin in scattering is

$$(\mathbf{I} \cdot \mathbf{r})(\mathbf{\sigma} \cdot \mathbf{r})g(r). \tag{50}$$

There are several other possibilities but these two, (49) and (50), encompass the physical possibilities which might be present because the target nucleus has a nonzero spin and which are not described by potential (48). The physical processes involved are the possible dependence of the scattering on channel spin and the possible spin-flip which can occur because of an interaction dependent upon the neutron and the target spin.

The importance of these various effects is not as yet known. A great deal of information could be gained if scattering and polarization experiments with separated isotopes and on neighboring elements were performed. One such experiment has been performed by Vanetsian, Klutcharev, and Fedtchenko (20). In experiments with 19.6-Mev protons incident upon $_{48}Cd^{111}$, $_{48}Cd^{113}$, $_{48}Cd^{116}$, $_{59}Sn^{116}$, the three cadmium isotopes gave very nearly the same results which, however, differed substantially from those obtained with tin. On the other hand $_{32}Ge^{74}$ (spin 0) and $_{32}Ge^{73}$ (spin 9/2) give rise to considerably different scattering beyond 100°. Fulmer (18) has scattered 22-Mev protons from $_{28}Ni^{64}$, $_{30}Zn^{64}$ and natural Cu. The shape of the angular distributions are completely similar.

Electromagnetic interactions of the incident nucleons with the nucleus must be added to Eq. (48). The simplest example is the Coulomb interaction of an incident charged particle with the target. In addition there is the interaction of the magnetic moment of the nucleon with the Coulomb field **E** of the target nucleus. For an incident neutron, the case of greatest interest, the interaction is given by

$$V_{em} = \mu_n(e\hbar/2M^2C^2)\boldsymbol{\sigma} \cdot (\mathbf{E} \times \mathbf{p}) \tag{51}$$

where μ_n is 1.91, the neutron magnetic moment in nuclear magnetons. For spherical nuclei

$$\mathbf{E} = -\mathbf{r}\frac{1}{r}\frac{d\phi}{dr}$$

where ϕ is the electrostatic potential. Therefore

$$V_{em} = -\mu_n \frac{e\hbar}{2M^2C^2}\frac{1}{r}\frac{d\phi}{dr}\boldsymbol{\sigma} \cdot \mathbf{L}. \tag{52}$$

This potential, therefore, adds on a long-range tail to the nuclear spin orbit term in Eq. (48) affecting the small angle scattering and polarization. Another electromagnetic effect suggested by Thaler, stems from the possible polarization of the nucleon by the Coulomb field. It would contribute a potential energy of the form:

$$V_{pol} = -\tfrac{1}{2}\alpha E^2, \tag{53}$$

where α is the polarizability. An experimental observation of the polarizability effect has not yet been made.

b. Scattering by Spin Zero Nuclei

We give here the formulae for the cross sections $<\sigma_t>$, $<\sigma_{el}>$, and $<\sigma_r>$ which are obtained from the potential [Eq. (48)] in terms of the complex phase shifts of Eq. (19). We remind the reader that before $<\sigma_{el}>$ and $<\sigma_r>$ can be compared with experiment the cross section σ_{fl} must be added to $<\sigma_{el}>$ to obtain the average $\bar{\sigma}_{el}$ and subtracted from $<\sigma_r>$ to obtain $\bar{\sigma}_r$.

Many of the calculations have been made without the spin orbit term. In that event the scattering amplitude has the familiar form:

$$<f> = \frac{\lambda}{2i} \sum (2l+1)[\exp[2i(\xi_l + i\eta_l)] - 1]P_l(\cos\vartheta). \qquad (54)$$

From this expression it follows that

$$<\sigma_{el}> = \int |<f>|^2 d\Omega = 4\pi\lambda^2 \Sigma (2l+1)[\exp[-2\eta_l]\sin^2\xi_l \\ + \tfrac{1}{4}(1 - \exp[-2\eta_l])^2] \qquad (55)$$

$$<\sigma_t> = 4\pi\lambda^2 \Sigma(2l+1)[\sin^2\xi_l + \tfrac{1}{2}\cos 2\xi_l(1 - \exp[-2\eta_l])], \qquad (56)$$

and

$$<\sigma_r> = \pi\lambda^2 \Sigma(2l+1)(1 - \exp[-4\eta_l]). \qquad (57)$$

When the spin orbit forces are included in V, Eq. (48), the partial wave analysis must be made in terms of the eigenfunctions of the total angular momentum j which for a given value of the orbital angular momentum l can have two values:

$$j = l \pm \tfrac{1}{2}. \qquad (58)$$

Let the corresponding complex phase shifts be designated by $\zeta_l^{(+)}$ for $j = l + \tfrac{1}{2}$, and $\zeta_l^{(-)}$ for $j = l - \tfrac{1}{2}$ where

$$\zeta_l^{(+)} = \xi_l^{(+)} + i\eta_l^{(+)}.$$

Following the method of Schwinger (21,22) we write the asymptotic form of the wave function for the neutron as follows:

$$\psi \to \exp[i\mathbf{k}_i \cdot \mathbf{r}]\chi + \frac{\exp[ikr]}{r}F\chi, \qquad (59)$$

where χ is a spin wave function and $\hbar\mathbf{k}_i$ is the incident momentum of the neutron. Then F, the scattering amplitude, is now an operator which includes simultaneously the description of the scattering when the initial spin is up or down. From general invariance arguments it follows that

$$F = <f> = A(\vartheta) + B(\vartheta)\mathbf{\sigma} \cdot \mathbf{n}, \qquad (60)$$

where **n** is a unit vector perpendicular to the plane of scattering:

$$\mathbf{n} = (\mathbf{k}_f \times \mathbf{k}_i)/|\mathbf{k}_f \times \mathbf{k}_i|. \qquad (61)$$

The vector $\hbar\mathbf{k}_f$ is the final momentum. The functions A and B can be expressed in terms of the phase shifts ζ as follows:

$$A = \frac{1}{2ik}\sum_{0}^{\infty}[(l+1)(\exp[2i\zeta_l^{(+)}] - 1) + l(\exp[2i\zeta_l^{(-)}] - 1)]P_l(\cos\vartheta) \qquad (62)$$

$$B = -\frac{1}{2k}\sum_{1}^{\infty}[\exp(2i\zeta_l^{(+)}) - \exp(2i\zeta_l^{(-)})]P_l^{(1)}(\cos\vartheta), \qquad (63)$$

where

$$P_l^{(1)} = \sin\vartheta\,\frac{d}{d(\cos\vartheta)}\,P_l(\cos\vartheta). \qquad (64)$$

For an unpolarized beam the differential cross section is given by

$$\left\langle\frac{d\sigma_{el}}{d\Omega}\right\rangle = \frac{1}{2}\,\text{tr}\,F\dagger F = |A|^2 + |B|^2, \qquad (65)$$

where tr signifies trace and $F\dagger$ is the adjoint of F.
The total shape elastic cross section is just a simple generalization of Eq. (55).

$$<\sigma_{el}> = 4\pi\lambda^2\sum_l\{(l+1)[\exp[-2\eta_l^{(+)}]\sin^2\xi_l^{(+)} + \tfrac{1}{4}(1 - \exp[-2\eta_l^{(+)}])^2]$$
$$+ l[\exp[-2\eta_l^{(-)}]\sin^2\xi_l^{(-)} + \tfrac{1}{4}(1 - \exp[-2\eta_l^{(-)}])^2]\}. \qquad (66)$$

The total cross-section and reaction cross section are corresponding analogues of Eq. (56) and Eq. (57).

Since spin 1/2 particles can be polarized, further information on the interaction of these particles with spin zero nuclei can be obtained from double and triple scattering experiments (22). A description of the double and triple scattering experiments together with the relevant formulae is given in Chapter V.A. The angular distribution predicted by Eq. (60) for an unpolarized beam striking a polarizer P and then being scattered by S is

$$\left\langle\left(\frac{d\sigma}{d\Omega}\right)_{PS}\right\rangle = \left(\frac{d\sigma}{d\Omega}\right)_P\left[\left\langle\frac{d\sigma}{d\Omega}\right\rangle_S + <2\,\text{Re}\,A_S^*B_S>P_P\mathbf{n}_P \cdot \mathbf{n}_S\right], \qquad (67)$$

where P_P is the polarization of the beam incident on the scatterer. The polarization is defined as the expectation value of the spin component in the direction of $\mathbf{n}_P$. If now we consider the angular distribution for first a scattering by a polarizer P, then by a scatterer S, and finally by an analyzer A, we find that form (60) predicts:

$$\left\langle \left(\frac{d\sigma}{d\Omega}\right)_{PSA} \right\rangle = \left(\frac{d\sigma}{d\Omega}\right)_P \left(\frac{d\sigma}{d\Omega}\right)_A \left[\left\langle \frac{d\sigma}{d\Omega} \right\rangle_S + P_P P_A \mathbf{n}_P \cdot \mathbf{n}_A \right.$$
$$+ P_A \langle 2\,\mathrm{Re}\,A_S{}^* B_S \rangle \mathbf{n}_A \cdot \mathbf{n}_S + P_P \langle 2\,\mathrm{Re}\,A_S{}^* B_S \rangle \mathbf{n}_P \cdot \mathbf{n}_S$$
$$+ P_P P_A \{ \langle 2\,\mathrm{Im}\,A_S B_S{}^* \rangle (\mathbf{n}_P \times \mathbf{n}_S) \cdot \mathbf{n}_A$$
$$\left. + \langle 2|B_S|^2 \rangle (\mathbf{n}_P \times \mathbf{n}_S) \cdot (\mathbf{n}_S \times \mathbf{n}_A) \} \right], \quad (68)$$

where P_A is the polarization in the direction $\mathbf{n}_A$ which would be produced if an unpolarized beam struck the analyzer.

By suitably choosing appropriate directions $\mathbf{n}_P$, $\mathbf{n}_S$, $\mathbf{n}_A$, it is possible to determine $2\,\mathrm{Re}\,A_S{}^* B_S$, $2\,\mathrm{Im}\,A_S B_S{}^*$, $|B_S|^2$. The angular distribution $(d\sigma/d\Omega)_S$ gives $|A_S|^2 + |B_S|^2$. It thus follows that by experiment it is possible (though at the present stage not feasible) to determine the amplitudes A_S and B_S except, of course, for an all over phase. Of course formulae (67) and (68) give only the results predicted by Eq. (48). We need to correct them for the compound elastic scattering before we can compare with the experimentally observed averages.

These formulae, Eq. (59) to Eq. (68) hold rigorously when the target has spin zero. They are also valid for target nuclei with nonzero spin if the complex potential is given by Eq. (48) and does not include terms of the type described in Eq. (49) and Eq. (50).

c. Fluctuation Cross Section and Elastic Scattering

In this section the calculation of the fluctuation cross section, σ_{fl}, in terms of the phase shifts of the complex model will be discussed. It follows from Eq. (10) that σ_{fl} must be added to $\langle \sigma_{el} \rangle$ before we can obtain the experimentally observable average elastic scattering cross section.[b] Our primary interest is the effect of σ_{fl} on the angular distribution for elastic scattering. The expression for $d\sigma_{fl}/d\Omega$ is obtained from Eq. (5) of Section V.A and Eq. (35) by taking their difference:

$$\frac{d\sigma_{fl}}{d\Omega} = \frac{\lambda^2}{(2i+1)(2I+1)} \sum P_L(\cos\vartheta) \frac{(-)^{s-s'}}{4}$$
$$\bar{Z}(l_1 J_1 l_2 J_2; sL) \bar{Z}(l_1' J_1 l_2' J_2; s'L)$$
$$\mathrm{Re}\,\{ \langle \mathfrak{J}(\alpha s' l_1' | \alpha s l_1 | J_1 \Pi_1) \mathfrak{J}^*(\alpha s' l_2' | \alpha s l_2 | J_2 \Pi_2) \rangle$$
$$- \langle \mathfrak{J}(\alpha s' l_1' | \alpha s l_1 | J_1 \Pi_1) \rangle \langle \mathfrak{J}^*(\alpha s' l_2' | \alpha s l_2 | J_2 \Pi_2) \rangle \}. \quad (69)$$

[b] We remind the reader that σ_{fl} at low energies is approximately equal to the cross section for compound elastic scattering.

VI.D. THE COMPLEX POTENTIAL MODEL

Only the resonance parts of $\mathfrak{J}$ will survive in the fluctuation term in the curly brackets. We shall now determine under what conditions there will be an absence of interference among the various possible transitions. Note that only if $s = s'$, and $l_1' = l_1$ will $\mathfrak{J}(\alpha s' l_1' | \alpha s l_1 | J_1 \Pi_1)$ have a phase which varies smoothly with energy. We make the *statistical assumption* that when the phase fluctuates rapidly with energy that the average values indicated in Eq. (69) are zero. Second, we assume that the resonance levels do not overlap which leads to the requirement that $J_1 = J_2$ and that the parity of l_1 be the same as that of l_2. From this last result we can immediately conclude that only *even* L's will appear in (69) so that $d\sigma_{fl}/d\Omega$ is symmetric about 90°. Generally, however, there will be several different values of l which permit a transition from a given value of the channel spin s to a particular value of the total angular momentum J. There will, therefore, be interference terms in Eq. (69) between these various possible l-values. These interference terms are present only if the spin of the target nucleus differs from zero. If it is zero only one value of l can occur for a transition from the target nucleus ($s = 1/2$) to a particular J of a particular parity. Actually for the spin zero case the only assumption we need to make in order to prove the absence of interference terms is that the various resonances do not overlap. The statistical assumption is not required.

Because of these additional complexities we shall discuss the target spin zero case only. By virtue of the preceding discussion, Eq. (69) reduces to

$$d\sigma_{fl}/d\Omega = \tfrac{1}{2}\lambda^2 \Sigma P_L(\cos\vartheta) \sum_{l,J} (\tfrac{1}{4}) \bar{Z}^2(lJlJ;\tfrac{1}{2}L)[<|\mathfrak{J}(\alpha\tfrac{1}{2}l|\alpha\tfrac{1}{2}l|J\Pi)|^2> \\ - |<\mathfrak{J}(\alpha\tfrac{1}{2}l|\alpha\tfrac{1}{2}l|J\Pi>|^2], \quad (70)$$

where for a given l, J can take on the value $l \pm 1/2$. Finally, it is necessary to evaluate the quantity in square brackets. Let us do this first for the simpler case of no inelastic scattering. Then from $\sigma_{fl} = <\sigma_r>$ it follows that[e]

$$d\sigma_{fl}/d\Omega = \tfrac{1}{2}\lambda^2 \Sigma P_L(\cos\vartheta) \bar{Z}^2(lJlJ;\tfrac{1}{2}L)\tfrac{1}{4}(1 - \exp[-4\eta_{lJ}]) \quad (71)$$
target nucleus spin = 0, and $\sigma_r = 0$

where

$$\eta_{l,l+\frac{1}{2}} = \eta_l^{(+)} \quad \text{and} \quad \eta_{l,l-\frac{1}{2}} = \eta_l^{(-)}. \quad (72)$$

[e] It may be convenient to employ another form for $d\sigma_{el}/d\Omega$ which follows directly Eq. (71). We need only realize that the compound elastic scattering will in this case affect only the noninterfering terms in $|A|^2$ and $|B|^2$. Then

$$|A|_{fl}^2 = \tfrac{1}{4}\lambda^2 \Sigma\{(l+1)^2(1 - \exp[-4\eta_l^{(+)}]) + l^2(1 - \exp[-4\eta_l^{(-)}])\}(P_l)^2$$
$$|B|_{fl}^2 = \tfrac{1}{4}\lambda^2 \Sigma\{(1 - \exp[-4\eta_l^{(+)}]) + (1 - \exp[-4\eta_l^{(-)}])\}(P_l^{(1)})^2.$$

Adding Eq. (71) to the shape elastic cross section Eq. (65) gives us then the experimentally observable average elastic cross section. It is important to notice that this final result is expressed completely in terms of the phase shifts of the complex potential model.

At this point we discuss the effect of compound elastic scattering on the interpretation of polarization experiments, considering again only the case for which Eq. (70) is valid, that is, target nucleus spin zero and zero reaction cross section. Consider 2 Re $A_s^* B_s$ of Eq. (67) first. The term as one readily discovers by using the explicit expressions for A and B involves only interference terms. For the case being considered as we have shown just above the interference terms in the compound elastic scattering vanish so that

$$<\text{Re } A_s^* B_s> = \overline{A_s^* B_s} \tag{73}$$

where again we have employed angular brackets to indicate the complex potential amplitudes and the bar to indicate energy averages.

Turning now to 2 Im $A_s^* B_s$ of Eq. (68), the situation is not so simple. Compound elastic scattering will contribute to Im $A_s^* B_s$ since it contains squared terms in addition to interference terms. We note that the noninterfering part of $<\text{Im } A_s^* B_s>$ is

$$\frac{\lambda^2}{4} \sum \{(l+1)|\exp[2i\zeta_l^{(+)}] - 1|^2 - l|\exp[2i\zeta_l^{(-)}] - 1|^2\} P_l P_l^{(1)}. \tag{74}$$

As we can see by comparing Eqs. (70) and (71) we must add a compound elastic contribution to this term equal to:

$$(\text{Im } A_s^* B_s)_{fl} = \frac{\lambda^2}{4} \sum [(l+1)(1 - \exp[-4\eta_l^{(+)}]) - l(1 - \exp[-4\eta_l^{(-)}])] P_l P_l^{(1)}. \tag{75}$$

The expression for 2 Im $A_s^* B_s$ is now

$$2 \overline{\text{Im } A_s^* B_s} = 2 <\text{Im } A_s^* B_s> + 2 (\text{Im } A_s^* B_s)_{fl}. \tag{76}$$

This expression can now be compared directly with experiment. An entirely similar discussion gives the following result for $|B_s|^2$

$$\overline{|B_s|^2} = <|B_s|^2> + (|B_s|^2)_{fl} \tag{77}$$

where

$$(|B_s|^2)_{fl} = \frac{\lambda^2}{4} \sum [(1 - \exp[-4\eta_l^{(+)}]) + (1 - \exp[-4\eta_l^{(-)}])][P_l^{(1)}]^2. \tag{78}$$

With this formula we conclude the discussion of the case in which the target nucleus has spin zero and there is zero reaction cross section.

VI.D. THE COMPLEX POTENTIAL MODEL

The next case to be considered is that of target nucleus still with spin zero but with a nonzero reaction cross section. In this situation, more assumptions and more information are necessary before a final formula for $d\sigma_{fl}/d\Omega$ can be obtained and used. We assume that $\langle\sigma_r\rangle$ gives the cross section for the formation of a compound nucleus. More specifically, the cross section for formation of a compound nucleus of angular momentum J with a neutron of orbital angular momentum l is

$$\sigma_C^{(lJ)} = \pi\lambda^2(2J+1)(1 - \exp[-4\eta_{lJ}]). \tag{79}$$

Second, the assumption [see discussion of statistical theory of nuclear reactions (Section V.A)] is made that the branching ratio for decay of the compound nuclear state to the ground state of the target nucleus with emission of a neutron of an angular momentum l is

$$w_{lJ} = \left\langle \frac{\Gamma(\alpha l|J\Pi)}{\Gamma_{J\Pi}} \right\rangle \tag{80}$$

$$w_{lJ} = \frac{[T_l^{(J)}(E)]_{CN}}{\Sigma[T_{l'}^{(J)}(E')]_{CN}}, \tag{81}$$

where $T_l^{(J)}$ is given by Eq. (37) and the energies and angular momenta are those of the emitted particles. With these assumptions

$$d\sigma_{fl}/d\Omega = \tfrac{1}{2}\lambda^2 \Sigma P_L(\cos\vartheta)\bar{Z}^2(lJlJ;\tfrac{1}{2}L)\tfrac{1}{4}(1 - \exp[-4\eta_{lJ}])w_{lJ}. \tag{82}$$

The difficulties in employing these formulae are fairly obvious. In order to calculate w_{lJ} we need to know the energies, spins, and parities of all the levels of the target nucleus which can be excited by the incident neutron.

Polarization experiments are important because

$$2\,\overline{\text{Re}\,A_\text{s}^*B_\text{s}},\; 2\,\overline{\text{Im}\,A_\text{s}^*B_\text{s}},\; \overline{|B_\text{s}|^2}$$

together with the angular distribution $\overline{d\sigma}/d\Omega$ and total cross section $\bar{\sigma}_t$ are sufficient to determine the complex phase shifts of the complex potential model. We may see this most easily by discussing a specific case. Suppose that the energy is low enough so that only $l = 0$ and $l = 1$ neutrons need to be considered so that three complex phase shifts or six numbers are to be obtained from experiment. Suppose for the moment that the scattering is pure elastic. From $\overline{d\sigma}/d\Omega$ we obtain three relations between the phase shifts; from the angular distribution of $2\,\overline{\text{Re}\,A_\text{s}B_\text{s}^*}$ two more, and similarly two from $2\,\overline{\text{Im}\,A_\text{s}^*B_\text{s}}$ and two from $2\overline{|B_\text{s}|^2}$. Altogether we have nine relations which over-determine the three complex phase shifts. When reactions are possible, three more constants, the three giving the fraction of $<\sigma_r>$ which leads to elastic scattering, are involved. It is important that these can also be obtained from experiment. If the w's are added to the complex phase shifts as additional unknowns we are left with nine quantities to be determined. We have, however, ten relations between these constants, nine as described above and one from $\bar{\sigma}_t$ (or $\bar{\sigma}_r$). It is thus possible not only to obtain these quantities but also to check on the statistical hypothesis which forms the basis for the expressions for w_{lJ} given in Eq. (81).

We shall not discuss the case for which the target nucleus has a nonzero spin. Assumptions of a type similar to those developed immediately above are required even when there is no inelastic scattering.

d. Empirical Values for the Complex Potential Parameters

Various authors have worked with special cases of this general form Eq. (48). For example, many have omitted the spin orbit term, sometimes because of the absence of polarization data which would help to determine the magnitude of this term in the potential, or because it was thought that this term is small so that on a first over-all survey it could be neglected. Another difference has been in the relation between the shape of the real part of central potential, ρ_c and the shape of the imaginary part ρ_c'. Some authors have equated ρ_c and ρ_c'. Others have concentrated ρ_c' at the surface of the nucleus. On the other hand, most authors have placed ρ_{so} and ρ_{so}' equal to ρ_c, a form which is suggested by the multiple scattering approximation of Riesenfeld and Watson (*23*) and of Fernbach, Heckrotte, and Lepore (*23*). This choice for the spin orbit potential has the consequence that it is concentrated in the surface region since ρ_c is generally taken to be constant in the interior of the nucleus. In the earliest papers (*3*) on this subject ρ_c was taken to be a square well; that is, it was constant for r less than R and zero for r greater than R. The results of these calculation showed that this form was adequate for a qualitative description of neutron total cross sections and angular distributions but that it gave much too small a reaction cross section. The high reflectivity of such a potential well form was found to be responsible (*3,24*). In other words the interaction in the surface region of the target was not described properly by the square well. This was also demonstrated (*25*) by the failure of the model to describe proton angular distributions correctly as might be expected since the scattering of protons is sensitive to the nature of the surface interaction by virtue of the Coulomb barrier which inhibits penetration into the nuclear interior. The cure was to use a tapered well for ρ_c, consisting of an interior region in which ρ_c is essentially constant and a finite surface region over which ρ_c drops from its interior value to zero. A typical tapered well is the Woods-Saxon potential (*26*):

$$(\rho_c)_{\text{WS}} = \frac{1}{1 + \exp\left[(r - R)/a\right]}. \tag{83}$$

The width of the surface region is determined by the parameter a, being smaller for smaller a. The parameter R is the value of r at which ρ_c has approximately (1/2) of its interior value and is therefore a measure of the nuclear radius. Another form which has similar properties is that employed by Emmerich and Amster (*27*).

$$(\rho_c)_{\text{DS}} = \begin{matrix} 1 & r < R \\ \dfrac{1}{e^x - x} & r > R. \end{matrix} \quad x = (r - R)/a \tag{84}$$

There is obviously a wide variety of possible ρ_c, etc. Those which have actually been used are tabulated in Table I of ref. *6*. For the purposes of this chapter it is sufficient to limit ourselves to three forms given in Table I.

TABLE I. POTENTIAL WELL TYPE. THE SYMBOLS ρ_c, ρ_c', ρ_{so}, ρ_{so}' ARE DEFINED BY EQ. (48); $x = (r - R)/a$

Type	Designation	ρ_c	ρ_c'	$\rho_{so} = \rho_{so}'$
Woods-Saxon	V_{WS}	$\left[1 + \exp\left(\dfrac{r-R}{a}\right)\right]^{-1}$	ρ_c	0
Derivative surface absorption	V_{DS}	$[\exp x - x]^{-1}$ $\begin{matrix} r \leq R \\ r \geq R \end{matrix}$	$-\dfrac{d\rho_c}{dx}$	0
Bjorklund-Fernbach	V_{BF}	$\left[1 + \exp\left(\dfrac{r-R}{a}\right)\right]^{-1}$	$\exp\left[-\dfrac{(r-R)^2}{b^2}\right]$	ρ_c

The qualitative features of the cross sections determined by these potentials and their differences and similarities are listed below. These results have been obtained for the most part by direct numerical calculation so that their domain of validity is not known.

(1) At low energies the cross sections, particularly the position of minima and maxima, depend on the parameters V_c and R through the combination $V_c R^2$. As the energy increases the combination involved is written $V_c R^n$ where n slowly varies from two to three in going from low to high energy. The criterion that V_c and R should be independent of the energy leads to a determination of V_c and R separately. At lower energies it is possible in principle to determine these parameters from the measurement of the strength function and the scattering length (*27a*).

(2) As the imaginary part of the potential W_c is increased, the oscillations in the various cross sections are damped, the peaks broadening. Concentrating the imaginary part of the potential in the surface reduces the slope of $<\sigma_r>$ and $<\sigma_t>$ as a function of mass number A.

(3) Increasing a, the length of the taper, increases $<\sigma_r>$ but decreases $<\sigma_{el}>$. Potentials of similar shape will yield nearly identical results if the parameter a is adjusted in each case so as to give identical values to the thickness t of the surface region, defined to be the distance over which the potential drops from $\frac{9}{10}$ to $\frac{1}{10}$ of its value at the origin.

(4) Including a spin orbit term dampens the oscillations in the angular distribution of elastically scattered nucleons as well as in the total cross section.

We now turn to the determination of the parameters of these potentials. At "zero neutron energy" the neutron scattering length a and the ratio $<\Gamma/D>$ can be measured (see Fig. 1 and Fig. 2) as we mentioned earlier [see discussion following Eq. (46)] or in other words the zero energy limits of the complex phase shift can be obtained directly from experiment. At higher energies it is not possible to determine these phase shifts directly because of the lack of data. The total cross section plays an important role and one can use its distinctive features, such as the location of the maximum cross section, and the dependence on A of the various single-particle resonances. Finally, attempts are made to match the experimental angular distributions. As we have emphasized exact comparison is generally not possible because of the difficulty of taking account of the compound elastic scattering precisely. We do know however that Eq. (71) gives the maximum of the compound elastic scattering for spin 0 target nuclei and we can require that the experimental curve fall between $<d\sigma_{el}/d\Omega>$ and $<d\sigma_{el}/d\Omega>$ plus the $d\sigma_{fl}/d\Omega$ of Eq. (71). Since generally the compound elastic scattering is relatively small except for large angles, this requirement is fairly severe. Again at higher energies (say >10 Mev; the exact figure will certainly vary with the target nucleus) fluctuation scattering is negligible so that the computed complex well angular distribution may be compared directly with experiment.

It is rather remarkable that it is possible to fit the experimental data over a wide range in energies and over the entire periodic table with the potentials listed in Table I. It is of course this fact which makes the complex potential model not only a useful model but also a valid approximation to the exact description of the nucleon-nucleus interaction.

In Tables II and III the values of the parameters are listed for the Woods-Saxon and the derivative surface absorption potentials. Bjorklund and Fernbach (32) place $R = 1.24 A^{\frac{1}{3}} f$; $a = 0.65 f$; $b = 0.98 f$.

TABLE II. PARAMETERS FOR THE WOODS-SAXON POTENTIAL
$[R = (r_0 A^{\frac{1}{3}} + r_1)]$

Energy (Mev)	Particle	V_c (Mev)	W_c (Mev)	a (fermi)	r_0 (fermi)	r_1 (fermi)	Ref.
0–4	neutron	52	3.1	0.52	1.15	0.4	28*
10	proton	53	7	0.50	1.3	—	29
17	proton	47	8.5	0.49	1.33	—	29
17	proton	50	8	0.50	1.3	—	30
31.5	proton	35.5	15.5	0.53	1.33	—	30
40	proton	36	15 ± 5	0.66 ± 0.1	1.3	—	29
90	proton	26	10 ± 5	0.66 ± 0.1	1.3	—	29

* Tables of the phase shifts for this case will be published shortly as a technical report of the Laboratory of Nuclear Science, Massachusetts Institute of Technology.

TABLE III. PARAMETERS FOR DERIVATIVE SURFACE ABSORPTION POTENTIAL
$\bar{R} = R + 1.36a$; $\bar{R} = (1.25 A^{\frac{1}{3}} + 0.5)f$; $a = 0.84f$

Energy (Mev)	V_c (Mev)	W_c/V_c	Ref.
0.1–1.0	44	0.34	31*
1.5–3.0	43	0.37	
4.0–6.0	42	0.40	
7.0–10.0	41	0.43	
12.0–14.0	40	0.46	

* Tables of transmission coefficients are available in this reference.

With these parameters kept constant for all nuclei and energies, the best values of V_c, W_c, V_{so}, W_{so} were evaluated by fitting a variety of experimental results involving neutron and proton scattering and polarizations.

Of course the agreement of the cross sections calculated from these potentials with experiment is not precise. In the next few paragraphs we shall discuss the deficiencies of the fit obtained with these potentials but it should be always borne in mind that the qualitative over-all fit is excellent and in some cases exact.

The results at zero energy are given in Fig. 1 and Fig. 2. In Fig. 1 we note that the peak at $A = 155$ is much broader and more irregular than the calculated result. Bohr and Mottelson (33) suggested that this discrepancy is a consequence of the asphericity of the target nuclei in this mass number region. This surmise has been verified by a number of authors. For example, Margolis and Troubetzkoy (34) pointed out that closed shell nuclei did fall on the calculated curve very nicely as in fact can be seen in Fig. 1 on the low A side of the peak. Moreover, they showed the affect of asphericity is to break up the giant resonance into a number of resonances depending upon the size of the deformation. One observes that this structure together with the variation in deformation with target nucleus is sufficient to account for the structure of Fig. 2 near

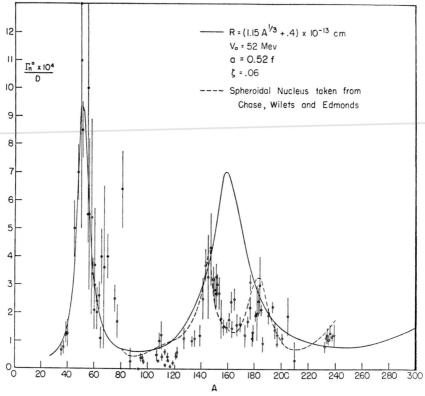

FIG. 1. The ratio Γ_n^0/D of the neutron width to level spacing. Here $\Gamma_n^{(0)}/D = (E_0/E)^{\frac{1}{2}}\Gamma_n/D$ where $E_0 = 1$ ev. The solid curve is taken from Ref. 28 where the spherical potential V_{WS} was employed with the constants noted in the figure. The dotted line is that obtained in Ref. 35 in which the effect of the aspherical shape of the target nuclei on neutron scattering was evaluated. The experimental points are reported by the groups at Argonne, Brookhaven, Duke, and Wisconsin.

$A = 155$. The quantitative situation is quite satisfactory as is indicated by the calculations of Chase, Wilets, and Edmonds (35).

Of course, this asphericity effect is not restricted to zero energy scattering. The calculated results for $\bar{\sigma}_t$ are shown in Fig. 3. These are in qualitative agreement with the data. However difficulties in matching the data with spherical potentials are experienced near the D-wave maximum which occurs for $A \sim 140$ at an energy of a few hundred thousand electron volts.

A second difficulty at low energy revealed in Fig. 1 is that the valley between the two giant resonances is much deeper than the theory predicts. This discrepancy has not been resolved. There is, of course, always the possibility that these elements have a genuine paucity of levels near zero energy. Lane et al. (36) suggest that this is an effect associated with

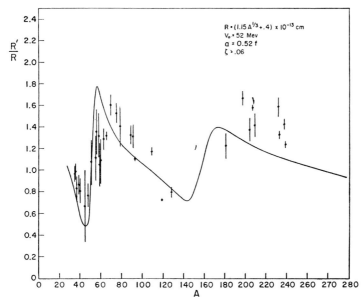

Fig. 2. Ratio of the potential scattering length to nuclear radius. The solid curve is taken from Ref. 28 where the spherical potential V_{WS} was employed with the constants noted in the figure. The experimental points are reported by the groups at Brookhaven and Duke. We are indebted to R. L. Zimmerman and K. K. Seth for informing us of the results obtained by these groups prior to their publication.

the magic number 50 which occurs at $A = 90$ and 120. They also point out that, for these nuclei the imaginary potential will be not only smaller but also more concentrated at the surface.

At higher energies the principal qualitative discrepancies appear in the angular distributions. Generally it is possible to match the experimental data quite beautifully for relatively small angles through several diffraction minima and maxima. However, at back angles the experimental curves do not always have the structure of the calculated curves. This is particularly serious when spin orbit terms are omitted as in the V_{WS} and V_{DS} potentials. Introducing the spin orbit potential reduces the discrepancy considerably and indeed in many cases to within the experi-

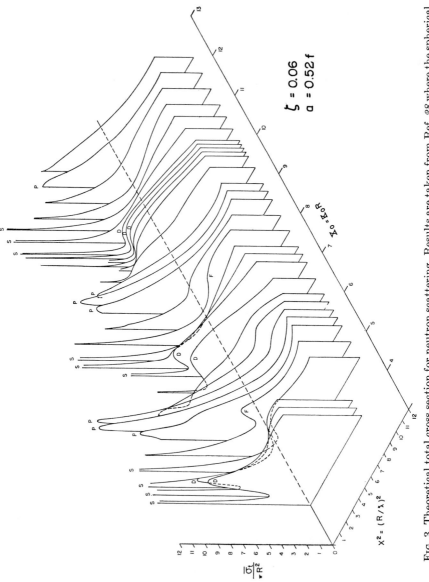

FIG. 3. Theoretical total cross section for neutron scattering. Results are taken from Ref. 28 where the spherical potential V_{WS} was employed with the constants noted in the figure. The letters at the "giant" resonances indicate the partial wave which is involved.

mental error. The remaining discrepancy may be a consequence of the nonlocality of the true complex potential.

Optical model analyses have also been made of alpha particle elastic scattering for alpha particles (37,38,39) 20 Mev and above and for deuterons of about 15 Mev (40). In both cases it is found that the imaginary term, that is, the absorption, is so large that these experiments are sensitive only to the nature of the potential in the nuclear surface region. For example, Igo finds that all fits of the data have identical forms for r greater than that distance at which the real potential is -10 Mev.

REFERENCES

1. H. A. Bethe, Phys. Rev. **57**, 1125 (1940).
2. H. H. Barschall, Phys. Rev. **86**, 431 (1952).
3. H. Feshbach, C. E. Porter, and V. F. Weisskopf, Phys. Rev. **96**, 448 (1954).
4. H. Feshbach and V. F. Weisskopf, Phys. Rev. **76**, 1550 (1949).
5. A. M. Lane and R. G. Thomas, Revs. Modern Phys. **30**, 257 (1958).
6. H. Feshbach, Ann. Rev. Nuclear Sci. **8**, 49 (1958).
7. F. L. Friedman and V. F. Weisskopf, *Niels Bohr and the Development of Physics* (Pergamon Press, London, 1955), p. 134.
8. H. Feshbach, Ann. Phys. (N.Y.) **5**, 357 (1958).
9. G. E. Brown, C. T. de Dominicis, and J. E. Langer, Ann. Phys. (N.Y.) **6**, 209 (1959).
10. R. K. Adair, S. E. Darden, and R. E. Fields, Phys. Rev. **96**, 503 (1954).
11. J. M. Blatt and V. F. Weisskopf, *Theoretical Nuclear Physics* (John Wiley and Sons, New York, 1952).
12. S. Yoshida, Proc. Phys. Soc. **A69**, 668 (1956).
13. H. Feshbach, D. C. Peaslee, and V. F. Weisskopf, Phys. Rev. **71**, 145 (1947).
14. A. E. S. Green and P. C. Sood, Phys. Rev. **111**, 1147 (1958).
15. T. H. R. Skyrme, Nuclear Phys. **9**, 615 (1959).
16. A. M. Lane, Compt. rend. congr. intern. physique nucléaire, Paris, 1958, p. 32 (1959).
17. J. P. Schiffer, Proc. Intern. Conf. Nuclear Optical Model, Florida Univ. Studies (Tallahassee) **32**, 254 (1959).
18. C. B. Fulmer, Proc. Intern. Conf. Nuclear Optical Model, Florida Univ. Studies (Tallahassee) **32**, 260 (1959).
19. H. A. Bethe and J. Goldstone, Proc. Roy. Soc. **A238**, 551 (1956).
20. R. A. Vanetsian, A. P. Klutcharev, and E. D. Fedtchenko, Compt. rend. congr. intern. physique nucléaire, Paris, 1958, p. 607 (1959).
21. J. Schwinger, Phys. Rev. **73**, 407 (1948).
22. See reviews in L. Wolfenstein, Ann. Rev. Nuclear Sci. **6**, 43 (1956), and R. van Wageningen, Some Aspects of the Theory of Elastic Scattering of Spin 0 and in Spin 1/2 Particles by Spin 0 Targets, Thesis, University of Groningen (1957).
23. W. B. Riesenfeld and K. M. Watson, Phys. Rev. **102**, 1107 (1956); S. Fernbach, W. Heckrotte, and J. V. Lepore, Phys. Rev. **97**, 1059 (1955).
24. M. Walt and J. R. Beyster, Phys. Rev. **98**, 655 (1955); M. Kawai and H. Ui, Progr. Theoret. Phys. (Kyoto) **14**, 263 (1955).
25. R. E. Lelevier and D. S. Saxon, Phys. Rev. **87**, 40 (1952); D. M. Chase and F. Rohrlich, *ibid.* **94**, 81 (1954).

26. R. D. Woods and D. S. Saxon, Phys. Rev. **95**, 577 (1954).
27. W. S. Emmerich and H. J. Amster, Physica **22**, 11,3 (1956).
27a. K. K. Seth, D. J. Hughes, R L. Zimmerman, and R. C. Garth, Phys. Rev. **110**, 692 (1958).
28. H. Feshbach, C. E. Porter, and V. F. Weisskopf, unpublished.
29. A. E. Glassgold and P. J. Kellogg, Phys. Rev. **109**, 1291 (1958); A. E. Glassgold, Revs. Modern Phys. **30**, 419 (1958).
30. M. A. Melkanoff, J. Nodvik, D. S. Saxon, and R. D. Woods, Phys. Rev. **106**, 793 (1957).
31. W. S. Emmerich, Westinghouse Research Lab. Research Report 6-94511-6-R19 (1958).
32. F. E. Bjorklund and S. Fernbach, University of California Radiation Lab. Report UCRL 5028 (1958).
33. Quoted in H. H. Barschall, Phys. Rev. **96**, 448 (1954).
34. B. Margolis and E. S. Troubetzkoy, Phys. Rev. **106**, 105 (1957).
35. D. M. Chase, L. Wilets and A. R. Edmonds, Phys. Rev. **110**, 1080 (1958).
36. A. M. Lane, J. E. Lynn, M. A. Melkanoff, and E. R. Rae, quoted by A. M. Lane, Proc. Intern. Conf. Nuclear Optical Model, Florida Univ. Studies (Tallahassee) **32**, 168 (1959).
37. W. B. Cheston and A. E. Glassgold, Phys. Rev. **106**, 1215 (1957).
38. G. Igo and R. M. Thaler, Phys. Rev. **106**, 126 (1957).
39. G. Igo, Phys. Rev. Letters **1**, 12 (1958).
40. M. A. Melkanoff, Proc. Intern. Conf. Nuclear Optical Model, Florida Univ. Studies (Tallahassee) **32**, 207 (1959).

See also general references listed at the end of Section V.A of the present volume.

Some Tables of Transmission Factors

Neutrons

E. Campbell, H. Feshbach, C. E. Porter, and V. F. Weisskopf, M. I. T. Lab. Nuclear Sci. Tech. Report to be issued shortly.
W. Schrandt, J. R. Beyster, M. Walt, and E. N. Salmi, Los Alamos Laboratory Report LA-2099 (1956).
J. E. Monahan, L. C. Biedenharn, and J. P. Schiffer, Argonne National Laboratory Report ANL-5846 (1958).

Charged Particles

A. Tubis, Los Alamos Laboratory Report LA-2150 (TID-4500) (1958).
Tables of Coulomb Wave Functions National Bureau of Standards, Washington, D.C., 1952, Vol. I.
C. E. Froberg and P. Rabinowitz, National Bureau of Standards Report 3033 (1954).
R. F. Christy and R. Latter, Revs. Modern Phys. **20**, 185 (1948).
I. Bloch, M. H. Hull, Jr., A. A. Broyles, W. G. Bouricius, B. E. Freeman, and G. Breit, Revs. Modern Phys. **23**, 147 (1951).
H. Feshbach, M. M. Shapiro, and V. F. Weisskopf, Atomic Energy Commission Report NYO3077 (1953).
W. D. Barfield and A. A. Broyles, Phys. Rev. **88**, 892 (1952).
W. T. Sharp, H. E. Gove, and E. B. Paul, Atomic Energy Commission Lab. Report TPI-70 AECL No. 268 (1953).

See also list of tables in *Coulomb Wave Functions* by M. H. Hull, Jr., and G. Breit, in *Handbuch der Physik* Vol. XLI/1 (Springer Verlag, Berlin, 1959).

Appendixes

Appendix I. Constants and Conversion Factors*

Velocity of light $c = 299793.0 \pm 0.3$ km-sec^{-1}

Planck's constant $h = (6.62517 \pm 0.00023) \times 10^{-27}$ erg-sec

Avogadro's constant   (physical scale)
$N = (6.02486 \pm 0.00016) \times 10^{23}$ (gm-mole)$^{-1}$

Electronic charge $e = (4.80286 \pm 0.00009) \times 10^{-10}$ esu
$e/c = (1.60206 \pm 0.00003) \times 10^{-20}$ emu

Electron rest mass $m = (9.1083 \pm 0.0003) \times 10^{-28}$ gm

Fine structure constant $\alpha = e^2/\hbar c = (7.29729 \pm 0.00003) \times 10^{-3}$

Bohr magneton $\mu_0 = he/4\pi mc = (0.92731 \pm 0.00002) \times 10^{-20}$ erg-gauss^{-1}

Nuclear magneton $\mu_n = he/4\pi m_p c = (0.505038 \pm 0.000018) \times 10^{-23}$ erg-gauss^{-1}

$$1 \text{ gm} = (5.61000 \pm 0.00011) \times 10^{26} \text{ Mev}$$
$$1 \text{ electron mass} = 0.510976 \pm 0.000007 \text{ Mev}$$
$$1 \text{ atomic mass unit} = 931.141 \pm 0.010 \text{ Mev}$$
$$1 \text{ ev} = (1.60206 \pm 0.00003) \times 10^{-12} \text{ erg}$$

* E. R. Cohen, J. W. M. DuMond, T. W. Layton, and J. S. Rollett, Revs. Modern Phys. **27**, 363 (1955). Because of more recent determinations and corrections, these values are likely to require revisions by small amounts: see discussions in E. R. Cohen, K. Crowe and J. W. M. DuMond, "The Fundamental Constants of Physics" (Interscience Publishers, New York, 1957), E. R. Cohen and J. W. M. DuMond, Phys. Rev. Letters **1**, 291, 382 (1958), J. W. M. DuMond, Ann. Phys. (N.Y.) **7**, 365 (1959).

Appendix II. Table of Isotopes

D. STROMINGER

The following table is an abridged version of the Table of Isotopes, compiled by D. Strominger, J. M. Hollander, and G. T. Seaborg, which was published in Revs. Modern Phys. **30,** 585 (1958). The unabridged version of the Table of Isotopes should be consulted for more detail and for references to the original literature.

The first column contains the atomic numbers, chemical symbols, and mass numbers of the nuclear species. Metastable states are listed only when the half-life is longer than approximately 0.01 sec.

The second column lists the half-lives of the isotopes.

The third column lists the type of decay for each nuclide. The word "stable" appears in this column for those nuclides which occur in nature and have not been observed to decay. Other symbols used are:

β^- negative beta-particle (negatron) emission.
β^+ positive beta-particle (positron) emission.
α alpha-particle emission.
EC orbital electron capture.
IT isomeric transition (transition from upper to lower energy level of same nucleus).
n neutron emission following a β^- decay.

The fourth and fifth columns list relative isotopic abundances. The fourth column lists the abundances as they occur in nature while the fifth lists the enriched abundances available.*

* The maximum enrichment listed is that available as of July 1958 at

  (a) Isotopes Division, Oak Ridge National Laboratory, P. O. Box X, Oak Ridge, Tennessee.
  (b) Electromagnetic Separation Group, Atomic Energy Research Establishment, Harwell, Berks., England.
  (c) Isomet Corporation, P. O. Box 34, Palisades Park, New Jersey.

Many suppliers are available for enriched isotopes marked (d), and Harwell is willing to consider special enrichment problems and to fabricate certain targets.

The availability of enriched isotopes is time dependent, and the numbers given here are just meant to give the reader some idea of what is obtainable. We are grateful to Dr. P. V. Arow of Oak Ridge and to Dr. M. L. Smith of Harwell for sending us their latest values for the enrichments.

Columns 6 through 8 list the nuclear moments. The spin J is given in units of $\hbar$. The magnetic dipole moment μ is given in units of nuclear magnetons. The value is given without the diamagnetic correction. The electric quadrupole moment q is given in units of barns (10^{-24} cm^2).

The ninth column lists the isotopic mass excesses $(M - A)$ in million electron volts (Mev). These are taken from the tables by A. H. Wapstra [Physica **21**, 367, 385 (1955)] and J. R. Huizenga [Physica **21**, 410 (1955)], except for a few values from F. Ajzenberg-Selove and T. Lauritsen, Nuclear Phys. **11**, 1 (1959).

APPENDIX II. TABLE OF ISOTOPES

Isotope		Half-Life	Type of Decay	Abundance (per cent)		Nuclear Moments			Isotopic Mass Excess
Z	A			Natural	Enriched	Spin J	Magnetic μ	Quadrupole q	
$_0n^1$		12.8 min	β^-			1/2	-1.91316		8.368
$_1H^1$		Stable		99.99		1/2	$+2.79270$		7.585
H^2		Stable		0.014	High^d	1	$+0.857393$	$+0.00273$	13.726
H^3		12.262 yr	β^-			1/2	$+2.7988$		15.835
$_2He^3$		Stable		10^{-4}–10^{-5}	High^d	1/2	-2.1274		15.817
He^4		Stable		~ 100		0			3.607
He^6		0.81 sec	β^-						19.398
$_3Li^6$		Stable		7.4	99.999^a	1	$+0.82193$		15.862
Li^7		Stable		92.6	99.998^a	3/2	$+3.2560$		16.977
Li^8		0.841 sec	$\beta^-, 2\alpha$						23.310
Li^9		0.168 sec	β^-, n						28.1
$_4Be^7$		53.3 days	EC						17.840
Be^8		$\sim 10^{-15}$ sec	2α						7.309
Be^9		Stable		100		3/2	-1.1773	$\sim \pm 0.2$	14.010
Be^{10}		2.5×10^6 yr	β^-						15.566
$_5B^8$		0.78 sec	$\beta^+, 2\alpha$						25.287
B^{10}		Stable		19	99.9^a	3	$+1.8006$	$+0.074$	15.010
B^{11}		Stable		81	99.8^a	3/2	$+2.6880$	$+0.0355$	11.914
B^{12}		0.019 sec	β^-, α 1.3%						16.917
$_6C^{10}$		19.1 sec	β^+						18.79
C^{11}		20.4 min	β^+ 99+%, EC 0.19%						13.895
C^{12}		Stable		98.89	99.9^a	0			3.541
C^{13}		Stable		1.11	69.2^a	1/2	$+0.702205$		6.963
C^{14}		5570 yr	β^-			0			7.157
C^{15}		2.25 sec	β^-						14.305
$_7N^{12}$		0.0125 sec	$\beta^+, 3\alpha$						21.00
N^{13}		10.05 min	β^+						9.185
N^{14}		Stable		99.64		1	$+0.40357$	$+0.02$	7.002
N^{15}		Stable		0.365	$>95^c$	1/2	-0.2830		4.528
N^{16}		7.35 sec	β^-						10.402
N^{17}		4.14 sec	β^-, n						13.022
$_8O^{14}$		72 sec	β^+						12.149
O^{15}		124 sec	β^+						7.233
O^{16}		Stable		99.76		0			0
O^{17}		Stable		0.037	1.4^c	5/2	-1.8930	-0.027	4.222
O^{18}		Stable		0.204	90^c	0			4.521
O^{19}		29.4 sec	β^-						8.931
$_9F^{17}$		66 sec	β^+						6.989
F^{18}		112 min	β^+ 97%, EC 3%						6.188
F^{19}		Stable		100		1/2	$+2.6275$		4.142
F^{20}		10.7 sec	β^-						5.904
$_{10}Ne^{18}$		1.6 sec	β^+						10.415

TABLE OF ISOTOPES (Continued)

Isotope Z	A	Half-Life	Type of Decay	Abundance (per cent) Natural	Abundance (per cent) Enriched	Spin J	Magnetic μ	Quadrupole q	Isotopic Mass Excess
Ne	19	17.7 sec	β^+						7.398
Ne	20		Stable	90.92		0	$< 2 \times 10^{-4}$		−1.146
Ne	21		Stable	0.257		3/2	−0.66140		0.465
Ne	22		Stable	8.82			~ 0		−1.533
Ne	23	40.2 sec	β^-						1.643
Ne	24	3.38 min	β^-						1.113
$_{11}$Na	20	0.39 sec	β^+, α						14.187
Na	21	22 sec	β^+						3.987
Na	22	2.58 yr	β^+ 89% EC 11%			3	+1.746		1.307
Na	23		Stable	100		3/2	+2.2161	+0.10	−2.744
Na	24m	$\sim$0.02 sec	IT, β^-						
Na	24	15.0 hr	β^-			4	+1.69		−1.336
Na	25	60 sec	β^-						−2.07
$_{12}$Mg	23	12 sec	β^+						1.353
Mg	24		Stable	78.7	>99.9[b]		~ 0		−6.853
Mg	25		Stable	10.1	97.5[a]	5/2	−0.8547		−5.818
Mg	26		Stable	11.2	99.0[a,b]		~ 0		−8.569
Mg	27	9.45 min	β^-						−6.641
Mg	28	21.2 hr	β^-						−6.784
$_{13}$Al	24	2.1 sec	β^+ $\alpha \sim 10^{-2}\%$						7.16
Al	25	7.3 sec	β^+						−1.57
Al	26m	6.7 sec	β^+						
Al	26	8×10^5 yr	β^+ 84% EC 16%						−4.544
Al	27		Stable	100		5/2	+3.6385	+0.149	−9.236
Al	28	2.27 min	β^-						−8.594
Al	29	6.56 min	β^-						−9.38
$_{14}$Si	27	4.3 sec	β^+						−4.409
Si	28		Stable	92.18	99.4[a]			~ 0	−13.246
Si	29		Stable	4.71	83.0[a]	1/2	±0.5548	$<10^{-4}$	−13.353
Si	30		Stable	3.12	72.5[a]			~ 0	−15.595
Si	31	2.62 hr	β^-						−13.825
Si	32	$\sim$710 yr	β^-						−14.77
$_{15}$P	28	0.28 sec	β^+						0.55
P	29	4.4 sec	β^+						−8.386
P	30	2.6 min	β^+						−11.280
P	31		Stable	100		1/2	+1.1305		−15.308
P	32	14.2 days	β^-			1	−0.2523		−14.873
P	33	24.4 days	β^-						−16.615
P	34	12.4 sec	β^-						−14.79
$_{16}$S	31	2.7 sec	β^+						−9.87
S	32		Stable	95.0	99.1[a]	0			−16.579
S	33		Stable	0.76	25.1[a]	3/2	+0.6427	−0.064	−16.864

TABLE OF ISOTOPES (Continued)

Isotope Z	A	Half-Life	Type of Decay	Abundance (per cent) Natural	Abundance (per cent) Enriched	Spin J	Magnetic μ	Quadrupole q	Isotopic Mass Excess
S^{34}			Stable	4.22	44.6[a]			$<2 \times 10^{-3}$	−19.890
S^{35}		87 days	β^-			3/2	±1.0	+0.045	−18.545
S^{36}			Stable	0.014	2.4[a]			<0.01	−20.08
S^{37}		5.04 min	β^-						−16.71
S^{38}		2.87 hr	β^-						
$_{17}Cl^{32}$		0.31 sec	β^+ $\alpha\ 10^{-2}\%$						−3.55
Cl^{33}		2.8 sec	β^+						−11.41
Cl^{34m}		32.4 min	$\beta^+ \sim 50\%$ IT $\sim 50\%$						
Cl^{34}		1.53 sec	β^+						−14.37
Cl^{35}			Stable	75.5	96.0[a]	3/2	+0.82091	−0.0789	−18.712
Cl^{36}		3.1×10^5 yr	β^- 98% EC 2%			2	+1.2839	−0.0168	−18.914
Cl^{37}			Stable	24.5	75.2[a]	3/2	+0.6833	−0.0621	−20.914
Cl^{38}		37.3 min	β^-						−18.66
Cl^{39}		56 min	β^-						−18.79
Cl^{40}		1.4 min	β^-						
$_{18}A^{35}$		1.83 sec	β^+						−13.302
A^{36}			Stable	0.337			~ 0		−19.628
A^{37}		35 days	EC						−20.098
A^{38}			Stable	0.063					−23.475
A^{39}		~ 265 yr	β^-						−21.75
A^{40}			Stable	99.60			~ 0		−23.232
A^{41}		110 min	β^-						−20.92
A^{42}		>3.5 yr	β^-						
$_{19}K^{37}$		1.2 sec	β^+						−13.97
K^{38}		7.7 min	β^+						−17.60
K^{38}		0.95 sec	β^+						
K^{39}			Stable	93.08	99.9[a,b]	3/2	+0.3909	+0.11	−22.31
K^{40}		1.25×10^9 yr	β^- 89% EC 11%	0.012	6.2[a]	4	−1.2964		−21.74
K^{41}			Stable	6.91	99.2[a]	3/2	+0.2151	±0.1	−23.50
K^{42}		12.5 hr	β^-			2	−1.137		−22.51
K^{43}		22.3 hr	β^-						−23.90
K^{44}		22 min	β^-						−22.44
K^{45}		34 min	β^-						
$_{20}Ca^{38}$		0.66 sec	β^+						
Ca^{39}		1.0 sec	β^+						−15.46
Ca^{40}			Stable	96.97	99.997[a]		~ 0		−23.07
Ca^{41}		1.1×10^5 yr	EC						−23.07
Ca^{42}			Stable	0.64	88[b]				−26.18
Ca^{43}			Stable	0.145	74[b]	7/2	−1.3153		−25.75
Ca^{44}			Stable	2.06	98[b]				−28.55
Ca^{45}		164 days	β^-						−27.61

TABLE OF ISOTOPES (Continued)

Isotope Z	A	Half-Life	Type of Decay	Abundance (per cent) Natural	Abundance (per cent) Enriched	Nuclear Moments Spin J	Nuclear Moments Magnetic μ	Nuclear Moments Quadrupole q	Isotopic Mass Excess
	Ca^{46}		Stable	0.0033	9.5[a]				
	Ca^{47}	4.7 days	β^-						−28.44
	Ca^{48}		Stable	0.185	85.7[b]				−30.08
	Ca^{49}	8.8 min	β^-						−26.86
$_{21}Sc^{40}$		0.22 sec	β^+						−9.08
	Sc^{41}	0.87 sec	β^+						−17.11
	Sc^{42}	0.68 sec	β^+						
	Sc^{43}	3.92 hr	β^+						−23.53
	Sc^{44m}	2.44 days	IT						
	Sc^{44}	3.92 hr	β^+ 93% EC 7%						−24.90
	Sc^{45}		Stable	100		7/2	+4.749		−27.87
	Sc^{46m}	20 sec	IT						
	Sc^{46}	83.9 days	β^-						−28.41
	Sc^{47}	3.43 days	β^-						−30.51
	Sc^{48}	44 hr	β^-						−30.35
	Sc^{49}	57 min	β^-						−32.09
	Sc^{50}	1.5 min	β^-						
$_{22}Ti^{43}$		0.6 sec	β^+						
	Ti^{44}	~10^3 yr	EC						
	Ti^{45}	3.09 hr	β^+, EC						−25.82
	Ti^{46}		Stable	7.99	86.4[a]				−30.77
	Ti^{47}		Stable	7.32	85.6[a]	5/2	−0.7871		−31.19
	Ti^{48}		Stable	73.99	99.2[a]				−34.34
	Ti^{49}		Stable	5.46	84.8[a]	7/2	−1.1023		−34.09
	Ti^{50}		Stable	5.25	84.7[a]				−36.71
	Ti^{51}	5.79 min	β^-						−34.77
$_{23}V^{46}$		0.40 sec	β^+						−23.37
	V^{47}	31.1 min	β^+						−28.28
	V^{48}	16.0 days	β^+ 56% EC 44%						−30.31
	V^{49}	330 days	EC			7/2	±4.46		−33.48
	V^{50}	4×10^{14} yr	EC	0.25	13.2[b]	6	+3.3413		−34.34
	V^{51}		Stable	99.75	>99.9[b]	7/2	+5.139	~+0.3	−37.21
	V^{52}	3.76 min	β^-						−36.14
	V^{53}	2.0 min	β^-						
	V^{54}	55 sec	β^-						
$_{24}Cr^{48}$		23 hr	EC						
	Cr^{49}	41.8 min	β^+						−30.92
	Cr^{50}		Stable	4.31	94.5[a]				−35.72
	Cr^{51}	27.8 days	EC						−36.46
	Cr^{52}		Stable	83.76	99.97[a]				−40.05
	Cr^{53}		Stable	9.55	96.5[a]	3/2	−0.47354		−39.61
	Cr^{54}		Stable	2.38	94.3[a]				−40.95

TABLE OF ISOTOPES (Continued)

Isotope Z	A	Half-Life	Type of Decay	Abundance (per cent) Natural	Abundance (per cent) Enriched	Spin J	Magnetic μ	Quadrupole q	Isotopic Mass Excess
	Cr^{55}	3.52 min	β^-						−38.71
$_{25}Mn^{50}$		0.28 sec	β^+						−27.94
	Mn^{51}	45 min	β^+						−33.24
	Mn^{52m}	21 min	β^+						
	Mn^{52}	5.60 days	EC 67% β^+ 33%						−35.32
	Mn^{53}	$\sim 2 \times 10^6$ yr	EC			7/2	±5.050		−39.02
	Mn^{54}	291 days	EC						−39.75
	Mn^{55}		Stable	100		5/2	+3.4614	+0.4	−41.53
	Mn^{56}	2.576 hr	β^-			3			−40.42
	Mn^{57}	1.7 min	β^-						
$_{26}Fe^{52}$		8.3 hr	β^+, EC						−33.33
	Fe^{53}	8.9 min	β^+						−35.15
	Fe^{54}		Stable	5.84	96.7a				−40.38
	Fe^{55}	2.6 yr	EC						−41.30
	Fe^{56}		Stable	91.68	99.93a				−44.10
	Fe^{57}		Stable	2.17	84.1a	1/2	<0.05		−43.37
	Fe^{58}		Stable	0.31	78.4a				−45.19
	Fe^{59}	45.1 days	β^-						−43.19
	Fe^{60}	$\sim 3 \times 10^5$ yr	β^-						
	Fe^{61}	5.5 min	β^-						
$_{27}Co^{54}$		0.18 sec	β^+						−31.5
	Co^{55}	18.2 hr	$\beta^+ \sim 60\%$ EC $\sim 40\%$						−37.84
	Co^{56}	77.3 days	EC 80% β^+ 20%			4	±3.86		−39.47
	Co^{57}	270 days	EC			7/2	±4.65		−42.85
	Co^{58m}	9.2 hr	IT						
	Co^{58}	71.3 days	EC 85% β^+ 15%			2	±4.05		−42.89
	Co^{59}		Stable	100		7/2	+4.639	+0.5	−44.75
	Co^{60m}	10.5 min	IT 99+% β^- 0.3%						
	Co^{60}	5.24 yr	β^-			5	+3.80		−43.89
	Co^{61}	99 min	β^-						−45.36
	Co^{62}	1.6 min	β^-						
	Co^{62}	13.9 min	β^-						−43.82
$_{28}Ni^{56}$		6.2 days	EC						
	Ni^{57}	36 hr	β^+ 50% EC 50%						−39.61
	Ni^{58}		Stable	67.8	99.9a				−43.05
	Ni^{59}	1×10^5 yr	EC						−43.68
	Ni^{60}		Stable	26.2	99.2b				−46.71
	Ni^{61}		Stable	1.25	83.1a		~ 0		−46.85
	Ni^{62}		Stable	3.66	97.8a				−48.82
	Ni^{63}	125 yr	β^-						−47.02

TABLE OF ISOTOPES (Continued)

Isotope		Half-Life	Type of Decay	Abundance (per cent)		Nuclear Moments			Isotopic Mass Excess
Z	A			Natural	Enriched	Spin J	Magnetic μ	Quadrupole q	
Ni	64		Stable	1.16	95.9[a]				−48.30
Ni	65	2.564 hr	β−						−45.95
Ni	66	54.8 hr	β−						
29Cu	58	9.5 min	β+						−33.5
Cu	58	3.0 sec	β+						
Cu	59	81.5 sec	β+						
Cu	60	23 min	β+ 93% EC 7%			2			−40.43
Cu	61	3.32 hr	β+ 68% EC 32%			3/2			−44.62
Cu	62	9.73 min	β+						−44.89
Cu	63		Stable	69.1	99.8[a]	3/2	+2.221	−0.16	−47.08
Cu	64	12.80 hr	EC 42% β− 39% β+ 19%			1	±0.40		−46.62
Cu	65		Stable	30.9	99.4[b]	3/2	+2.380	−0.15	−48.05
Cu	66	5.10 min	β−						−46.78
Cu	67	58.5 hr	β−						−47.38
Cu	68	32 sec	β−						
30Zn	60	2.1 min	β+, EC						
Zn	61	1.48 min	β+						
Zn	62	9.33 hr	EC ~90% β+ ~10%						−43.19
Zn	63	38.3 min	β+ 93% EC 7%						−43.74
Zn	64		Stable	48.89	98.5[a]		~0		−47.19
Zn	65	245 days	EC 99% β+ 1.5%						−46.70
Zn	66		Stable	27.81	97.8[a]		~0		−49.41
Zn	67		Stable	4.11	81.0[a]	5/2	+0.8735	+0.18	−47.96
Zn	68		Stable	18.56	96.8[a]		~0		−49.80
Zn	69m	13.8 hr	IT						
Zn	69	57 min	β−						−47.92
Zn	70		Stable	0.62	59.3[a]				−48.95
Zn	71m	3 hr	β−						
Zn	71	2.2 min	β−						−46.71
Zn	72	49 hr	β−						
31Ga	64	2.6 min	β+						−40.0
Ga	65	15.2 min	β+, EC						−43.58
Ga	65	8.0 min	β+, EC						
Ga	66	9.45 hr	β+ 66% EC 34%			0	<4 × 10−5		−44.24
Ga	67	77.9 hr	EC			3/2	+1.84	+0.21	−46.95
Ga	68	68 min	β+ 85% EC 15%						−46.89
Ga	69		Stable	60.2	98.4[a]	3/2	+2.011	+0.189	−48.82
Ga	70	21.1 min	β−						−48.29

TABLE OF ISOTOPES (Continued)

Isotope Z	A	Half-Life	Type of Decay	Abundance (per cent) Natural	Abundance (per cent) Enriched	Spin J	Magnetic μ	Quadrupole q	Isotopic Mass Excess
	Ga^{71}		Stable	39.8	98.1^a	3/2	+2.5550	+0.119	−49.01
	Ga^{72}	14.3 hr	β$^-$			3	±0.12		−47.58
	Ga^{73}	5.0 hr	β$^-$						−48.47
	Ga^{74}	7.8 min	β$^-$						
$_{32}$Ge^{66}		~150 min	β$^+$						
	Ge^{67}	21 min	β$^+$						−42.53
	Ge^{68}	280 days	EC						
	Ge^{69}	40.4 hr	EC ~67% β$^+$ ~33%						−46.59
	Ge^{70}		Stable	20.55	93.7^a			<0.007	−49.94
	Ge^{71}	11.4 days	EC						−48.76
	Ge^{72}		Stable	27.37	94.9^a			<0.007	−51.59
	Ge^{73m}	0.53 sec	IT						
	Ge^{73}		Stable	7.67	78.0^a	9/2	−0.8768	−0.2	−49.86
	Ge^{74}		Stable	36.74	95.2^a			<0.007	−51.60
	Ge^{75m}	48 sec	IT						
	Ge^{75}	82 min	β$^-$						−49.71
	Ge^{76}		Stable	7.67	80.6^b			<0.007	−50.91
	Ge^{77m}	54 sec	β$^-$, IT						
	Ge^{77}	11.3 hr	β$^-$						−48.40
	Ge^{78}	86 min	β$^-$						
$_{33}$As^{68}		~7 min	β$^+$						
	As^{69}	15 min	β$^+$						
	As^{70}	52 min	β$^+$						
	As^{71}	62 hr	EC ~70% β$^+$ ~30%						−46.85
	As^{72}	26 hr	EC, β$^+$						−47.23
	As^{73}	76 days	EC						−49.49
	As^{74}	17.5 days	EC 38% β$^-$ 33% β$^+$ 29%						−49.03
	As^{75}		Stable	100		3/2	+1.4349	+0.3	−50.84
	As^{76}	26.4 hr	β$^-$			2	−0.903		−49.79
	As^{77}	38.7 hr	β$^-$						−51.13
	As^{78}	91.0 min	β$^-$						−49.83
	As^{79}	9.0 min	β$^-$						−50.24
	As^{80}	~36 sec	β$^-$						
$_{34}$Se^{70}		~44 min	β$^+$						
	Se^{71}	4.5 min	β$^+$						
	Se^{72}	8.40 days	EC						
	Se^{73}	7.1 hr	β$^+$, EC						−46.72
	Se^{73}	44 min	β$^+$						
	Se^{74}		Stable	0.87	33.1^a			<0.002	−50.39
	Se^{75}	121 days	EC			5/2		+1.1	−49.97
	Se^{76}		Stable	9.02	88.5^a		~0	<0.002	−52.76

TABLE OF ISOTOPES (Continued)

Isotope Z A	Half-Life	Type of Decay	Abundance (per cent) Natural	Abundance (per cent) Enriched	Spin J	Magnetic μ	Quadrupole q	Isotopic Mass Excess
Se^{77m}	17.5 sec	IT						
Se^{77}		Stable	7.58	86.6[a]	1/2	+0.5325	<0.002	−51.81
Se^{78}		Stable	23.52	96.5[a]	0			−53.92
Se^{79m}	3.91 min	IT						
Se^{79}	$<7 \times 10^4$ yr	β^-			7/2	−1.015	+0.9	−52.54
Se^{80}		Stable	49.82	98.4[a]	0			−54.03
Se^{81m}	56.8 min	IT						
Se^{81}	18.2 min	β^-						−52.54
Se^{82}		Stable	9.19	89.9[a]		∼0	<0.002	−53.44
Se^{83}	70 sec	β^-						
Se^{83}	25 min	β^-						
Se^{84}	3.3 min	β^-						
Se^{85}	40 sec	β^-						
$Se^{86,87}$	17 sec	β^-						
$_{35}Br^{74}$	36 min	β^+, EC						
Br^{75}	1.6 hr	β^+, EC						−47.25
Br^{76}	17.2 hr	β^+						−48.17
Br^{77}	57 hr	EC 99% β^+ 1%						−50.45
Br^{78m}	6.4 min	IT						
Br^{78}	<6 min	β^+						−50.42
Br^{79}		Stable	50.52	95.1[a]	3/2	+2.0992	+0.33	−52.69
Br^{80m}	4.38 hr	IT						
Br^{80}	17.6 min	β^- 92% β^+ ∼3% EC ∼5%						−52.14
Br^{81}		Stable	49.48	96.8[a]	3/2	+2.2625	+0.28	−53.93
Br^{82}	35.87 hr	β^-			5	±1.6	±0.7	−53.37
Br^{83}	2.30 hr	β^-						−54.57
Br^{84}	6.0 min	β^-						
Br^{84}	31.8 min	β^-						−52.90
Br^{85}	3.00 min	β^-						−53.44
Br^{87}	55.6 sec	β^-, n 2%						−46.7
Br^{88}	15.5 sec	β^-, n						
Br^{89}	4.48 sec	β^-, n						
Br^{90}	1.4 sec	n						
$_{36}Kr^{76}$	9.7 hr	EC						
Kr^{77}	1.1 hr	EC, β^+						−47.57
Kr^{78}		Stable	0.354					−51.32
Kr^{79m}	55 sec	IT						
Kr^{79}	34.5 hr	EC 95% β^+ 5%						−51.07
Kr^{80}		Stable	2.27					−54.14
Kr^{81m}	13 sec	IT						
Kr^{81}	2.1×10^5 yr	EC						−53.78

TABLE OF ISOTOPES (Continued)

Isotope		Half-Life	Type of Decay	Abundance (per cent)		Nuclear Moments			Isotopic Mass Excess
Z	A			Natural	Enriched	Spin J	Magnetic μ	Quadrupole q	
	Kr⁸²		Stable	11.56			∼0		−56.43
	Kr⁸³ᵐ	114 min	IT						
	Kr⁸³		Stable	11.55		9/2	−0.96706	+0.22	−55.55
	Kr⁸⁴		Stable	56.90			∼0		−57.68
	Kr⁸⁵ᵐ	4.36 hr	β⁻ 77% IT 23%						
	Kr⁸⁵	10.3 yr	β⁻			9/2	−1.001	+0.25	−56.24
	Kr⁸⁶		Stable	17.37	99.6ᵃ		∼0		−57.54
	Kr⁸⁷	78 min	β⁻						−54.68
	Kr⁸⁸	2.77 hr	β⁻						−53.57
	Kr⁸⁹	3.18 min	β⁻						−51.41
	Kr⁹⁰	33 sec	β⁻						
	Kr⁹¹	9.8 sec	β⁻						
	Kr⁹²	3.0 sec	β⁻						
	Kr⁹³	2.0 sec	β⁻						
	Kr⁹⁴	1.4 sec	β⁻						
	Kr⁹⁷	∼1 sec	β⁻						
₃₇	Rb⁷⁹	24 min	β⁺						
	Rb≤⁸⁰	8 days	EC						
	Rb⁸¹ᵐ	31.5 min	β⁺, IT			9/2			
	Rb⁸¹	4.7 hr	EC 87% β⁺ 13%			3/2	+2.05		−51.58
	Rb⁸²ᵐ	6.3 hr	EC 94% β⁺ 6%			5	+1.50		
	Rb⁸²	1.25 min	β⁺						−52.56
	Rb⁸³	83 days	EC			5/2	+1.42		
	Rb⁸⁴ᵐ	23 min	IT, EC						
	Rb⁸⁴	33.0 days	EC 78% β⁺ 19% β⁻ 2.5%			2	−1.32		−55.03
	Rb⁸⁵		Stable	72.15	99.1ᵃ	5/2	+1.3482	+0.30	−56.91
	Rb⁸⁶ᵐ	1.02 min	IT						
	Rb⁸⁶	18.66 days	β⁻			2	−1.69		−57.25
	Rb⁸⁷	5.0 × 10¹⁰ yr	β⁻	27.85	95.9ᵃ	3/2	+2.7414	+0.14	−58.78
	Rb⁸⁸	17.8 min	β⁻						−56.52
	Rb⁸⁹	15.4 min	β⁻						−55.41
	Rb⁹⁰	2.74 min	β⁻						−54.10
	Rb⁹¹	14 min	β⁻						
	Rb⁹¹	1.67 min	β⁻						
	Rb⁹²	80 sec	β⁻						
₃₈	Sr⁸¹	29 min	EC, β⁺						
	Sr⁸²	25.5 days	EC						
	Sr⁸³	33 hr	EC, β⁺						
	Sr⁸⁴		Stable	0.56	63.7ᵃ				−56.00
	Sr⁸⁵ᵐ	70 min	IT 86% EC 14%						

TABLE OF ISOTOPES (Continued)

Isotope		Half-Life	Type of Decay	Abundance (per cent)		Nuclear Moments			Isotopic Mass Excess
Z	A			Natural	Enriched	Spin J	Magnetic μ	Quadrupole q	
	Sr^{85}	64.0 days	EC						−55.87
	Sr^{86}		Stable	9.86	88.6[a]		∼0		−59.02
	Sr^{87m}	2.80 hr	IT						
	Sr^{87}		Stable	7.02	60.2[a]	9/2	−1.0893		−59.06
	Sr^{88}		Stable	82.56	99.8[a]		∼0		−61.69
	Sr^{89m}	∼10 days	IT						
	Sr^{89}	50.5 days	β^-						−59.92
	Sr^{90}	27.7 yr	β^-						−59.81
	Sr^{91}	9.67 hr	β^-						−57.11
	Sr^{92}	2.60 hr	β^-						−56.20
	Sr^{93}	8.2 min	β^-						
	Sr^{94}	1.3 min	β^-						
	Sr^{95}	∼0.7 min	β^-						
$_{39}$	Y^{82}	70 min	EC						
	Y^{83}	3.5 hr	EC						
	Y^{84}	3.7 hr	β^+, EC						
	Y^{85}	5 hr	EC						
	Y^{86}	14.6 hr	β^+						−54.82
	Y^{87m}	14 hr	IT						
	Y^{87}	80.0 hr	EC 99+% β^+ ∼0.3%						−57.37
	Y^{88}	104 days	EC, β^+						−57.95
	Y^{89m}	16.1 sec	IT						
	Y^{89}		Stable	100		1/2	−0.13683		−61.38
	Y^{90}	64.2 hr	β^-						−60.34
	Y^{91m}	50.3 min	IT						
	Y^{91}	57.5 days	β^-						−59.78
	Y^{92}	3.60 hr	β^-						−58.12
	Y^{93}	10.4 hr	β^-						−57.12
	Y^{94}	16.5 min	β^-						−54.38
	Y^{95}	10.5 min	β^-						
$_{40}$	Zr^{86}	17 hr	EC						
	Zr^{87}	94 min	β^+, EC						−53.86
	Zr^{88}	85 days	EC						
	Zr^{89m}	4.4 min	IT 93% EC 5.6% β^+ 1.8%						
	Zr^{89}	79.3 hr	EC ∼75% β^+ ∼25%						−58.54
	Zr^{90m}	0.83 sec	IT						
	Zr^{90}		Stable	51.46	98.6[a]				−62.54
	Zr^{91}		Stable	11.23	86.9[a]	5/2	−1.29803		−61.33
	Zr^{92}		Stable	17.11	95.8[a]				−61.62
	Zr^{93}	1.1×10^6 yr	β^-						−60.22
	Zr^{94}		Stable	17.40	97.9[a]				−59.78

TABLE OF ISOTOPES (Continued)

Isotope Z	A	Half-Life	Type of Decay	Abundance (per cent) Natural	Abundance (per cent) Enriched	Spin J	Magnetic μ	Quadrupole q	Isotopic Mass Excess
	Zr^{95}	65 days	β^-						−57.82
	Zr^{96}		Stable	2.80	85.2[a]				−57.24
	Zr^{97}	17.0 hr	β^-						−54.55
	Zr^{99}	30 sec	β^-						
41	Nb^{89m}	0.8 hr	β^+						
	Nb^{89}	1.9 hr	β^+						
	Nb^{90m}	24 sec	IT						
	Nb^{90}	14.60 hr	EC, β^+						−58.10
	Nb^{91m}	64 days	IT 90% EC 10%						
	Nb^{91}	long	EC						−59.94
	Nb^{92m}	13 hr	EC						
	Nb^{92}	10.1 days	EC						−59.96
	Nb^{93m}	12 yr	IT						
	Nb^{93}		Stable	100		9/2	+6.144	−0.4	−60.28
	Nb^{94m}	6.6 min	IT 99+% β^- ~0.1%						
	Nb^{94}	1.8×10^4 yr	β^-						−59.11
	Nb^{95m}	90 hr	IT						
	Nb^{95}	35 days	β^-						−58.94
	Nb^{96}	23.35 hr	β^-						−57.49
	Nb^{97m}	60 sec	IT						
	Nb^{97}	72.1 min	β^-						−57.22
	Nb^{98}	26 min	β^-						
	Nb^{99}	3.8 min	β^-						
42	Mo^{90}	5.7 hr	β^+, EC						−55.31
	Mo^{91m}	66 sec	IT, β^+, EC						
	Mo^{91}	15.5 min	β^+						−56.35
	Mo^{92}		Stable	15.86	95.5[a]		~0		−60.33
	Mo^{93m}	6.95 hr	IT						
	Mo^{93}	>2 yr	EC						−59.85
	Mo^{94}		Stable	9.12	84.9[a]		~0		−61.15
	Mo^{95}		Stable	15.70	97.2[a]	5/2	−0.9290		−59.87
	Mo^{96}		Stable	16.50	93.2[a]		~0		−60.62
	Mo^{97}		Stable	9.45	93.8[a]	5/2	−0.9485		−59.26
	Mo^{98}		Stable	23.75	96.4[a]		~0		−59.07
	Mo^{99}	66.0 hr	β^-						−55.87
	Mo^{100}		Stable	9.62	97.9[a]		~0		−57.47
	Mo^{101}	14.61 min	β^-						
	Mo^{102}	11.5 min	β^-						
	Mo^{105}	<2 min	β^-						
43	Tc^{92}	4.3 min	β^+, EC						−53.9
	Tc^{93m}	43.5 min	IT ~80% EC ~20%						

TABLE OF ISOTOPES (Continued)

Isotope Z	A	Half-Life	Type of Decay	Abundance (per cent) Natural	Enriched	Spin J	Magnetic μ	Quadrupole q	Isotopic Mass Excess
Tc	93	2.75 hr	EC 88% β^+ 12%						−56.71
Tc	94	53 min	β^+ ∼75% EC ∼25%						−56.82
Tc	95m	60 days	EC 96% IT ∼3% β^+ ∼0.4%						
Tc	95	20.0 hr	EC						−58.24
Tc	96m	51.5 min	IT β^+ ∼0.01%						
Tc	96	4.20 days	EC						−57.5
Tc	97m	91 days	IT						
Tc	97	2.6×10^6 yr	EC						
Tc	98	1.5×10^6 yr	β^-						
Tc	99m	6.04 hr	IT						
Tc	99	2.12×10^5 yr	β^-			9/2	+5.657	+0.3(?)	−57.3
Tc	100	15.8 sec	β^-						
Tc	101	14.0 min	β^-						
Tc	102	5 sec	β^-						
Tc	102	4.5 min	β^-						
Tc	104	18 min	β^-						
Tc	105	10 min	β^-						
44 Ru	93	50 sec	β^+ (?)						
Ru	94	57 min	EC						
Ru	95	1.65 hr	EC, β^+						−56.15
Ru	96		Stable	5.57	95.5a				−57.81
Ru	97	2.88 days	EC						
Ru	98		Stable	1.86	34.2a				−58.5
Ru	99		Stable	12.7	91.2a	5/2	−0.6		−57.5
Ru	100		Stable	12.6	88.9a				
Ru	101		Stable	17.0	91.1a	5/2	−0.7		
Ru	102		Stable	31.6	97.2a				−59.21
Ru	103	39.8 days	β^-						−57.08
Ru	104		Stable	18.5	98.2a				−57.9
Ru	105	4.5 hr	β^-						−55.08
Ru	106	1.00 yr	β^-						−55.65
Ru	107	4 min	β^-						
Ru	108	∼4 min	β^-						
45 Rh	97	35 min	β^+						
Rh	98	8.7 min	β^+						
Rh	99	15.0 days	β^+						
Rh	99	4.7 hr	EC 90% β^+ 10%						
Rh	100	20.8 hr	EC ∼95% β^+ ∼5%						
Rh	101	4.7 days	EC						

TABLE OF ISOTOPES (Continued)

Isotope		Half-Life	Type of Decay	Abundance (per cent)		Nuclear Moments			Isotopic Mass Excess
Z	A			Natural	Enriched	Spin J	Magnetic μ	Quadrupole q	
	Rh^{101}	5 yr							
	Rh^{102}	210 days	β^+, β^-, EC						-57.03
	Rh^{103m}	57 min	IT						
	Rh^{103}	Stable		100		1/2	-0.0879		-57.83
	Rh^{104m}	4.4 min	IT 99+% β^- ~0.1%						
	Rh^{104}	44 sec	β						-56.24
	Rh^{105m}	45 sec	IT						
	Rh^{105}	36.5 hr	β^-						-57.09
	Rh^{106}	130 min	β^-						-55.69
	Rh^{106}	30 sec	β^-						
	Rh^{107}	24 min	β^-						-55.65
	Rh^{108}	18 sec	β^-						
$_{46}$	Pd^{98}	17.5 min	EC						
	Pd^{99}	21.6 min	β^+						
	Pd^{100}	4.0 days	EC						
	Pd^{101}	8.5 hr	EC 96% β^+ 4%						
	Pd^{102}		Stable	0.96	50.9^a				-58.17
	Pd^{103}	17.0 days	EC						-57.27
	Pd^{104}		Stable	10.97	78.3^a				-58.78
	Pd^{105}		Stable	22.2	78.2^a	5/2	-0.57		-57.66
	Pd^{106}		Stable	27.3	82.3^a				-59.22
	Pd^{107m}	21.3 sec	IT						
	Pd^{107}	~7×10^6 yr	β^-						-56.85
	Pd^{108}		Stable	26.7	94.7^a				-57.92
	Pd^{109m}	4.75 min	IT						
	Pd^{109}	13.5 hr	β^-						-55.40
	Pd^{110}		Stable	11.8	91.4^a				-56.24
	Pd^{111m}	5.5 hr	IT 75% β^- 25%						
	Pd^{111}	22 min	β^-						-53.18
	Pd^{112}	21 hr	β^-						-53.29
	Pd^{113}	1.4 min	β^-						
	Pd^{114}	2.4 min	β^-						
	Pd^{115}	45 sec	β^-						
$_{47}$	Ag^{103}	66 min	β^+, EC						
	Ag^{104}	1.2 hr	β^+, EC						-55.06
	Ag^{104}	27 min	β^+			2			
	Ag^{105}	40 days	EC			1/2			-55.4
	Ag^{106}	24.0 min	β^+, EC, β^- (?)			1			
	Ag^{106}	8.2 days	EC			6			-56.25
	Ag^{107m}	44.3 sec	IT						
	Ag^{107}		Stable	51.35	98.8^a	1/2	-0.11301		-56.89

TABLE OF ISOTOPES (Continued)

Isotope Z	A	Half-Life	Type of Decay	Abundance (per cent) Natural	Abundance (per cent) Enriched	Spin J	Magnetic μ	Quadrupole q	Isotopic Mass Excess
	Ag^{108}	2.3 min	β^- 98% EC 1.6% β^+ 0.1%						−55.79
	Ag^{109m}	39.2 sec	IT						
	Ag^{109}		Stable	48.65	99.5[a]	1/2	−0.12992		−56.48
	Ag^{110m}	253 days	β^- 95% IT 5%			6			
	Ag^{110}	24.2 sec	β^-						−54.58
	Ag^{111m}	74 sec	IT						
	Ag^{111}	7.6 days	β^-			1/2	−0.145		−55.33
	Ag^{112}	3.20 hr	β^-						−53.59
	Ag^{113m}	1.2 min	β^-						
	Ag^{113}	5.3 hr	β^-						−53.75
	Ag^{114}	5 sec	β^-						
	Ag^{114}	2 min	β^-						
	Ag^{115m}	∼20 sec	β^-						
	Ag^{115}	21.1 min	β^-						−50.95
	Ag^{116}	2.5 min	β^-						
	Ag^{117}	1.1 min	β^-						
$_{48}$	Cd^{104}	59 min	EC						−53.06
	Cd^{105}	55 min	EC, β^+						−52.4
	Cd^{106}		Stable	1.22	36.0[b]				−56.32
	Cd^{107}	6.7 hr	EC 99+% β^+ 0.3%						−55.45
	Cd^{108}		Stable	0.88	68.6[a]				−55.55
	Cd^{109}	470 days	EC						−56.33
	Cd^{110}		Stable	12.39	70.0[a]		∼0		−57.45
	Cd^{111m}	48.6 min	IT						
	Cd^{111}		Stable	12.75	89.9[a]	1/2	−0.5922		−56.38
	Cd^{112}		Stable	24.07	96.5[a]		∼0		−57.52
	Cd^{113m}	5.1 yr	β^-, IT						
	Cd^{113}		Stable	12.26	87.3[a]	1/2	−0.6195		−55.61
	Cd^{114}		Stable	28.86	98.2[a]		∼0		−56.29
	Cd^{115m}	43 days	β^-						
	Cd^{115}	53 hr	β						−53.95
	Cd^{116}		Stable	7.58	93.8[a]		∼0		−54.15
	Cd^{117m}	3.0 hr	IT						
	Cd^{117}	∼50 min	β^-						−51.4
	Cd^{118}	50 min	β^-						
	Cd^{119}	10 min	β^-						
	Cd^{119}	2.9 min	β^-						
$_{49}$	In^{107}	30 min	β^+						−52.12
	In^{108m}	55 min	β^+, IT						
	In^{108}	40 min	β^+						
	In^{109m}	< 2 min	IT						

TABLE OF ISOTOPES (Continued)

Isotope Z	A	Half-Life	Type of Decay	Abundance (per cent) Natural	Abundance (per cent) Enriched	Spin J	Magnetic μ	Quadrupole q	Isotopic Mass Excess
	In^{109}	4.3 hr	EC 94% β^+ 6%						−54.56
	In^{110m}	5.0 hr	EC 99+% IT 0.6%						
	In^{110}	66 min	β^+, EC						−53.52
	In^{111}	2.81 days	EC						−55.5
	In^{112m}	20.7 min	IT						
	In^{112}	15 min	β^- 44% β^+ 24% EC 32%						−54.98
	In^{113m}	104 min	IT			1/2	±0.217		
	In^{113}		Stable	4.23	65.4a	9/2	+5.496	+0.75	−55.76
	In^{114m}	50.0 days	IT 96.5% EC 3.5%			5	+4.7		
	In^{114}	72 sec	β^- 98% EC 1.9% β^+ 4 × 10^{-3}%						−54.4
	In^{115m}	4.50 hr	IT 95% β^- 5%						
	In^{115}	6 × 10^{14} yr	β^-	95.77	99.9a	9/2	+5.508	+0.76	−55.40
	In^{116m}	53.99 min	β^-			5	+4.21		
	In^{116}	13 sec	β^-						−53.51
	In^{117m}	1.90 hr	β^- 78% IT 22%						
	In^{117}	1.1 hr	β^-						−54.20
	In^{118}	4.5 min	β^-						
	In^{118}	5.5 sec	β^-						
	In^{119}	~2 min							
	In^{119}	17.5 min	β^-						
$_{50}$Sn^{108}		9 min	EC						
	Sn^{109}	18.1 min	EC, β^+						
	Sn^{110}	4.0 hr	EC						
	Sn^{111}	35.0 min	EC ~71% β^+ ~29%						−53.0
	Sn^{112}		Stable	0.95	72.5a				−55.63
	Sn^{113}	119 days	EC						
	Sn^{114}		Stable	0.65	50.0a				−56.4
	Sn^{115}		Stable	0.34	17.6a	1/2	−0.91320		−55.91
	Sn^{116}		Stable	14.24	94.3b		~0		−56.83
	Sn^{117m}	14.0 days	IT						
	Sn^{117}		Stable	7.57	83.8a	1/2	−0.9949		−55.69
	Sn^{118}		Stable	24.01	96.2a		~0		−56.52
	Sn^{119m}	~250 days	IT						
	Sn^{119}		Stable	8.58	83.6a	1/2	−1.0409		−54.94
	Sn^{120}		Stable	32.97	98.2a		~0		−55.81
	Sn^{121m}	>400 days	β^-						
	Sn^{121}	27.5 hr	β^-						−53.63

TABLE OF ISOTOPES (Continued)

Isotope		Half-Life	Type of Decay	Abundance (per cent)		Nuclear Moments			Isotopic Mass Excess
Z	A			Natural	Enriched	Spin J	Magnetic μ	Quadrupole q	
	Sn^{122}		Stable	4.71	88.9[a]				−53.91
	Sn^{123}	39.5 min	β^-						−51.57
	Sn^{123}	136 days	β^-						
	Sn^{124}		Stable	5.98	95.0[a]				−51.64
	Sn^{125}	9.5 min	β^-						−48.96
	Sn^{125}	9.4 days	β^-						
	Sn^{126}	~50 min	β^-						
	Sn^{127}	2.1 hr	β^-						
	Sn^{128}	57 min	β^-						
	Sn^{130}	2.6 min	β^-						
	Sn^{131}	3.4 min	β^-						
	Sn^{132}	2.2 min	β^-						
$_{51}$	Sb^{115}	60 min	β^+						
	Sb^{116}	60 min	β^+, EC						
	Sb^{116}	16 min	β^+						−52.11
	Sb^{117}	2.8 hr	EC 97+% β^+ 2.6%						
	Sb^{118m}	3.5 min	β^+, IT (?)						
	Sb^{118}	5.1 hr	EC						−52.42
	Sb^{119}	38.0 hr	EC						
	Sb^{120}	5.8 days	EC						
	Sb^{120}	16.4 min	β^+, EC						−53.10
	Sb^{121}		Stable	57.25	99.4[a]	5/2	+3.3418	−0.5	−54.01
	Sb^{122m}	3.5 min	IT						
	Sb^{122}	2.80 days	β^- 97% EC 3% β^+ 0.01%			2	−1.90		−52.44
	Sb^{123}		Stable	42.75	96.7[a]	7/2	+2.533	−0.7	−52.98
	Sb^{124m_2}	21 min	IT, β^-						
	Sb^{124m_1}	1.3 min	IT, β^-						
	Sb^{124}	60.9 days	β^-						−51.00
	Sb^{125}	2.0 yr	β^-						−51.31
	Sb^{126}	9 hr	β^-						
	Sb^{126}	28 days	β^-						
	Sb^{126}	19 min	β^-						
	Sb^{127}	88 hr	β^-						
	Sb	6.2 days	β^-						
	Sb^{128}	10.3 min	β^-						
	Sb^{128}	9.6 hr	β^-						
	Sb^{129}	4.2 hr	β^-						
	Sb^{130}	7.1 min	β^-						
	Sb^{130}	33 min	β^-						
	Sb^{131}	23.1 min	β^-						
	Sb^{132}	2.1 min	β^-						
	Sb^{133}	4.4 min	β^-						

TABLE OF ISOTOPES (Continued)

Isotope		Half-Life	Type of Decay	Abundance (per cent)		Nuclear Moments			Isotopic Mass Excess
Z	A			Natural	Enriched	Spin J	Magnetic μ	Quadrupole q	
Sb	134, 135	~50 sec	β⁻						
₅₂Te		1.4 hr	β⁺						
Te	<118	2.5 hr	β⁺						
Te	118	6.0 days	EC						
Te	119	4.5 days	EC						
Te	120		Stable	0.089	32.0ᵇ				−53.52
Te	121m	154 days	IT						
Te	121	17 days	EC						
Te	122		Stable	2.46	82.1ᵇ				−54.42
Te	123m	104 days	IT						
Te	123		Stable	0.87	68.3ᵇ	1/2	−0.7320		−52.8
Te	124		Stable	4.61	90.7ᵇ				−53.91
Te	125m	58 days	IT						
Te	125		Stable	6.99	85.4ᵇ	1/2	−0.8824		−52.07
Te	126		Stable	18.71	96.4ᵇ		~0		−52.52
Te	127m	105 days	IT 98% β⁻ 2%						
Te	127	9.4 hr	β⁻						−50.60
Te	128		Stable	31.79	98.0ᵇ		~0		−50.19
Te	129m	33.5 days	IT						
Te	129	72 min	β⁻						−48.81
Te	130		Stable	34.49	98.8ᵇ		~0		−48.61
Te	131m	30 hr	β⁻ 78% IT 22%						
Te	131	24.8 min	β⁻						−46.54
Te	132	77.7 hr	β⁻						−46.32
Te	133m	63 min	β⁻ 87% IT 13%						
Te	133	2 min	β⁻						−43.9
Te	134	44 min	β⁻						
₅₃I	118	~10 min							
I	119	18 min	β⁺						
I	120	1.4 hr	EC						
I	121	2.0 hr	β⁺						
I	122	3.5 min	β⁺						−50.27
I	123	13.0 hr	EC			5/2			
I	124	4.5 days	EC ~70% β⁺ ~30%			2			−50.68
I	125	60.0 days	EC			5/2	±3.0	−0.66	−51.92
I	126	2.6 hr							
I	126	13.3 days	EC 55% β⁻ 44% β⁺ 1.3%						−50.42
I	127		Stable	100		5/2	+2.7935	−0.69	−51.40
I	128	24.99 min	β⁻ 94% EC 6.4%			1			−49.66

TABLE OF ISOTOPES (Continued)

Isotope		Half-Life	Type of Decay	Abundance (per cent)		Nuclear Moments			Isotopic Mass Excess
Z	A			Natural	Enriched	Spin J	Magnetic μ	Quadrupole q	
I	129	1.72×10^7 yr	β^-			7/2	+2.603	−0.49	−50.51
I	130	12.6 hr	β^-						−48.60
I	131	8.08 days	β^-			7/2		−0.35	−48.74
I	132	2.26 hr	β^-						−46.73
I	133	20.8 hr	β^-						−46.9
I	134	52.2 min	β^-						−45.4
I	135	6.68 hr	β^-						
I	136	86 sec	β^-						−40.11
I	137	22.0 sec	β^-, n 6%						
I	138	5.9 sec	β^-, n						
I	139	2.7 sec	β^-						
54Xe	121	40 min	β^+						
Xe	122	19.5 hr	EC						
Xe	123	2.1 hr	EC, β^+						
Xe	124		Stable	0.096					−50.85
Xe	125m	55 sec	IT						
Xe	125	18.0 hr	EC						
Xe	126		Stable	0.090					−51.68
Xe	127m	75 sec	IT						
Xe	127	36.41 days	EC						−50.4
Xe	128		Stable	1.919					−51.68
Xe	129m	8.0 days	IT						
Xe	129		Stable	26.44		1/2	−0.7725		−50.70
Xe	130		Stable	4.08					−51.57
Xe	131m	12.0 days	IT						
Xe	131		Stable	21.18		3/2	+0.68680	−0.12	−49.71
Xe	132		Stable	26.89			∼0		−50.28
Xe	133m	2.3 days	IT						
Xe	133	5.270 days	β^-						−48.7
Xe	134		Stable	10.44			∼0		−48.77
Xe	135m	15.6 min	IT						
Xe	135	9.13 hr	β^-						
Xe	136		Stable	8.87			∼0		−46.51
Xe	137	3.9 min	β^-						−41.7
Xe	138	17 min	β^-						
Xe	139	41 sec	β^-						
Xe	140	16.0 sec	β^-						
Xe	141	1.7 sec	β^-						
Xe	143	1.0 sec	β^-						
Xe	144	∼1 sec	β^-						
55Cs	123	6 min	β^+						
Cs	125	45 min	β^+, EC						
Cs	126	1.6 min	β^+ 82% EC 18%						−46.9

TABLE OF ISOTOPES (Continued)

Isotope Z A	Half-Life	Type of Decay	Abundance (per cent) Natural	Abundance (per cent) Enriched	Spin J	Magnetic μ	Quadrupole q	Isotopic Mass Excess
Cs^{127}	6.3 hr	EC, β^+			1/2	±1.41		−48.3
Cs^{128}	3.8 min	β^+ 75% EC 25%						−47.6
Cs^{129}	30.7 hr	EC			1/2	±1.47		−49.6
Cs^{130}	30 min	β^+, β^-, EC			1	+1.32		−48.58
Cs^{131}	9.6 days	EC			5/2	+3.48		−49.36
Cs^{132}	6.2 days	EC			2	+2.20		−48.5
Cs^{133}		Stable	100		7/2	+2.5642	−0.003	−49.2
Cs^{134m}	3.2 hr	IT β^- ~1%			8	+1.10		
Cs^{134}	2.07 yr	β^-			4	+2.973		−47.5
Cs^{135}	3.0×10^6 yr	β^-			7/2	+2.7134		
Cs^{136}	12.9 days	β^-						
Cs^{137}	26.6 yr	β^-			7/2	+2.8219		−45.7
Cs^{138}	32.2 min	β^-						−42.9
Cs^{139}	9.5 min	β^-						
Cs^{140}	66 sec	β^-						
Cs^{142}	~1 min	β^-						
$_{56}Ba^{126}$	97 min	EC						
Ba^{127}	12 min	β^+						
Ba^{128}	2.4 days	EC						
Ba^{129}	2.45 days	β^+						−47.0
Ba^{130}		Stable	0.101	23.3[a]				−49.02
Ba^{131}	11.5 days	EC						
Ba^{132}		Stable	0.097	12.9[b]				
Ba^{133m}	38.8 hr	IT						
Ba^{133}	7.2 yr	EC						
Ba^{134}		Stable	2.42	64.3[b]		~0		−49.6
Ba^{135m}	28.7 hr	IT						
Ba^{135}		Stable	6.59	67.3[a]	3/2	+0.8323		
Ba^{136}		Stable	7.81	63.9[b]		~0		
Ba^{137m}	2.60 min	IT						
Ba^{137}		Stable	11.32	45.1[a]	3/2	+0.9311		−46.9
Ba^{138}		Stable	71.66	98.7[a]		~0		−47.8
Ba^{139}	84.0 min	β^-						−44.6
Ba^{140}	12.80 days	β^-						−42.5
Ba^{141}	18 min	β^-						
Ba^{142}	6 min	β^-						
$_{57}La^{131}$	58 min	β^+						
La^{132}	4.5 hr	β^+						
La^{133}	4.0 hr	EC, β^+						
La^{134}	6.5 min	β^+ ~44% EC ~56%						−45.9
La^{135}	19.5 hr	EC						
La^{136}	9.5 min	EC ~67% β^+ ~33%						

TABLE OF ISOTOPES (Continued)

Isotope Z A	Half-Life	Type of Decay	Abundance (per cent) Natural	Abundance (per cent) Enriched	Spin J	Magnetic μ	Quadrupole q	Isotopic Mass Excess
La^{137}	6×10^4 yr	EC						
La^{138}	1.0×10^{11} yr	EC, β^-	0.089	1.74^a	5	+3.685	±0.9	−46.5
La^{139}		Stable	99.911	99.99^a	7/2	+2.761	+0.27	−47.0
La^{140}	40.22 hr	β^-						−43.8
La^{141}	3.8 hr	β^-						−42.5
La^{142}	77 min	β^-						
La^{143}	~19 min	β^-						
$_{58}$Ce^{131}	30 min	β^+						
Ce^{132}	4.2 hr	β^+						
Ce^{133}	6.30 hr	EC, β^+						
Ce^{134}	72.0 hr	EC						
Ce^{135}	22 hr	EC						
Ce^{136}		Stable	0.193	29.9^a				
Ce^{137m}	34.5 hr	IT 99+% EC ~0.1%						
Ce^{137}	8.7 hr	EC						
Ce^{138}		Stable	0.250	13.1^a				−47.5
Ce^{139m}	55 sec	IT						
Ce^{139}	140 days	EC			3/2	±0.84		−46.9
Ce^{140}		Stable	88.48	99.6^a				−47.6
Ce^{141}	33.1 days	β^-			7/2	±0.89		−45.0
Ce^{142}	5×10^{15} yr	α	11.07	90.1^a				−43.7
Ce^{143}	33 hr	β^-						−40.5
Ce^{144}	285 days	β^-						−38.7
Ce^{145}	3.0 min	β^-						
Ce^{146}	13.9 min	β^-						−33.2
$_{59}$Pr^{135}	22 min	β^+, EC						
Pr^{136}	70 min	β^+						
Pr^{137}	1.4 hr	EC 83% β^+ 17%						
Pr^{138}	2.0 hr	EC ~90% β^+ ~10%						−44.0
Pr^{139}	4.5 hr	EC ~94% β^+ ~6%						
Pr^{140}	3.4 min	β^+ ~54% EC ~46%						−44.3
Pr^{141}		Stable	100		5/2	+3.8	−0.054	−45.5
Pr^{142}	19.2 hr	β^-			2	±0.15		−43.0
Pr^{143}	13.76 days	β^-						−41.9
Pr^{144}	17.27 min	β^-			0			−39.0
Pr^{145}	5.95 hr	β^-						
Pr^{146}	24.4 min	β^-						−34.3
$_{60}$Nd^{138}	22 min	β^+						
Nd^{139}	5.50 hr	EC ~90% β^+ ~10%						
Nd^{140}	3.3 days	EC						−44.2

TABLE OF ISOTOPES (Continued)

Isotope Z A	Half-Life	Type of Decay	Abundance (per cent) Natural	Abundance (per cent) Enriched	Spin J	Magnetic μ	Quadrupole q	Isotopic Mass Excess
Nd^{141}	2.42 hr	EC 98% β^+ 1.9%						−43.8
Nd^{142}		Stable	27.13	93.9^a				−45.2
Nd^{143}		Stable	12.20	85.6^b	7/2	−1.1	±1	−42.8
Nd^{144}	5 × 10^{15} yr	α	23.87	95.7^a				−42.0
Nd^{145}		Stable	8.29	78.6^a	7/2	−0.62	±1	
Nd^{146}		Stable	17.18	95.6^a				−38.5
Nd^{147}	11.06 days	β^-			5/2	±0.56		−35.8
Nd^{148}		Stable	5.72	93.0^b				−33.5
Nd^{149}	2.0 hr	β^-						−31.0
Nd^{150}		Stable	5.60	94.7^a				−29.9
Nd^{151}	15 min	β^-						
$_{61}$Pm^{141}	20 min	β^+						
Pm^{142}	~30 sec	β^+, EC						
Pm^{143}	270 days	EC						
Pm^{144}	300 days	EC						
Pm^{145}	18 yr	EC						
Pm^{146}	~1 yr	β^-						−38.2
Pm^{147}	2.64 yr	β^-						−36.7
Pm^{148}	5.3 days	β^-						−33.3
Pm^{148}	42 days	β^-						
Pm^{149}	54 hr	β^-						−32.7
Pm^{150}	2.7 hr	β^-						−28.8
Pm^{151}	27.5 hr	β^-						
Pm	12.5 hr	β^-						
$_{62}$Sm^{142}	72 min	β^+, EC						
Sm^{143}	9.0 min	β^+						
Sm^{144}		Stable	3.16	92.9^b				−41.0
Sm^{145}	340 days	EC						
Sm^{146}	5 × 10^7 yr	α						−38.9
Sm^{147}	1.3 × 10^{11} yr	α	15.07	92.5^b	7/2	−0.76	<±0.7	−37.0
Sm^{148}		Stable	11.27	86.0^b				−36.0
Sm^{149}		Stable	13.84	88.8^a	7/2	−0.64	<±0.7	−34.0
Sm^{150}		Stable	7.47	92.8^b				−34.1
Sm^{151}	~93 yr	β^-						
Sm^{152}		Stable	26.63	98.5^b				−30.4
Sm^{153}	47.1 hr	β^-						
Sm^{154}		Stable	22.53	99.1^a				−27.5
Sm^{155}	23.5 min	β^-						−24.7
Sm^{156}	9.0 hr	β^-						−23.2
$_{63}$Eu^{144}	18 min	β^+						
Eu^{145}	5 days	EC						
Eu^{146}	38 hr	EC						
Eu^{147}	24 days	EC 99+% α ~10^{-3}%						

TABLE OF ISOTOPES (Continued)

Isotope Z	A	Half-Life	Type of Decay	Abundance (per cent) Natural	Abundance (per cent) Enriched	Spin J	Magnetic μ	Quadrupole q	Isotopic Mass Excess
	Eu^{148}	54 days	EC						
	Eu^{149}	120 days							
	Eu^{150}	15.0 hr	β^-						-31.5
	Eu^{151}		Stable	47.77	97.8^a	5/2	+3.4	+1.2	
	Eu^{152}	9.2 hr	EC 25% β^- 75%						
	Eu^{152}	12.7 yr	EC 73% β^- 27%			3	±2.0		
	Eu^{153}		Stable	52.23	95.0^a	5/2	+1.5	+2.5	
	Eu^{154}	16 yr	β^-			3	±2.1		-25.2
	Eu^{155}	1.7 yr	β^-						-26.9
	Eu^{156}	15.4 days	β^-						-24.1
	Eu^{157}	15.4 hr	β^-						-25.1
	Eu^{158}	60 min	β^-						
$_{64}$Gd^{147}		29 hr	EC						
	Gd^{148}	~130 yr	α						-34.1
	Gd^{149}	9.3 days	EC 99+% α 7 × 10^{-4}%						
	Gd^{150}	>10^5 yr	α						-32.5
	Gd^{151}	150 days	EC						
	Gd^{152}		Stable	0.20	14.9^a				
	Gd^{153}	236 days	EC						
	Gd^{154}		Stable	2.15	33.2^a				-28.2
	Gd^{155}		Stable	14.7	72.3^a	3/2	-0.24	+1.1	-27.1
	Gd^{156}		Stable	20.5	80.2^a				-26.5
	Gd^{157}		Stable	15.7	69.7^a	3/2	-0.32	+1.0	-26.8
	Gd^{158}		Stable	24.9	92.9^a				-24.8
	Gd^{159}	18.0 hr	β^-						
	Gd^{160}		Stable	21.9	95.4^a				-20.7
	Gd^{161}	3.6 min	β^-						
$_{65}$Tb		>17 hr	β^+						
	Tb^{149}	4.1 hr	EC ~85% α ~15%						
	Tb^{151}	19 hr	EC 99+% α 3 × 10^{-4}%						
	Tb^{153}	62 hr	EC						
	Tb	17 hr	β^-						
	Tb^{154}	8 hr	EC, β^+						
	Tb^{154}	17.2 hr	EC 99+% β^+ ~0.5%						
	Tb^{155}	5.6 days	EC						
	Tb^{156m}	5.5 hr	IT, EC, β^-						
	Tb^{156}	5.6 days	EC, β^-						
	Tb^{158m}	11.0 sec	IT						
	Tb^{159}		Stable	100		3/2	±1.5		
	Tb^{160}	72.3 days	β^-						-21

TABLE OF ISOTOPES (Continued)

Isotope Z	A	Half-Life	Type of Decay	Abundance (per cent) Natural	Abundance (per cent) Enriched	Spin J	Magnetic μ	Quadrupole q	Isotopic Mass Excess
	Tb161	6.88 days	β−						
	Tb163	6.5 hr							
	Tb164	23 hr							
66Dy152		2.3 hr	α						
	Dy153	5.0 hr	α						
	Dy154	13 hr	α						
	Dy155	10 hr	EC						
	Dy156		Stable	0.0524	13.8a				
	Dy157	8.2 hr	EC						
	Dy158		Stable	0.0902	10.6a				
	Dy159	134 days	EC						
	Dy160		Stable	2.29	67.0a				−23
	Dy161		Stable	18.88	76.6a	5/2	−0.37	+1.1	
	Dy162		Stable	25.53	87.2a				−21
	Dy163		Stable	24.97	75.6a	5/2	+0.51	+1.3	
	Dy164		Stable	28.18	90.0a				−18
	Dy165m	1.25 min	IT, β−						
	Dy165	139 min	β−						−16
	Dy166	82 hr	β−						
67Ho156		∼1 hr							
	Ho159	33 min	EC						
	Ho160m	5.0 hr	IT						
	Ho160	28 min	EC 99+% β+ ∼0.5%						
	Ho161	2.5 hr	EC						
	Ho162	67 min	EC						
	Ho163m	0.8 sec	IT						
	Ho164	36.7 min	β− 53% EC 47%						−16.0
	Ho165		Stable	100		7/2	±3.3	±2	−18
	Ho166	>30 yr	β−						
	Ho166	27.3 hr	β−						
	Ho167	3.0 hr	β−						
68Er160		29 hr	EC						
	Er161	3.1 hr	EC						
	Er162		Stable	0.136	14.1a				
	Er163	75 min	EC						
	Er164		Stable	1.56	35.1a				−17.0
	Er165	10.0 hr	EC						
	Er166		Stable	33.41	72.9a				
	Er167m	2.5 sec	IT						
	Er167		Stable	22.94	58.8a	7/2	±0.5	±10	
	Er168		Stable	27.07	76.9a				−15.0
	Er169	9.4 days	β−						
	Er170		Stable	14.88	87.3a				−10

TABLE OF ISOTOPES (Continued)

Isotope Z	A	Half-Life	Type of Decay	Abundance (per cent) Natural	Abundance (per cent) Enriched	Spin J	Magnetic μ	Quadrupole q	Isotopic Mass Excess
	Er^{171}	7.8 hr	β$^-$						
	Er^{172}	50 hr	β$^-$						
$_{69}$Tm^{165}		29 hr	EC						
	Tm^{166}	7.7 hr	EC 99+% β$^+$ ~0.5%						
	Tm^{167}	9.6 days	EC						
	Tm^{168}	85 days	EC						
	Tm^{169}		Stable	100		1/2	−0.21		
	Tm^{170}	129 days	β$^-$ 99+% EC 0.15%						
	Tm^{171}	680 days	β$^-$						
	Tm^{172}	63.6 hr	β$^-$						
$_{70}$Yb^{166}		54 hr	EC						
	Yb^{167}	74 min	β$^+$						
	Yb^{167}	19 min	EC						
	Yb^{168}		Stable	0.140					
	Yb^{169}	31.8 days	EC						
	Yb^{170}		Stable	3.03					
	Yb^{171m}	short	IT						
	Yb^{171}		Stable	14.31		1/2	+0.45		
	Yb^{172}		Stable	21.82					−15
	Yb^{173}		Stable	16.13		5/2	−0.67	+2.4	
	Yb^{174}		Stable	31.84					−18
	Yb^{175}	101 hr	β$^-$						
	Yb^{176}		Stable	12.73					
	Yb^{177}	1.9 hr	β$^-$						
$_{71}$Lu^{169}		~2 days	EC						
	Lu^{170}	1.7 days	EC						
	Lu^{171}	8.5 days	EC						
	Lu^{171}	~600 days	EC						
	Lu^{172}	6.70 days	EC						
	Lu^{172}	4.0 hr	β$^+$, EC						
	Lu^{173}	1.4 yr	EC						
	Lu^{174}	165 days	EC ~80% β$^-$ ~20%						
	Lu^{175}		Stable	97.40		7/2	+2.0	+5.6	
	Lu^{176m}	3.71 hr	β$^-$						
	Lu^{176}	2.4 × 10^{10} yr	β$^-$	2.60		6	+2.8	+8.0	−2
	Lu^{177}	6.75 days	β$^-$						
	Lu$^{178, 179}$	18.7 min							
$_{72}$Hf^{170}		112 min	β$^+$						
	Hf^{171}	16.0 hr	EC						
	Hf^{172}	~5 yr	EC						
	Hf^{173}	23.6 hr	EC						
	Hf^{174}		Stable	0.163	10.1[a]				

TABLE OF ISOTOPES (Continued)

Isotope Z A	Half-Life	Type of Decay	Abundance (per cent) Natural	Abundance (per cent) Enriched	Spin J	Magnetic μ	Quadrupole q	Isotopic Mass Excess
Hf^{175}	70 days	EC						
Hf^{176}		Stable	5.21	59.5^a				−3
Hf^{177}		Stable	18.56	62.2^a	7/2	+0.61	+3	
Hf^{178m}	4.8 sec	IT						
Hf^{178}		Stable	27.1	84.8^a		∼0		0
Hf^{179m}	19 sec	IT						
Hf^{179}		Stable	13.75	53.3^a	9/2	−0.47	+3	
Hf^{180m}	5.5 hr	IT						
Hf^{180}		Stable	35.22	93.3^a		∼0		2
Hf^{181}	44.6 days	β$^-$						4
Hf^{183}	64 min	β$^-$						
$_{73}$Ta^{176}	8.0 hr	EC						
Ta^{177}	53 hr	EC 99+% β$^+$ ∼2 × 10^{-3}%						
Ta^{178}	2.1 hr	EC ∼97% β$^+$ ∼3%						
Ta^{178}	9.35 min	EC 98% β$^+$ 2%						
Ta^{179}	∼600 days	EC						
Ta^{180m}	8.15 hr	EC ∼79% β$^-$ ∼21%						
Ta^{180}		Stable	0.0123	0.25^a				2
Ta^{181}		Stable	99.9877		7/2	+2.1	+6	3
Ta^m	0.33 sec	IT						
Ta^{182m}	16.5 min	IT						
Ta^{182}	115.1 days	β$^-$						5
Ta^{183}	5.0 days	β$^-$						
Ta^{184}	8.7 hr	β$^-$						
Ta^{185}	50 min	β$^-$						
Ta^{186}	10.5 min	β$^-$						
$_{74}$W^{176}	80 min	EC 99+% β$^+$ 0.5%						
W^{177}	130 min	EC						
W^{178}	21.5 days	EC						
W^{179}	30 min	EC						
W^{179}	5.2 min	EC or IT						
W^{180}		Stable	0.135	6.95^a				1
W^{181}	145 days	EC						
W^{182}		Stable	26.4	94.2^a				3
W^{183m}	5.5 sec	IT						
W^{183}		Stable	14.4	86.2^a	1/2	+0.115		5
W^{184}		Stable	30.6	95.7^a				6
W^{185m}	1.62 min	IT						
W^{185}	75.8 days	β$^-$						
W^{186}		Stable	28.4	97.9^a				10

TABLE OF ISOTOPES (Continued)

Isotope		Half-Life	Type of Decay	Abundance (per cent)		Nuclear Moments			Isotopic Mass Excess
Z	A			Natural	Enriched	Spin J	Magnetic μ	Quadrupole q	
	W^{187}	24.0 hrs	β^-						12
	W^{188}	69.5 days	β^-						
75	Re^{177}	17 min	β^+						
	Re^{178}	15 min	β^+						
	Re^{180}	2.4 min	β^+, EC						
	Re^{180}	18 min	EC						
	Re^{180}	20 hr	β^+						
	Re^{181}	20 hr	EC						
	Re^{182}	12.7 hr	EC						
	Re^{182}	64.0 hr	EC						
	Re^{183}	71 days	EC						
	Re^{184}	50 days	EC						
	Re^{185}		Stable	37.07	96.0[a]	5/2	+3.144	+2.8	
	Re^{186}	88.9 hr	β^- 92% EC 8%						10
	Re^{187}	$\sim 5 \times 10^{10}$ yr	β^-	62.93	98.8[a]	5/2	+3.176	+2.6	10
	Re^{188m}	18.7 min	IT						
	Re^{188}	16.7 hr	β^-						15
	Re^{189}	150 days	β^-						
	Re	9.8 min	β^-						
	Re^{190}	2.8 min	β^-						
76	Os^{181}	23 min	EC						
	Os^{182}	21.9 hr	EC						
	Os^{183m}	10 hr	IT, EC						
	Os^{183}	13.5 hr	EC						
	Os^{184}		Stable	0.018					
	Os^{185}	93.6 days	EC						
	Os^{186}		Stable	1.59					9
	Os^{187}		Stable	1.64		1/2	+0.12		10
	Os^{188}		Stable	13.3					13
	Os^{189m}	5.7 hr	IT						
	Os^{189}		Stable	16.1		3/2	+0.6507	+0.6	17
	Os^{190m}	9.5 min	IT						
	Os^{190}		Stable	26.4					16
	Os^{191m}	14 hr	IT						
	Os^{191}	16.0 days	β^-						20
	Os^{192}		Stable	41.0					21
	Os^{193}	30.6 hr	β^-						25
	Os^{194}	~ 700 days	β^-						
	Os^{195}	6.5 min	β^-						
77	Ir^{185}	15 hr	EC						
	Ir^{186}	15 hr	EC, β^+						
	Ir^{187}	13 hr	EC						

TABLE OF ISOTOPES (Continued)

Isotope Z	A	Half-Life	Type of Decay	Abundance (per cent) Natural	Abundance (per cent) Enriched	Spin J	Magnetic μ	Quadrupole q	Isotopic Mass Excess
Ir	188	41.5 hr	EC 99+% β^+ ~0.3%						
Ir	189	11 days	EC						
Ir	190	3.2 hr	EC ~90% β^+ ~10%						
Ir	190	11 days	EC						
Ir	191m	4.9 sec	IT						
Ir	191	Stable		38.5	85.9[a]	3/2	+0.16	+1.5	20
Ir	192m	1.42 min	IT 99.9% β^- 0.1%						
Ir	192	74.37 days	β^- 95.5% EC 4.5%						23
Ir	193m	11.9 days	IT						
Ir	193	Stable		61.5	89.1[a]	3/2	+0.17	+1.5	23
Ir	194	19.0 hr	β^-						24.6
Ir	195	2.3 hr	β^-						26.7
Ir	196	9.7 days	β^-						
Ir	197	7 min	β^-						
Ir	198	50 sec	β^-						31
$_{78}$Pt	186	2.5 hr	EC						
Pt	188	10.0 days	EC						
Pt	189	11 hr	EC						
Pt	190	8×10^{11} yr	α	0.0127	0.76[a]				
Pt	191	3.00 days	EC						
Pt	192	~10^{15} yr	α	0.78	13.9[a]				21.5
Pt	193m	4.3 days	IT						
Pt	193	>74 days	EC						23.6
Pt	194	Stable		32.9	65.1[a]		~0		22.3
Pt	195m	3.5 days	IT						
Pt	195	Stable		33.8	60.1[a]	1/2	+0.6004		24.6
Pt	196	Stable		25.2	65.9[a]		~0		25.0
Pt	197m	78 min	IT						
Pt	197	18 hr	β^-						27
Pt	198	Stable		7.19	60.9[a]				27
Pt	199	31 min	β^-						
Pt	200	11.5 hr	β^-						
$_{79}$Au	183-187	4.3 min	EC, β^+ α ~0.01%						
Au	186	~15 min	EC						
Au	188	4.5 min	EC						
Au	189	42 min	EC						
Au	191	3.0 hr	EC			3/2	~+0.14		
Au	192	4.7 hr	EC 99% β^+ ~1%			1			
Au	193m	3.9 sec	IT						
Au	193	15.8 hr	EC			3/2	~+0.14		

TABLE OF ISOTOPES (Continued)

Isotope		Half-Life	Type of Decay	Abundance (per cent)		Nuclear Moments			Isotopic Mass Excess
Z	A			Natural	Enriched	Spin J	Magnetic μ	Quadrupole q	
	Au^{194}	39.5 hr	EC ~97% β^+ ~3%			1	±0.068		
	Au^{195m}	30.6 sec	IT						
	Au^{195}	180 days	EC						
	Au^{196}	14.0 hr	EC or IT						
	Au^{196}	5.55 days	EC ~95% β^- ~5%						26
	Au^{197m}	7.2 sec	IT						
	Au^{197}		Stable	100		3/2	+0.136	+0.56	27
	Au^{198}	2.697 days	β^-			2	±0.50		28
	Au^{199}	3.14 days	β^-			3/2	±0.24		29
	Au^{200}	48 min	β^-						32
	Au^{201}	26 min	β^-						33
	Au^{203}	55 sec	β^-						
$_{80}$Hg^{189}		23 min	EC						
	Hg^{191}	57 min	EC						
	Hg^{192}	5.7 hr	EC, β^+						
	Hg^{193m}	10.0 hr	EC 84% IT 16% β^+						
	Hg^{193}	~6 hr	EC						
	Hg^{194}	~130 days	EC						
	Hg^{195m}	40 hr	EC 50% IT 50%						
	Hg^{195}	9.5 hr	EC						
	Hg^{196}		Stable	0.146	1.9[a]	0			25
	Hg^{197m}	24 hr	IT 97% EC 3%						
	Hg^{197}	65 hr	EC			1/2	+0.52		
	Hg^{198}		Stable	10.02	79.1[a]		~0		27
	Hg^{199m}	42 min	IT						
	Hg^{199}		Stable	16.84	73.1[a]	1/2	+0.4993		28
	Hg^{200}		Stable	23.13	91.3[a]		~0		30
	Hg^{201}		Stable	13.22	71.8[a]	3/2	−0.607	+0.42	32
	Hg^{202}		Stable	29.80	98.3[a]		~0		32.9
	Hg^{203}	46.9 days	β^-						33.9
	Hg^{204}		Stable	6.85	89.2[a]		~0		34.75
	Hg^{205}	5.5 min	β^-						37.58
$_{81}$Tl^{195m}		3.5 sec	IT						
	Tl^{195}	1.2 hr	EC						
	Tl^{196}	1.8 hr	EC						
	Tl^{197m}	0.54 sec	IT						
	Tl^{197}	2.7 hr	EC			1/2			
	Tl^{198m}	1.90 hr	EC ~60% IT ~40%			7			
	Tl^{198}	5.3 hr	EC						

TABLE OF ISOTOPES (Continued)

Isotope Z	A	Half-Life	Type of Decay	Abundance (per cent) Natural	Abundance (per cent) Enriched	Spin J	Magnetic μ	Quadrupole q	Isotopic Mass Excess
Tl	199	7.4 hr	EC			1/2			
Tl	200	26.1 hr	EC, β^+			2			
Tl	201	72 hr	EC			1/2			
Tl	202	12.0 days	EC						33.9
Tl	203		Stable	29.50	92.6[a]	1/2	+1.5960		33.5
Tl	204	3.6 yr	β^- 98% EC 2%			2	±0.089		35.08
Tl	205		Stable	70.50	99.0[a]	1/2	+1.6117		35.83
Tl	206	4.19 min	β^-						37.663
Tl	207	4.79 min	β^-						39.237
Tl	208	3.10 min	β^-						43.770
Tl	209	2.2 min	β^-						47.19
Tl	210	1.32 min	β^-, n						51.787
82 Pb	195	17 min	EC						
Pb	196	37 min	EC						
Pb	197m	42 min	EC ~80% IT ~20%						
Pb	198	25 min	EC						
Pb	198	2.4 hr	EC						
Pb	199m	12.2 min	IT						
Pb	199	90 min	EC, β^+						
Pb	200	21.5 hr	EC						
Pb	201m	61 sec	IT						
Pb	201	9.4 hr	EC, β^+						
Pb	202m	3.62 hr	IT 90% EC 10%						
Pb	202	~3 × 10^5 yr	EC						34.0
Pb	203m	6.1 sec	IT						
Pb	203	52.1 hr	EC						34.8
Pb	204m	66.9 min	IT						
Pb	204		Stable	1.40	43.6[b]		~0		34.32
Pb	205	~5 × 10^7 yr	EC						35.886
Pb	206		Stable	25.1	80.4[a]		~0		36.153
Pb	207m	0.80 sec	IT						
Pb	207		Stable	21.7	71.5[a]	1/2	+0.584		37.787
Pb	208		Stable	52.3	96.7[b]		~0		38.774
Pb	209	3.30 hr	β^-						43.27
Pb	210	19.4 yr	β^-						46.400
Pb	211	36.1 min	β^-						50.99
Pb	212	10.64 hr	β^-						54.166
Pb	214	26.8 min	β^-						62.00
83 Bi	198	7 min	EC 99+% α 0.05%						
Bi	199	~25 min	EC 99+% α ~0.01%						

TABLE OF ISOTOPES (Continued)

Isotope		Half-Life	Type of Decay	Abundance (per cent)		Nuclear Moments			Isotopic Mass Excess
Z	A			Natural	Enriched	Spin J	Magnetic μ	Quadrupole q	
Bi^{200}		35 min	EC						
Bi^{201}		62 min	EC 99+% $\alpha\ 3 \times 10^{-3}$%						
Bi^{201}		1.85 hr	EC						
Bi^{202}		95 min	EC						
Bi^{203}		12.3 hr	EC, β^+ $\alpha \sim 10^{-5}$%			9/2			
Bi^{204}		11.6 hr	EC			6			
Bi^{205}		14.5 days	EC, β^+						
Bi^{206}		6.4 days	EC			6			39.75
Bi^{207}		8.0 yr	EC						40.19
Bi^{208}		$\sim 3 \times 10^4$ yr	EC						41.70
Bi^{209}			Stable	100		9/2	+4.0388	−0.4	42.64
Bi^{210}		5.013 days	β^-			1			
Bi^{210}		2.6×10^6 yr	α 99.6% β^- 0.4%						46.336
Bi^{211}		2.16 min	α 99.7% β^- 0.3%						49.595
Bi^{212}		60.5 min	β^- 64% α 36%						53.584
Bi^{213}		47 min	β^- 98% α 2%						56.77
Bi^{214}		19.7 min	β^- 99+% α 0.04%						61.008
Bi^{215}		8 min	β^-						64.17
$_{84}$Po^{196}		1.9 min	α						
Po^{197}		~ 4 min	α						
Po^{198}		~ 6 min	α						
Po^{199}		~ 11 min	α						
Po^{200}		11 min	EC, α						
Po^{201}		18 min	EC, α						
Po^{202}		51 min	EC 98% α 2%						
Po^{203}		42 min	EC						
Po^{204}		3.8 hr	EC ~ 99% $\alpha \sim 1$%						
Po^{205}		1.8 hr	EC 99+% α 0.074%						
Po^{206}		8.8 days	EC 95% α 5%						
Po^{207}		5.7 hr	EC 99+% $\beta^+ \sim 0.2$% $\alpha \sim 10^{-2}$%						
Po^{208}		2.93 yr	α, EC						43.14
Po^{209}		103 yr	α 99+% EC ~ 0.5%			1/2			44.47
Po^{210}		138.401 days	α						45.166
Po^{211m}		25 sec	α						

TABLE OF ISOTOPES (Continued)

Isotope		Half-Life	Type of Decay	Abundance (per cent)		Nuclear Moments			Isotopic Mass Excess
Z	A			Natural	Enriched	Spin J	Magnetic μ	Quadrupole q	
Po	211	0.52 sec	α						48.984
Po	212	3.04×10^{-7} sec	α						51.334
Po	213	4.2×10^{-6} sec	α						55.38
Po	214	1.64×10^{-4} sec	α						57.838
Po	215	1.83×10^{-3} sec	α 99+% β^- 5×10^{-4}%						62.12
Po	216	0.158 sec	α						64.680
Po	217	<10 sec	α						69.0
Po	218	3.05 min	α 99+% β^- 0.02%						71.72
$_{85}$At	<202	43 sec	α, EC						
At	<203	1.7 min	α, EC						
At	203	7 min	α, EC						
At	204	~25 min	EC						
At	205	25 min	α, EC						
At	206	2.9 hr	EC						
At	207	1.8 hr	EC ~90% α ~10%						
At	208	6.3 hr	EC						
At	208	1.6 hr	EC 99+% α 0.5%						
At	209	5.5 hr	EC ~95% α ~5%						
At	210	8.3 hr	EC 99+% α 0.17%						48.99
At	211	7.20 hr	α 41% EC 59%						49.773
At	212	0.22 sec	α						53.02
At	214	~2×10^{-6} sec	α						58.90
At	215	~10^{-4} sec	α						61.36
At	216	~3×10^{-4} sec	α						65.135
At	217	0.018 sec	α						67.52
At	218	2 sec	α 99+% β^- 0.1%						71.37
At	219	54 sec	α ~97% β^- ~3%						74.17
$_{86}$Rn	204	3 min	α						
Rn	206	6.5 min	α, EC						
Rn	207	11 min	EC 96% α 4%						
Rn	208	23 min	EC ~80% α ~20%						
Rn	209	30 min	EC 83% α 17%						
Rn	210	2.7 hr	α ~96% EC ~4%						
Rn	211	16 hr	EC 74% α 26%						

TABLE OF ISOTOPES (Continued)

Isotope		Half-Life	Type of Decay	Abundance (per cent)		Nuclear Moments			Isotopic Mass Excess
Z	A			Natural	Enriched	Spin J	Magnetic μ	Quadrupole q	
	Rn^{212}	23 min	α						53.13
	Rn^{215}	$\sim 10^{-6}$ sec	α						61.36
	Rn^{216}	$\sim 10^{-4}$ sec	α						63.105
	Rn^{217}	$\sim 10^{-3}$ sec	α						66.88
	Rn^{218}	0.019 sec	α						68.709
	Rn^{219}	3.92 sec	α						72.66
	Rn^{220}	51.5 sec	α						74.689
	Rn^{221}	25 min	β^- $\sim 80\%$ α $\sim 20\%$						78.69
	Rn^{222}	3.8229 days	α						80.92
$_{87}$	Fr^{212}	19.3 min	EC 56% α 44%						
	Fr^{218}	5×10^{-3} sec	α						70.51
	Fr^{219}	0.02 sec	α						72.41
	Fr^{220}	27.5 sec	α						75.56
	Fr^{221}	4.8 min	α						77.55
	Fr^{222}	14.8 min	β^- 99+% α $\sim 0.03\%$						81.08
	Fr^{223}	22 min	β^- 99+% α $\sim 4 \times 10^{-3}$ %						83.43
$_{88}$	Ra^{213}	2.7 min	α						
	Ra^{219}	$\sim 10^{-3}$ sec	α						73.12
	Ra^{220}	0.03 sec	α						74.286
	Ra^{221}	30 sec	α						77.32
	Ra^{222}	38 sec	α						78.994
	Ra^{223}	11.68 days	α						82.24
	Ra^{224}	3.64 days	α						84.084
	Ra^{225}	14.8 days	β^-						87.42
	Ra^{226}	1622 yr	α						89.39
	Ra^{227}	41.2 min	β^-						93.38
	Ra^{228}	6.7 yr	β^-						95.50
	Ra^{230}	1 hr	β^-						101.63
$_{89}$	Ac^{222}	5.5 sec	α						81.21
	Ac^{223}	2.2 min	α 99% EC 1%						82.78
	Ac^{224}	2.9 hr	EC $\sim 90\%$ α $\sim 10\%$						85.45
	Ac^{225}	10.0 days	α						87.06
	Ac^{226}	29 hr	β^- $\sim 80\%$ EC $\sim 20\%$						90.22
	Ac^{227}	21.6 yr	β^- $\sim 99\%$ α 1.2%			3/2	+1.1	-1.7	92.07
	Ac^{228}	6.13 hr	β^-						95.46
	Ac^{229}	66 min	β^-						97.29
$_{90}$	Th^{223}	~ 0.1 sec	α						84.42
	Th^{224}	~ 1 sec	α						85.16

TABLE OF ISOTOPES (Continued)

Isotope Z A	Half-Life	Type of Decay	Abundance (per cent) Natural	Abundance (per cent) Enriched	Spin J	Magnetic μ	Quadrupole q	Isotopic Mass Excess
Th^{225}	8.0 min	α ~90% EC ~10%						87.62
Th^{226}	30.9 min	α						89.06
Th^{227}	18.17 days	α						91.99
Th^{228}	1.910 yr	α						93.215
Th^{229}	7340 yr	α						96.14
Th^{230}	8.0×10^4 yr	α						97.77
Th^{231}	25.64 hr	β$^-$						101.14
Th^{232}	1.39×10^{10} yr	α	100					103.17
Th^{233}	22.12 min	β$^-$						106.45
Th^{234}	24.10 days	β$^-$						108.74
$_{91}$Pa^{226}	1.8 min	α						91.75
Pa^{227}	38.3 min	α ~85% EC ~15%						92.97
Pa^{228}	22 hr	EC ~98% α ~2%						95.26
Pa^{229}	1.5 days	EC 99+% α 0.25%						96.46
Pa^{230}	17.7 days	EC ~85% β$^-$ ~15% α ~0.003%						99.28
Pa^{231}	3.43×10^4 yr	α			3/2			100.81
Pa^{232}	1.31 days	β$^-$						103.53
Pa^{233}	27.0 days	β$^-$						105.22
Pa^{234m}	1.175 min	β$^-$ 99+% IT 0.6%						
Pa^{234}	6.66 hr	β$^-$						108.54
Pa^{235}	23.7 min	β$^-$						110.81
Pa^{237}	11 min	β$^-$						116.53
$_{92}$U^{227}	1.3 min	α						94.95
U^{228}	9.3 min	α ~80% EC ~20%						95.56
U^{229}	58 min	EC ~80% α ~20%						97.76
U^{230}	20.8 days	α						98.66
U^{231}	4.3 days	EC 99+% α 6×10^{-3} %						101.15
U^{232}	74 yr	α						102.237
U^{233}	1.62×10^5 yr	α			5/2	±0.51	±3.4	104.66
U^{234}	2.48×10^5 yr	α	0.0056					106.22
U^{235m}	26.5 min	IT						
U^{235}	7.1×10^8 yr	α	0.720	High^d	7/2	±0.34	±4.0	109.41
U^{236}	2.39×10^7 yr	α						111.36
U^{237}	6.75 days	β$^-$						114.30
U^{238}	4.51×10^9 yr	α	99.28	High^d				116.60
U^{239}	23.54 min	β$^-$						120.27
U^{240}	14.1 hr	β$^-$						122.73

TABLE OF ISOTOPES (Continued)

Isotope		Half-Life	Type of Decay	Abundance (per cent)		Nuclear Moments			Isotopic Mass Excess
Z	A			Natural	Enriched	Spin J	Magnetic μ	Quadrupole q	
$_{93}$Np^{231}		~50 min	α						102.97
	Np^{232}	~13 min	EC						104.89
	Np^{233}	35 min	EC 99+% α ~10^{-3}%						105.70
	Np^{234}	4.40 days	EC 99+% β^+ ~0.05%						108.34
	Np^{235}	410 days	EC 99+% α 1.6 × 10^{-3}%						109.57
	Np^{236}	>5 × 10^3 yr							112.21
	Np^{236}	22 hr	EC 43% β^- 57%						
	Np^{237}	2.20 × 10^6 yr	α			5/2	±6		113.79
	Np^{238}	2.10 days	β^-			2			116.73
	Np^{239}	2.346 days	β^-			5/2			118.97
	Np^{240}	7.3 min	β^-						122.37
	Np^{240}	60 min	β^-						
$_{94}$Pu^{232}		36 min	α >2% EC <98%						105.87
	Pu^{233}	20 min	EC 99+% α 0.1%						107.87
	Pu^{234}	9.0 hr	EC 94% α 6%						108.57
	Pu^{235}	26 min	EC 99+% α 3 × 10^{-3}%						110.71
	Pu^{236}	2.85 yr	α						111.70
	Pu^{237m}	0.18 sec	IT						
	Pu^{237}	45.6 days	EC 99+% α 3.3 × 10^{-3}%						114.00
	Pu^{238}	86.4 yr	α						115.42
	Pu^{239}	24,360 yr	α			1/2	~±0.02		118.26
	Pu^{240}	6,580 yr	α						120.22
	Pu^{241}	13.0 yr	β^- 99+% α 4 × 10^{-3}%			5/2	~±1.4		123.05
	Pu^{242}	3.79 × 10^5 yr	α						125.19
	Pu^{243}	4.98 hr	β^-						128.58
	Pu^{244}	~7.6 × 10^7 yr	α						130.97
	Pu^{245}	10.1 hr	β^-						134.2
	Pu^{246}	10.85 days	β^-						137.0
$_{95}$Am^{237}		~1.3 hr	EC 99+% α 5 × 10^{-3}%						115.43
	Am^{238}	1.9 hr	EC						117.92
	Am^{239}	12 hr	EC 99+% α 4 × 10^{-3}%						119.04
	Am^{240}	51 hr	EC						121.57
	Am^{241}	458 yr	α			5/2	+1.4	+4.9	123.03
	Am^{242m}	16.01 hr	β^- 81% EC 19%						

TABLE OF ISOTOPES (Continued)

Isotope		Half-Life	Type of Decay	Abundance (per cent)		Nuclear Moments			Isotopic Mass Excess
Z	A			Natural	Enriched	Spin J	Magnetic μ	Quadrupole q	
	Am²⁴²	~100 yr	β^- 90% EC 10% α 1%						125.88
	Am²⁴³	7.95 × 10³ yr	α			5/2	+1.4	+4.9	128.02
	Am²⁴⁴	26 min	β^- 99+% EC 0.04%						131.22
	Am²⁴⁵	1.98 hr	β^-						133.00
	Am²⁴⁶	25.0 min	β^-						136.70
₉₆	Cm²³⁸	2.5 hr	EC < 90% α > 10%						118.79
	Cm²³⁹	2.9 hr	EC						120.82
	Cm²⁴⁰	26.8 days	α						121.67
	Cm²⁴¹	35 days	EC 99% α 1%						123.91
	Cm²⁴²	162.5 days	α						125.25
	Cm²⁴³	35 yr	α 99+% EC 0.3%						128.02
	Cm²⁴⁴	17.9 yr	α						129.72
	Cm²⁴⁵	8 × 10³ yr	α						132.31
	Cm²⁴⁶	6600 yr	α						134.21
	Cm²⁴⁷	> 4 × 10⁷ yr	α						137.34
	Cm²⁴⁸	4.7 × 10⁵ yr	α 89% Spont. fission 11%						139.76
	Cm²⁴⁹	64 min	β^-						143.0
	Cm²⁵⁰	2 × 10⁴ yr	Spont. fission						145.86
₉₇	Bk²⁴³	4.5 hr	EC 99+% α 0.15%						129.48
	Bk²⁴⁴	4.4 hr	EC 99+% α 6 × 10⁻³%						131.80
	Bk²⁴⁵	4.98 days	EC 99+% α 0.11%						133.08
	Bk²⁴⁶	1.8 days	EC						135.69
	Bk²⁴⁷	~10⁴ yr	α						137.48
	Bk²⁴⁸	16 hr	β^- 70% EC 30%						140.3
	Bk²⁴⁹	314 days	β^- 99+% α 2 × 10⁻³%						142.10
	Bk²⁵⁰	3.13 hr	β^-						145.86
₉₈	Cf²⁴⁴	25 min	α						132.55
	Cf²⁴⁵	44 min	EC 70% α 30%						134.62
	Cf²⁴⁶	35.7 hr	α						135.72
	Cf²⁴⁷	2.5 hr	EC						138.25
	Cf²⁴⁸	350 days	α						139.70
	Cf²⁴⁹	360 yr	α						142.02
	Cf²⁵⁰	10.9 yr	α						143.96

TABLE OF ISOTOPES (Continued)

Isotope Z	A	Half-Life	Type of Decay	Abundance (per cent) Natural	Abundance (per cent) Enriched	Spin J	Magnetic μ	Quadrupole q	Isotopic Mass Excess
	Cf^{251}	~800 yr	α						147.13
	Cf^{252}	2.2 yr	α 97% Spont. fission 3%						149.60
	Cf^{253}	17 days	β⁻						152.6
	Cf^{254}	56 days	Spont. fission						155.28
$_{99}E^{246}$		7.3 min	EC, α						139.84
	E^{248}	25 min	EC 99+% α ~0.3%						142.58
	E^{249}	2 hr	EC 99+% α 0.13%						143.57
	E^{250}	8 hr	EC						145.90
	E^{251}	1.5 days	EC 99+% α 0.53%						147.71
	E^{252}	~140 days	α						150.59
	E^{253}	20.03 days	α						152.44
	E^{254}	480 days	α						155.98
	E^{254}	37 hr	β⁻ 99+% EC ~0.1%						
	E^{255}	24 days	β⁻						158.1
$_{100}Fm^{250}$		30 min	α, EC						146.83
	Fm^{251}	7 hr	EC ~99% α ~1%						149.01
	Fm^{252}	22.7 hr	α						150.51
	Fm^{253}	~4.5 days	EC 89% α 11%						152.86
	Fm^{254}	3.24 hr	α						154.88
	Fm^{255}	21.5 hr	α						157.95
	Fm^{256}	~3–4 hr	Spont. fission						160.21
$_{101}Mv^{256}$		~30 min	EC						161.87
102	254	~3.1 sec	α						158.80

Author Index

The numbers in parentheses are reference numbers and are included to assist in locating references when the authors' names are not mentioned at the point of reference in the text. Numbers in italics refer to the page on which the reference is listed.

A

Abramowitz, M., 775(35 and 36 see footnote u), 801(35), *810*
Adair, R. K., 896(10), *930*, 933, 952(7), *959*, 1037, *1061*
Ajzenberg-Selove, F., 698, 895(6 see footnote g), *930*, 940, 949, 954(57), *960*, 1067
Alaga, G., 880(33), 882(33), *889*
Alburger, D. E., 978(32), *982*, 997(8), *1008*, 1023(8), 1024(8 see footnote e), *1032*
Alcock, G., 679, *694*
Alder, K., 754(22, 24), 764(22), 778(24), 783(22), 784(22), 801(22, 24), *805*, *809*, 880(33), 882(33), *889*, 1009(see footnote a), 1023(see footnote c), *1032*
Allen, J. S., 821(9, 10 see footnote l), *832*
Almqvist, E., 698, 922(60), *931*, 940, 949 (58), *960*
Altman, A., 950(62), *960*
Amado, R. D., 877(30), *889*
Ambler, E., 817(4), 829(4), *832*
Amster, H. J., 1055, *1062*
Arfken, G. B., 753(19 see footnote i), *809*
Arima, A., *805*, 979(36), 980(39), *982*
Auerbach, T., 725(10), *731*, 906, 913, *931*
Austern, N., 671(3), 675(7), 679, 680(12), *694*

B

Bacher, R. F., 904, *931*, 968(9), *981*
Bair, J. K., 895(9), *930*, 949(59), *960*
Band, I. M., 839(8), 840, 842, *851*
Banerjee, M. K., 680, 685(14, 19), 686(1, 7, 19, 23), 687, 688, *694*, *695*, 729, *731*, 975(see footnote e)
Baranger, E., 899(21), *930*
Barfield, W. D., *1062*

Barker, F. C., 877(29), *889*, 942(45), 955, *960*
Barschall, H. H., 637, 896(10), *930*, 1033, 1057(33), *1061*, *1062*
Bartholomew, G. A., 919(53), *931*
Bayman, B., 993(6), *1008*
Bender, R. S., 686(9), 688, 909(40), *931*
Bennett, E., 686(7, 25), 688
Benoist, P., 679, *694*
Benveniste, J., 686(13), 688
Bernal, M. J. M., *806*
Bethe, H. A., 671(2), 689(24), *694*, *695*, 724, *731*, 861(14), 884(see footnote q), *888*, *889*, 897, *930*, 968(9), 975 (25), *981*, *982*, 1033, 1034(1), 1046 (19), *1061*
Beyster, J. R., 1054(24), *1061*, *1062*
Bhatia, A. B., 893(3), *930*
Biedenharn, L. C., 626, 631(see footnote b), 658(9), *668*, 742(8), 743(see footnote e), 753(19 see footnote i, 20), 754(20, 21), 755(20), 762(30 see footnote n), 764(21), 767(33), 768(20), 771(20 see footnote s), 772(20 see footnote t), 775(36 see footnote u), 778(20), 783(21), 786(11), 787(see footnote v), 788(45 see footnote w), 793(49 see footnote x), 801(20, 21), *805*, *806*, *807*, *809*, *810*, 903(see footnote j), *931*, *1062*
Bjørnholm, S. B., 991(5), 992, *1008*
Bjorklund, F. E., 1056, *1062*
Blair, J. S., 682, 683, 684, *694*, *695*
Blampied, W., 692, *695*
Blatt, J. M., 626, 631(see footnote b), 633, 658(9), *668*, 689(23), *695*, 708 (6), *731*, 743(see footnote e), 767(33), 772(34), 786(11), *805*, *807*, *809*, 854 (2), 855, 879, *888*, 891, 892, 903(29 see footnote j), 920(see footnote x), *930*, *931*, 1006(see footnote g), *1008*, 1042(11), *1061*

Bleuler, E., 687(33), 688
Blin-Stoyle, R. J., 764(32), *807*, *809*, 815(3), *832*, 877(30), *889*, 978, 979(35, 37), *982*
Bloch, C., 666, *669*
Bloch, I., 775(35 see footnote u), 801(35), *810*, *1062*
Bloom, S. D., 933(21, 22, 23), 952(21, 22, 23), *959*
Blumberg, S., 662(21), 663, *669*
Bockelman, C. K., 637, 896(10), 898(17), 917(17), *930*
Bohr, A., 676, *694*, 754(22), 764(22), 783(22), 784(22), 801(22), *809*, 876(27), 877(31), 880(33), 882(33), *888*, *889*, 919(55), *931*, 980(see footnote f), 1009(1 see footnote a), 1023(1 see footnote c), *1032*, 1057
Bonner, T. W., 949
Bouricius, W. G., 775(35 see footnote u), 801(35), *810*, *1062*
Boys, S. F., *806*
Braams, C. M., 895(7 see footnote g), 898(17), 901(7), 917(17), *930*, 975(24), *982*
Brady, F. P., 686(4), 687, 688
Braid, T. H., 821(9), *832*
Breit, G., 775(see footnote u), 790(46), 801(35), *810*, 932, 937(3, 4), *959*, *1062*
Bromley, D. A., 882(35), *889*, 922, *931*, 949(58), *960*, 1029(10), *1032*
Brown, G. E., 1034(9), *1061*
Brown, H., 824(19), *833*
Browne, C. P., 898(17), 917(17), *930*, 953, *960*
Broyles, A. A., 775(35 see footnote u), 801(35), *810*, *1062*
Brueckner, K. A., 861(14), *888*, 981(43), *982*
Brussaard, P. J., 742(9), *809*
Brysk, H., 825(21), 832, *833*
Buechner, W. W., 898(17), 917(17), *930*
Bullock, M. L., 698, 940
Burcham, W. E., 858, *888*, 939(42), 956(71), *959*(see footnote f), *960*, 1023(see footnote c), *1032*
Burgy, M. T., 829(25), *833*
Burhop, E. H. S., 856(4), 857, *888*
Burman, R. L., 821(9), *832*
Burrows, H. B., 898(19), *930*

Butler, S. T., 671(3), 675(7), 679, *694*, 696, 724, *731*, 893(3), 897, 922, *930*, *931*, 975(23, 25), *982*

C

Caine, C. A., 978, *982*
Calvert, J. M., 916(46, 47 see footnote t), *931*
Carlson, B. C., 943(47), *960*
Cassen, B., 932, 936(2), *959*
Chase, D. M., 676, 677, 682, *694*, 1054(25), 1058, *1061*, *1062*
Chen, S., 686(2, 19), 687, 688
Cheston, W. B., 1061(37), *1062*
Christy, R. F., 736, *808*, 909, *931*, 933, *959*, *1062*
Church, E., 835(see footnote b), 841, *851*
Clarke, R. L., 690, *695*
Clegg, A. B., 933(16, 17, 19), 952(16, 17, 19), *959*
Cliff, M. J., *806*
Coester, F., *807*
Cohen, B. L., 691, *695*
Cohen, E. R., 1065
Cohn, H. O., 895(9), *930*, 949(59), *960*
Condon, E. U., 710(8), 722(8), *731*, *806*, *807*, 862(15), *888*, 903, *931*, 932, 936(2), 937(3), 942(44), *959*, *960*, 986(1), *1008*
Conzett, H. E., 686(8), 687(6), 688
Cook, C. F., 949
Cowan, C. L., Jr., 812, *832*
Cox, J. A. M., 746(13), *809*
Crowe, K., 1065

D

Daitch, P. B., 893(3), *930*
Darden, S. E., 1037(10), *1061*
Davis, L., 953(64), *960*
Day, R., 997(8), *1008*
Dayton, I. E., 687(42), 688
de Dominicis, C. T., 1034(9), *1061*
de Groot, S. R., *808*
Demeur, M., *668*
Deutsch, M., *808*
Devons, S., 754(23), 801(23), *808*, *809*
Dirac, P. A. M., 936, *959*
Dismuke, N. M., 824(16), 825(16), *833*
DuMond, J. W. M., 1065

E

Edmonds, A. R., 676, 682(9), *694*, 742 (10), *808*, *809*, 885(40), 886, 887, *889*, 903(see footnote j), 904, *931*, 981(40), *982*, 1001(13), *1008*, 1058, *1062*
Ehrman, J. B., 920, *931*
Eisberg, R. M., 691, *695*
Eisenbud, L., 626, 631(8), *668*, 787, 788, *810*
El Bedewi, F. A., 687(30), 688
Elliott, J. P., 672(4), 676(4), *694*, 864(20), 867(22), 877, 885(22), *888*, 904(34), *931*, 1003(14), *1008*, 1025(9), *1032*
Elsasser, W. J., 965(2), 968, *981*
Elwyn, A. J., 917(48), *931*
Emmerich, W. S., 1055, 1057(31), *1062*
Endt, P. M., *668*, 895(7 see footnote g), 901(7), *930*
Enge, H. A., 917, *931*
Ericson, T., 665(23), 666(23), *669*, 736(see footnote c), *808*
Evans, N. T. S., 719, 730, *731*
Ewing, D. H., 662(20), 667(20), *669*, 671 (2), *694*

F

Falkoff, D. L., *806*
Fano, U., 735, 746, 760, 762(30 see footnote n), 763(27), *808*, *809*, 903(see footnote j), *930*
Farwell, G. W., 686(24), 688
Fedtchenko, E. D., 1047, *1061*
Feenberg, E., 825(22), 826, 827, 828, 830, 831, 832, *833*, 932, 933, 937(4), 939 (8), *959*, 968, 976(28), *981*, *982*, 1005 (15), 1007, *1008*
Feldman, D., 937(37), *959*
Ferentz, M., *806*
Ferguson, A. J., 787(see footnote v), *806*, 922(60), *931*, 949(58), *960*
Fermi, E., 981(42), *982*
Fernbach, S., 1054, 1056, *1061*, *1062*
Ferrell, R. A., 680, 899(23), *930*, 948, *960*
Feshbach, H., 632(13), 641, 660(6), 665 (23), 666(23), *668*, *669*, 671(1, 2), 672, 675, *694*, 775(35 see footnote u), 801(35), *810*, 922, *931*, 1033, 1034(3, 4, 6, 8), 1036(3), 1043(13),

1045(6), 1054(3), 1057(28), 1058(28), 1060(28), *1061*, *1062*
Fields, R. E., 1037(10), *1061*
Finke, R. G., 686(13), 688
Fischer, G. E., 686(17), 688
Flowers, B. H., 867(22), 877, 885(22, 40), 886, 887, *888*, *889*, 904, *931*, 981(40), *982*, 1001(13), 1003(14), *1008*
Foldy, L., 938(41), 953(41), *960*
Ford, K. W., 981(41), *982*, 991(4), 992, 994(7), 995, 999(10), *1008*, 1023(8), 1024(8 see footnote e), *1032*
Fowler, T. K., 675, 680, *694*
Fowler, W. A., 643, 920, *931*
Francis, N. C., 675, *694*
Frauenfelder, H., 733, *808*
Freeman, B. E., 775(35 see footnote u), 801(35), *810*, *1062*
Freemantle, R. G., 686(22, 28), 687(40), 688
Freier, G., 971(17), *981*
French, A. P., 730, *731*
French, J. B., 725, *731*, 893(3), 895(5), 898(18), 917(18), 919, *930*, 981(41), *982*, 1000(11), *1008*
Friedman, F. L., 893(3), *930*, 1034(7), 1037, *1061*
Froberg, C. E., 775(see footnote u), *810*, *1062*
Fujii, A., 725(10), *731*
Fujimoto, Y., 893(3), *930*
Fulmer, C. B., 691, *695*, 1046, 1047, *1061*

G

Galonsky, A. I., 896(11), *930*
Gammel, J. L., 938(40), *960*, 981(43), *982*
Gardner, J. W., 753, *809*
Garth, R. C., 1055(27a), *1062*
Geer Illsley, E. H., 686(27), 688, 701
Gell-Mann, M., 858(8), 868(8), 884(see footnote q), *888*, 953, *960*
Gerhart, J. B., 933(27, 28), 947(27, 28), 949(51, 52, 53), *959*, *960*
Ghiorso, A., 965(23), *981*
Gibson, W. M., 686(22), 688
Glassgold, A. E., 1057(29), 1061(37), *1062*
Glendenning, N. K., 681, 685(15), *694*
Gluckstern, R. L., 775(36 see footnote u), *810*
Goertzel, G. H., 838(5), 842(16), *851*

Goldfarb, L. J. B., 754(23), 801(23), *808, 809*
Goldhaber, M., 829(24), *833*, 874(26), 880(26), *888*
Goldstein, M., 754(21), 764(21), 783(21), 801(21), *806, 809*
Goldstein, S., 981(41), *982*
Goldstone, J., 1046(19), *1061*
Gomes, L. C., 981(43), *982*
Gottfried, K., 1020(5), 1021(5), *1032*
Goudsmit, S., 904, *931*
Gove, H. E., 745(see footnote f), 775(35 see footnote v), 787(see footnote u), 801(35), *810*, 882(35), *889*, 919(53), 922(60), *931*, 949(58), *960*, 1029(10), *1032, 1062*
Grace, M. A., 764(32), *807, 809*, 815(3), *832*
Graves, E. R., 692, *695*
Grayson, W. C., Jr., 977, *982*
Green, A. E. S., 970(14), *981*, 1046, *1061*
Green, D., 949, *960*
Green, L. L., 896, *930*
Green, T. A., 839(9), 841(9), *851*
Green, T. S., 686(14), 688, 898(19), *930*
Greuling, E., 821, 824, *833*
Grodzins, L., 829(24), *833*
Gugelot, P. C., 687(34, 36, 42), 688, 692, *695*

H

Haffner, J. W., 686(5, 10), 687, 688
Halban, H., 815(3), *832*
Halbert, E. C., 895(5), 903(see footnote j), *930*
Halpern, I., 736(6 see footnote c), *808*
Halpern, O., 762(31), *809*
Hamburger, A. I., 899(21), 927(see footnote d), 928(see footnote e), *930*
Hamilton, D., 746(13), *809*
Hammack, K. C., 968, *981*
Harr, J., 842(16), *851*
Harrison, F. B., 812(1), *832*
Harvey, J. A., 997(8), *1008*
Hauser, I., 778(38), *810*
Hauser, W., 665(23), 666(23), *669*, 671(2), *694*
Haxel, O., 969, *981*
Hayward, R. W., 817(4), 829(4), *832*
Heckrotte, W., 1054, *1061*
Heisenberg, W., 932, *959*

Heitler, W., 854(3 see footnote c), *888*
Henley, E., 684, *695*
Hermannsfeldt, W. B., 821(9, 10 see footnote l), *832*
Herzberg, G., 1014(see footnote b), *1032*
Heusinkveld, M., 971(17), *981*
Heydenburg, N. P., 687(32), 688, 1023 (see footnote c), *1032*
Hinds, S., 687(35, 38), 688, 898(19), *930*
Hintz, N. M., 686(2), 687
Hofstadter, R., 859(12), 868(12), *888*
Holladay, G. S., *806*
Hollander, J. M., 1066
Holmes, D. K., 824(20), *833*
Holmgren, H. D., 686(26, 27), 688, 699, 701
Holt, J. R., 898(16), 914(43), *930, 931*
Hootan, P., 679, *694*
Hope, J., *806*, 904(34), *931*
Hoppes, D. D., 817(4), 829(4), *832*
Horie, H., 805, *807*, 949(53), *960*, 979(36), 980(39), *982*
Hornyak, W., 685, 686(1, 23), 687, 688, *695*, 949(53), *960*
Horowitz, J., 893(4), 894(4), *930*
Hossain, A., 687(40), 688
Hoyle, M. G., 793(48 see footnote x), 805(50), *807, 810*
Huang, K., 893(3), *930*
Huby, R., 626, *668*, 685, *695*, 743(see footnote e), 769, *809*, 893(3), *930*
Hudson, R. P., 817(4), 829(4), *832*
Hughes, D. J., 636, 644, 968(8), *981*, 1055(27a), *1062*
Huizenga, J. R., 965, *981*, 1067
Hull, M. H., Jr., 775(see footnote u), 801(35), *810, 1062*
Hulthen, L., 937, *960*
Hunt, S. E., 949
Hunting, C. E., 687(39), 688, 699
Huus, T., 754(22), 764(22), 783(22), 784(22), 801(22), *809*, 1009(1 see footnote a), 1023(1 see footnote c), *1032*

I

Igo, G., 691, 692, *695*, 1061(38, 39), *1062*
Inglis, D. R., 868(25), 885(25), *888*, 899(22), 909(22), *930*, 970(16), *981*, 1007(17), *1008*
Irving, J., 927, *931*
Ishidzu, T., *807*

J

Jackson, H. L., 896(11), *930*
Jackson, J. D., 819(see footnote j), *832*
Jacob, M., 630(11 see footnote a), *668*
Jacobsohn, B., 937(37), *959*
Jaffe, A. A., 916(46, 47 see footnote t), *931*
Jahn, H. A., *806*, 885(41), 886, 887, *889*, 904, *931*
Jauch, J. M., *807*
Jeffries, C. D., 975(26), *982*
Jensen, J. H. D., 964(see footnote a), 969, 973(1), 976(1), 979(1), *981*, 998(9), *1008*
Johnson, C. H., 821(9), *832*
Johnsrud, A., 997(8), *1008*
Johnston, R. L., 686(27), 688
Jones, G. A., 933(14, 18), 952(14, 18), *959*

K

Kalos, M. H., 893(4), 894(4), *930*
Kawai, M., 1054(24), *1061*
Kazek, C., Jr., *806*
Kearsley, M. J., 981(41), *982*, 994(7), *1008*, 1023(8), 1024(8 see footnote e), *1032*
Kellog, P. J., 1057(29), *1062*
Kennedy, J. M., 793(48 see footnote x), 805(50), *806*, *807*, *810*, 862(18), *888*
Kerman, A. K., 1009(see footnote a), *1032*
Kikuchi, K., 685, 687(41), *695*, 893(3), *930*
King, J. S., 976, *982*
King, R. W., 895(8 see footnote g), *930*, 949, *960*
Kington, J. D., 895(9), *930*, 949, *960*
Kinsey, B. B., 879(32), *889*, 1023(see footnote c), *1032*
Klein, P., 936, *959*
Kline, R. M., 949, *960*
Klutcharev, A. P., 1047, *1061*
Kobayaski, S., 685, 687(41), *695*
Kofoed-Hansen, O., 821(8), 827(8), *832*
Konopinski, E. J., 821(see footnote n), 824(17), *833*, 948(49), *960*
Kraus, A. A., Jr., 793(see footnote x), *810*
Krohn, V. E., 829(25), *833*
Kroll, M., 938(41), 953(41), *960*
Kruse, H. W., 812(1), *832*
Kurath, D., 672(4), 676(4), *694*, 867(21), 885(21), *888*, 972(see footnote d), 981(40), *982*, 1007(19), *1008*

L

Landau, L., 937(33), *959*
Lane, A. M., *668*, 786(42 see footnote v), 788(42 see footnote w), 790(42), *810*, 863(19), 865(19), 867(22), 885(22, 39), 887(42), *888*, *889*, 892, 896(12), 906, 907(37), 908, 922, *930*, *931*, 1007(17), *1008*, 1025(9), *1032*, 1034(5), 1046(16), 1059, *1061*, *1062*
Lang, J. M. B., 668(26), *669*
Langer, J. E., 1034(9), *1061*
Lascoux, J., 922, *931*
Latter, R., *1062*
Laubitz, M. J., 818(5), *832*
Lauritsen, C. C., 643.
Lauritsen, T., 643, 895(6 see footnote g), *930*, 949, 954(57), *960*, 1067
Lawson, R. D., 970(15), 974(15), 981, *982*
Layton, T. W., 1065
LeCouteur, K. J., 668(26), *669*
Lee, K., 970(14), *981*
Lee, T. D., 813(2), 829, *832*, 947(48), *960*
Lelevier, R. E., 1054(25), *1061*
Lepore, J. V., 654(18), *669*, 1054, *1061*
Levine, S. H., 686(9), 688, 909(40), *931*
Levinger, J. S., 884(see footnote q), *889*
Levinson, C. A., 680, 685(14, 19), 686(1, 7, 19, 23), 687, 688, *694*, *695*, 975(see footnote e), 981(41, 43), *982*, 999(10), *1008*
Liberman, D. A., 909, *931*
Likely, J. G., 686(4), 687, 688
Lind, D., 997(8), *1008*
Lippman, B. A., 762(31), *809*
Litherland, A. E., 745(see footnote f), 882(35), *889*, 916(47 see footnote t), 919, 922(60), *931*, 949(58), *960*, 1029, *1032*
Lloyd, S. P., 746(13), 753, *809*
Longmire, C., 824(19), *833*
Lubitz, C. R., 928, *931*
Lynn, J. E., 1059(36), *1062*

M

McCarthy, I. E., 680, *694*
MacDonald, W. M., 858(9), 869(9), *888*, 933(13, 24, 25, 29), 942(24, 25). 948

(29), 950(62), 952(13), 953(24), *959*, *960*
McGinnis, C. L., 895(8 see footnote g), *930*
McGowan, F. K., 779(39), 784(40), *810*
McGruer, J. N., 686(9), 688, 697, 698, 899(21), 909(40), 917(51), *930*, *931*
McGuire, A. D., 812(1), *832*
McHale, J. L., 754(21), 764(21), 783(21), 801(21), *809*
McManus, H., 675(7), 694, 882(35), *889*, 1029(10), *1032*
Madansky, L., 725, 728, *731*
Mahmoud, H. M., 824(17), *833*, 948(49), *960*, 981(43), *982*
Mann, A. K., 955, *960*
Margolis, B., 1057, *1062*
Marion, J. B., 900(25), 920, *930*, *931*, 949, 953(64), *960*
Maris, T. A. J., 686(20), 688
Mark, H., 970(15), 974(15), *981*
Marshak, R. E., 937(39), *960*
Marsham, T. N., 898(16), 914(43), *930*, *931*
Martinelli, E. A., 686(13), 688
Marty, C., 679, *694*
Maslin, E. E., 916(46, 47 see footnote t), *931*
Matsuda, K., 685, 687(41), *695*
Matsunobu, H., *806*
Maxson, D. R., 686(7, 25), 688, 821(10 see footnote l), *832*
Mayer, M. G., 964(see footnote a), 969, 973, 976(1), 979(1), *981*, 998(9), *1008*
Melkanoff, M. A., 1057(30), 1061(40), *1062*
Melvin, M. A., *807*
Meshkov, S., 899(21), 900(25), *930*
Messiah, A. M. L., 893(4), 894(4), *930*
Meyer, P., 679, *694*
Middleton, R., 686(14), 687(35), 688, 898(19), *930*
Miller, D. W., 896(10), *930*
Miller, J., 948(49a), 949, *960*
Monahan, J. E., *1062*
Moore, W. E., 899(21), *930*
Morinaga, H., 949, 956(70), *960*
Morita, M., 778(38), 807, *810*, 986(1), *1008*
Morpurgo, G., 870, 933, 951, 952, *959*
Moszkowski, S. A., 860, 876(27), *888*, 978(33), *982*

Mott, N. F., 854(1), *888*
Mottelson, B. R., 676, *694*, 754(22), 764(22), 783(22), 784(22), 801(22), *809*, 876(27), 877(31), 880(33), 882 (33), *888*, *889*, 980(see footnote f), 1009(see footnote a), 1023(1 see footnote c), *1032*, 1057
Muether, H. R., 949(50), *960*

N

Nagahara, Y., 685(22), 687(41), *695*
Nagarajan, M. A., 729, *731*
Nakada, M. P., 693, *695*
Neudachin, V. G., 942(46), *960*
Neuert, H., 909(39), *931*
Newns, H. C., 685, *695*, 893(3), *930*
Newton, T. D., 668, *669*, 793, *810*
Nielsen, O. B., 991(5), 992, *1008*
Nilsson, S. G., 841, *851*, 882(36), *889*, 1009(see footnote a), 1020(5),· 1021, *1032*
Nodvik, J., 1057(30), *1062*
Nordheim, L. W., 968, 977, *981*, *982*
Novey, T. B., 829(25), *833*

O

Obi, S., *807*
Oda, Y., 685(22), 687(41), *695*
Ofer, S., 772(see footnote t), *809*
Okai, S., 919, *931*
Oppenheimer, J. R., 695, *731*
Owen, G. E., 725, 728, *731*, 824(18), *833*

P

Pandya, S. P., 981(41), *982*
Parkinson, W. C., 719, *731*, 897, 914(see footnote q), *930*, 976, *982*
Parry, G., 687(35), 688
Paul, E. B., 690, *695*, 775(35 see footnote u), 801(35), *810*, 882(35), *889*, 919 (53), 922(60), *931*, 949(58), *960*, 1003(14), *1008*, 1029(10), *1032*, *1062*
Pearson, C., 671(3), 679, *694*
Peaslee, D. C., 947(48), 954, *960*, 1043 (13), *1061*
Peele, R. W., 686(18), 688
Perks, M. A., 979, *982*
Perlman, I., 965(3), *981*, 1023(see footnote c), *1032*

Perry, C. L., 824(16), 825(16), *833*
Peterson, R. E., 637
Phillips, M., 695, *731*, 1005(15), *1008*
Phillips, P. R., 687(34), 688
Pičman, L., 672(4), 676(4), *694*
Pieper, G. F., 686(16), 687(16, 32), 688
Pleasonton, F., 821(9), *832*
Porter, C. E., 632(13), 641, 662, 663, *668, 669*, 922(63), *931*, 1033, 1034(3), 1036(3), 1054(3), 1057(28), 1058(28), 1060(28), *1061, 1062*
Present, R. D., 932, 937(3), *959*
Primakoff, H., 824(18), *833*
Prosser, F. W., 788(45 see footnote w), 793(49 see footnote x), *810*
Prowse, D. J., 686(28), 687(40), 688
Pryce, M. H. L., 997(8), *1008*, 1023(8), 1024(8 see footnote e), *1032*
Pursey, D. L., 821, *833*

R

Rabinowitz, P., 775(35 see footnote u), *810, 1062*
Rac, 1059(36), *1062*
Racah, G., 735, 742(7), 756(25), 760(27), 763(27), *808, 809*, 903(see footnote j), 904, 915(28 see footnote r), *930, 931*, 973, *982*, 1001, *1008*
Radicati, L. A., 858(8, 9), 863(19), 865(19), 868(8), 869(9), *888*, 933, 942, *959, 960*, 1007(17), *1008*
Radtke, M. G., *807*
Rae, R. E., 1059(36), *1062*
Rainwater, J., 676(8), *694*
Rasmussen, J. O., 841, *851*, 881(34), *889*, 1023(see footnote c), *1032*
Rasmussen, S. W., 686(11), 688
Raz, B. J., 725(10), *731*, 898(18), 917(18), 919, *930*, 981(41), *982*, 1000(11), *1008*
Redlich, M., 672(4), 676(4), *694*
Reiner, A. S., 841(see footnote e), *851*, 993(6), *1008*
Reines, F., 812(1), *832*
Reitz, J., 824(19), *833*, 839(8), 850(8), *851*
Reynolds, J. B., 697
Richardson, J. R., 949, *960*
Rickey, M., 686(24), 688
Riesenfeld, W. B., 1054, *1061*
Ringo, G. R., 829(25), *833*

Roberson, P. C., 686(15), 687(15), 688
Rohrlich, F., 1054(25), *1061*
Rollett, J. S., 1065
Rose, M. E., 753(19 see footnote i, 20), 754(20), 755(20), 768(20), 771(20 see footnote s), 772(20 see footnote t), 778(20), 801(20), *805, 806, 807, 808, 809*, 820(7), 824(16, 19, 20), 825(16, 21), *832, 833*, 834(see footnote a), 836(2), 839(6, 9), 840(10), 841, 842 (15, 16, 17), *851*, 857(6), 862(16, 17), *888*, 903(29, 30 see footnote j), *931*, 987(see footnote c), *1008*
Rosen, L., 692, 693, *695*
Rosenfeld, L., 912, *931*, 936, *959*
Rosenzweig, N., *806*
Ross, A. A., 968(7), 970(15), 974(15), *981*
Rotblat, J., 686(22, 28), 687(40), 688
Rubin, A. G., 696
Rutledge, A. R., *806*

S

Sachs, R. G., *668*
Saito, R., *807*, 986(1), *1008*
Salmi, E. N., *1062*
Saltpeter, E., 937(36), *959*
Sano, M., 919, *931*
Satchler, G. R., 685, 695, *731*, 746(13), 753(see footnote h), 772, *809*, 906, 920(56), *931*
Sato, M., *807*
Sawicki, J., 920(56), *931*
Saxon, D. S., 1054(25), 1055(26), 1057(30), *1061, 1062*
Scharff-Goldhaber, G., 877(28), 881(28), *889*, 967, *981*
Schiff, L. I., 696(5), *731*, 977(18), *981*
Schiffer, J. P., 793(49 see footnote x), *810*, 1046, *1061, 1062*
Schmidt, T., 979, *982*
Schrandt, W., *1062*
Schrank, G., 687(42), 688
Schwartz, C., 900(24), *930*, 980(38), *982*
Schwartz, R. B., 636, 644
Schwinger, J., 760(28), *808, 809*, 937(34, 35), *959*, 1048, *1061*
Seaborg, G. T., 965(3), *981*, 1066
Sears, B. J., 793(48 see footnote x), 805(50), *806, 807, 810*
Sehni, R. C., *806*
Seidlitz, L., 687(33), 688

Sells, R. E., *806*
Serber, R., 696, *731*
Seth, K. K., 1055(27a), 1059, *1062*
Shapiro, D., 775(35 see footnote u), 801 (35), *810*
Shapiro, M. M., 633(14), *668*, *1062*
Sharp, R. R., 898(17), 917(17), *930*
Sharp, W. T., 775(35 see footnote u), 793(see footnote x), 801(35), 805, *806*, *807*, *808*, *810*, 862(18), *888*, *1062*
Sheline, R. K., 981(41), *982*, 991(5), 992, 993(6), *1008*
Sherman, D., 968(8), *981*
Sherr, R., 685, 686(1, 23, 24, 25), 687, 688, 692, *695*, 922, *931*, 933(27), 947(27), 949(50, 51, 53), *959*, *960*
Shook, G., 686(15), 688
Shortley, G. H., 710(8), 722(8), *731*, *806*, *807*, 862(15), *888*, 903, *931*, 942(44), *960*, 986(1), *1008*
Shull, F. B., 917(48), *931*
Signell, R. S., 937(39), *960*
Silver, R., 686(8), 688
Simon, A., 652, 659(17, 19), *669*, *807*, 986(1), *1008*
Singh, J. J., 896(13), *930*
Skyrme, T. H. R., 864(20), *888*, 1046(15), *1061*
Sliv, L. A., 839(8), 840, 850, *851*
Smith, A., 821(see footnote n), *833*
Smith, K., *807*
Smorodinskii, J., 937(33), *959*
Sneddon, I. N., 854(1), *888*
Snell, A. H., 821(9), *832*
Sogo, P. B., 975(26), *982*
Sood, P. C., 1046, *1061*
Sperduto, A., 898(17), 917(17), *930*
Spiers, J. A., *808*
Spinrad, B. I., 842(16), *851*
Squires, E. J., 679, *694*
Stähelin, P., 821(9, 10 see footnote l), *832*
Standing, J. G., 697
Standing, K. G., 899(20), *930*
Stech, B., 754(24), 778(24), 801(24), *809*, 978(33), *982*
Steffen, R. M., 733, *808*
Stelson, P. H., 779(39), 784(40), *810*
Sternheimer, R. M., 971(17), *981*
Stevenson, J. W., *807*
Stewart, L., 692, *695*
Strominger, D., 881(34), *889*, 1066
Strong, P., 842(16), *851*

Strutinsky, V., 665(23), 666(23), *669*, 736 (see footnote c), *808*, *809*
Suess, H. E., 969, *981*
Sugawara, M., 937, *960*
Summers-Gill, R. G., 686(12), 688
Sunyar, A. W., 829(24), *833*, 874(26 see footnote m), 880(26), *888*
Sutton, D. C., 948(49a), 949, *960*
Swamy, N. V. V. J., *807*

T

Takano, N., 685(22), 687(41), *695*
Takebe, H., *806*
Takeda, M., 685(22), 687(41), *695*
Talmi, I., 943(47), *960*, 981(40, 41), *982*, 989(3), 1001(12), *1008*
Tanabe, Y., 805, *807*
Teichman, T., 634(15), *668*, 916(45 see footnote s), *931*
Telegdi, V. L., 829(25), *833*, 858(8), 868(8), 884(see footnote q), *888*, 953, 954, *960*
Temmer, G. M., 1023(see footnote c), *1032*
Tendam, D. J., 687(33), 688
Thaler, R. M., 754(21), 764(21), 783(21), 801(21), *809*, 938(40), *960*, 1061(38), *1062*
Thieberger, R., 943(47), *960*
Thomas, L. H., 970(16), *981*
Thomas, R. G., 662, *669*, 786(42 see footnote v), 788(see footnote w), 790(42), *810*, 920, 922(64), *931*, 1034(5), *1061*
Tobocman, W., *731*, 893(3, 4), 894(4), *930*
Tolhoek, H. A., 742(9), 746(13), *808*, *809*
Toppel, B. J., 933(21, 22), 952(21, 22), *959*
Trainor, L. E. H., 933, 951, *959*
Tralli, N., 838(5), *851*
Treiman, S. B., 819(6 see footnote j), *832*
Trigg, G., 825(22), 826, 827, 828, 830, 831, 832, *833*
Troubetzkoy, E. S., 1057, *1062*
True, W. W., 981(41), *982*, 994(7), 995, *1008*, 1023(8), 1024(8 see footnote e), *1032*
Trumpy, G., 974(22), *982*
Tubis, A., *1062*
Tyren, H., 686(20), 688

AUTHOR INDEX 1113

U

Ufford, C. W., 900(25), *930*
Ui, H., 1054(24), *1061*
Umeda, K., 839(7), *851*
Uretsky, J. L., 981(41), *982*

V

Vander Sluis, J. H., *807*
Vanetsian, R. A., 1047, *1061*
van Lieshout, R., 895(8 see footnote g), *930*
van Wageningen, R., 654, *669*, 1048(22), 1049(22), *1061*
van Wieringen, H., 885(41), *889*, 904(34), *931*
Vaughn, F. J., 686(21), 687(21), 688
Visscher, W. M., 899(23), *930*, 948, *960*
Vogelsang, W. F., 697, 698
Vogt, E., 634(16), *668*, 922, *931*
von Herrmann, P., 686(16), 687(16), 688

W

Walecka, J. D., 981(43), *982*
Wall, N. S., 687(39), 688, 699
Walt, M., 1054(24), *1061*, *1062*
Wapstra, A. H., 965, *981*, 1067
Warburton, E. K., 917(51), *931*
Watson, K. M., 674, *694*, 1054, *1061*
Watters, H. J., 681, 686(3), 687
Way, K., 895(8 see footnote g), *930*
Weaner, D. H., 917(50), *931*
Weber, G., 953, *960*
Weisskopf, V. F., 632(13), 633, 641, 662 (20), 667, *668*, *669*, 671(2), 689(23), *694*, *695*, 708(6), *731*, 772(34), 775 (35 see footnote u), 801(35), *809*, *810*, 854(2), 855, 859(11), 879, *888*, 891, 892, 920(see footnote x), 922 (63), *930*, *931*, 978(33), 981(42, 43), *982*, 1006(see footnote g), *1008*, 1033, 1034(3, 4, 7), 1036(3), 1037, 1042 (11), 1043, 1054(3), 1057(28), 1058 (28), 1060(28), *1061*, *1062*
Welton, T. A., 659(19), *669*
Weneser, J., 746(13), *809*, 835(see footnote b), 841, *851*, 874(26 see footnote m), 877(28), 880(26), 881(28), *888*, *889*

White, M. G., 949(50), *960*
Wick, G. C., 630(see footnote a), *668*
Wigner, E. P., 626, 631(8), 632(12), 634 (15), 662, 663, *668*, *669*, 742(10), 747(15), 756(25), 786(see footnote v), 787, 788, *808*, *809*, *810*, 915(45 see footnote s), 922(64), *931*, 932, 933, 939(8), 955(6), 956(69), *959*, *960*, 976(29), *982*, 1005(15), 1007, *1008*
Wilets, L., 676, 682(9), *694*, 1058, *1062*
Wilkinson, D. H., *731*, 735, *808*, 867(23, 24), 868(24), 869(24), 870(23, 24), 871 (24, see footnote l), 884(37), 885(39), *888*, *889*, 907(37), 915, *931*, 933, 952, 955, 956, 957, 958, *959*, *960*
Willard, H. B., 787(see footnote v), 790 (46), *810*, 895(9), *930*, 949(59), *960*
Willmott, J. C., 896(13), *930*
Wilson, R., 863(6), *888*
Winther, A., 754(22, 24), 764(22), 778 (24), 783(22), 784(22), 801(22, 24), *809*, 821(8), 827(8), *832*, 1009(1 see footnote a), 1023(1 see footnote c), *1032*
Wolfenstein, L., 654(18), 665(23), 666 (23), *669*, 671(2), *694*, 1048(22), 1049(22), *1061*
Wolicki, E. A., 686(27), 688, 701
Woods, R. D., 1055(26), 1057(30), *1062*
Wu, C. S., 764, 817(4), 821(see footnote m), 829(4), *832*, *833*
Wyld, H. W., Jr., 819(6 see footnote j), *832*

Y

Yamazaki, T., 685(22), 687(41), *695*
Yanagawa, S., *807*
Yang, C. N., 813(2), 829, *832*, 947(48), *960*
Yoshida, S., 893(3), 919, *930*, *931*, 1042, *1061*
Yoshiki, H., 687(37), 688

Z

Zaffarano, D. J., 949, *960*
Zimmerman, R. L., 1055(27a), 1059, *1062*

Subject Index

This is the complete subject index for Parts A and B.

A

A^{35}, 821
$A^{40}(p,p')A^{40}$ reaction, 687
Absolute stripping width, 923
Absorption cross section
  effective, 1040
Absorption curve for continuous β-spectrum, 29
Absorption of electrons
  in matter, 27–28
Absorption of nucleons, 1033
Accelerator
  positive ion, 102–103
  pulsed beam from, 372
Accelerator energy calibration, use of narrow resonances for, 286
Adiabatic approximation, 676, 682
  Hamiltonian for, 677
Ag^{107}, energy levels of, 1027
Al^{27}, 692
$Al^{27}(d,d')Al^{27}$ reaction, 687
$Al^{27}(d,n)Si^{28}$ reaction, 696, 916
$Al^{27}(d,p)Al^{28}$ reaction, 318
Al^{25}, level spectra for, 1029
$Al^{27}(n,\gamma)Al^{28}$ reaction, 318
Al^{26}, odd parity level in, 896
$Al^{27}(p,\gamma)Si^{28}$ reaction, 268
Al^{27}, quadrupole moment of, 1029
Aligned nuclei, 447
Alkali halides, 36, 38
Allowed beta decay
  matrix elements in nonrelativistic limit, 977
Allowed beta spectra
  analysis of shapes, 820
Allowed positron decay, 947
Allowed transitions,
  in beta decay, 141, 147, 819
(α,α') processes, 682
(α,α') scattering, 680
Alpha decay, 170–204
  as function of mass number, 182
  Coulomb barrier against, 180
  dependence on isobaric spin selection rule, 494
  energy levels populated in, 185
  excited states in, 185
  favored, 199
  hindrance factors for, 187–189
  occurrence in periodic table, 170, 180
  of odd-mass nuclei, 196
  unfavored, 200–201
Alpha decay energy, 180–183, 965
  half-life for, 184
Alpha decay theory, "one-body," 183
Alpha emission
  conversion electrons associated with, 178
  gamma rays associated with, 178, 179
Alpha emitter
  in rare earth elements, 181
  lifetime considerations, 183–185
Alpha-gamma ray coincidences, 191
Alpha-neutron reaction, 400
Alpha particle, 678, 681
Alpha-particle model, 921
Alpha particle "pile up," 355
Alpha particle spectrograph, 172–175
  capabilities of, 175
  double-focusing, 173
  important characteristics of, 172
  with permanent-magnet, 174
Alpha particle spectrum
  half-widths achieved, 177
  of Th^{227}, 176
  of U^{235}, 177
Alpha spectra
  experimental techniques, 171–180
  systematic study of, 173
Am^{241}, 198–200
Amplitude factors, 631
Analysis of beta decay data, 811–832
Angular correlation, 139, 732–805
  application to specific cases, 769–799
  in beta decay, 819
  intermediate state, 761
  involving quadrupole transition, 773

SUBJECT INDEX

nonrelativistic case, 747
of gamma rays by inelastic scattering
  of protons, 777
over two helicities, 749
relativistic case, 748
summary of formulas and notations,
  800–805
Angular correlation analysis, 249
Angular-correlation experiments, 311
Angular correlation, final
  classical, 740
Angular correlation measurements
  geometrical corrections to, 301
Angular correlation pattern
  predicted by Born approximation, 685
Angular-correlation process, 736–769
  from a semiclassical view, 736
  limiting form, 736
  with intrinsic spins, 739
Angular distribution
  formula for, 738
  from observation of recoils, 469
  interference maxima and minima in, 679
  of alpha particles scattered from C^{12}, 681
  of reaction products, 646
  use in resolving resonances, 456
Angular distribution of alpha particles
  scattered from S^{32}, 683
Angular distribution, of elastically scattered neutrons, experimental determination of, 469
Angular momenta
  coupling in shell model, 902
  for $(j)^3$ configurations, 999
Angular momenta of resonances, 447
Angular momenta, unsharp
  Jacobi polynomial, 742
Angular momentum
  change in nuclear transition, 190
  conservation of, 723
  coupling, 728, 890, 919
  dependence of level spacing on, 436
  in semiclassical approach to scattering, 678
  of nucleus, 250
  of quadrupole mode, 1013
  selection rules on, 727
Angular momentum barrier, 305
Angular momentum functions
  references to tabulations of, 805–807

Angular momentum of the target, 671
Angular momentum transfer, 672, 675
Angular momentum triangle, 744
Angular momentum vector
  of a nuclear state, 550
Anomalously long lifetimes
  accidental cancellation of matrix elements, 814
Antihermitian, operator, 816
Antineutrino, 812, 819
Antineutrino state, 816
Antiparticle, 811, 829
Antisymmetrization, 907
  by circular arc, 903
  neglect of, 926
Argonne time of flight spectrometer, 345
Atomic beam refocusing method, 559
Atomic beam resonance arrangement, 559
Atomic beams, 559
Atomic fractional parentage coefficient, 886
Atomic screening, 223
Atomic spectroscopy, 733
Atomic stopping cross section
  dependence on material, 7
  formula for, 7
Attenuation coefficient, total, 224
  linear, 212
Au^{196}, 252
Auger electrons, 812
Average elastic cross section
  experimentally observed, 1052
Average field in nucleus, 1010
Average nuclear field
  motion in, 1019
Axial-focusing spectrometer, transmission of, 237
Axial vector interaction, 817
Axially symmetric field
  two fold degeneracy in, 1022
Axially symmetric nuclei, 1015
Azimuthal averaging, 741

B

B^{11}, 795
Ba^{137}, 974
$B^{11}(d,n)C^{12}$ reaction, 916
  angular distribution of neutrons from, 728

$B^{10}(d,p)B^{11}$ reaction, 730, 731
  angular distribution of ground state protons, 720
  selection rules in, 724
$B^{10}(He^3,n)N^{12}$ reaction, 698
$B^{10}(n,\alpha)Li^7$ reaction, 351
$B^{11}(p,\gamma)C^{12}$ reaction, 289
Barber's rule for focusing, 61
Barium, spectral distribution for, 324
Barrier height
  for neutrons, 640
Barrier penetration, 183, 272, 895
Barschall giant resonances, 1034
Be^8, 910
$Be^9(\alpha,\alpha')Be^9$ reaction, 684
$Be^9(d,\alpha)Li^7$ reaction
  Doppler shift in γ-transition of Li^7 from, 531
$Be^9(d,d')Be^9$ reaction, 684
$Be^9(d,p)Be^{10}$ reaction
  angular distribution of protons from, 729
$Be^9(d,n)B^{10}$ reaction
  neutron threshold for, 484
  time spectrum for, 374
$Be^9(p,d)Be^8$ reaction, 953
$Be^9(p,\gamma)B^{10}$ reaction, 276
$Be^9(p,\alpha\gamma)Li^6$ reaction, 264
Bent crystal, shaping of, 242
Bent-crystal spectrometer
  useful range of, 244
Bessel function, spherical, 676, 817
Beta decay, 310, 811
  angular distribution of electrons in, 817
  angular momentum selection rules, 814
  as tool in nuclear spectroscopy, 813
  constant for, 825
  Coulomb field in, 817
  energy distribution, 819–820
  energy resolution correction, 824
  ft values for, 147, 825
  gamma-gamma coincidences in, 163
  half-life, 827
  inclusion of Coulomb field, 822
  influence of lepton angular momentum on, 814
  matrix element for, 946
  momentum distributions for, 145
  nonunique transitions, 823
  outline of theory, 815–819
  predictions of shell model, 976
  relativistic corrections, 948
  role in nuclear physics, 813
  second forbidden transitions in, 818
  table of selection rules for, 818
  unique forbidden, 154
  ft values for, 154
  unique transitions, 822
Beta decay data
  analysis of, 811–832
Beta decay operators
  possible choices, 816
Beta decays of I^{126}, 153–167
  β-γ coincidence spectrum, 160
  branch ratio analysis for, 158–160
  composite spectrum of β^+ for, 161–163
  decay scheme for, 164
  γ-ray spectrum for, 163
  Kurie plot analysis of, 155–158
  total beta-spectrum for, 155
Beta decay spectral shapes
  corrections to, 824
  neutrino mass from, 821
Beta decay spectrum
  corrections for finite nuclear size, 824
  shape, 814
Beta emission
  energy conditions for, 813
Beta-gamma cascade
  circular polarization of photons, 778
Beta-gamma correlations, 754
Beta-gamma correlation measurements, 815
Beta interaction
  coupling constants in, 139
  formulation of, 816
Beta particle, interaction with matter, 15–30
Beta-process, 811
  cross section of, 812
Beta-ray spectra
  measurement of, 70–97
  measurement of momentum, 71
  measurement of total ionization, 71
  resolution corrections for, 74
  sources of error in measuring, 72
  use of magnetic spectrometer in, 79
Beta-ray spectrograph, uniform field type, 174
Beta-ray spectrometer
  axial-focusing, 234
  double-focusing, 235
  efficiency of intermediate-image, 241
  prismatic, 234
  variable-field, 236

Beta-recoil, 536
Beta-spectrum
  precision limits of, 80–81
  shape of, 141
Beta transitions
  classed according to ft values, 148
  experimental information, 140–141
  forbidden, 151
  order of forbiddenness, 140, 814
  super-allowed, 147
β_z matrices for $Li^7 + p$ reactions, 910
Betatron, 348, 491
  bremsstrahlung photons from, 494
  energy scale of, 501
Bethe-Goldstone equation, 1046
BF_3 gas proportional counter, 351
Bi^{206}, 258, 1024
Binding energies
  of even-even nuclei, 1010
  of odd A nuclei, 1010
Binding energy, 717, 1046
  of last neutron, 720
Binding energy of electrons, 6
  related to orbital capture, 831
Binding energy of odd neutron
  table of, 966
Binding energy of odd nucleon
  table of, 965
Binding energy of odd proton, 966
Binding field
  spherically symmetric, 1009
Bipartition angle, 219
Bismuth β-ray spectrum, 71
Black sphere scattering, 460
Blatt-Biedenharn formula, 804
Bohr compound nucleus theory, 695
Bohr magneton, 569
Bohr-Mottelson strong coupling model, 677
Bohr straggling, 384
  formula for, 385
Boltzmann constant, 689
Boltzmann distribution, 561, 567
Born approximation, 6, 922
  distorted, 674
  for relativistic particles, 541
  plane wave, 675
  surface, 685
Boron reactions
  use in emulsions for high-energy neutrons, 392
Bose statistics, 1013

Bound nuclear states, 514
  experimental procedures, 405
  gamma decay of, 246–259
Bragg reflection, 229, 242
  higher order, 243
Bragg relationship, 343
Branching ratio, 640
  collective model, 882
  for decay of compound nucleus, 1053
  in beta processes, 812
  shell model, 882
Breathing mode configuration, 899
Breit-Wigner amplitudes, 280, 660
  phase difference between, 790
Breit-Wigner cross section
  Doppler broadened, 494
Breit-Wigner dispersion terms, 767
Breit-Wigner formula
  for single-level, 418
Breit-Wigner level shapes, 495
Breit-Wigner relations, 661
Bremsstrahlung, 491
Bremsstrahlung energy
  maximum, 498
Bremsstrahlung spectrum
  weak internal, 830
Bromine, isomerism of, 247
Brookhaven crystal
  spectrometer, 343, 420
Bubble chambers, 51–52, 375–379
  hydrogen pressure-volume equilibrium curve for, 376
  operation of, 376
Bubbles, charged and uncharged
  pressure-radius curves for, 377
Butler radius, 923

C

C^{10}, 821
C^{12}, 725
  giant resonance in, 955
C^{13}, 899
C^{14}, 787, 814
Carbon, 675, 681
$C^{12}(d,p)C^{13}$ reaction, 724
$C^{13}(d,p)C^{14}$ reaction, 913
$C^{12}(\gamma,n)C^{11}$ cross-section, 500
$C^{12}(He^3,\alpha)C^{12}$ reaction
  differential cross section for, 700
$C^{12}(He^3,p)N^{14}$ reaction
  differential cross section for, 701

SUBJECT INDEX

$C^{12}(p,\gamma)N^{13}$ reaction, 282
$C^{12}(p,\gamma p)C^{12}$ reaction
    gamma rays from, 285
$C^{12}(p,p')C^{12*}$ reaction, 680, 685
C^{12} and O^{16} emission from nuclei, 180
$C^{14}(\alpha,\gamma)O^{18}$ reaction, 291
$C^{14}(p,\gamma)N^{15}$ reaction, 279
$C^{14}(p,n)N^{14}$ reaction, 281
C^{12} wave functions, 680
Ca^{40}, 685, 857
Ca^{43}, 917
$Ca^{40}(d,p)Ca^{41}$ reaction, 897
$Ca^{42}(d,p)Ca^{43}$ reaction, 897
Ca isotopes, 1000
Cadmium
    total cross section for neutrons on, 644
Canonically conjugate momenta, 704
Capture gamma rays
    polarization correlation, 785
Carbon, inelastic scattering of electrons on, 542
    neutron interaction with, 370
Carbon contamination of targets, 270
    reduction of, 272
Carbon reactions
    use in emulsions for high-energy neutrons, 392
Cascades, 744–746
    over-all parity change, 781
    parallel angular momenta, 746
    with unobserved intermediate gamma rays, 773
Cascade feeding of low-lying states, 448
Cascade gamma-rays
    geometrical arrangements for study of, 300
    polarization-directional correlations, 311
Cascade transitions
    of vibrational levels, 1014
Cathode sputtering
    use in target preparation, 112
Cd^{111}, proton scattering on, 1047
Cd^{113}, resonance in, 419
Cd^{114}, neutron-capture γ-ray decay scheme of, 322
Center-of-mass corrections, 792
Center of mass energy, 626
Center of mass system, 5, 125, 696, 701
Central interaction parameters, 912
Central potential
    nonlocal, 1046

Central potential scattering
    formula for cross section, 647
Centrifugal barrier, 480, 853
Cerenkov counter, 25
Cerenkov radiation, 519
Cesium iodide, 36
Channel spin, 626, 652
    potential causing, 1046
Channel spin formalism, 794
Channel-spin ratio, 909
Charge, collection of, 117
Charge conjugation invariance
    in beta decay, 829
Charge conservation, 811
Charge dependence
    of transmission factors, 640
Charge dependence of energy loss, 6
Charge distribution in nucleus
    eccentricity of, 1018
Charge independence, 147, 149, 932, 937–938
    experimental evidence for, 937
Charge independent potential
    containing a spin-orbit interaction, 937
Charge measurement, 117–118
    current integration for, 117
    electron suppressor for, 117
Charge parity, 938
Charge parity selection rules
    application to absorption of γ rays by C^{12}, 954
Charge symmetry, 147, 938, 958
Charged particle deflection, 105
    electrostatic, 105–106
    experimental methods, 101–125, 136
Charged particle detectors, 31–52
Charged particle emission, 690
Charged particle reactions, 99–136
Charged particle scattering, 540
Charged particle source, 101–105
    auxiliary equipment for, 105–107
    ideal, 105
Charged particles
    angular distribution in scattering of, 649
    elastic scattering of, 649–652
Charge-spin fractional parentage coefficient, 885
Chemical binding
    effect of, 420
Chirality, 630
Chi-squared distribution, 428

SUBJECT INDEX

Cl^{34}, 940
Clebsch-Gordon coefficient, 627, 653, 862
Clebsch-Gordon summation, 925
Clebsch-Gordon transformation, 925
Closed geometries, 410
Closed shell configurations
  high stability of, 1009
Closed shells, 186, 434
  interaction with, 1009
  level density near, 318
  related to alpha decay, 181
Cloud chambers, 51–52, 308, 379–380
  solid, 43
Cloudy crystal ball model of nucleus, 440
Cm^{243}, alpha decay of, 1031
Co^{55}, 797
Co^{60}, 233
  decay of, 772
Cockcroft-Walton accelerator, 261
  current from, 409
Coherent channel spin interference, 799
Coherent effects in nuclei
  deformation, 1010
Coherent interference, 763
Coincidence analyzer
  use in determining directional correlation, 575
Coincidence spectrometry, 96
Collective flow in nucleus
  related to moment of inertia, 1018
Collective flow patterns, 1019
  irrotational flow, 1019
  rigid rotation, 1019
Collective mode
  excitation energy of, 678
  in nucleus, 676
  permanent deformation, 877
Collective model
  coupled equations for, 682
  intermediate coupling region of, 319
  of nucleus, 141
  particle surface interaction, 676
  pear shaped deformation, 877
  properties of, 319
  reduced widths from, 919
  strong coupling, 676
  transitions in, 876
  vibrational distortion, 881
Collective model of nucleus, 672
  special selection rules, 880
Collective motion in the nucleus, 250

Collective motions, 1009–1032
  excitation of, 667
  period of, 678
  separation into rotational and vibrational modes, 1011
Collimators, design of, 412
Column vector, 910
Commuting operators, 902
Complex phase shift
  determination of, 1053
Complex potential
  condition for resonances, 1041
  determination of, 1034
  empirical determination of, 1045–1061
  form of, 1045
  formula for, 1045
  imaginary part, 1037
  requirements of, 1033
Complex potential cross section, 1035
Complex potential model, 632, 1033–1061
  average total cross section, 1040
  parameters of, 1039
  validity of, 1034
Complex potential parameters
  empirical values for, 1054
Compound elastic cross section
  average, 1036
Compound elastic scattering
  cross section formula for, 639
Compound nuclear levels
  polarization measurement, 654
Compound nuclear processes
  radiative capture, 642
Compound nuclear resonances, 1033
Compound nuclear states, 670
Compound nucleus, 136, 272, 284, 285, 305, 625–668, 691
  angular momentum of, 418, 453
  decay of, 403
  energy levels of, 335
  lifetime of, 625
  magnetic moment of, 625
  overlapping levels, 660–661
  resonances, 338, 451–459
  transmission coefficient for formation, 1042
Compound nucleus resonance, 674
  total cross-sections for, 451–455
Compound state, sharp, 739
Compound width
  energy dependence, 1037

Pages 1–582 are in Part A. Pages 623–1103 are in Part B.

SUBJECT INDEX

Compounds, stopping cross section of, 14
Compton cross section, integrated, 221
Compton differential scattering cross section, 221
Compton effect, 211, 219–224
 energy distribution of recoil electrons, 222
Compton electrons, 222, 230
Compton polarimeter, 296
Compton scattered photons, energy profile of, 220
Compton scattering, 231, 263, 288
 for determining degree of polarization of gamma rays, 296
 formula for, 220
Compton spectrometer, 232
 two crystal, 308
Cone effect, 480–481
 relation to target thickness, 481
Configuration
 definition of, 984
Configuration admixtures
 effects on gamma-transition probabilities, 996
Configuration interaction, 995
 evaluated by perturbation theory, 995
Configuration selection rules, 895
Configuration space, 706, 713
Configuration states, pure
 linear combinations, 996
Conservation of angular momentum, 811
Conservation of parity, 723
Conservation of statistics, 811
Constants and conversion factors, 1064
Contamination build-up on deflecting plates, 106
Conversion coefficient, absolute, 239
Conversion electron energy, determination of, 79
Conversion electrons, 835
Conversion ratios, 837
Converter, 241
Coordinate system
 fixed in nucleus, 1015
Core oscillation, 919
Correlation functions, 799
Correlation processes
 complex, 754
Correspondence principle, 742, 853
Cosmic rays, correction for, 287

Coulomb barrier, 180, 307, 366, 640
Coulomb barrier effect, 690
Coulomb correction
 to ratio of ft values for O^{14}, 950
Coulomb coupling, 753
Coulomb excitation, 255–257, 266, 285, 580, 683
 of odd A elements, 782
Coulomb excitation correlation, 753
Coulomb excitation-gamma ray correlation, 783
Coulomb field
 polarization of nucleon by, 1047
Coulomb field Dirac waves, 839
Coulomb field of nucleus, 224
 pair production in, 224
Coulomb force, 696
 relaxation of charge-independence, 858
Coulomb function tables, 775
Coulomb interaction
 between low-lying configurations, 942
 in isotopic spin notation, 940
Coulomb interference
 formula in scattering, 651
Coulomb perturbation
 light nuclei, 869
 matrix elements of, 869
 of isotopic spin states, 940
Coulomb scattering, 122
 amplitude, 649
Counter, gas, 45
 resolution of, 45
Counter ratio
 at resonance, 487
Counter ratio data
 fluctuations in, 486
Counter telescope, 47, 365–368
 efficiency of, 367
 energy resolution of, 366
 for neutron work, 365
Counters, total energy, 44–47
Coupled equations, 674
 for direct interaction elastic scattering, 676
Coupling, intermediate
 in light nuclei, 1002
Coupling, weak
 of two odd groups, 909
Coupling constants
 in beta processes, 813

SUBJECT INDEX

Coupling of L and S
  by Clebsch-Gordon coefficients, 906
Coupling of three nucleons, 997–1000
  experimental evidence, 999
  pertinent configurations, 997–998
Coupling of two angular momenta, 985
Coupling potential, 673, 676
Coupling procedure, fundamental, 984
Coupling scheme(s)
  extrapolation to more complex nuclei, 1000
  for deformed nuclei, 1016
Coupling strength
  of electron-neutrino field, 811
Cr, 685
Cr^{53}, 917
$Cr^{52}(p,p')Cr^{52}$ reaction, 687
Cross section ratio
  related to widths, 640
Cross section(s)
  average, 1034–1045
  determinations of, 412–414
    corrections to, 413
    multiple scattering corrections, 414
  energy dependence for (α,α') reactions, 644
  expansion in associated Legendre polynomials, 660
  fluctuations in, 1050
  for Coulomb excitation, 255
  for exciting channels, 674
  for initially unpolarized beam, 655
  for a thin scatterer, 412
  in the laboratory system, 126
  maximum value of, 645
  transformation to center of mass system, 126–127
Crystal diffraction spectroscopy, 229, 242–244
Crystalline field, 556
Cs^{135}, γ-transition in, 524
Cu^{64}, branching ratio in decay of, 812
Current conservation, 631
Current density in nuclei, 855
Current integration, 117–118
Cut-off approximation
  in stripping theory, 730
Cut-off radius
  in stripping, 715
Cyclotron, 348
  current from, 372
Cyclotron target, 241

D

Damped wave function, 718
Data processing, 126–136
  correction for resolution, 128
  for target thickness, 128
  estimation of uncertainties, 127
Data taking techniques, optimal, 125
de Broglie wave length, 691, 824
Decay constant, 183, 521
  measurements, accuracy of, 523
  obtained from prompt coincidence curve, 522
  self-comparison method, 524
Decay scheme, of Am^{241}, 198
  Cm^{242}, 195
  Cm^{243}, 201
  I^{126}, 164, 165
  multiplicity of, 329
  Ra^{226}, 194
Decay times of scintillators, 520
Decoupling parameter, 1017
  expression for, 1022
Deflection analyzers, 59–67
  precision of, 69
Deflection spectrometry, 531
Deformation parameter
  of collective model, 683
Deformation parameter ϵ, 1020
Deformed nuclei, 1014–1022
  rotational spectra, 1014–1019
Degeneracy, 556
Degenerate states, 995
Degrees of freedom
  coupling between, 1014
Delay cable, 522
Delbruck scattering, 211
Densitometer, 255
Density distribution
  in nucleus, 1010
Density matrix, 760–764
  Fano's parametrization, 762
  for axially symmetric states, 761
  operator form, 784
Density of levels, 509, 668, 689, 968
  formulas for, 668, 924
Detailed balancing, 689
Deuteron, 695, 726
  photodisintegration of, 296
  quadrupole moment of, 549
Deuteron-helium scattering, 135

Pages 1–582 are in Part A. Pages 623–1103 are in Part B.

1122 SUBJECT INDEX

(d,n) reactions
  experimental setup, 406
  level structure from, 406
Deuteron stripping reactions
  related to reduced widths, 891
Diagrammatic wave functions, 903
Differential cross section, 671
  absolute, 691
  contributions to, 646–647
  for incident and emergent particle spin greater than 1/2, 658
  for inelastic channel, 674
  for $P^{31}(d,p)P^{32}$ reaction, 897
  for polarized beam, 656
  for triple scattering, 657
  for unpolarized beam, 655
  formula for, 708
  sensitivity to collective admixtures, 672
Differential elastic cross section
  for complex potential model, 1041
Diffraction effects, 541
Diffraction maxima and minima
  in scattering, 682
Diffraction spectrometer flat crystal, 310
Diffusion coefficient, 23
Diffusion of electrons, 22
Diffusion thickness for electrons, 23
Digital computer, 136
  in phase shift calculation, 136
Dipole absorption
  in heavy nuclei, 496
Dirac electron
  parity operator for, 822
  theory, 548, 838
Dirac matrices, 816
Dirac spinors, 948
Direct interactions, 670–731
  in elastic scattering, 670–695
  theoretical considerations, 672–680
Direct interaction amplitude, 671
Direct-interaction inelastic scattering, 922
Direct interaction model
  inelastic scattering in, 671
Direction-direction correlations, 800
  gamma rays, 770, 800
  heavy particles, 774
  nuclear particle emission, 800
  summary for pure states, 740

Directional correlation function, 572
Directional correlation pattern of γ,γ cascade
  rotation of, 577
Directional distribution
  of a gamma transition, 565
  polar diagram of, 566
Directional distribution function, 564
Directional relations
  diagramming, 738
Discrete angles, 742
Dispersion, 662
Distorted Born approximation, 680
Distribution of reduced transition speeds
  light elements, 871
Doppler angle, 497
Doppler broadening
  due to thermal motion, 537
  in nuclear recoil, 529
Doppler effect, 127, 257, 283, 866
Double focusing, 62
Double scattering experiments, 1044
Double scattering geometry, 656
Doubly magic nuclei, 990
Duality transformation, 750
Du Mond spectrometer, 242
Dy^{161}, level spectrum for, 1030
Dynamic distortion, 944

E

E1 and M1 transitions
  comparison of, 871
E1 transitions, slow, 877
E2 moments, 1013
Effective charge factor, 864, 879
Effective potential
  for nucleons, 988
Eigenphases, 631
Elastic channel, 670
Elastic collisions with atoms, 18
Elastic photon scattering
  experimental arrangement for, 494
Elastic potential scattering amplitudes, 629–631
Elastic scattering, 132, 336, 338, 670, 677
  amplitude, 678
  at $\theta \simeq \pi$, 132
  nonresonant, 1039
  of photons, in giant resonance region, 496
  by nuclei, 492

of spin (1/2) particles, 655
width for, 646
Elastic scattering cross-section, average for 7-Mev photons, 495
Elastic scattering data, 136
Electric dipole matrix element retardation terms, 951
Electric dipole selection rule violations of, 951
Electric dipole transitions, 192
Electric field, axially symmetric, 550
Electric monopole transitions, 251
Electric multipoles
  odd order, 836
  parity of, 251
  transition, 195
Electric quadrupole absorption, 499
Electric quadrupole moment, 550, 555
  determination of, 576
  for quadrupole deformations, 1013
Electric quadrupole transitions, 165, 246, 250
Electric transition matrix elements
  formula for, 860
Electromagnetic field
  coupling to the magnetization of nucleus, 855
Electromagnetic interactions, 341, 1047
Electromagnetic transitions, 514
  general selection rules for, 853–857
  methods of observation, 865
Electron
  binding energy of, 813
  energy imparted to, 4
  energy loss in collisions with, 4–5
  inelastic collisions with, 15
  momentum imparted to, 4
  orbital capture of, 812
  radiative energy loss of, 15
Electron capture, 7, 164–166
  intensity of, 164–166
Electron correlation
  internal conversion, 801
Electron dynamics, 838
Electron elastic scattering, 540
Electron-electron collisions, 27
Electron-neutrino field, 811
Electron penetration of nucleus, 840
Electron positron pair
  emission of, 834
Electron scattering
  inelastic experimental results, 545
Electron spin, flip of, 570
Electronic energy levels, 963
Electrostatic accelerator, tandem, 104
Electrostatic analyzers, 59–60, 106
  cylindrical geometry, 59
  focusing properties of, 59
  spherical geometry, 59–60
Electrostatic energy selector, see Electrostatic analyzer
Ellipsoidal deformations of nuclei, 878
Emission of gamma rays
  factors influencing, 852
Empirical range-energy relationship for electrons, 29
Emulsion
  lithium in, 391
  nuclear, analysis of, 381
    energy resolution obtainable with carbon boron reactions in, 393
  processing of, 381
  stopping power variation with humidity, 383
  water content of, 383
Energy absorption, requirement for, 5
Energy average
  formula for, 1034, 1035
Energy calibration standards
  absolute, 107
Energy conservation, 811
Energy eigenfunctions, 672
Energy from single particle levels
  in Pb^{206}, 994
Energy levels
  angular momentum of, 436
  determination, 178
  in heavy elements, 170
  of an atom in $^2S_{\frac{1}{2}}$ state, 570
  of even-even nuclei, 194
  of intrinsic spectrum, 1020
  rotational, 255
  selection rule for, 186
  spacings, 250
Energy loss, 9–14
  for ions heavier than alpha particles, 11–13
  of alpha particles, 11
  of high-velocity ions, 9
  of low-velocity ions, 11
  rate of, 9
  to electrons, 5
    quantum mechanical formula for, 5

SUBJECT INDEX

Energy matrix
  Rosenfeld interaction, 911
Energy measurement
  of particles from accelerators, 261
  relativistic corrections, 108
Energy-modulated radiations
  coincidences between, 527
Energy-range measurements
  effect of small angle scattering, 384
Energy-range variation
  causes of, 386
Energy resolution
  in emulsions, 394
Energy separation
  states of different J, 988
  with no configuration interaction, 997
Energy spectrum
  for single-particle motion in an ellipsoidal potential, 1021
  of low-energy neutrons, 403
Energy splitting
  in a magnetic field, 555
Energy straggling, 41
Energy to light conversion efficiency, 77
Entropy, 689
Erbium, neutron-capture γ-ray spectrum of, 320
Eu, 363
Euler angles, 677
Even-even core
  vibrational excitations, 1014
Even-even nuclei, 183, 186, 306, 552
  angular momentum of, 984
  capture of alpha particles by, 291
  elastic scattering from, 421
  excited state energy of, 189
  ground state of, 186
  heavy, 190
  inelastic proton scattering by, 298
  proton capture by, 293
  quadrupole vibrational spectrum for, 1013
  systematics of the spectra, 1025
Even-odd and odd-even nuclei, 554
Even-odd nuclei, 306
Excitation cross-section
  of vibrational levels, 1013
Excitation curve
  thin target, 643
Excitation energies
  low, 1011
Excitation processes, 16

Excitation spectra of nuclei
  associated with reorientation in space, 1014
Excited states
  low-lying
    due to breaking pairs, 1022
  of target nucleus, 673
Exchange effect
  estimate of, 731
  in stripping reactions, 725
Exclusion principle, 719, 724
Expansion
  order of forbiddenness, 822
External perturbing fields, 733
Extra-nuclear absorption, 534
Extreme angular distributions, 765

F

f for β^+ emission
  high energy scale, 828
  low-energy scale, 827
f for β^- emission
  high-energy scale, 826
  medium-energy scale, 826
F^{19}, 265, 702, 881
  contaminant in targets, 270
  determination of nuclear g-factor for, 578
  inelastic scattering of photons on, 578
$F^{19}(p,\alpha)O^{16}$ reaction, 700
$F^{19}(d,d')F^{19}$ reaction, 687
$F^{19}(d,p)F^{20}$ reaction
  angular distribution of protons in, 729
Fano's statistical tensors, 784
Fano's X-coefficient, 759, 803
$F^{19}(p,\alpha\gamma)O^{16}$ reaction, 265
  as a function of proton energy, 271
$F^{19}(p,\gamma)Ne^{20}$ reaction, 265
Fe, 685
Fe^{54}, 797
Fe^{56}, energy levels of, 1026
$Fe^{56}(p,p')Fe^{56}$ reaction, 687
Fe^{56}, resonances in, 422
Fermi coupling constant, 149
Fermi energy, 438
Fermi function, 819
Fermi gas, 690
  model of nucleus, 435
Fermi integral function, 140

SUBJECT INDEX

Fermi interaction, 827
Fermi matrix element, 819
  formula for, 947
Fermi plot, 820
Fermi theory of beta decay, 141
Fermi-Thomas model, 9
Fermions, 811
Fictitious channel spin, 776
Fierz interference, 948
  in beta decay, 820
Figure of merit for spectrometer, 83
Fine structure constant, 819, 853
First forbidden beta spectra, 824
Fission, 340, 355, 358, 417
  application of collective model to, 426
  channels, 425
  fragments, 355
  resonances
    shape of, 426
  widths, 431
Fluctuation cross section, 1035
  at low energies, 1036
  limit of, 1036
Fluorescence decay times, 33
Fluorescence radiation, 73
Fluorescent scattering, 534–540
Forbidden spectra, 821
Forbidden transition
  estimate of speed, 870
  nonunique, 142
  unique, 144
Forces
  delta function approximation of, 973
  pairing, 1010
  spin orbit, 963
  velocity dependent, 883
Fourier transform, 709
Fractional parentage coefficients, 722, 852, 863, 884–887
Fractional parentage expansion
  for L-S coupling, 863
Fractional polarization, 784
Fraunhofer optics
  approximations, 684
Frequency of vibration
  of nucleus, 1012
ft values, 167–169, 825–828
  calculation of, 153, 168
  from shell model, 976
"Full-energy-loss" peak, 231

G

Gamma decay
  hindrance factor in, 185
  multipole order of, 250–252
Gamma-energy shift
  from physicochemical sources, 538
Gamma-gamma cascades, 572
Gamma-gamma coincidence, 191, 232, 311
Gamma-gamma correlation
  for radiations of mixed multipolarity, 751
  with definite multipolarity (pure-pure case), 770
Gamma-radiating levels
  width measurement of, 533–545
Gamma radiation
  absorption of, 534–540
Gamma-ray
  absorption coefficient, 211
  abundance measurements, 179
  annihilation, 231
  branching ratios, 290
  Doppler shift of, 289
  double correlations of, 291
  energies of, 242
  energy range of interest, 211
  formula for absorption, 212
  ground state, table of, 314
  interaction with matter, 211–227
  linear polarization of, 298
  mixed transitions, 771
  monochromatic, 492, 499
  neutron capture, 304–329
  nonresonant yields of, 284–286
  partial width, 274
  probability of emission, 255
  relative intensities of, 315–316
Gamma ray absorption and emission
  sensitivity to multipole moment, 652
  spectral distribution of, 323
  unbound nuclear states from, 260–302
Gamma-ray correlations
  triple, 293
Gamma ray data
  analysis of, 852–887
Gamma ray detection, 179, 228, 234
  efficiency of, 179
Gamma ray emission, 642
  from nuclei
    modes of de-excitation, 852

1126 SUBJECT INDEX

Gamma-ray scattering
  from water, 493
Gamma-ray sources
  standard, 290
Gamma-ray spectra, 178
  at resonance, 788
  coincident, 288
  from capture of charged particles, 287
  from resonant capture, 448
  measurement of, 234–244
  neutron capture, 240
  properties of, 312
  shell model properties of, 313
Gamma-ray spectrometer, 262–266
  bent-crystal, 242
Gamma-ray spectroscopy, 211–332
Gamma-ray transitions, 228
  half-life of, 255–257
  inhibited by large angular momentum change, 978
Gamma-ray width
  measurement of, 283
Gamma-transition probabilities
  estimate of absolute value of, 515
Gamma transitions
  absolute
    methods of measuring, 517
  highly forbidden, 812
Gamow-Teller interactions, 142, 827
Gamow-Teller matrix element, 819
  formula for, 947
Gamow-Teller selection rules, 151
Gas counter
  He^3 filled, 360
  hydrogen filled, 361
    pulse-height spectrum for, 362
    resolution of, 363
    useful range, 363
  limitations of, 360
Gas multiplication factor, 229
Gas scattering chamber
  slit geometry for, 122
  small volume, 124
Gas targets, 269
Gating circuit, 502
Gaussian distribution
  multiple scattering angles, 21–22
  of matrix elements of width Γ, 663
Gd, 1028
Ge^{72}, 857
Ge^{74}, 1047

Geiger-Müller counters, 50–51, 308
  for detection of gamma rays, 229
Gettering, 113
Giant resonance, 338, 459–464, 1034
  angular distribution of elastically scattered neutrons from, 462
  characteristics of, 502
  distinguished from compound nucleus resonance, 460
  polarization by elastic scattering from, 464
  strength function for, 461
  total cross-sections for, 459
Good geometry, 211
Green's function
  in solution of Schroedinger equation, 705
Green's theorem, 891
Ground state rotational band
  in even-even nuclei, 1022

H

H^3, 979
Hamiltonian
  charge independent, 944
  charge symmetric, 938
  for target, 673
  interaction, 919
  single particle, 1019
Hankel function, 717
Harmonic oscillator
  selection rule for, 1014
Harmonic oscillator level sequence, 970
Harmonic oscillator model
  energy separations in, 943
Harmonic oscillator potential, 864, 892
Harmonic spectrum
  corrections to, 1013
He^3 counter, 352
He^6, angular correlation measurements on, 821
Heavy nuclei
  distribution of magnetic transitions, 875
  distribution of matrix elements for E1 transitions, 873
  distribution of matrix elements for E2 transitions, 873
  E3, E4, and E5 transitions, 874
  empirical data, 872–879

SUBJECT INDEX

location of energy levels, 435
radiative widths, 874
Heavy particle emission, 853
Heavy particle stripping, 727, 922
Helium-3 spectrometer
  response to neutrons, 361
Hermetian conjugate, 816
Hf^{180}, 880
Ho^{161}, K capture in, 1030
Holes in a shell, 993
Hulthén wave function, 714, 719, 926
  deuteron function, 927
Huygens construction, 26
Hydrodynamic theories, 503
Hyperfine spectra, 549
Hyperfine splitting of an atomic term, 558
Hyperfine structure of spectral lines, 557

I

I^{129}
  beta spectrum of, 79
  decay scheme, *see* Decay scheme of I^{126}
  related nuclear states, 164, 165
Ilford C2 emulsion, 380
  proton range-energy in, 383
Impact parameter, 17
  minimum value of, 4
  upper limit of, 5
In^{115}, 430
In^{111}–Cd^{111} cascade, 573
Incident beam
  energy spread in, 664
Incident particle energy, measurement of, 107
Incident particle sources, 101–108
Independent particle approximation
  degeneracy of solutions, 1009
Independent particle model, 250, 255, 681, 863–865
  equivalent-particle formulation, 863
  intermediate coupling, 866, 868
  related to gamma ray emission, 852
Independent particle shell model, 680
Inelastic alpha scattering, 775
Inelastic collisions
  with atomic electrons, 16
Inelastic potential scattering amplitudes, 629–631

Inelastic scattering, 285
  direct interaction model of, 670
  involving nuclear electric transitions, 541
  involving transitions between bound states, 542
  of photons, 496
  semiclassical approach, 678
  W.K.B. calculations of, 679
Inelastic scattering calculations, 680–684
Inglis spin orbit model, 912
Initial link, 803
Intensity rules, 1017
Interaction
  spin-dependent, 676
Interaction fluctuations
  average over, 1033
Interaction potential, 673
Interaction radius, 696
Interconfigurational mixing, 875
Interference
  coherent and incoherent, 764–768
Interfering operators, 778
Intermediate coupling, 672
Intermediate nuclear states
  mixed, 804
  sharp angular momentum and parity, 800–804
Internal conversion, 70, 228, 252–255, 815, 857
  angular correlation in, 753
  by pair creation, 857
  discussion of the process, 834
  parity selection rules for, 837
  part photoelectric, 857
Internal conversion coefficients, 253
  calculation of, 838–850
  definition of, 834
  dynamic structure effects on, 850
  finite nuclear size effect, 839
  K shell, 248
  measurement of, 254
  nonstructure dependent, 840
  nuclear structure and screening model, 839
  numerical values, 850
  relationship to charge, 517
  structure dependent, 841
  unscreened, 839
Internal conversion data
  analysis of, 834–850
Internal-conversion electrons, 234–239

Pages 1–582 are in Part A. Pages 623–1103 are in Part B.

1128 SUBJECT INDEX

Internal conversion ratio, 253
  measurement of, 254
Internal conversion sources, 74
Internal-coupling angular
  momenta, 902
Intrinsic angular momentum
  quantum number K, 1019
Intrinsic angular momentum
  vanishing of, 1015
Intrinsic modes of excitation, 1011
Intrinsic properties
  associated with odd nucleon in odd-A
    nucleus, 1022
Intrinsic spectra, 1019
Intrinsic spin
  classical flipping, 862
Intrinsic states of the nucleus, 1018
Invariance arguments, 1049
Ion detectors, solid state, 51–52
Ion energy loss, 4, 6
  dependence on charge, 6
  velocity dependence of, 6
Ion identification, 47
Ion-optical aberrations, 62–63
Ionization
  energy to produce, 44
  per unit path length, 18
Ionization chamber, 179
  for detection of gamma rays, 229
  pulse, 44–47
Ionization detection, gaseous, 43–51
Ionization detector, specific, 47–50
Ionization potential, 6, 8
  of an atom, 5
Ionization processes, 16
Ionized particle
  trajectory of, 375
Irrotational flow in the nucleus
  related to moment of inertia, 1018
Irving's wave function, 927
Isobaric nuclei, 939
Isochromat, 498
Isolated resonances, 272
  scattering from, 647
Isomeric lifetimes, 874
Isomeric state, 247, 812
Isotopic spin, 147–148, 294, 672, 885
  characterization of nuclear states, 933
  commutation relations for, 935
  Coulomb corrections to, 939
  eigenfunctions, 935
  introduction of, 932

Pauli principle for, 936
  quantum number, 939
  raising and lowering operators, 935
  selection rules for, 858
  single particle functions, 925
Isotopic spin Clebsch-Gordon factor, 915
Isotopic spin components, 934
Isotopic spin coupling factor, 892, 916, 924
Isotopic spin functions
  symmetry of, 935
Isotopic spin impurity, 941
  coefficients, 945
  estimates of, 953
Isotopic spin of nuclear states, 934
Isotopic spin quantum number
  validity of, 957
Isotopic spin selection rules, 307, 932–959
  derivation for electric dipole transitions, 933
  on E1 transitions, 868
  relating cross sections, 955
  violation, 867
Isotopic spin space, 934
Isotopic spin states
  admixture of, 958
  Coulomb perturbation, 940
  formula for, 956
  mixing, 940
Isotopic spin triplet, 938
Isotopic spin vector operator, 933

J

j-forbidden reactions, 880, 899
j-j charge fractional percentage coefficient, 886
j-j coupling, 295, 884
  closed shell, 896
  independent particle model, 863
  in shell model, 723, 880, 972
  sum rules for, 916
  total isotopic spin impurity in, 943
j-j representation, 1005
j-selection rule, 898
Jacobi polynomial, 767

K

K^{39}, 917, 979
K^{40}, 72
K absorption edge, 213

SUBJECT INDEX 1129

K/β^+ branching ratio
  energy scale, 830, 831
K-capture, 166
  branching ratio to β^+ emission, 812
  decay schemes, 213
K conversion coefficients, 839
K-edge jump ratio, 218
K-electrons, 9, 812, 831
  stopping power of, 6
K-hindering, 841, 842
K/L ratio, 248–249
K-shell conversion coefficients
  table of, 843–846
K-shell photopeaks, 238
K-shell matrix elements, 842
Kicksorter, 264
Kinetic energies in nucleus, 677
Kinetic energy, end-point
  for positron decay, 949
Kinetic energy operator, 673
Klein-Nishina formula, 220
Klystron bunching of beam, 372
Knock-out process, 727
Kurie plot, 76, 141–142, 143, 153, 820

L

L-electrons, 9
  stopping power of, 6
l-forbiddenness
  in shell model, 879
L/K branching ratio
  formula for, 832
LS and jj coupling
  transition between, 913
LS coupling, 295, 723
  independent particle model, 863
  in shell model, 880
  selection rules, 880
LS coupling, extreme, 906
LS-coupling shell model state, 904
LS model of nucleus
  of Wigner and Feenberg, 1007
LS representation, 905
LS selection rules, 900
L shell conversion coefficients
  table of, 847–850
l-unfavored amplitude, 919
Laboratory coordinate system, 1018
Laboratory cross sections, 126
  calculation of, 126
Larmor frequency, 556

Larmor precession frequency, 554
Lead isotopes, 319
  spectra of, 1023
  spins and parities of levels in, 1023
Lead, mass attenuation
  coefficients for, 226
Legendre coefficient, 745
Legendre functions, 742
Legendre polynomials, 627, 646, 817
  addition-theorem for, 738
Legendre series
  correlation of directions of motion, 737
Lens spectrometer, 90–93
  long lens type, 92
  resolution of, 90
  short lens type, 91
  spherical aberrations in, 90
Lepton, 749
  circular polarization measurements on, 778
  total angular momentum of, 813
Lepton field, 816
Level
  shell model identification, 985
Level crossovers, 974
Level density, 689
  dependence on angular momentum, 431
  for independent particle model, 435
  formula for, 435
  in target nucleus, 401
  observed by thermal neutrons, 668
  on excitation energy, 437
  on nuclear mass, 437
Level sequence
  near neutron shells, 971
Level shifts
  and reduced widths, 920
Level spacings, 432, 632
  average values of, 433
  even-even nuclei in, 432
  experimental determination of, 432
  formula for, 668
  from slow neutron resonances, 434
  in medium and heavy nuclei, 495
  shell effects in, 437
  statistical study of, 335
  table of average, 444–446
Level widths
  determined by inelastic electron scattering, 545
  electromagnetic contribution to, 514
  table of, 545

Pages 1–582 are in Part A. Pages 623–1103 are in Part B.

1130 SUBJECT INDEX

Lifetimes
  coincidence method, 518
  experimental methods for determining, 517
  measurement
    in heavy radiative nuclei, 530
    of vibrational level, 1013
    pulse method, 518
    using coincidence technique, 521
    using electronic methods, 517
  table of, determined by electronic timing after detection, 543
  very short, 512–545
Li^6 mirror reactions, 956
$Li^6(n,t)He^4$ reaction, 391
$Li^6(p,p')Li^6$ reaction, 686
$Li^6(p,He^3)He^4$ reaction, 686
$Li^6(\alpha,\alpha')Li^6$ reaction, 686
Li^7, 886
$Li^7(\gamma,n)Li^6$ reaction
  in support of selection on charge independence, 954
$Li^7 + p$ reactions, 909
$Li^7(p,p')Li^{7*}$ reaction, 684
$Li^7(p,n)Be^7$ reaction, 409
  threshold as an energy standard, 478
$Li^7(p,d)Li^6$ reaction, 697
$Li^7(d,n)Be^8$ reaction, 484
Light fermion, 811
Light nuclei
  deformations in, 1011
  electric dipole transitions, 867
  electric quadrupole transitions, 871–872
  empirical data, 867–872
  higher multipole transitions, 872
  intermediate coupling in, 1007
  magnetic dipole transitions, 870
Line broadening, 237
Linear accelerator, 348
Linear polarization correlation, 780
Liquid drop model, 1010
  surface vibrations, 1012
Liquid hydrogen, 375
Liquid scintillator, boron-loaded, 351
Lithium iodide
  pulse height response of, 364
  scintillator, 353
Long counter, 477
  modified, 483
Long wave length approximation
  electric dipole matrix element for, 951

Longitudinal polarization
  of electrons, 829
Lorentz transformation
  extended group, 816
Low energy nuclear spectra, 1012
  survey of, 1023–1032
Low energy spectra of nuclei
  collective motions associated with, 1011
Low-lying states
  spins of, 981
Lu^{176}, 72
Lubitz' tables, 929

M

Magic numbers, 248, 638
  evidence for, 964–968
  from grouping of levels, 968
  list of, 963
Magnetic analyzer, 60–67, 106
  calibration of, 67–68
  focusing properties of, 61, 62
  interpretation of results from, 69
  180 degree, 62
  target preparation for, 68
Magnetic bunching, 372
Magnetic dipole absorption, 509
Magnetic field calibration points, 68
Magnetic moment, 548
  of neutron, 1047
  of nucleus, coupling to electron, 569
Magnetic multipoles
  even order, 837
  parity of, 251
  radiation, 252
Magnetic octopole, 551
Magnetic pair spectrometer, 239–241, 309
Magnetic spectrograph, 174
  resolution and transmission of, 174
Magnetic spectrometer
  flat-type, 83
  focusing properties of, 80
  lens-type, 83
  third-order focusing in, 89
Magnetic quantum number, 554, 569, 710, 743
Magnetic quantum number sums, 734
Magnetic radiation, 652
Magnetic transition matrix elements
  formula for, 860

Pages 1–582 are in Part A. Pages 623–1103 are in Part B.

Magnetized iron analyzers, 311
Manganese
  interference of slow neutron resonances in, 424
  total neutron cross section for, 424
Many body problem
  relativistic, 838
Many particle spectra, 997–1002
Mass attenuation coefficients, 212
  as function of photon energy, 224
  particular, 227
Mass exchange reaction
  angular distribution in, 709–721
  Born approximation in, 702
  cross section of, 702–709
  dynamics of, 699–702
  formula for cross section, 705
  momentum transfer in, 703
  Schroedinger equations for, 704
Mass 12 isobaric triad, 940
Matrix elements
  absolute value of, 513
  for electromagnetic transitions, from charged particle scattering, 540
  of operators, 512
Matrix formulation
  of multipole operators, 762
Maximal techniques
  of magnetic quantum number sums, 745
Maxwellian distribution
  of neutrons in a pile, 344
  of target atoms, 419
Mayer anomaly, 900
Mechanical choppers, 347
Mesic units, 701
Method of self self-consistent field, 1010
Mg, 685
$Mg^{24}(d,d')Mg^{24}$ reaction, 687
$Mg^{24}(p,\gamma)Al^{25}$ reaction, 269, 282, 287
Mg^{25}, collective model characteristics of, 320
Microwave spectroscopy, 558
Mirror nuclei, 147, 939
  super allowed transitions between, 827
Mixed radiations, 765
Mixed-mixed correlations, 789
Mixing coefficients, 771
Mixing parameter, 800
Mixing ratios, 789
Mn^{56}, 449
$Mn^{55}(d,p)Mn^{56}$ reaction, 900

Moderator, 481
Møller's formula for electron-electron scattering, 18
Molecular beams, 559
  deflection of, 559
Molecular beam magnetic resonance, 560
Molecular stopping cross section, 14
Moment of inertia
  effective, for nucleus, 1015
Momentum space
  volume of, 820
Momentum transfer, 701
Monoenergetic γ-radiation
  production of, 534
Moszkowski units, 860
Mott scattering, 16
  cross section for, 18
Multiple scattering, 15, 17, 21, 414
  approximation, 1054
Multiple scattering formalism, 675
Multiplicity, 557
  coincidence method for measuring, 329
  formula for, 329
Multiparticle levels, 898
Multipole mixtures, 836
Multipole operator
  gamma ray, 771
Multipole radiation
  in beta decay, 140
Multipole spin operator, 784

N

N^{13}, 896
N^{14}, ground state, 953
$N^{14}(d,d')N^{14}$ reaction, 686
$N^{14}(d,p)N^{15}$, stripping cross section in, 895
N^{15}, 307
$N^{15}(\gamma,p)C^{14}$ cross-section, 507
N^{16}, beta decay of, 241
Na^{23}, 901
$Na^{23}(\alpha,p\gamma)Mg^{26}$ reaction, 284
$Na^{23}(\alpha,\alpha'\gamma)Na^{23}$ reaction, 284
$Na^{23}(d,p)Na^{24}$ reaction, 697
$Na^{23}(d,t)Na^{22}$ reaction, 698
Naked nucleus, 835
Naphthalene, phosphor of, 75
Naturally radioactive
  elements, 642
Natural radioactivity
  correction for, 287

SUBJECT INDEX

Nb^{90}, positron decay of, 990
Nb^{145}, 974
Ne, 685
Ne^{19}, angular correlation
    measurements on, 821
Ne$^{20}(\alpha,\alpha')$Ne^{20} reaction, 687
Ne$^{20}(p,\gamma)$Na^{21} reaction, 921
Negative potential phase shift, 636
Neutral particles
    angular distribution of, 646–649
    elastic scattering of, 646–649
Neutrino, 170, 811
    helicity, 754
    mass, 816
    monoenergetic, 830
Neutron(s), 390, 682
    angular, distribution of scattered, 411
    binding energy of, 437
    detection efficiency for, 371
    elastically scattered angular distribution of, 455
    energy resolution of, 359
    energy spectra of, 402
    epithermal, 361
    fast, sources of, 464
    high energy, 392
    inelastically scattered, spectrum of, 411
    interactions with matter, 335–341
    types of, 336–346
Neutrons, low energy, 662
    moderated in lead, 349
    potential scattering of, 336
    production of, 359
    pulsed
        from accelerators, 348
        sources of, 347
    from reactions, 464–465
    resonant-direct, 504
    resonant scattering of, 338
    slow radiation widths for, 442
    transmission factors for, 641
Neutron absorption, 339–341
Neutron beam
    attenuation by emulsion and glass, 388
Neutron camera, 395
Neutron capture, 246, 305, 339, 724
    direct, 306
    thermal, 304
Neutron-capture cross-sections
    radiative, 354
Neutron capture gamma-rays, 304–329

    experimental methods for study of, 308, 309, 312
    heavy elements, 318
    intensities, of low-energy, 326
        relative, 315
    measurements, 307
    multiplicities of, 328
    reactor experiments on, 308
    statistical properties of, 322
Neutron chopper, 342
Neutron cross section
    average
        measurement of, 1034
    correction for back-ground in measuring, 468
    for fission, 354
    for iron, 337
    measurement of, 468–473
        techniques of, 350
    total, 350, 468
        experimental arrangement for measuring, 469
        for helium, 456
        for nitrogen, 457
        for oxygen, 453
        for sulfur, 452
    using reactions, 472
    variation with energy, 336
Neutron detection reactions, 363–364
Neutron-deuteron scattering, 395
Neutron emission, 403, 641
    near threshold, cross section of, 480
    threshold(s) for, 479
Neutron energy
    "bright-line" method, 352
    discrimination of, using the cone effect, 480
    from time-of-flight, 347
Neutron energy spread
    factors determining, 466
Neutron excess, 966
Neutron flux, absolute, 413
    formula for, 407
    for scattering experiments, estimation of, 407
    in emulsion formula for, 388
Neutron magnetic moment, 549
Neutron moderator, 348
Neutron monochromator, mechanical, 344
$(n,2n)$ process, 403
    permissible energy values for, 404

SUBJECT INDEX 1133

Neutron producing reactions
  characteristics of, 467
Neutron-proton mass difference, 952
Neutron-proton scattering, 338, 388
  cross-section, 370
Neutron resonances
  fast, 451–475
    experimental techniques, 464
  in sulfur, 636
  partial widths of, 417
  shapes of, 417
  single levels, 417–422
  slow, 417–449
Neutron scattering
  angular resolution, 408
  cross-sections, 352
    problem in measuring, 353
    theoretical, 1060
  elastic, corrections for multiple scattering, 470
  inelastic, 374, 403, 471, 991
Neutron separation energy, 304, 305
  distribution, 326
Neutron source
  energy resolution, 466
  monoenergetic, 407
Neutron spectra
  difficulties in measuring, 359
  fast, 358–396
    measurement of, 360–396
    from time-of-flight measurements, 369
    use of scintillators in measuring, 363–365
Neutron spectrometer, 343–350
  comparison of slow, 349
  crystal, 343–345
  gas counter type, 360–363
  properties of various, 344
Neutron spectroscopy, 342–355
  limitations of, 400
Neutron spectrum measurements
  with cloud chamber, 379
Neutron strength function for heavy nuclides, 441
Neutron threshold(s), 477–487
  accuracy of determining, 478
  experimental results, 483
  for $Cu^{63}(p,n)Zn^{63}$ reaction, 485–486
  for $Li^7(d,n)Be^8$ reaction, 485
  in endothermic reactions, 477–479
  table of, for (p,n) reactions, 478
  weak, counter-ratio method, 482–483

Neutron transmission, 350
  measurement of, 350
Neutron widths
  distribution of, 662
  for slow neutron resonances of U^{238}, 429
Newton's rules, 443
Ni, 685
$Ni^{60}(p,\gamma)Cu^{61}$ reaction resonances in, 278
$Ni(p,p')Ni$ reaction, 687
Ni^{64}, proton scattering from, 1046
Noble gases, 36
  electron structure of, 963
  use in counters of, 36
Nonfissionable nuclides, 427
Nonlocal potential, 1046
Nonlocality
  effects of, 1046
Nonresonant cross section, 629
Nonresonant interactions, 629
Normal modes of oscillation, 1012
Normal modes of vibration
  labelled by multipole order, 1012
Normalization, 744
Np^{239}, beta decay of, 1031
Nuclear angular momentum
  precession of, 557
  in a magnetic field, 555
Nuclear Bohr magneton, 860
Nuclear charge distribution, 549
Nuclear core, 693
Nuclear coupling schemes, 983–1007
Nuclear current operator, 772
Nuclear deformation(s), 441, 503
  quadrupole type, 1012
Nuclear deformation parameter
  formula for, 1011
Nuclear degrees of freedom, 1014
Nuclear density, 675
Nuclear density distribution, 712, 1010
Nuclear dipole polarizability
  tensor nature of, 496
Nuclear dissociation energy, 514
Nuclear distortion
  influence on projectile wave function, 694
  treated as a perturbation, 1020
Nuclear electromagnetic moments
  measurement of, 548–580
Nuclear emulsion(s), 51–52, 171, 380–396
  as both radiator and detector, 386
  range-energy relations for, 382

Pages 1–582 are in Part A. Pages 623–1103 are in Part B.

1134 SUBJECT INDEX

range straggling in, 384
recoil hydrogen nuclei in, 386
stopping power of, 382
use as detector only, 394
Nuclear energy levels, 335, 732
  decay schemes of, 257–259
Nuclear equilibrium shape, 1010
  influence of pairing forces, 1011
Nuclear exchange potentials, 932
Nuclear excitation
  particles with spin, 746
  within same configuration, 908
Nuclear excitation energy, 246
Nuclear field
  oblate distortion of, 1020
  quadrupole distortion, 1020
Nuclear forces, 261, 335, 341, 540, 649
Nuclear g-factor, 550, 574
  arrangement for determination of, 578
Nuclear ground state, 676
Nuclear gyromagnetic ratio, 550
Nuclear Hamiltonian, 933
Nuclear index of refraction, 672
Nuclear induction, 561
Nuclear isomerism, 247–250, 978–980
  historical summary, 247
Nuclear isotopic spin assignment, 939
Nuclear levels
  from β-decay, 814
  spacings, 32, 434
  spin and parity of, 147
Nuclear lifetimes
  from crystal diffraction of X-rays, 533
  from measurement of internal conversion coefficient, 533
Nuclear magnetic moment, 550–554
  measurement of, 549
Nuclear magnetic resonance fluxmeter, 68
Nuclear magneton, 548, 980
Nuclear matrix elements
  related to transition probabilities, 514
Nuclear matter
  resonant structure of, 670
Nuclear model(s), 550–554, 672
  in beta decay, 828
  shell, 248
  simplest, 968
  unified, 186, 192
  wave function of, 514
Nuclear moment of inertia, 1015
Nuclear moments, 548–580, 978–980
  general rules concerning, 551–554

interaction with external fields, 554–557
measurement of, 557–562
  by radiation pattern method, 563
  methods based on emission of nuclear radiation, 562
  using pure quadrupole resonance, 561
  using resonance methods, 560
Nuclear motion
  intrinsic and rotational, 1015
Nuclear orientation, 563
Nuclear pairs, emission of, 228
Nuclear parity
  for quadrupole deformation, 1013
Nuclear parity change, 653
Nuclear particle emission
  channel spin scheme, 800
  Coulomb excitation, 801
  j_p scheme, 801
  unobserved intermediate radiations, 801
Nuclear particles, 3
  interaction with matter, 5
  spectra, 55–70
Nuclear photodisintegration
  giant resonance in, 884
Nuclear plates
  examination of, 380
  technique, 692
Nuclear potential
  central, 650
  harmonic oscillator, 968
  static isotropic, 1010
  variations in shape, 1010
Nuclear property
  related to that of other state in band, 1017
Nuclear radiation
  emission of, 562
  successive directional correlation of, 571
Nuclear radius, 678, 696, 818, 1044
Nuclear reaction experiments
  table of, 686
Nuclear reactions, 401
  angular distribution for, 646
  beta decay, 946
  compound statistical model of, 671
  data processing for, 126–136
  directional distribution in, 577
  for spinless particles, 737

SUBJECT INDEX 1135

gamma rays in, 652–654
initiated by p-wave protons, 798
involving fast neutrons, 457
isotopic spin impurity determined by, 957
parameters, 101
resonant and nonresonant, 100
single step model, 699
statistical theory of, 685
targets, 108–117
unified theory of, 672
use of isotopic spin quantum numbers in, 946
Nuclear reactor, 343
Nuclear recoil, 257
from direct distance measurement, 527
from Doppler-shift measurement, 528
geometrical arrangement for measurement, 529
in gamma ray absorption, 283
Nuclear relaxation time, 571
Nuclear resonance, 568
in liquids and solids, 560
Nuclear resonance absorption and induction, 560
Nuclear resonance scattering, 211
Nuclear restoring force, 1012
Nuclear scattering
Coulomb interference term, 651
Nuclear shape, 188
ellipsoidal, 1011
for minimum energy, 1011
Nuclear shell model, 963–981
two nucleons in same level, 986
Nuclear shell structure, 554
Nuclear softening
collective vibration frequencies, 1011
Nuclear spectra, 1009–1032
rotational band structure, 1016
systematic variations, 1010
Nuclear spin, 247, 648, 672
Nuclear stability, 967
Nuclear states
bound, 101, 399, 414
spin and parity of, 399
study using continuous spectra, 401
study using discrete spectra, 400–401
study using neutron spectroscopy, 399

characteristics of bound excited, 246
determination of properties of, 291
identification of, 792
low-lying, 513
metastable, 247
neutron thresholds leading to, 479–487
observed in alpha decay, 170–204
observed in beta decay, 139–168
resonant compound, 671
several intermediate, 785
spins of, 513
study of, 245–332
unbound, 417–475
from charged particle capture, 260
Nuclear states expansion
into isotopic spin eigenstates, 944
Nuclear structure
unified description, 1011
Nuclear surface, 675
inelastic processes in, 681
oscillations in, 255
Nuclear temperature, 503, 668, 690
Nuclear transition energies
from gamma-ray measurements, 258
Nuclear transitions, 811
absolute, measurement of, 516
cascade relations between, 259
of high multipolarity, 514
parity of, 251
Nuclear volume, 675
Nuclear wave function, 512, 861
dependence of cross section on, 721
Nuclei
effective interaction in, 918
multipolarities of, 314
near magic numbers, 841
odd-mass, 196
orientation of
dynamic method, 567
static method, 567
oriented, emission of beta particles from, 566
self-conjugate, 294, 858
spherical, 491
stable, 400
Nucleon
binding energy, 246
coupling, 732
energy
as function nuclear field shape, 1011
interaction
first approximation, 1009

magnetic moment of, 860
oscillations, 246
Nucleon-nucleon interaction, 675
  average, 1033
  from reduced width analysis, 891
Nucleon-nucleon resonances
  fluctuations in interaction, 1033
Nucleonic orbits, 1010
Nucleons
  correlation among, 719
  correlation of motion, 1019
  emission of, 246
  evaporated, 491
  motion relative to body fixed coordinate system, 1014
  spin of, 854
  unpolarized, 690
Nucleons, two identical ($1p$)
  jj description, 1003
  LS description, 1005
Nucleus
  as dynamic system, 1012
  charge and current distribution in, 978
  completely absorbing black, 682
  excited, radiative decay of, 443
  orientation in space, 1014
  residual, energy levels of, 399

O

O^{14}, 821
$O^{16}(d,\alpha)N^{14}$ reaction
  application of isotropic spin selection rules to, 952
$O^{16}(d,p)O^{17}$ reaction, 700, 716, 725
  angular distribution of protons in, 729
$O^{16}(p,\gamma)F^{17}$ reaction, 286
$O^{16}(p,p')O^{11}$ reaction, 681, 687
O^{17}, 716, 896, 971
O^{18}, energy level diagram for, 292
Oblate deformation
  of nucleus, 1010
Observed magnetic moments, 979
Odd-A nuclei
  low-lying intrinsic states, 1011
  nonspheroidal regions, 202
  spheroidal region, 196
Odd-even nuclei, 900
Odd-odd nuclei, 306, 307
180 degree spectrometer, 83
  constant-field type, 84
  constant radius, 83

One-level approximation, 797
One-particle model
  low-energy spectra for even-even nuclei, 1022
Open geometries, 410
Operator expansion
  for density matrix, 761
Operators, 512
  relativistically invariant, 816
Optical atomic spectra, 558
Optical model, 670
  of nuclear reaction, 440
  of nucleus, 463
    assumptions of, 440
    diffuse edge of, 440
    in explanation of giant resonances, 463
  scattering amplitude in, 670
  square well parameters for, 681
Optical potential, 675
  real part of, 630
  spin-orbit coupling term in, 464
Orange type spectrometer, 89, 234
Orbital angular momenta, 131, 186, 279, 696
  in alpha decay, 186
  in shell model, 676
  of compound state, 131
  relative reduced width for, 905
  scattering, 630
Orbital capture
  energy condition for, 813
  observation of, 812
Orbital capture of electron, 812, 830–832
Orbital fractional parentage coefficient, 885
Order of forbiddenness
  in beta decay, 814
Organic contamination of targets, 109
Oriented nuclei, 567–571
  angular distribution of radiations from, 764
  beta emission from, 815
Oscillations
  independent, 1012
Overlapping levels, 660
  many, 661–667
  statistical theory, 661
Overlapping resonances
  empirical analysis of, 661

SUBJECT INDEX

P

P^{29}, 971
$P^{31}(\alpha,p)S^{31}$ reaction, 687
$P^{31}(d,p)P^{32}$ reaction, 318, 914
$P^{31}(n,\gamma)P^{32}$ reaction, 318
P^{32}, 814
(p,α) process, 791
Pb^{206}, application of coupling scheme to, 993
Pb^{206}, comparison of experiment with theory, 994, 996
Pb^{207}, single particle levels from, 993
Pb^{208}, 965
Pb^{208}, doubly closed shell at, 877
$Pb(p,p')Pb$ reaction, 691
$(p,p'\gamma)$ process, 766
p-wave neutrons, 637
p-wave resonances, 900
Pair coincidence spectra
  in $F^{19}(p,\alpha)O^{16}$ reaction, 241
Pair emission, 856
Pair production, 211, 224–226
  angular distribution of negatron-positron pairs, 225
  atomic electron effect, 224
  internal, 228
Pair spectrometer, intermediate-image, 236
Pairing effects, 1010
Pairing energy, 312, 327, 437, 973
  for odd number of particles in level, 973
Pairing forces, 1010
Parallel angular momenta, 746, 774
Parametrization, 750
Parentage coefficients, 884–887
Parentage rule, 723
Parity, 131, 132, 140, 239, 625, 665, 682
  change in target nucleus, 676
Parity assignments
  relative, 814
Parity change
  in beta decay, 814
  of the nucleus, 250
Parity conservation, 744
  effect on electric multipole transitions, 653
  effect on magnetic multipole transitions, 653
  failure of, 829
Parity conserving coupling, 947

Parity determination
  in beta decay, 815
Parity-favored transition, 855
Parity of a state
  defined, 975
Parity restrictions, 724
Parity rule for radiation emission
  derivation of, 854
Parity selection rule
  related to multipole order, 837
Partial lifetime, 857
Partial wave, lth
  logarithmic, 891
Partial wave analysis, 133–136, 455, 460, 1048
  graphical method of, 133–135
  in potential scattering, 452
Partial waves, 696, 1040
  parity of, 646
Partial width, 130–131, 277
  related to peak cross-section, 427
Particle absorption by the nucleus, 679
Particle accelerators, 261–262
Particle beam
  unpolarized, 649
Particle, localization of, 41
  step-wedge arrangement for, 42
Particle orbits
  in nucleus, 1010
Particle parameters, 800
  for spinless particles, 752
Particle reactions
  isotopic spin from, 956
  selection rules, 952
Particle-surface coupling, 919
Particle trajectory, 380
Pauli principle, 672, 864, 972, 986
  generalized, 936
Pauli spin operators, 989, 1045
Pauli spin theory, 932
Pb, low-lying states, 1024
Pb^{206}, 258
Pb^{207}, 1024
Pb^{208}, ground state of, 1023
Pd, 691
Pd^{106}, energy levels of, 1027
Peak cross section σ_0, 426
Penetrability, 632
  of centrifugal barrier, 454
Penetration factor
  modified by single-particle resonance, 640

1138   SUBJECT INDEX

Perturbation theory, 941
  amplitude calculation, 918
  time dependent, 853
  use in calculating internal conversion coefficients, 838
Phase factor, 760
Phase problem
  in angular correlation, 769
Phase shift, 629
  complex, 631, 1038
  for potential (hard sphere) scattering, 788
  imaginary part of, 632
  machine calculation of, 136
  repulsive sphere, 630
Phase shift analysis
  ambiguities in, 1045
Phase space, 689
Phonon, 920
Phosphors
  characteristics of, 36–39
  organic, 36
  plastic, 36
"Phoswich," 43
Photo-peak, 36
Photo stars, 508
Photoactivation, 492, 497–501
  use in determining thresholds of (γ,n) reactions, 500
  yield curve, breaks in, 498
Photoalphas, 508–509
Photoelectric conversion, 857
Photoelectric cross section, 218
Photoelectric effect, 213–219
  relative strength for atomic shells, 218
Photoelectric efficiency, 37, 520
Photoelectrons, 234
  angular distribution of, 219
Photon difference method of analysis, 502
Photon energy distribution
  in bremsstrahlung, 508
Photon excitation of nuclei, 358
Photon total absorption, 496
Photon transition
  between $J = 0$ states, 855
  in lead, 507
Photoneutron(s), 492, 501–505
  from betatron beam, arrangement for detection of, 502
  thresholds, table of, 504

Photons, 491
  circular-circular polarization correlation for, 780
  circularly polarized, 750
  spin, 652
Photonuclear absorption
  at giant resonance, 491
Photonuclear calibration points
  table of, 500
Photonuclear effect, 228
Photonuclear measurements
  in elastic scattering, 492–496
Photonuclear reaction, 491–509
Photoproton(s), 492, 505–508
  emission, effect of Coulomb barrier, 505
  energy, measurement of, 506
  from lead, energy distribution of, 506
Pi meson, 628
  Compton wave length, 1045
  emission of, 853
Pickup reaction, 698
Plane of scattering, 1049
Plane wave
  expansion of, 710
Plastifluor B, 371
Plural scattering, 21
Pm^{147}, gamma transition in
  prompt coincidence curve for, 523
Po^{214}, 857
Poincaré vector, 754
Poisson distribution, 520
Polarizability effect, 1048
Polarization, 564, 649
  circular, 566
  correlations, 296
  defined, 458
  dependence on energy, 459
  in statistical model, 665
  of emitted particle, 626
  of gamma-rays, 296, 782
  of neutrons by elastic scattering, 458
  observation of, 473
  of scattered neutrons, experimental arrangement for measuring, 474
Polarization, measurement of
  on outgoing projectiles, 671
Polarization analyzer, 654
Polarization-direction correlations, 778, 779, 802
  for photons, 780
Polarization independent correlation, 780
Polarization of pure gamma ray, 782

Polarized beam, 654
Polarized γ-rays from capture of polarized neutrons, 311
Polarized neutrinos, 829
Polarized neutrons, 311, 447
  sources of, 473
Polarized nuclei, 817
  by dipole (hyperfine) coupling, 764
Polarizing power of scatterer, 473
Polyethylene radiator, 359
Porter-Thomas distribution, 428
Positive helicity, 749
Positrons, 20, 811
  decay
    experimental information, 949
  differential cross section for creation of, 226
  Mott scattering of, 20
  spectrum, 832
Potential
  complex, 1033–1061
  effective distorting, 675
  imaginary part of, 463
  intermediate between square well and harmonic oscillator, 970
Potential scattering, 100
  cross section, 421
  interference with, 423
  partial wave series for, 649
  phase shift, 1038
Potential well, 859, 1055
  phase change of wave over, 1034
  resonances, 338
Prismatic spectrometer resolution of, 237
Processes, short-lived, 375
Projectile wave function
  absorptive distortion of, 682
Prompt coincidence wave, 522
Proportional counter, 40–47
  for detection of gamma rays, 229
  gases for, 74
  rise time of, 50
  spectroscopy, 72–75
Proton
  capture, 246
  emission, 695
  exchange, 726
  magnetic moment, 549
  recoil, 365, 370
    geometrical considerations, 386
    spectrum, produced by neutrons, 367
Proton spectroscopy, 399

Proton stopping cross-section, 13
  in various materials, 11
  ratio to alpha stopping cross section, 12
Proton(s), 8, 675, 678
  range-energy relations in liquid hydrogen, 377
  stopping cross section in various materials, 8
  weakly-coupled to C^{12}, 896
Pseudoscalar interaction, 816
Pt, 1028
Pt^{196}, multipole radiations in decay of, 252
Pu^{238}, decay scheme, 188
Pu^{239}, level spectrum of, 1031
Pu^{239}, 426
Pulse-height analyzer, multichannel, 232
Pulse spectrum, analysis of, 288
Pulsed activation, 255
Pure nuclear states
  direction-direction correlation, 770
Pure state correlations, 789

Q

Q-value of reaction, 704
Quadrupole deformation, 1012
Quadrupole excitation
  energy levels graphed, 1013
Quadrupole interaction, 556
  in various media, typical values of, 574
Quadrupole modes
  formula for energy, 1012
  harmonic oscillator equivalence, 1012
Quadrupole moment
  related to intrinsic quadrupole moment, 1018
Quadrupole moments
  as function of odd nucleon number, 553
  near closed shells, 980
  theoretical, 980
Quadrupole vibrations
  independent types, 1012
Quantization of angular momentum, 742
Quantum electrodynamics, 25, 838
Quantum mechanical angle functions, 766
Quantum mechanical excitation spectrum, 1013
Quenching of discharge, 230

R

Ra^{224}, 191–192
Racah coefficients, 627, 862
  interpretation of, 735
  normalized, 903
Radial boundary conditions, 893
Radial wave function, 633, 1040
  for bound state, 709
  in beta decay, 824
  table of, 712
Radiation
  emitted in cascade, 733
  from accelerated charge, 853
Radiation emission
  bond rupture by, 537
Radiation, nuclear, polarization of, 563
Radiation pattern
  from oriented nuclei, 564
Radiation width(s), 430–431, 442–446, 858
  dependence on nuclear size, 442
  distribution for individual transitions, 431
  for gamma-rays, 535
  formula for, 430
  table of, 444–446
Radiationless transitions, 835
Radiations, unpolarized
  directional correlations on, 735
Radiative capture, 417, 425, 625, 636
  cross-section, 418
    formula for, 639
    measurement of, 310
  of α-particles, 774
  of neutron, 417–421
  of fast neutrons, 458
Radiative channel, 425
Radiative collision of electrons with atomic nuclei, 25
Radiative transitions
  Hamiltonian for, 950
  mean lifetime, 858
  special selection rules, 879–882
Radio-frequency field, 570
Radioactive sources
  extremely long-lived, 72
  line shapes from, 288
  photo-induced, 497
RaE, 814
Raising and lowering operators
  for isotopic spin, 935

Raman scattering, 211
Range-energy tables, 9
Range ratio for charged particles, 6
Range straggling
  Gaussian distribution of, 385
Rare earth elements, 170
  spectra of, 319
Rate of energy loss for electrons, 16
Ratio of neutron width to level spacing
  for various mass number, 1058
Ratio of potential scattering length to nuclear radius
  for various mass number, 1059
Rayleigh scattering, 211
Rb^{85}, 974
Reaction amplitude, 678
  for mass exchange reaction, 710
Reaction channels, 673
Reaction cross sections, 639–645, 1036
Reaction radii, 719
Reactions
  competition of, 637
Reactions for detecting neutrons, 472
Reactions initiated by charged particles
  gamma-rays from, 284
Reciprocity relation
  for cross section, 628
Recoil-distance measurements
  table of, 543
Recoil Doppler shifts
  table of, 544
Recoil nuclei
  slowing-down in solid material, 530
  slowing-down time of, 532
Recoil protons in emulsion
  energy distribution of, 389
  total number of, 389
Recoupling rules, basic, 903
Reduced mass, 16, 131, 702
Reduced matrix elements
  standard form, 772
Reduced transition speeds
  function of level spacing, 878
Reduced width, 128, 130, 274
  analysis by selection rules, 894–901
  analysis of, 890–930
  from Born approximation, 922
  maximum value of, 454
  quantitative analysis via the shell model, 901–914
  relative, 893, 914, 915
  statistical factors, 915

stripping reaction, 922
unfavored, 918
variations, 915–922
Reflection symmetry
in nuclei, 1016
Reflectivity of a potential, 1054
Refractive index, complex, 100
Relativistic correction
of scalar matrix element, 950
Relativistic effects, 747
Relativistic invariance
number of coupling constants in
β-decay, 813
Relativistic rate of energy loss for
electrons, 17
Relaxation methods, 561
Repulsive sphere scattering, 1043
Residual interactions
compared to spin-orbit splitting, 1003
Resolution of double focusing
spectrometer, 86
Resolution of a line, 74
Resolved final states
experiments with, 684
Resonance(s)
Doppler broadened, 420, 439
Doppler smearing of, 493
energies, 625
interferences between, 279
isolated, fission component of cross
section in, 422
observed shapes of, 275
overlapping, 1033
p-wave neutrons for, 448
peak scattering cross section for, 422
self-shielding effect in, 439
Resonance amplitudes
for isolated resonances, 631
Resonance characteristics
special techniques for obtaining, 281
Resonance energy, 127, 280
Resonance fluorescence, 492, 866
Resonance γ-scattering
table of physical parameters in, 539
Resonance interference, 286
Resonance parameters
average properties, 1039
average values of, 433
distribution of, 428
extraction of, 127
neutron widths, 428
Resonance scattering, 283, 535

elastic cross section for, 535
interference in, 421
interference with potential, 339, 454
problems associated with, 538
Resonance scattering and absorption
arrangement for detection of, 538
Resonance shape(s)
effects of energy straggling on, 275
multi-level effects, 422
single from scattering, 421
Resonance yields, theoretical expressions
for, 272–274
Resonant absorption
cross section for, 535
Resonant γ-scattering and absorption,
544
Resonant partial wave, 134
graphical representation of, 134
Resonant reactions, 125, 625
interference with nonresonant, 129
Resonant widths, 894
as a function of mass number, 1044
derivation of, 426–428
Rh^{103}, 692
Rhodium, total neutron cross section of,
420
Ritz combination principle, 258
Rotational band, 187, 676
even-parity, 186, 191
for $K = \frac{1}{2}$, 1017
odd-parity, 190
Rotational energy of nucleus
formula for, 1015
Rotational levels, 667
Rotational mode, 671
Rotational spectra
for an $s_{\frac{1}{2}}$ particle coupled to an even-
even core, 1017
for odd J values, 1017
in collective model, 877
Rotations
of isotopic spin space, 941
Row vector, 910
Rutherford scattering, 125, 132, 257, 784
formula for, 18, 650
graphical treatment of, 134

S

$S^{32}(\alpha,\alpha')S^{32*}$ reaction, 683
S matrix, 768
S-wave interaction, 695

1142 SUBJECT INDEX

S-wave scattering, 132
S-waves, 660
   graphical treatment of, 134
Saxon potential well, 641
Scalar interaction, 817
Scalar product, 816
Scaling circuit, 229
Scattered neutron spectrum, 358
Scattering amplitude, 674
   dependence on spin, 648
   for double scattering, 655
   for incident and emergent spins greater than 1/2, 658
Scattering amplitude, exact
   average over, 1033
Scattering angle distribution, 1036
Scattering cross-section
   for particular angular momentum, 423
   multi-level, 425
Scattering, elastic, 211
Scattering experiments
   flux requirements, 405
   neutron sources for, 405
Scattering from spin zero target, 648
Scattering lengths
   sign of, 1043
Scattering matrix, 787
   formalism, 786
   formula for, 788
   vicinity of a resonance, 790
Scattering of electrons
   by electrons, 21
   by nuclei, 18
   by positrons, 23
Scattering of particles having spin
   from nuclei having spin, 1037–1045
Scattering of spinless particles
   by spinless nuclei, 1035
Scattering, semiclassical approach
   "active cylinder" in, 679
Schiff bremsstrahlung spectrum, 507
Schmidt diagram, 554, 980
   for odd neutron nuclei, 552
   for odd proton nuclei, 551
Schroedinger equation, 673, 677
   for extra nucleon, 944
   two body, 629
Scintillation counter, 32–43
   applications of, 39–43
   basic circuit description, 33–36
   for detection of gamma rays, 230–234
   resolution considerations in, 36–39

   resolving time of, 33
   spectroscopy, 75–79
   use in coincidence measurements, 179
Scintillation spectrometry, 531
Scintillator delay, 519
Scintillator-photomultiplier detector, 519
Scintillators
   organic, 234
Screening correction
   in beta decay, 824
Screening effect, 839
   for heavy elements, 839
   of atomic electrons, 25
   radii, 788
Second-order focusing, 63
Secondary electrons, 229
Selection rules, 723, 836–838
   miscellaneous, 901
Selection rules for electric radiation
   with L-S coupling, 864
Self absorption of γ-ray, 538
Self-comparison method, 525
Self-conjugate nuclei, 858
   isotopic spin change, 870
   sum rule for, 883
Self-consistent field, 677, 1010
Self-consistent method
   for nucleon-nucleon potential, 1046
Semiclassical approach to scattering
   ray optics in, 680
Seniority concept, 898, 1001–1002
   application of, 1001
   selection rules, 901
Separated isotopes
   experiments with, 1047
Separation energy
   near magic numbers, 967
   Z dependence, 967
Serber exchange factor, 680
Shell model, 442, 672, 878, 963
   branching ratios, 882
   calculation of isotopic spin impurities, 943
   configuration mixing, 1009
   degenerate levels, 972
   experimental confirmation from deuteron stripping, 975
   ground state spins, 971
   intermediate coupling, 676
   level spacing, 438
   light elements, 313

parity of nuclear levels, 975
predictions of, 976
restriction on l, 724
special selection rules, 879
spherical-wall, 203
spherically symmetrical binding field, 1009
theoretical foundation, 989
use of square well potential in, 969
Shells, major
separation of, 879
Shells, unfilled
in nucleus, 1010
Shielding material
arrangements, 410
neutrons scattered by, 371
$Si^{28}(p,p')Si^{28}$ reaction, 687
$Si^{30}(p,\gamma)P^{31}$ reaction, 293
Single-channel analyzer, 288
Single particle levels, 724, 916–918
experimental evidence for, 991
Single-particle magnetic dipole moment, 979
Single-particle model, 250, 554
l-forbidden selection rule in, 841
mean lifetimes for, 860
of transition probabilities, 430
radiative width, 860
Single-particle orbital angular momentum rule, 895
Single particle reactions
widths measured in, 893
Single particle wave functions, 985
Single-particle widths, 891–894, 908
expressions for, 865
Single photon transitions, 853
Size resonances, 1040
Slow neutron spectroscopy, 342–355
Small amplitude assumption, 1013
Small oscillations
in nucleus, 1012
normal modes for, 1012
Sn, 693
Sn^{116}, proton scattering on, 1047
Sodium iodide, 36
efficiency in various geometries, 263
mass attenuation coefficients for, 225
Solenoid spectrometer, 92
luminosity of, 92
transmission of, 92
Space exchange force
attractive, 973

Space exchange operator, 680
Space-symmetry selection rules, 901
Specific ionization, 51
Spectra measurements, 55–97
Spectra of even-even nuclei from Gd to Pt
systematics of, 1028
Spectra of lead isotopes, 1023
Spectral distribution
computation of, 323
energy for centroid of, 327
J-dependence of, 324
Spectrograph
broad-range, 63
geometry of, 64
multiple-gap, 64–65
quadrupole ens for, 67
double-focusing, 179
Spectrometer(s)
aberrations, 82
as momentum measuring device, 238
bent-crystal transmission type, 242
characteristics of, 94
charged particle, 39–41
resolution of, 40–41
classification, 83
comparison of, 94
dispersion of, 82
double focusing, 65–66, 86
having sector-shape magnets, 89
hollow crystal, 77–78
intermediate-image, 93, 235
iron free, 88
lens type, 90–93
luminosity of, 82
magnetic gamma-ray, 234
neutron slowing-down time, 348
optimum conditions for operation of, 82–83
resolution of, 82
semicircular focusing pair, 240
split crystal, 76
three-crystal pair, 233
time of flight, 342
transmission of, 81
Spectrometry, high-precision, 97
Spectrum
for two identical ($1p$) nucleons, 1004
Spectrum measurements, integrated, 408
Spectrum of $(j_1)^2(j_2)$, 998
Spectrum of $(j_1)^3$, 998

1144 SUBJECT INDEX

Spectrum ordering
  limits of weak or dominant spin-orbit coupling, 1006
Sphere method for inelastic cross-sections, 471
Spherical Bessel function, 696
Spherical harmonics, 710
  related to nuclear density distribution, 1016
  transformation like, 659
Spherical mapping, 738
  for triple correlation, 758
Spherical nucleus, 1045
Spherical oscillating system
  normal modes, 1012
Spherical potential
  degeneracy in, 1020
Spherical waves
  converging, 674
  outgoing, 672
Spherically symmetric density distribution, 1017
Spheroidal shape of nucleus, 1011
Spin assignments, 981
Spin change of target nucleus, 678
Spin decoupling
  for $K = \frac{1}{2}$ bands, 1017
Spin-flip, 132, 630, 649, 1047
  effect on electric transition, 862
Spin-lattice interaction, 561
Spin measurements, 916–918
Spin-orbit coupling, 464
  formula for, 969
Spin-orbit interaction, 666
Spin-orbit splitting, 835
Spin-orbital fractional parentage coefficient, 887
Spin polarization
  measurement of, 754, 760
Spin-spin interaction, 561, 666
Spin-2 phonon
  excitation of, 919
Spin wave function
  for incident particle, 655
Spin zero nuclei
  scattering by, 1048
Spinless particle parameter
  formula for, 774
Spins of even-even nuclei, 973
Square-well potential, 435, 690
  in stripping, 716
  neutron scattering by, 892

Stable configuration, 984
Standard gamma-gamma correlation, 751
Standard triple correlation, 803
Static nuclear size effects
  on internal conversion coefficients, 841
Statistical assumption, 401, 1051
  results from, 667
Statistical factor, 418
Statistical model of inelastic scattering
  compared with direct interaction, 689–693
Statistical model of nucleus, 503, 672
Statistical tensor formulation, 760–764
"Statistical-separation," 240
Statistical weight factors, 861
Stilbene scintillator, 368
Stokes polarization description, 779
Stopping cross section, 8
  of compounds, 14
  of heavy ions, 12
Stopping processes, 3, 7
  experimental information, 7
Straggling effects for electrons, 27
Strength function, 438–441, 1043
  determination of, 438
  experimental measurement of giant resonance, 462
Strength of coupling
  relative, 811
Stripping and pickup reactions, 316, 339, 695–731
  angular distribution in, 698
  radial wave functions for, 713
  theory of, 695–731
  widths, 891
Subchannels
  interference between, 660
Sulfur
  neutron resonance in, 637
Sum rule(s), 503, 861, 882–884
Super-allowed transitions, 149–150, 827
Supermultiplet theory, 932, 951
  beta decay, 932
Surface absorption potentials, 1055–1056
  parameters for, 1057
Surface reactions, 718
Symmetry axis of nuclei
  angular momentum of, 1015
Symmetry, azimuthal, 711
Symmetry selection rule, 910
Symplectic symmetry, 904
Synchrocyclotron, 346, 348

Pages 1–582 are in Part A. Pages 623–1103 are in Part B.

SUBJECT INDEX

T

$T(d,n)He^4$ reaction, 413
$T(d,n)He^4$ reaction, 690
$T(p,\gamma)He^4$ reaction, 269
Ta^{181}, 841
Tb^{161}, β-decay of, 1030
Tb^{156}, decay of, 772
Th^{228}, 692
Th^{234}, 247
Tl^{204} beta spectrum, 146–147
Tl^{208}
  pulse spectrum of gamma rays from, 263
Table of isotopes, 1068–1103
Tantalum, neutron-capture γ-ray spectrum of, 321
Tantalum target backing, 269
Target
  Coulomb barrier in, 695
  stopping power of, 276
Target assembly
  multiple, 119
Target backing, 266
  thick, 110
Target chamber design, 118–125, 266
Target foils, 111
Target materials and techniques, 269
Target nuclei
  thermal motion of, 419, 493
Target nucleus
  zero spin, 650
Target preparation
  by electroplating, 112
  by vapor plating, 113
Target thickness
  choice of, 108–109, 267, 784
  measurement of, 114–115, 267
Target wave function
  for ground state, 672
Targets, 266–272
  contamination of, 109, 370
  gas, 115–116, 120
    chambers for, 120–125
  preparation of, 108
  rotating, 270
  solid, 118–120
    chambers for, 118
  temperature
    problems in, 110
  thin solid, 111–113
Tensor force, 549

Tensor interaction, 817
Tensor parameters, 761
Thallium, 230
  as an impurity, 36
Thermal neutron(s), 304–306
  capture, 305, 878
Thermodynamic equilibrium, 567
Thin target data from thick targets, 115
Thomas-Fermi-Dirac screening function, 839
Thomas-Fermi model, 824
Thomas term, 970
Thomson scattering, 211
Three-stage correlations, 790
Threshold counter, 481–482
Threshold energy for reaction, 641
Ti, 49, 685
$Ti(p,p')Ti$ reaction, 687
Time-measuring technique
  vernier principle, 373
Time-of-flight measurement
  contributions to time resolution, 369
Time-of-flight neutron spectrometer, 345–348
Time reversal invariance, 552
Total angular momentum, 626
Total cross section, 645–646
  elastic, 634, 1035
  formula for, 645, 1035
  information derived from, 453
  in terms of the strength function, 439
  per unit energy, 666
Tracer technique
  in atomic beam experiments, 562
$trans$-stilbene, 525
Transformation properties, 816
Transition
  between neighboring levels in harmonic oscillator, 1014
  parity unfavored, 856
Transition amplitude, 662
Transition, cross-over, 1014
Transition density, 541
Transition lifetimes $\lesssim 10^{-10}$ sec, 526–533
  by direct timing, 526
  by recoil methods, 527
Transition matrix, 626, 1037
  for scattering, 649
  formula for, 1038
Transition probability, 252, 819
  measurement of, 513

related to hindrance factor, 203
upward, 256
Transition strength
  logarithmic scale, 868
  units of, 858
Transitions
  l-forbidden, 814
  mixtures of, 251
Transitions, fast, 515
  multipolarity of, 515
Transitions from closely spaced levels, 878
Transmission
  average cross-section from, 439
Transmission coefficients, 222, 632, 665
  for complex potential model, 1041
  for various l, 642
Transmission data
  area method of analysis, 426
Transmission factor, 666
Transmission measurements, 447
Transmutation, 625, 654
Transuranium elements, 187, 193
Triple coincidences, 792
Triple correlation(s), 299–301, 791–799, 803–804
  channel spin case, 793
  classical, 757
  coefficients, 300
  experimental arrangement, 757
  for mixed intermediate states, 805
  general formulas, 793–795
  inelastic proton scattering (p,p',γ), 797
  measurements, interpretation of, 301
  of spinless particles, 755
  (p,γ,γ) process, 795
Triple correlation angle functions
  table of, 757
Triple correlation formulas
  mixed, 769
Triple correlation in direction
  interference between parity states, 792
Triple scattering experiments, 1044, 1049
Triple scattering geometry, 657
Triton, 698, 702
Two-body forces, 673
Two body forces in shell model
  first order correction to energy of state, 973
  treated by perturbation methods, 972

Two-body system, 336
Two channel approximation, 682

U

U, 853
U^{234}, 191
U^{235}, 177–178
  decay scheme of, 178
U^{238}, 400
  resonances in, 432
U functions, 903
Unbound states, 246
Uncertainty principle, 246, 711, 860
Unidirectional radiation
  density matrix, 741
Unified model of nucleus, 250
Uniform-field spectrograph, 175
Uniform model of nucleus, 942
Unitary condition, 768
Unitary sum rule
  formula for, 903
Universal constants
  in beta decay, 828
Unpaired particles in the nucleus, 1011
Unstable nuclei
  ground states of, 812

V

V^{51}, 999
Van de Graaff generator, 241, 257, 261, 342, 478
  current from, 409
Vector addition coefficients, 721, 742, 986
Vector-coupling, 904
Vector interaction, 817
Vector potential
  standing wave, 772
Velocity dependence of energy loss, 6
Vernier chronotron, 373
Vibrational collective motion, 682, 877
Vibrational description of nuclei
  accuracy of, 1014
Vibrational excitation
  use of Coulomb excitation in studying, 1025
Vibrational levels
  Coulomb excitation of, 1013
  decay of, 1013, 1014
  E2 transitions between, 1013
  lifetime of, 1013

SUBJECT INDEX

Vibrational mode, 671
Vibrational spectra, 1012–1014
Virtual mesons
  recoil effects caused by emission and absorption, 952

W

W.K.B. approximation, 678, 694, 920, 1041
W.K.B. level-shift formula, 920
Wave function, 705
  boundary conditions on, 705
  with isotopic spin, 869
Wave function, closed channel, 673
  many body, 336
Wave lengths
  average, 1034–1045
Wave number, 924
Wave packet, 1037
Weak interactions, 139, 829
Weight function
  energy
    width of, 1035
  for energy average, 1034
Weisskopf units of transition strength, 858–860
  energy limitation, 861
Weizäcker-Williams approximation, 753
Widths
  gross-energy dependence, 645
  properties of, 631–634
Widths of γ-radiating levels
  from charged particle scattering, 540
Wigner coefficients
  symmetrical form, 756
Wigner-Eckart theorem, 733

Wigner-Eisenbud nuclear reaction theorem, 734
Wigner $9j$ coefficients, 659, 803
Wigner-limit value, 892
Wigner superallowed transition, 976
Wigner supermultiplet theory, 932
Wigner-Teichmann sum rule, 634
Wigner two-level formula, 281
"Window" curve, 237
Woods-Saxon potential, 1055
  parameters for, 1057

X

X-functions, 903
X ray critical-absorption and emission for elements, 214–217

Y

Yield-curve measurements, 282
Yukawa two-body operator, 680

Z

Z coefficients, 752, 805
$\bar{Z}$ coefficient, 665
Zeeman effect, 560, 733
  in nuclear spectroscopy, 733
Zero reaction cross section, 1052
Zero-zero radiationless transition, 835
Zero-zero transitions, 228, 856–857
  allowed, 819
Zn^{64}, proton scattering from, 1046
Zr^{90}, 857, 990
  interpretation of spectrum, 992

PURE AND APPLIED PHYSICS

A Series of Monographs and Textbooks

Consulting Editors

H. S. W. Massey
University College, London, England

Keith A. Brueckner
University of California, San Diego
La Jolla, California

1. F. H. Field and J. L. Franklin, Electron Impact Phenomena and the Properties of Gaseous Ions.
2. H. Kopfermann, Nuclear Moments, English Version Prepared from the Second German Edition by E. E. Schneider.
3. Walter E. Thirring, Principles of Quantum Electrodynamics. Translated from the German by J. Bernstein. With Corrections and Additions by Walter E. Thirring.
4. U. Fano and G. Racah, Irreducible Tensorial Sets.
5. E. P. Wigner, Group Theory and Its Application to the Quantum Mechanics of Atomic Spectra. Expanded and Improved Edition. Translated from the German by J. J. Griffin.
6. J. Irving and N. Mullineux, Mathematics in Physics and Engineering.
7. Karl F. Herzfeld and Theodore A. Litovitz, Absorption and Dispersion of Ultrasonic Waves.
8. Leon Brillouin, Wave Propagation and Group Velocity.
9. Fay Ajzenberg-Selove (ed.), Nuclear Spectroscopy. Parts A and B.
10. D. R. Bates (ed.), Quantum Theory. In three volumes.
11. D. J. Thouless, The Quantum Mechanics of Many-Body Systems.
12. W. S. C. Williams, An Introduction to Elementary Particles.
13. D. R. Bates (ed.), Atomic and Molecular Processes.
14. Amos de-Shalit and Igal Talmi, Nuclear Shell Theory.
15. Walter H. Barkas. Nuclear Research Emulsions. Part I.
    Nuclear Research Emulsions. Part II. *In preparation*
16. Joseph Callaway, Energy Band Theory.
17. John M. Blatt, Theory of Superconductivity.
18. F. A. Kaempffer, Concepts in Quantum Mechanics.
19. R. E. Burgess (ed.), Fluctuation Phenomena in Solids.
20. J. M. Daniels, Oriented Nuclei: Polarized Targets and Beams.
21. R. H. Huddlestone and S. L. Leonard (eds.), Plasma Diagnostic Techniques.
22. Amnon Katz, Classical Mechanics, Quantum Mechanics, Field Theory.
23. Warren P. Mason, Crystal Physics in Interaction Processes.
24. F. A. Berezin, The Method of Second Quantization.
25. E. H. S. Burhop (ed.), High Energy Physics. In four volumes.

26. L. S. Rodberg and R. M. Thaler, Introduction to the Quantum Theory of Scattering.
27. R. P. Shutt (ed.), Bubble and Spark Chambers. In two volumes.
28. Geoffrey V. Marr, Photoionization Processes in Gases.
29. J. P. Davidson, Collective Models of the Nucleus.
30. Sydney Geltman, Topics in Atomic Collision Theory.
31. Eugene Feenberg, Theory of Quantum Fluids.
32. Robert T. Beyer and Shephen V. Letcher, Physical Ultrasonics.

In preparation
    J. Killingbeck and G. H. A. Cole, Physical Applications of Mathematical Techniques.